Basic Mathematics

FUNDAMENTALS AND ALGEBRA

Basic Mathematics

FUNDAMENTALS AND ALGEBRA

SECOND EDITION

Richard Williams
City Colleges of Chicago

Scott, Foresman/Little, Brown College Division

Scott, Foresman and Company
Glenview, Illinois
Boston
London

This book is dedicated in loving memory to Bernis Williams

Library of Congress Cataloging-in-Publication Data
Williams, Richard (Richard W.)
  Basic mathematics.

  Includes index.
  1. Arithmetic—1961–   . 2. Algebra. I. Title.
QA107.W57   1988   513′.122   87-12683

ISBN 0-673-18338-6

Copyright © 1984, 1988 Scott, Foresman and Company.
All Rights Reserved.
Printed in the United States of America.

2 3 4 5 6 VIK 93 92 91 90 89 88

Preface

Ever since the first edition of *Basic Mathematics,* I have tried to write a book that is both useful to students and helpful for instructors. For the student, the text is written in an informal, nonrigorous, and nonthreatening style that guides the students through the instructional material like an effective tutor. Students who successfully complete this textbook will be prepared for a subsequent algebra course, a technical mathematics course, an allied health mathematics course, or a liberal arts mathematics course. For the instructor, the text is designed for a course in arithmetic and algebra or an elementary algebra course. Moreover, the text can be utilized in a traditional lecture-discussion format, a competency-based program, or in a mathematics laboratory.

All the basic arithmetic topics, except whole numbers, and all the topics required for a course in elementary algebra are covered in this text. This combination of topics was chosen because many college freshmen have forgotten most of their computational skills with fractions, decimals, and percents, but few of them need a complete course in arithmetic. Many of the students in an elementary algebra class need to review fractions, decimals, or percents in order to successfully complete the course.

Changes in the Second Edition

Motivating Text
Basic concepts, rules, definitions, and procedures are motivated by using carefully constructed illustrative examples.

New Design
Discussions of key ideas are indicated by numbered boxes (IDEA 1). Helpful annotations to examples are displayed in color. Important properties are screened in gray, while step-by-step procedures are screened in blue.

Watch Your Step
To help eliminate misconceptions and to prevent the development of bad habits, numerous warnings that point out the most common errors found in homework and test papers have been inserted throughout the text. Look for the *!* symbol.

Calculator Problems
Calculator problems have been inserted into the exercise sets and are indicated by the symbol.

Chapter Summaries
All summaries have been rewritten. These highlight every definition, property, skill, and procedure presented in the chapter.

Chapter Tests
Each chapter ends with a chapter test. These problems will offer the student more practice in choosing the appropriate method for solving a variety of problems.

Rational Expressions
A chapter on rational expressions has been added.

Review Exercises	Additional problems have been added to all review exercises.
Special Features	Based upon the results of a three-year investigation to improve achievement scores and attitudes toward mathematics of college students, this book is designed with the following special features: Each chapter begins with a list of objectives along with sample problems for each objective. The sample problems are keyed to the appropriate section and *Idea*. Each new *Idea* is indicated by a boxed number in the left-hand margin. These Ideas are keyed to the problems in the objectives. Basic **computational skills** with fractions, decimals, and signed numbers are continually **reinforced** once introduced. **Skill procedures are written in a step-by-step format** followed by several examples. Before examples are solved the rule is restated, and a color annotation or other pedagogical technique is used in guiding students through critical steps. Each chapter ends with a summary of Important Terms and Important Skills referenced by section and Idea.
Exercises	An average of 371 exercises per chapter. An average of 50 *Practice Problems* per chapter. These problems allow the students to practice each new skill before proceeding to the next skill or concept. **Review Exercises** are randomly arranged so that students must select the proper solution strategy. *Fill in the blank* exercises. This type of problem helps develop the student's vocabulary and serves as an excellent review of the rules and procedures presented in the chapter. **Verbal statements** of problems. These problems will sharpen the skills the students need to solve word problems. They will also help to develop the vocabulary needed for most mathematics courses. There are over **4800 exercises, 650 Practice Problems,** and **770 worked examples** in the text. **Word problems** have been integrated into practically every section of the book and an entire chapter is devoted to the algebraic solution of word problems. Moreover, these word problems include applications to business, consumer mathematics, science, and technical mathematics.
Flexible Organization	Integers are discussed in Chapter 1 and reinforced in the last section of Chapters 2 and 3 on fractions and decimals. This approach was chosen to stimulate students with a new concept while reinforcing basic skills with whole numbers. However, if preferred, Chapter 1 along with Section 2.9 (Signed Fractions) and 3.7 (Signed Decimals) can be taught directly after Chapter 3. Then Chapters 4–13 or the remaining material can be selected to meet the abilities and needs of the students.
Instructor's Guide and Solutions Manual	To help teachers use *Basic Mathematics,* an Instructor's Guide and Solutions Manual is available. It contains answers to the even-numbered exercises and worked-out solutions to every third exercise in the text; two quizzes for every section and three tests for every chapter of the text; two final exams; and teaching suggestions.
Computer-Assisted Testing System (CATS)	CATS software can be used with either Apple or IBM computers to construct and print tests. Over 3000 questions arranged according to topic are included, as well as an option to create new questions.
Acknowledgments	The author wishes to extend his gratitude to the individuals whose expertise and generous assistance made this book possible. First, I am indebted to the students, staff, tutors, and faculty of Malcolm X College who offered helpful comments and

suggestions in the development of this text. Secondly, I am indebted to the reviewers, Bonnie M. Hodge, Laura J. Shaprio, Frances Smith, Anna Jo Ruddel, Patricia Wilkinson, Douglas Robertson, John G. Michaels, and Phillip Stoddard, who offered helpful suggestions and criticisms that helped to improve this edition. In particular, I wish to thank Laurie Golson for her detailed review of the manuscript and Steve Quigley for his support throughout preparation of the manuscript. A special thanks and sincere appreciation are extended to Eloise Green who typed this manuscript and to Brenetta McGee who prepared the index.

A final and loving thanks to my parents Bernis and Willie Ruth Williams whose sacrifices provided me with the opportunities to develop the expertise and whose guidance provided me the tenacity required to initiate and complete this project.

To the Student

If your instructor wishes, you can proceed at your own pace. Each unit begins with a list of objectives and sample problems. Quickly look them over.

If you feel you are not ready to work these sample problems, we suggest you skip them. Instead, turn the page to the first section of the unit and work through the entire unit.

If you feel you can work the sample problems, do so. Then check your answers with those at the back of the book.

If you missed any problems and feel you need just a *quick review*, read the Ideas that explain the problems you missed.

If you missed any problems and feel you need a *more thorough review*, read the sections containing the problems you missed and work the *Practice Problems*. Then work the exercises at the end of each section.

Answers to the practice problems are at the end of the section; answers to odd-numbered exercises and to all sample problems and review problems are at the back of the book. We encouarge you, however, to work each problem before looking at the answer. Only through working math will you learn to do math.

Contents

Chapter 1: INTEGERS — 1

    1.1 Integers and Absolute Value   2
    1.2 Adding Integers   4
    1.3 Subtracting Integers   11
    1.4 Adding and Subtracting Integers   15
    1.5 Multiplying Integers   19
    1.6 Dividing Integers   23
    1.7 Order of Operations   25
    1.8 Algebraic Expressions and Formulas   30
         Summary   35
         Review Exercises   37
         Test   39

Chapter 2: FRACTIONS AND MIXED NUMBERS — 41

    2.1 Finding Prime Factorizations   42
    2.2 Reducing Fractions   47
    2.3 Improper Fractions and Mixed Numbers   52
    2.4 Multiplying Fractions and Mixed Numbers   55
    2.5 Dividing Fractions and Mixed Numbers   62
    2.6 Adding Fractions   69
    2.7 Adding Mixed Numbers   77
    2.8 Subtracting Fractions and Mixed Numbers   81
    2.9 Signed Fractions   87
         Summary   94
         Review Exercises   96
         Test   98

Chapter 3: DECIMALS AND PERCENTS — 100

    3.1 Reading and Writing Decimals   101
    3.2 Adding and Subtracting Decimals   103
    3.3 Multiplying Decimals   107

3.4 Rounding Decimals 110
3.5 Dividing Decimals 113
3.6 Converting Fractions and Decimals 117
3.7 Signed Decimals 122
3.8 The Meaning of Percent 126
3.9 Writing Decimals and Fractions as Percents 130
Summary 134
Review Exercises 136
Test 138

Chapter 4: EXPONENTIAL EXPRESSIONS 140

4.1 The Meaning of Exponential Expressions 141
4.2 Multiplying and Dividing Exponential Expressions 143
4.3 Raising Products, Quotients, and Powers to Higher Powers 147
4.4 Scientific Notation 151
Summary 155
Review Exercises 157
Test 158

Chapter 5: POLYNOMIALS 160

5.1 Combining Like Terms 161
5.2 Evaluating Polynomials 166
5.3 Adding Polynomials 169
5.4 Subtracting Polynomials 171
5.5 Multiplying Polynomials 174
5.6 Special Products 179
5.7 Dividing by Monomials 184
5.8 Dividing by Polynomials 188
Summary 191
Review Exercises 192
Test 194

Chapter 6: LINEAR EQUATIONS AND INEQUALITIES IN ONE VARIABLE 196

6.1 Linear Equations 197
6.2 Solving Linear Equations Containing One Operation 200
6.3 Solving Linear Equations Containing More Than One Operation 208
6.4 Solving Linear Equations Containing Grouping Symbols or Fractions 212
6.5 Literal Equations 217

6.6 Solving Linear Inequalities Containing One Operation 220
6.7 Solving Linear Inequalities Containing More Than One Operation 227
Summary 231
Review Exercises 233
Test 235

Chapter 7: SOLVING WORD PROBLEMS BY USING LINEAR EQUATIONS IN ONE VARIABLE 236

7.1 Solving Word Problems 237
7.2 Ratio and Proportion Problems 246
7.3 Percent Problems 256
7.4 Simple Interest 265
7.5 Mixture Problems 270
7.6 Uniform Motion Problems 274
Summary 280
Review Exercises 282
Test 284

Chapter 8: GRAPHING 286

8.1 The Rectangular Coordinate System 287
8.2 Linear Equations in Two Variables 292
8.3 Slope-Intercept Form of Linear Equations 299
8.4 Linear Inequalities in Two Variables 303
Summary 310
Review Exercises 312
Test 314

Chapter 9: LINEAR SYSTEMS OF EQUATIONS 318

9.1 Intersecting, Parallel, and Equal Lines 319
9.2 Solving Linear Systems of Equations by Addition 326
9.3 Solving Linear Systems of Equations by Substitution 336
Summary 345
Review Exercises 346
Test 348

Chapter 10: FACTORING POLYNOMIALS 350

10.1 Common Monomial Factors 351
10.2 Factoring Trinomials 355

10.3 Special Factorization   363
Summary   369
Review Exercises   370
Test   371

Chapter 11: RATIONAL EXPRESSIONS   372

11.1 Simplifying Rational Expressions   373
11.2 Multiplying and Dividing Rational Expressions   377
11.3 Adding and Subtracting Rational Expressions   381
11.4 Complex Fractions   388
11.5 Solving Equations Containing Rational Expressions   393
11.6 Applications   398
Summary   403
Review Exercises   404
Test   406

Chapter 12: SQUARE ROOTS AND QUADRATIC EQUATIONS   408

12.1 Square Roots   409
12.2 Multiplying, Simplifying, and Dividing Square Roots   412
12.3 Adding and Subtracting Square Roots   420
12.4 Quadratic Equations   422
12.5 Solving Quadratic Equations by Factoring   428
12.6 The Quadratic Formula   435
Summary   439
Review Exercises   441
Test   443

Chapter 13: GEOMETRY   444

13.1 Perimeter   445
13.2 Area   453
13.3 Volume   466
13.4 Pythagorean Theorem   471
Summary   477
Review Exercises   479
Test   481

Appendix A: MEASUREMENT — 483

Appendix B: TABLE OF SQUARES, SQUARE ROOTS, AND PRIMES — 485

Answers — 487

Index — 509

1 Integers

Objectives The objectives for this chapter are listed below along with sample problems for each objective. By the end of this chapter you should be able to find the solutions to the given problems.

1. Find the absolute value of an integer *(Section 1.1/Idea 2)*.
   a. $|-10|$ b. $|30|$ c. $|+15|$

2. Add two or more integers *(Section 1.2/Ideas 1–4)*.
   a. $-30 + (-15)$ b. $30 + (-60)$ c. $-5 + 10 + (-6) + (-30) + 40$

3. Find the additive inverse of an integer *(Section 1.3/Idea 1)*.
   a. -10 b. 30 c. $+15$

4. Subtract one integer from another *(Section 1.3/Idea 2)*.
   a. $-30 - 15$ b. $-10 - (-45)$

5. Simplify an expression containing addition and subtraction *(Section 1.4/Idea 1)*.
   a. $-16 + 10 - 25 + 2$ b. $15 - 25 + (-8) - (-8)$

6. Multiply two or more integers *(Section 1.5/Idea 2)*.
   a. $5(-10)$ b. $(-2)(3)(-5)$

7. Divide one integer by another *(Section 1.6/Idea 1)*.
   a. $\dfrac{-60}{10}$ b. $\dfrac{-30}{-5}$

8. Simplify an expression containing more than one operation *(Section 1.7/Ideas 1–3)*.
   a. $5 - 6 \cdot 2 + 12 \div (-4) + 8$ b. $3[-6 + 8(-13 + 2 \cdot 5)]$

9. Evaluate an algebraic expression or a formula *(Section 1.8/Idea 1–2)*.
   $M = 6 - a(p + 5d)$ when $a = 4$, $p = -9$, and $d = 1$

1.1 Integers and Absolute Value

IDEA 1   The numbers

$$0, 1, 2, 3, 4, 5, 6, 7, 8, 9, 10, \text{ and so on}$$

are called **whole numbers.** Whole numbers can be shown by equally spaced points on a straight line called a **number line.** In Figure 1-1, the arrowhead shows that the number line continues to the right without ending. The three dots under the number line show that the numbers also continue without ending.

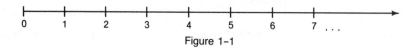

Figure 1-1

Whole numbers cannot be used in some situations. For example, no whole number can represent a temperature of 5° below zero or a loss of 10 yards in a football game. In such situations, integers must be used.

The numbers

$$\ldots -5, -4, -3, -2, -1, 0, 1, 2, 3, 4, 5, \ldots$$

are called **integers.** Integers can be shown on a number line as in Figue 1-2. The arrowheads show that the number line continues in both directions without ending.

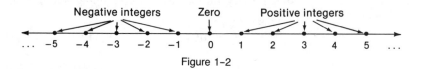

Figure 1-2

Integers to the right of zero are **positive integers** (greater than zero). A positive integer may have a plus sign (+) in front of it or it may have no sign at all. Integers to the left of zero are **negative integers** (less than zero). A negative integer has a minus sign (−) in front of it. Zero is the only integer that is neither positive nor negative. Also, the positive integers are called **natural numbers.**

Positive and negative integers can be used to represent opposite quantities. In football, for example, a gain of five yards can be expressed as 5 or +5 (positive 5), whereas a five-yard loss is expressed as −5 (negative 5). Some other examples are as follows:

1. A profit of $30 can be represented by 30 or +30; a $30 loss is expressed as −30.
2. A checking account containing $55 can be represented by 55 or +55; a checking account overdrawn $55 is represented by −55.

Practice Problem 1   *Express each statement as a positive or negative integer.*

    a. $50 profit   **b.** $35 loss

    c. 10-yard gain   **d.** 15° below zero

NOTE: Answers to Practice Problems are at the end of the section in which they appear.

IDEA 2

Until now, we have been concerned with the sign of an integer. However, frequently in mathematics we are interested in the numerical value of an integer without regard to its sign. This value is called the **absolute value** of an integer. Please note that in the following definition the letter x will be used to represent any integer.

> The **absolute value** of an integer x, written $|x|$, is the distance or number of units between 0 and x on the number line.

Example 1   Find the following absolute values.

  a. $|4|$ **b.** $|-4|$

Solution   **1a.** $|4| = 4$ since the number of units between 0 and 4 is 4 (see Figure 1-3).

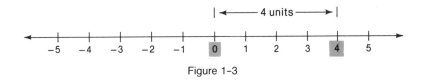

Figure 1-3

1b. $|-4| = 4$ since the number of units between 0 and -4 is 4 (see Figure 1-4).

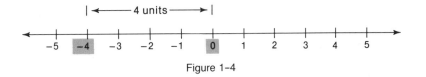

Figure 1-4

! Since the absolute value of a number simply represents a distance, it will never be negative.

When finding the absolute value of an integer, it may be inconvenient to draw a number line. Instead, we can use the fact that

Absolute Value Property

> If x is any positive number or zero, then the absolute value of x and the absolute value of $-x$ are both equal to x.
>
> or
>
> $|x| = x$ and $|-x| = x$
>
> where x is a positive number or zero.

Example 2   Find the following absolute values.

  a. $|-40|$ **b.** $|+50|$ **c.** $|727|$

Solution   **2a.** $|-40| = 40$ **2b.** $|+50| = 50$ **2c.** $|727| = 727$

Practice Problem 2   *Find the following absolute values.*

  a. $|-10|$ **b.** $|+75|$ **c.** $|3|$ **d.** $|0|$

1.1 Exercises

Express each statement as a positive or negative integer.

1. 300 feet above sea level
2. $13 debt
3. $100 profit
4. 10-pound gain
5. $39 loss
6. 5-pound loss
7. 300 feet below sea level
8. 8° above zero
9. having $35
10. 13° below zero

Find the following absolute values.

11. $|-9|$
12. $|-84|$
13. $|-95|$
14. $|25|$
15. $|35|$
16. $|-37|$
17. $|-104|$
18. $|+84|$
19. $|724|$
20. $|74|$
21. $|0|$
22. $|9|$
23. $|+5|$
24. $|+73|$
25. $|+324|$
26. $|622|$

Fill in the blank.

27. The absolute value of an integer is never _____.

Answers to Practice Problems 1a. 50, or +50   b. −35   c. 10, or +10   d. −15   2a. 10   b. 75   c. 3   d. 0

1.2 Adding Integers

IDEA 1 In this section we are going to develop three rules for adding integers. To develop these rules, we are first going to analyze how to add whole numbers by using a number line.

Let us first consider how to find the sum 2 + 4 by using the number line in Figure 1-5. To find the sum 2 + 4, we start at 0 and move 2 units to the right (the first arrow represents the number 2). Now, to add 4, we start at the end of the arrowhead representing 2 and we move 4 units to the right. Our final location is 6, which represents a total movement of 6 units to the right or the sum of the numbers 2 and 4. A careful analysis of the results in Figure 1-5 suggests that *the sum of two positive numbers is the sum of the absolute values of the numbers.* That is,

$$2 + 4 = |2| + |4| = 6$$

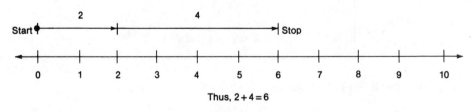

Thus, 2 + 4 = 6

Figure 1-5

Let us now try the same procedure with −2 + (−4), which is read "negative 2 plus negative 4." The "( )" around the −4 are parentheses and they are used to separate the integer −4 from the symbol for addition.

1.2 Adding Integers

To find the sum of -2 and -4, we start at 0 and move 2 units to the left (a negative sign means to move in the negative direction). To add -4, we move 4 more units to the left. Our final location is -6, which implies $-2 + (-4) = -6$ (see Figure 1-6). We can also use an example involving money to show that $-2 + (-4) = -6$. For example, if you lose \$2 (represented by -2) and you lose \$4 (-4), your combined loss would be \$6 (-6).

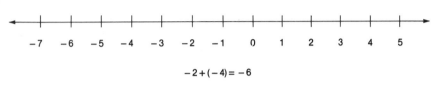

$-2 + (-4) = -6$

Figure 1-6

The fact that $-2 + (-4) = -6$ suggests that the *sum of two negative numbers is the negative of the sum of the absolute values of the addends*. That is,

$$-2 + (-4) = -(|-2| + |-4|) = -6$$

The parentheses in the expression $-(|-2| + |-6|)$ are used in a slightly different way than before. Here, they indicate that whatever is inside the parentheses is treated as one quantity or number. In the expression $-8 + 17$, for example, -8 is a negative number. In the expression $-(8 + 17)$, the negative of the sum of 8 and 17 is a negative number.

Based on the aforementioned statements, we can now state a rule for adding integers having the same sign.

To Add Two or More Integers Having the Same Sign:

1. Find the sum of their absolute values.
2. In front of this sum, place the common sign.

Example 3 Add.

  a. $-3 + (-10)$ **b.** $-100 + (-150)$ **c.** $-8 + (-10) + (-50)$

Solution To add two or more negative integers, find the sum of their absolute values and place a minus sign in front of this sum.

$$\text{Sign} \downarrow \quad \text{Sum of absolute values} \swarrow$$

3a. $-3 + (-10) = \boxed{-(3 + 10)}$ **3b.** $-100 + (-150) = -(100 + 150)$
$\qquad\qquad\qquad = -13$ $\qquad\qquad\qquad\qquad\qquad\qquad = -250$

3c. $-8 + (-10) + (-50) = -(8 + 10 + 50)$
$\qquad\qquad\qquad\qquad\quad = -68$

Remember that the "( )" used in Example 3 are used in two different ways.

1. They may indicate that whatever is inside the parentheses is treated as one quantity or number. For example, in the expression $-3 + 10$, -3 is a negative number. However, in the expression $-(3 + 10)$, the sum of 3 and 10 is a negative number.
2. They may separate a signed number from an operation symbol. For example, the parentheses in $-3 + (-10)$ separate the plus sign from the -10. You should *not* write $-3 + -10$.

Practice Problem 3 **Add.**

a. $-10 + (-15)$ b. $+37 + (+117)$

c. $-8 + (-15)$ d. $-2 + (-8) + (-11) + (-33)$

IDEA 2

Let us now consider a problem in which one number is positive and the other is negative. For example, to find the sum $5 + (-2)$, we start at 0 and move 5 units to the right. To add -2, we move 2 units to the left. Our final location is 3, which is the sum of 5 and -2 (see Figure 1-7).

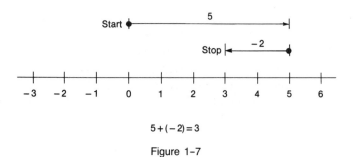

$$5 + (-2) = 3$$

Figure 1-7

In the above problem, it should be noted since we first moved *to the right* and then *to the left* that our final location (3) is the *difference* between the absolute values of the addends. Also, since we moved farther to the right (positive 5) than we did to the left (negative 2), the final location or result is positive. This implies that *the sign of the sum is the same as the sign of the number having the larger absolute value.* In other words,

$$5 + (-2) = |5| - |2| = 3$$

Based on the above statements, we have the following rule.

To Add Two Integers Having Different Signs:

1. Find the difference between their absolute values (subtract the smaller absolute value from the larger absolute value).
2. In front of this difference, place the sign of the integer with the larger absolute value.

Notice that the sum of a positive integer and a negative integer having *different absolute values* may be either positive or negative. It depends on which integer has the larger absolute value. However, the sum of the two integers having different signs and the same absolute value is always zero. For example,

$$6 + (-6) = 0 \quad \text{and} \quad -3 + 3 = 0$$

Example 4 Add.

a. $-60 + 45$ b. $-30 + 65$ c. $45 + (-60)$ d. $815 + (-175)$

e. $16 + (-16)$

Solution To add a positive integer and a negative integer, subtract their absolute values (the larger minus smaller). Place the sign of the number with the larger absolute value in front of this difference.

Sign of the integer with the larger absolute value ↓
Larger absolute value minus smaller ↓

4a. $-60 + 45 = -(|60| - |45|)$
$= -(60 - 45)$
$= -15$

4b. $-30 + 65 = +(65 - 30) = 35$

4c. $45 + (-60) = -(60 - 45)$
$= -15$

4d. $815 + (-175) = +(815 - 175) = 640$

4e. $16 + (-16) = 0$ Since 16 and -16 have the same absolute value.

! When adding a positive integer and a negative integer, be sure to distinguish between the larger integer and the integer with the larger absolute value. For example, 45 is the larger integer in the problem $-60 + 45$, since a positive integer is greater than a negative integer. However, -60 is the integer with the larger absolute value, since $|-60| = 60$ and $|45| = 45$.

Practice Problem 4 Add.

a. $-40 + 75$ b. $-80 + 10$ c. $50 + (-70)$ d. $85 + (-15)$

IDEA 3

There are two basic properties of addition. The first property is the **commutative law for addition.**

Commutative Law for Addition

> The order in which two numbers a and b are added does not affect the sum,
>
> or
>
> $a + b = b + a$

For example,

$$-45 + 10 = 10 + (-45)$$

The second property is the **associative law for addition.**

Associative Law for Addition

The order in which three (or more) numbers a, b, and c are added does not affect the sum,

or

$$(a + b) + c = a + (b + c)$$

For example,

$$(-2 + 3) + 5 = -2 + (3 + 5)$$

IDEA 4

Suppose you had to add more than two integers having different signs. You could add them two at a time. However, this can be inconvenient. Instead, use the associative and commutative laws to group the positive integers together and to group the negative integers together. For example,

$-2 + 3 + (-5) + 8$	Add three or more integers.
$= -2 + (-5) + 3 + 8$	Rearrange the addends by using the commutative and associative laws.
$= -7 + 11$	Add negative integers and add positive integers.
$= 4$	Add.

The above example suggests the following rule.

To Add Three or More Integers Having Different Signs:

1. Add all negative integers.
2. Add all positive integers.
3. Add the results obtained in steps 1 and 2.

Example 5 Add.

a. $-3 + 8 + (-10) + 40$
b. $-35 + (-10) + 15 + (-30) + 7$
c. $-10 + 18 + (-61) + 30 + (-65) + 42$

Solution To add integers having different signs, first add the negative integers, next add the positive integers, and then add the results.

5a. $-3 + 8 + (-10) + 40$
    $= -13 + 48$
    $= 35$

5b. $-35 + (-10) + 15 + (-30) + 7$
    $= -75 + 22$
    $= -53$

5c. $-10 + 18 + (-61) + 30 + (-65) + 42$
$= -136 + 90$
$= -46$

Practice Problem 5 **Add.**

a. $-2 + 5 + (-16)$ **b.** $8 + (-50) + 95 + (-18)$

c. $-16 + 50 + 8 + (-30) + (-118) + 3$

IDEA 5

Now that you have reviewed the basic skills required to add integers, read the following suggestions for solving word problems involving the addition of integers. These suggestions will also be useful when solving word problems involving the subtraction, multiplication, and division of integers.

To Solve a Word Problem:

1. **Identify knowns and unknowns.** Read the problem until you understand what information is being given and what must be found.

2. **Write a mathematical expression.** Reread the problem and then express the written statement as a mathematical expression. Sometimes it is helpful to first rewrite the original problem as a short statement.

3. **Find the answer.** Simplify the expression written in step 2 and answer the question asked in the original problem.

4. **Check the answer.** Determine (mentally) if the answer is reasonable.

Example 6 Express each statement as an addition problem and then solve.

a. A football player gained 11 yards on an off-tackle play and then lost 13 yards on a broken play. What is his final yardage gain or loss?

b. Ron weighed 260 pounds. Find his weight after the following changes: lost 10 pounds, gained 3 pounds, lost 15 pounds.

Solution

6a. **Think:** 11-yard gain + 13-yard loss = ?
Write: 11   + (-13)
       $= -(13 - 11)$
Find: $= -2$
The football player lost two yards.
Check: The football player lost more yardage than he gained. Therefore, a loss of two yards is a reasonable answer.

6b. **Think:** original weight + 10-lb loss + 3-lb gain + 15-lb loss
Write: 260   + (-10) +   3   + (-15)
       $= 263 + (-25)$
Find: $= 238$
Ron weighs 238 pounds.

Practice Problem 6 *Express each statement as an addition problem and then solve.*

    a. One day the temperature rose 9° from a previous reading of 15° below zero. What is the new temperature?

    b. A certain stock had an opening price of 95 (meaning $95). If the daily changes in the stock were listed as +2, −1, +5, −6, +1, and −3, find the final closing price of the stock.

1.2 Exercises

Add.

1. $-16 + (-10)$
2. $-10 + (-45)$
3. $16 + (-10)$
4. $-25 + 19$
5. $16 + (-30)$
6. $25 + (-45)$
7. $-100 + (-29)$
8. $-100 + 29$
9. $100 + (-30)$
10. $70 + (-85)$
11. $-70 + (-8) + (-13)$
12. $35 + 70$
13. $-35 + (-70)$
14. $-40 + (-10)$
15. $70 + 25$
16. $-3 + (-5) + (-8)$
17. $-6 + (-10) + (-118)$
18. $-30 + 17$
19. $16 + (-39)$
20. $-3 + (-4) + (-8) + (-35)$
21. $-16 + 400$
22. $-131 + (-151)$
23. $-601 + 101$
24. $-8 + (-5) + (-2)$
25. $10 + (-33) + 40$
26. $-6 + (-30) + 40$
27. $-10 + (-80) + 190 + (-70)$
28. $-35 + (-6) + 39 + 51$
29. $-350 + 500 + 600 + (-80)$
30. $-105 + 600 + (-800)$
31. $-350 + 600 + (-100) + 20$
32. $-61 + (-101) + 31 + (-81) + 131$
33. $-183 + 7 + (-190) + (-60) + 5$
34. $-61 + (-8) + 401 + (-71) + (-100) + 160 + (-3)$
35. $-3 + 8 + (-9) + (-6) + 13 + (-10) + 60 + 3$
36. $-8 + 13 + (-14) + (-11) + 18 + (-15) + 65 + 8$
37. $-8 + (-2) + (-10) + (-10) + (-20)$
38. $-5 + (-8) + (-15) + (-2) + (-8)$
39. $-5 + 8 + (-7) + 4$
40. $-10 + 13 + (-8) + 5$
41. $-15 + (-17)$
42. $-11 + (-12)$
43. $-18 + 8$
44. $-19 + 9$
45. $17 + (-20)$
46. $3 + (-9)$
47. $-8 + (-3) + (-4)$
48. $-3 + (-10) + (-5)$
49. $-400 + 100$
50. $-500 + 300$
51. $15 + (-10) + 45$
52. $10 + (-20) + 40$

53. $-8 + 2 + 5 + 13 + (-18)$
54. $-5 + 3 + 7 + (-18) + 2$
55. $-5 + 8 + (-9) + 3 + (-16)$
56. $-9 + 3 + (-15) + 25 + (-1)$
57. $-18 + 3$
58. $-17 + 11$
59. $323 + (-220)$
60. $831 + (-530)$
61. $15 + (-115)$
62. $13 + (-213)$
63. $-18 + 53$
64. $-19 + 81$

Fill in the blanks with one or more words.

65. If you add two or more negative integers, the sign of the sum is _____.

66. If you add two integers having different signs, the sign of the sum is determined by _____.

67. The _____ law allows us to change the order in which two integers are added without affecting the sum.

68. We can regroup a collection of integers without changing its sum by the _____ law.

Express each statement as an addition problem and then find the answer. Show your work.

69. If the entries in your financial records read $100 profit, $30 loss, $180 loss, and $50 profit, what is the amount of your final profit or loss expressed as an integer?

70. If Joe records the monthly changes in his weight on a chart that reads "gained 3, lost 4, gained 5, lost 7, and lost 2," what is his final weight gain or loss expressed as an integer?

71. If Joe's original weight was 190 pounds in problem 70, how much does he weigh after all the changes in his weight?

72. The opening price of a certain stock is listed as 55 (meaning $55). If changes in the stock are listed as $+1, -2, +5, -6, -2$, and $+10$ during the day, find the final closing price of the stock.

73. Find the sum of $-23{,}146$ and $-10{,}989$.
74. Find the sum of $-102{,}051$ and $-83{,}458$.
75. Find the sum of $-10{,}011$; $8{,}559$; $-7{,}369$; and $18{,}359$.
76. Find the sum of $-11{,}015$; $8{,}756$; $-6{,}348$; and $20{,}301$.

Answers to Practice Problems 3a. -25 b. 154 c. -23 d. -54 4a. 35 b. -70 c. -20 d. 70
5a. -13 b. 35 c. -103 6a. $-6°$ or $6°$ below zero b. 93

1.3 Subtracting Integers

IDEA 1 Before we discuss how to subtract one integer from another, let us consider the following special sums.

$$6 + (-6) = 0 \qquad 105 + (-105) = 0$$
$$-6 + 6 = 0 \qquad -105 + 105 = 0$$

In each case the sum of the two numbers is zero. Whenever the sum of two numbers is zero, the numbers are said to be *additive inverses* or *opposites* of each other.

The **additive inverse** or **opposite** of an integer x, denoted by $-x$, is that number that can be added to x so that the sum will be zero.

Based on this definition, the additive inverse of -3 is 3, since $-3 + 3 = 0$. Similarly, the additive inverse of 3 is -3. Please note that the additive inverse of any

12 Integers

integer x (except zero) is written as $-x$. Thus, $-(-3) = 3$ can be read "The additive inverse of -3 is equal to 3." Similarly, $-(-4) = 4$, $-(-5) = 5$, $-(-10) = 10$, and $-(-6) = 6$.

Example 7 Find the additive inverses of the following integers.

  a. -60 **b.** 160 **c.** 0

Solution To find the additive inverse of any integer except zero, change the sign of that integer.

7a. The additive inverse of -60 is 60, since $-60 + 60 = 0$.

7b. The additive inverse of 160 is -160, since $160 + (-160) = 0$.

7c. The additive inverse of 0 is 0, since $0 + 0 = 0$.

Practice Problem 7 *Find the additive inverses of the following integers.*

  a. 6 **b.** -13 **c.** $+8$ **d.** -101 **e.** 0

IDEA 2

The concept of additive inverses can be used to rewrite every subtraction problem as an equivalent addition problem. For example,

Subtraction problem	*Equivalent addition problem*	*Result*
$7 - 2$	$= \quad 7 + (-2)$	$= \quad 5$

In the statement $7 - 2 = 5$, 7 is called the **minuend**, 2 is the **subtrahend**, and 5 is the **difference**. In general, if a, b, and c are any three numbers and $a - b = c$, then a is the minuend, b is the subtrahend, and c is the difference.

The above example illustrates that subtracting a number from a given quantity is the same as adding the additive inverse of that number of the given quantity. We can state this rule as follows.

To Subtract One Integer from Another:

1. Keep the minuend the same.
2. Change the operation of subtraction to addition, and add the additive inverse of the subtrahend,

or

$$a - b = a + (-b)$$

where $-b$ is the additive inverse of b.

Example 8 Subtract.

  a. $-9 - (-15)$ **b.** $32 - 157$ **c.** $-10 - 15$ **d.** $9 - (-15)$

Solution To subtract one integer from another, keep the minuend the same and add the additive inverse of the subtrahend.

8a. $-9 - (-15) = -9 + 15 = 6$

Additive inverse of subtrahend
Change operation to addition

8b. $32 - 157 = 32 + (-157) = -125$

8c. $-10 - 15 = -10 + (-15) = -25$

8d. $9 - (-15) = 9 + 15 = 24$

Please note that the answer to a subtraction problem is correct if the subtrahend plus the difference is equal to the minuend. Thus, in Example 8 it is clear that:

$$-9 - (-15) = 6 \quad \text{if} \quad -15 + 6 = -9;$$
$$32 - 157 = -125 \quad \text{if} \quad 157 + (-125) = 32;$$
$$-10 - 15 = -25 \quad \text{if} \quad 15 + (-25) = -10; \text{ and}$$
$$9 - (-15) = 24 \quad \text{if} \quad -15 + 24 = 9$$

In general,

$$a - b = c \quad \text{if} \quad b + c = a$$

Practice Problem 8 **Subtract and check your answers.**

   a. $-35 - (-75)$ **b.** $-75 - 10$ **c.** $75 - 100$

The "−" sign can be used for three different purposes:

1. To indicate subtraction, as in $2 - 3$ (read as "two minus three");
2. To represent a negative number (for example, -3, which is read as "negative three"); and
3. To signify the additive inverse of a number (for example, -3 can be read as "the additive inverse of three").

The meaning of the sign is usually clear from the way in which the problem is written.

In Section 1.2, you learned that the order in which two numbers are added does not affect the answer (commutative law for addition). However, the order in which two numbers are subtracted *does* affect the answer. For example, $3 - 7$ is not equal to $7 - 3$, since $3 - 7 = -4$ and $7 - 3 = 4$.

IDEA 3

Since order is important in subtraction, be careful in identifying the subtrahend (number to be subtracted). This is especially true for word problems that use the word *difference*. "The difference between *a* and *b*" is written $a - b$, and *b* is the subtrahend. "Subtract *a* from *b*" is written $b - a$, and *a* is the subtrahend.

Example 9 Solve the following problems. Show all work.

   a. A person's net worth is determined by subtracting liabilities from assets. If your liabilities are $5500 and your assets are $4800, what is your net worth?

b. The highest temperature recorded in Chicago one winter day was $-12°$. The lowest temperature was $-20°$. Find the difference between the highest and lowest temperatures.

Solution

9a. Since the liabilities ($5500) must be subtracted from the assets to determine your net worth, think of the liabilities as the subtrahend.

$$\begin{align} \text{Net worth} &= \text{assets} - \text{liabilities} \\ &= 4800 - 5500 \\ &= 4800 + (-5500) \\ &= -700 \end{align}$$

This means that you are $700 in debt.

9b. Since you want to find the difference between the highest and the lowest temperatures, the lowest temperature (or the second number) mentioned is the subtrahend.

$$\begin{align} \text{Temperature difference} &= \text{highest} - \text{lowest} \\ &= -12 - (-20) \\ &= -12 + 20 \\ &= 8 \end{align}$$

The difference between the highest and the lowest temperatures is 8°.

"Subtract a from b" is written $b - a$, and a is the subtrahend. However, "find the difference between a and b" is written $a - b$, and b is the subtrahend.

Practice Problem 9   Solve. Show all work.

a. Subtract -5 from 37.

b. The average temperature in June for a certain city is 85°. On June 7, 1983, the temperature was 80°. Find the difference between the June 7 temperature and the average temperature in June.

1.3 Exercises

Find the additive inverses of the following integers.

1. -10	**2.** 5	**3.** -4	**4.** 10	**5.** $+8$
6. -100	**7.** 0	**8.** -35	**9.** 100	**10.** -16

Subtract and check your answers.

11. $-9 - 12$	**12.** $9 - (-12)$	**13.** $9 - 12$
14. $-9 - (-12)$	**15.** $50 - 75$	**16.** $75 - (-50)$
17. $-13 - (-16)$	**18.** $-16 - (-13)$	**19.** $-135 - 70$
20. $85 - 100$	**21.** $-16 - 32$	**22.** $-14 - 32$
23. $-32 - (-14)$	**24.** $81 - (-29)$	**25.** $-18 - 50$

26. 35 − 82	**27.** −15 − (−35)	**28.** −35 − (−15)
29. 85 − (−100)	**30.** 16 − 35	**31.** 0 − 15
32. −170 − 335	**33.** −911 − (−1100)	**34.** −600 − (−190)
35. −100 − (−2000)	**36.** 100 − (−615)	**37.** 851 − 1135
38. 150 − 250	**39.** −13 − 18	**40.** −19 − 35
41. −13 − (−18)	**42.** −50 − (−60)	**43.** −18 − (−10)
44. −28 − (−13)	**45.** 16 − (−16)	**46.** 10 − (−11)
47. −18 − (−18)	**48.** −13 − (−13)	**49.** −859 − (−359)
50. −789 − (−180)		

Fill in the blanks with one or more words.

51. In the problem 35 − 45 = ?, the subtrahend is _____.

52. To subtract one integer from another, we add the additive inverse of the _____ to the _____.

53. In problem 51, the additive inverse of the subtrahend is _____.

54. The additive inverse of a positive integer is a _____ integer. The additive inverse of a negative integer is a _____ integer.

55. _____ is its own additive inverse.

Solve. Show your work.

56. The highest temperature in New York one winter day was 15°, and the lowest temperature was −5°. Find the difference between the highest and the lowest temperatures.

57. Mount Whitney, California is 14,495 feet above sea level, and El Centro, California is 45 feet below sea level. Find the difference in their elevations by subtracting the lowest from the highest elevation.

58. You are playing a game in which you have a score of −68. How many points must you earn to obtain a score of 35? (*Hint:* see Idea 2.)

59. Subtract −16 from −50.

60. The latitude of Lima, Peru is 12° south of the equator (expressed as −12). The latitude of Perth, Australia is 32° south of the equator (expressed as −32). Find the difference between Lima's and Perth's latitudes.

61. Subtract −5 from −10.

62. Subtract −10 from −5.

63. Find the difference between −5 and −10.

64. Find the difference between −10 and −5.

65. Subtract 45,132 from −64,008.

66. Subtract −38,117 from 15,345.

67. Find the difference between −8437 and 19,145.

68. Find the difference between −10,745 and −17,647.

Answers to Practice Problems   7a. −6   b. 13   c. −8   d. 101   e. 0   8a. 40   b. −85   c. −25   9a. 42   b. −5°

1.4 Adding and Subtracting Integers

IDEA 1   Sometimes in mathematics you must find the value of an expression that contains both the operations of subtraction and addition. To compute this value, you could do the indicated operations in order from left to right. However, you could also proceed as follows.

16 Integers

> **To Simplify an Expression Containing Subtraction and Addition:**
>
> 1. Change every operation of subtraction to addition by using the fact that $a - b = a + (-b)$, where $-b$ is the additive inverse of b.
> 2. Solve the resulting addition problem.

Example 10 Find the value of each expression.

a. $-13 + 8 - 17$ b. $-35 - 75 + 14 - (-100)$

c. $-18 - 75 + (-10) - 8$

Solution To simplify an expression containing both the operations of subtraction and addition, first change all operations of subtraction to addition by using the fact that $a - b = a + (-b)$. Then, do the addition.

10a. $-13 + 8 - 17 = -13 + 8 + (-17)$

$= -30 + 8$
$= -22$

10b. $-35 - 75 + 14 - (-100) = -35 + (-75) + 14 + 100$
$= -110 + 114$
$= 4$

10c. $-18 - 75 + (-10) - 8 = -18 + (-75) + (-10) + (-8)$
$= -111$

! If you are having difficulty with a particular topic, it is probably because you have not mastered the skills developed *before* the new topic. For example, to find the value of an expression containing both the operations of subtraction and addition, you must first master how to add integers (Section 1.2) and how to subtract one integer from another (Section 1.3).

Practice Problem 10 *Find the value of the following expressions.*

a. $-81 - 17 + 100$ b. $35 - (-10) + (-18) - 15$

c. $-16 + (-30) - 17 - 8$

IDEA 2

The concepts of income and expenses in business produce problems involving both addition and subtraction.

Example 11 At the end of June, the ADT Paint Company's cash assets were $800. In July, the company's balance sheet indicated the following cash transactions:

1. Painted John Smith's house and collected $5000.

2. Paid $2500 in employees' salaries.

3. Sold paint supplies and collected $3000.

4. Paid Johnson's Can Company $4000.

5. Paid $500 in rent.

How much cash does the ADT Paint Company have at the end of July?

Solution Incoming cash increases the amount of available cash, and outgoing cash decreases it.

$$800 + 5000 - 2500 + 3000 - 4000 - 500$$
$$= 800 + 5000 + (-2500) + 3000 + (-4000) + (-500)$$
$$= 8800 + (-7000)$$
$$= 1800$$

This implies that the ADT Paint Company has $1800 in cash assets at the end of July.

Practice Problem 11 *Solve.*

Cazzie's Auto Supply Company has $5000 in cash. In January, his records indicate that he withdrew $600 for personal use, received $6000 for services rendered, paid $2000 to employees, collected $800 from Dr. Watson, and paid $600 for new equipment. If all of the above were cash transactions, what is Cazzie's cash balance?

IDEA 3 The rules used to find the value of an expression containing both the operations of subtraction and addition can be used to simplify an expression involving only the operation of subtraction.

Example 12 Find the value of each expression.

a. $5 - 7 - 8 - (-13)$ b. $-5 - 10 - 8 - 13 - 15$

Solution To find the value of an expression containing only the operation of subtraction, change every operation of subtraction to addition by using the subtraction rule. Next, do the resulting addition problem.

12a. $5 - 7 - 8 - (-13) = 5 + (-7) + (-8) + 13$
$$= 18 + (-15)$$
$$= 3$$

12b. $-5 - 10 - 8 - 13 - 15 = -5 + (-10) + (-8) + (-13) + (-15)$
$$= -51$$

Practice Problem 12 *Find the value of each expression.*

a. $8 - 15 - 10 - 3$ b. $-2 - 7 - 5 - 6 - 5$

1.4 Exercises

Find the value of each expression.

1. $-8 + 16 - 35$
2. $15 - 37 + (-18)$
3. $8 - (-35) + (-69)$
4. $-17 + (-40) - 6$
5. $-17 - (-35) + 15$
6. $-40 + 17 - (-6)$
7. $-16 - 10 + (-14)$
8. $-20 - 5 + 3$

18   Integers

9. $-17 - 8 + (-5) - 4$

10. $50 - 75 + (-8) - (-17)$

11. $60 - 70 - 40 - 110$

12. $-35 - (-15) + 81 - 3$

13. $-40 - (-50) + (-10) - 30$

14. $85 - 2 - 100 - 74$

15. $-14 + (-40) - (-5)$

16. $-117 - 501 - 171 - 34$

17. $-40 + 12 - 5$

18. $-65 - 401 - 101 - 807$

19. $-15 - 20 + 12 - (-8)$

20. $88 - 41 - 381 - 635$

21. $-8 - 8 + 2 - 5 + 1$

22. $-9 - 9 + 3 - 6 + 2$

23. $-3 - 4 - 1 - 5 - 7$

24. $-1 - 8 - 7 - 3 - 5 - 1$

25. $-10 + 15 - 30$

26. $-15 + 20 - 40$

27. $-10 - (-35) + 15 - 18$

28. $-8 - (-15) + 3 - 9$

29. $15 - 35 + (-8) + 5$

30. $8 - 13 + (-7) + 10$

31. $-40 + 15 - 8 + 2 - 10$

32. $-50 + 10 - 8 + 3 - 8$

33. $10 - 30 + (-15) + 18$

34. $8 - 20 + (-9) + 50$

35. $30 + (-10) + 13 + 17 - 18$

36. $20 + (-10) + 11 + 9 - 13$

37. $70 - 95 + (-5) + 31$

38. $60 - 85 + (-5) + 32$

39. $18 - 25 - 13 - 5$

40. $22 - 32 - 10 - 8$

41. $-18 + 10 - 17 + 5 - (-8) - 10 - 30 + (-3) - 8 + 3 - 50$

42. $-17 + 9 - 16 + 4 - (-7) - 20 - 35 + (-3) - 8 + 7 - 40$

Fill in the blanks.

43. To find the value of an expression containing both the operations of addition and subtraction, change every operation of _____ to _____ by using the fact that _____. Next, do the resulting _____ problem.

44. $5 - 18 + 15 - 8 - (-19) = 5 + ($ _____ $) + 15 + ($ _____ $) + ($ _____ $)$

45. The rule for solving a problem involving both addition and subtraction of integers is also used to find the value of an expression containing only the operation of _____.

Solve and show all work.

46. The PDQ Candy Company owes the Davis Chocolate Company $5000. PDQ's financial records indicate the following cash transactions:
    • Collected $3000 from Kirkwood & Sons.
    • Withdrew $900 for personal use.
    • Paid Davis Chocolate Company $1500.
    • Received $1125 due from Kathy Gross.
    What is PDQ's cash balance after the last transaction?

47. Guerin has $850 in his checking account at the end of April. In May, his records indicate that he wrote checks for $550 and $200, paid a $4 service charge, deposited $30, and wrote another check for $130. If the bank pays all of his checks, does he still have any available cash or is his account overdrawn?

48. In a series of card games, a gambler wins a total of $500. If he later loses $300, wins $50, and loses $250, how much cash does he have?

49. The ACT Moving Company has $30,000 in cash assets. ACT's balance sheet indicates that they paid $8000 for new equipment, paid $1900 for rent, collected $575 from a customer for services rendered, withdrew $15,000 for personal use, and received $3145 from the sale of used furniture. If ACT's business dealings are all cash transactions, how much available cash does the company have after selling the furniture?

Find the value of each expression.

50. $7435 + (-17{,}632) - 8652 + 878$

51. $-1645 - (-8799) + (-10{,}500) - 8762$

52. $-678 - (-8888) - 9567 + (-6542)$

53. $-679 - 784 + (-8111) - 6418$

Answers to Practice Problems   **10a.** 2   **b.** 12   **c.** -71 **11.** $8,600   **12a.** -20 **b.** -25

1.5 Multiplying Integers

IDEA 1

In arithmetic, multiplication is viewed as repeated addition. That is, $2 \cdot 5$ (read as "2 times 5") means the sum of 2 fives, or $2 \cdot 5 = 5 + 5 = 10$. Similarly, $5 \cdot 2$ means the sum of 5 twos, or

$$5 \cdot 2 = 2 + 2 + 2 + 2 + 2 = 10$$

This suggests that *the product of two positive integers is a positive integer*. Note that $2 \cdot 5 = 5 \cdot 2$, suggesting that there is a *commutative law for multiplication*.

Commutative Law for Multiplication

> The order in which two numbers a and b are multiplied does not affect the product,
>
> or
>
> $$a \cdot b = b \cdot a$$

For example,

$$(3)(-5) = (-5)(3)$$

is read as "3 times -5 is equal to -5 times 3."

Multiplication will be indicated by a dot or parentheses. Thus, 3 times 5 is written as $3 \cdot 5$, $3(5)$, or $(3)(5)$.

It is easy to see what happens when a positive integer is multiplied by a negative integer. For example, $(3)(-5)$ means the sum of 3 negative fives, or

$$(3)(-5) = (-5) + (-5) + (-5) = -15$$

However, when a negative integer is to be multiplied by a positive integer, you must be careful. For example, $(-3)(4)$ is meaningless since you cannot add the number 4 "negative three" times. But, since the order in which two numbers are multiplied does not affect the product, we know that

$$(-3)(4) = 4(-3) = -3 + (-3) + (-3) + (-3) = -12$$

Therefore, we can think of $(-3)(4)$ as $(4)(-3)$, or the sum of 4 negative threes. As these examples show, *the product of a positive integer and a negative integer is a negative integer.*

Now consider the following sequence:

$$4(-2) = -8$$
$$3(-2) = -6$$
$$2(-2) = -4$$
$$1(-2) = -2$$
$$0(-2) = 0$$

As the first multipliers decrease by 1 (from 4 to 3 to 2, etc.), the products or answers increase by 2 (from -8 to -6 to -4, etc.). If you continued this pattern for multipliers less than zero, you would obtain

$$(-1)(-2) = 2$$
$$(-2)(-2) = 4$$
$$(-3)(-2) = 6$$
$$(-4)(-2) = 8$$

This sequence suggests that *the product of two negative integers is always positive.* In fact, this is true in our number system.

Based on the above statements, we can state a general rule for multiplying two integers.

To Multiply Two Integers:

1. Multiply their absolute values.
2. If the signs of the two integers are the same, the product is positive. If the signs are different, the product is negative.

Example 13 Multiply.

a. $3(-7)$ **b.** $(-30)(7)$ **c.** $(-3)(-7)$ **d.** $(-10)(-9)$ **e.** $(-7)(0)$

Solution   To multiply two integers, multiply their absolute values. The product is a positive number if the signs are the same. The product is a negative number if the signs are different.

Signs different / Product is negative / Multiply absolute values

13a. $3(-7) = -(3 \cdot 7)$
$= -21$

13b. $(-30)(7) = -(30 \cdot 7)$
$= -210$

13c. $(-3)(-7) = +(3 \cdot 7)$
$= 21$

13d. $(-10)(-9) = +(10 \cdot 9)$
$= 90$

13e. $(-7)(0) = 0$ The product of zero and any integer is zero.

Practice Problem 13 **Multiply.**

a. $(-4)(8)$ **b.** $(-1)(9)$ **c.** $(5)(-7)$ **d.** $(-6)(-3)$

1.5 Multiplying Integers

IDEA 2

Suppose you wanted to multiply more than two integers. To do this, you could multiply them two at a time as follows:

$$(-2)(3)(-5)$$
$$= -(2 \cdot 3)(-5)$$
$$= (-6)(-5)$$
$$= +(6 \cdot 5)$$
$$= 30$$

$$(-2)(-3)(-5)$$
$$= +(2 \cdot 3)(-5)$$
$$= (6)(-5)$$
$$= -(6 \cdot 5)$$
$$= -30$$

However, it is easier simply to count the number of negative signs. The problems above indicate that a product involving two (or an even number of) negative signs results in a positive answer, and that a product involving three (or an odd number of) negative signs yields a negative answer. Using this concept, we can rewrite the rule for multiplying integers.

> **To Multiply Two or More Integers:**
>
> 1. Multiply their absolute values.
> 2. If the original expression contained an odd number of negative signs, the product is negative; if the original expression contained an even number of negative signs, the product is positive.

Example 14 Multiply.

    a. $(-2)(-3)(5)(-2)$ **b.** $(5)(-2)$

    c. $(-2)(3)(-5)(2)$ **d.** $(-2)(-3)(2)(-5)(-3)$

Solution   To multiply two or more integers, multiply their absolute values. The product is a positive number if the problem contains an even number of negative signs. The product is a negative number if the problem contains an odd number of negative signs.

              3 negative signs   Multiply absolute values
              ↓ ↓ ↓   ↓

14a. $(-2)(-3)(5)(-2) = -(2 \cdot 3 \cdot 5 \cdot 2)$
$$= -60$$

14b. $5(-2) = -(5 \cdot 2)$
$$= -10$$

14c. $(-2)(3)(-5)(2) = +(2 \cdot 3 \cdot 5 \cdot 2)$
$$= 60$$

14d. $(-2)(-3)(2)(-5)(-3) = +(2 \cdot 3 \cdot 2 \cdot 5 \cdot 3)$
$$= 180$$

IDEA 3

As for addition, there is an *associative law for multiplication*.

Associative Law for Multiplication

The order in which three (or more) numbers a, b, and c are multiplied does not affect the product,

or

$(a \cdot b) \cdot c = a \cdot (b \cdot c)$

For example,

$(2 \cdot 3) \cdot 5 = 2 \cdot (3 \cdot 5)$.

Practice Problem 14 **Multiply.**

a. $(-5)(-3)(2)$ c. $(-2)(-2)(-3)(5)(-2)$
b. $(-2)(-2)(3)(-2)$ d. $35(-3)$

1.5 Exercises

Multiply.

1. $(-5)(-2)$
2. $(-2)(-3)(5)(-2)$
3. $5(-3)$
4. $(-6)(-3)$
5. $4(-8)$
6. $(-7)(-2)(10)$
7. $(-6)(4)$
8. $(2)(-3)(5)(-2)$
9. $(-10)(-2)$
10. $(3)(-2)(-2)$
11. $(4)(3)(-2)$
12. $(-2)(-2)(-2)(-2)$
13. $(0)(-10)$
14. $(-10)(5)(-2)$
15. $(-5)(3)(-2)$
16. $17(-8)$
17. $(-5)(11)$
18. $(-6)(-9)$
19. $(-7)(-2)(-2)$
20. $(-8)(-9)$
21. $(-2)(-2)(3)(-2)(-5)$
22. $(15)(-3)(21)$
23. $(3)(-4)(-2)$
24. $(-2)(3)(-5)(-2)(3)(5)$
25. $(-1)(-1)(-4)(-1)(1)(-2)$
26. $(-1)(-1)(-2)(-1)(1)(-2)$
27. $3(-35)$
28. $4(-12)$
29. $(-4)(-7)$
30. $(-8)(-7)$
31. $(-2)(-2)(-2)$
32. $(-3)(-3)(-3)$
33. $(-2)(-2)(-2)(-2)$
34. $(-3)(-3)(-3)(-3)$
35. $(2)(-10)(3)$
36. $(3)(-8)(2)$
37. $(-3)(1)(-2)(-2)$
38. $(-3)(2)(-1)(-1)$
39. $(2)(-3)(-2)(5)(2)$
40. $(3)(-2)(-2)(5)(2)$

Fill in the blanks.

41. Multiplication is repeated _____.
42. $5(-6)$ can be thought of as the sum of 5 _____.
43. If you multiply a positive integer and a negative integer, the answer is always _____.
44. The product of two negative integers is always _____.
45. The product of two or more integers is a _____ number if the problem contains an odd number of negative signs.
46. The product of two or more integers is a _____ number if the problem contains an even number of negative signs.

47. To multiply two or more integers, you should multiply their _____ and then determine the _____.

Express each answer as an integer. Show all work.

48. Mr. Jones has 100 shares of a stock that has decreased $5 per share in value. If a $5 decrease in one share is expressed as a −5, what was his total loss?

49. Shirley lost three pounds per week for 10 consecutive weeks. If a three-pound weight loss is represented by −3, how much weight did she lose?

50. The ADT Paint Company laid off 90 employees who each earn $200 a week. If each person's salary decreases the company's weekly expenses by $200 (−200), then what is the total decrease in the company's weekly expenses?

51. Find the product of −3, −9, −10, and −10.
52. Find the product of −2, −2, −5, and −5.
53. Multiply −35 and −402.
54. Multiply −63 and 504.
55. Find the product of −10, 20, −18, and −13.
56. Find the product of −16, −18, −10, and −9.

Answers to Practice Problems **13a.** −32 **b.** −9 **c.** −35 **d.** 18 **14a.** 30 **b.** −24 **c.** 120 **d.** −105

1.6 Dividing Integers

IDEA 1

In arithmetic, to divide 12 by 6 $\left(\text{written as } \frac{12}{6} \text{ or } 12 \div 6\right)$, we must find an integer c such that $6 \cdot c = 12$. Clearly, c is equal to 2. In other words, $\frac{12}{6} = 2$ because $6 \cdot 2 = 12$. In general, $\frac{a}{b} = c$ if and only if $b \cdot c = a$ for a number c. In this example, a is called the **dividend,** b is the **divisor,** and c is the **quotient.**

Please note that zero divided by any nonzero number always equals zero. For example, $\frac{0}{22} = 0$ since $0 \cdot 22 = 0$. Division by zero, however is meaningless. For example, $\frac{5}{0}$ is said to be undefined since there is no number c such that $0 \cdot c = 5$. If a problem involves division by zero, write "undefined."

Let us now use the fact that $\frac{a}{b} = c$ if and only if $b \cdot c = a$ to help us determine how to divide one integer by another.

Example 15 Divide to find the quotient.

  a. $\frac{-6}{2}$ **b.** $\frac{-6}{-2}$ **c.** $\frac{6}{-2}$ **d.** $\frac{0}{-4}$ **e.** $\frac{5}{0}$

Solution

15a. $\frac{-6}{2} = -3$ since $2(-3) = -6$

15b. $\frac{-6}{-2} = 3$ since $-2(3) = -6$

15c. $\frac{6}{-2} = -3$ since $-2(-3) = 6$

15d. $\frac{0}{-4} = 0$ since $-4(0) = 0$

15e. $\frac{5}{0}$ = undefined   since division by zero is undefined

Based on the results in Example 15 we can state a rule for dividing one integer by another.

> **To Divide One Integer by Another:**
>
> 1. Divide their absolute values.
> 2. If the dividend and divisor have the same sign, the quotient is positive. If the signs are different, the quotient is negative.

Example 16 Divide.

a. $\frac{-30}{2}$ b. $\frac{-15}{-3}$ c. $\frac{75}{-5}$ d. $\frac{-60}{-10}$

Solution To divide one integer by another, divide their absolute values. The quotient is a positive number if the signs are the same. The quotient is a negative number if the signs are different.

16a. (Signs different, Divide absolute values) $\frac{-30}{2} = -\left(\frac{30}{2}\right) = -15$

16b. (Signs same, Divide absolute values) $\frac{-15}{-3} = +\left(\frac{15}{3}\right) = 5$

16c. $\frac{75}{-5} = -\left(\frac{75}{5}\right) = -15$

16d. $\frac{-60}{-10} = +\left(\frac{60}{10}\right) = 6$

! If you have divided correctly, the divisor times the quotient must be equal to the dividend. For example, $\frac{-30}{2} = -15$ since $2(-15) = -30$.

Practice Problem 15 *Divide.*

a. $\frac{-16}{2}$ b. $\frac{0}{-16}$ c. $\frac{-16}{-8}$ d. $\frac{-15}{0}$

1.6 Exercises

Divide and check your answers.

1. $\frac{-16}{-2}$ 2. $\frac{-36}{4}$ 3. $\frac{-15}{5}$ 4. $\frac{0}{-9}$ 5. $\frac{-35}{7}$

6. $\frac{16}{-2}$ 7. $\frac{-100}{2}$ 8. $\frac{-9}{0}$ 9. $\frac{-75}{-5}$ 10. $\frac{-28}{4}$

11. $\dfrac{-66}{-6}$ 12. $\dfrac{-728}{8}$ 13. $\dfrac{-35}{-35}$ 14. $\dfrac{-8118}{9}$ 15. $\dfrac{-35}{35}$

16. $\dfrac{255}{-5}$ 17. $\dfrac{65}{-5}$ 18. $\dfrac{0}{7}$ 19. $\dfrac{-70}{7}$ 20. $\dfrac{-800}{-20}$

21. $\dfrac{-88}{-4}$ 22. $\dfrac{19}{0}$ 23. $\dfrac{-64}{16}$ 24. $\dfrac{-32}{16}$ 25. $\dfrac{-16}{-2}$

26. $\dfrac{-18}{-9}$ 27. $\dfrac{64}{-8}$ 28. $\dfrac{32}{-2}$ 29. $\dfrac{500}{-5}$ 30. $\dfrac{600}{-2}$

31. $\dfrac{-26}{-13}$ 32. $\dfrac{-39}{-13}$ 33. $\dfrac{-65}{5}$ 34. $\dfrac{-75}{5}$ 35. $\dfrac{-622}{-2}$

36. $\dfrac{-844}{-4}$ 37. $\dfrac{0}{-1001}$ 38. $\dfrac{0}{-2001}$ 39. $\dfrac{-1001}{0}$ 40. $\dfrac{-2001}{0}$

Fill in the blanks.

41. If you divide a negative integer by a negative integer, the quotient or answer is _____.

42. If you divide a positive integer by a negative integer or a negative integer by a positive integer, the quotient or answer is _____.

43. To divide one integer by another, divide their _____ and then determine the _____ of your final answer.

44. Zero divided by a nonzero integer is _____.

45. Any integer divided by zero is _____.

Express each answer as an integer. Show all work.

46. A gambler lost a total of $1000 (−1000) in five consecutive hands of poker. Find his average loss per hand by dividing the amount he lost by the number of hands he played.

47. Twenty people lost a total $20,000 (−20,000) in a business deal. Find the average amount of each person's loss by dividing the total loss by the number of people involved in the deal.

48. One day in Sheboygan, Wisconsin, the temperature dropped 30° (−30) in 10 hours. Find the average temperature change per hour.

49. Divide −5535 by 123. 50. Divide −10,413 by −89.

51. Divide −77 into −8008. 52. Divide −63 into 9576.

53. Ten people lost a total of $1,000 (−1000) in a business deal. Find the average amount of each person's loss.

54. One day in Chicago, the temperature dropped 20° (−20) in five hours. Find the average temperature change per hour.

Answers to Practice Problems **15a.** −8 **b.** 0 **c.** 2 **d.** undefined

1.7 Order of Operations

IDEA 1 Suppose you wanted to simplify an expression containing more than one operation. For example, to simplify $3 + 4 \cdot 2$, you could add first and the answer would be:

$$3 + 4 \cdot 2 = 7 \cdot 2$$
$$= 14$$

26 Integers

However, if you multiplied first, your answer would be:

$$3 + 4 \cdot 2 = 3 + 8$$
$$= 11$$

Obviously, this situation could create many problems. To avoid confusion, mathematicians developed an order of operations rule. According to this rule we multiply and divide before adding and subtracting. Thus $3 + 4 \cdot 2 = 3 + 8 = 11$.

> **To Simplify an Expression Containing More than One Operation:**
>
> **1.** First, do all multiplication and/or division as you work from left to right.
>
> **2.** Next, do all addition and/or subtraction.

Example 17 Simplify.

a. $-3 + 4 \cdot 2$ **b.** $-7 - 5 \cdot 2$

c. $35 + (-50) \div 10$ **d.** $10 - 6 \cdot 5 + (-40) - 8 \div (-2)$

Solution To simplify an expression containing more than one operation do all multiplication and/or division before adding and/or subtracting.

17a. $-3 + \boxed{4 \cdot 2}$
$= -3 + 8$ Multiply
$= 5$ Add

17b. $-7 - \boxed{5 \cdot 2}$
$= -7 - 10$ Multiply
$= -7 + (-10)$ Subtraction rule
$= -17$

17c. $35 + \boxed{(-50) \div 10}$
$= 35 + (-5)$ Divide
$= 30$ Add

17d. $10 - \boxed{6 \cdot 5} + (-40) - \boxed{8 \div (-2)}$
$= 10 - 30 + (-40) - (-4)$ Multiply and divide
$= 10 + (-30) + (-40) + 4$ Subtraction rule
$= 14 + (-70)$
$= -56$ Add

If an expression contains the operations of multiplication and division multiply and divide in order from left to right. For example,

$$-6 \cdot 2 \div (-4) = -12 \div (-4)$$
$$= 3$$

Practice Problem 16 *Simplify.*

a. $8 + 5 \cdot 2$ **b.** $-16 - 30 \cdot 2$

c. $-12 - 6(-3) + 40 \div (-10)$ **d.** $-60 \div 15 \cdot 7$

IDEA 2

In Idea 1 we said to multiply and/or divide before adding and/or subtracting. In some cases, however, a problem may call for adding or subtracting *before* multiplying or dividing. When this happens, parentheses ( ), brackets [ ], or braces { }, are placed around the part of the problem that must be computed out of the order stated in Idea 1. The ( ), [ ], and { } are *grouping symbols*. **Grouping symbols** are pairs of symbols used to indicate that whatever is inside the pairs is to be treated as one quantity or number.

To simplify an expression containing more than one operation and only one grouping symbol, we will simplify inside that symbol first. For example,

$$5(2 - 6) = 5(-4) = -20$$

We read $5(2 - 6)$ as "five times the quantity two minus six."

To Simplify an Expression Containing One Pair of Grouping Symbols:

1. First, do all operations inside the grouping symbols. Remember to multiply and/or divide before adding and/or subtracting.
2. Next, multiply and/or divide as you work from left to right.
3. Finally, add and/or subtract.

Example 18 Simplify.

a. $40 + 3(8 - 15)$

b. $3 - 6(-5 + 7 \cdot 2)$

c. $-16 + (-80 + 6 \cdot 6) \div 4$

d. $-80 - 20(-8 + 16 \div 8) + 4(-12 - 2 \cdot 5)$

Solution To simplify an expression containing more than one operation and one pair of grouping symbols, first do the operations inside the grouping symbols. Next, follow the order of operation rule.

18a. $40 + 3(8 - 15)$
$= 40 + 3(-7)$ Simplify inside the ( )
$= 40 + (-21)$ Multiply
$= 19$ Add

18b. $3 - 6(-5 + 7 \cdot 2)$
$= 3 - 6(-5 + 14)$ Simplify inside the ( )
$= 3 - 6(9)$
$= 3 - 54$ Multiply
$= -51$ Subtract

18c. $-16 + (-80 + 6 \cdot 6) \div 4$
$= -16 + (-80 + 36) \div 4$ Simplify inside the ( )
$= -16 + (-44) \div 4$
$= -16 + (-11)$ Divide
$= -27$ Add

18d.
$$-80 - 20(-8 + 16 \div 8) + 4(-12 - 2 \cdot 5)$$
$$= -80 - 20(-8 + 2) + 4(-12 - 10) \quad \text{Simplify inside each ( )}$$
$$= -80 - 20(-6) + 4(-22)$$
$$= -80 - (-120) + (-88) \quad \text{Multiply}$$
$$= -80 + 120 + (-88) \quad \text{Subtraction Rule}$$
$$= -168 + 120 \quad \text{Add}$$
$$= -48$$

Earlier we used a bar to indicate division. For example, $\frac{-6}{2}$ means $-6 \div 2$. A bar can also be used as a grouping symbol. For example,

$$\frac{-40 + 12}{-7 + 3}$$

means $(-40 + 12) \div (-7 + 3)$. To simplify this expression, simplify the expression above the bar, simplify the expression below the bar, and then divide. That is,

$$\frac{-40 + 12}{-7 + 3} = \frac{-28}{-4}$$
$$= 7$$

Practice Problem 17 **Simplify.**

a. $-20 + 3(-5 - 2)$ **b.** $-4 + (-16 + 3 \cdot 4) \div 2$

c. $-16 + 3(-7 + 2 \cdot 5) - 2(-30 + 8 \cdot 3)$ **d.** $\frac{(-4 + 7)(-2)}{-6}$

IDEA 3

Sometimes an expression may contain a grouping symbol within a grouping symbol (nested symbols). To simplify this type of expression, we simplify within the innermost grouping symbol first and work outwards. For example, $4[(2 - 3) + 6] = 4[-1 + 6] = 4[5] = 20$.

> **To Simplify an Expression Containing Nested Grouping Symbols:**
>
> 1. First, simplify inside the innermost grouping symbol. Remember to multiply and/or divide before adding and/or subtracting.
>
> 2. Next, repeat step 1 until all symbols of grouping have been eliminated.
>
> 3. Finally, multiply and/or divide before adding and/or subtracting in the final expression.

Example 19 Simplify.

a. $3[-10 + 4(2 - 3)] + 2$ **b.** $-6 + 4[(-2 - 3) + (-5 + 3 \cdot 2)]$

c. $-2\{1 + 3[-2 + 5(-6 + 9)]\}$

Solution To simplify an expression containing more than one operation and nested grouping symbols, simplify inside the innermost grouping symbol first.

19a. $3[-10 + 4(2 - 3)] + 2$
$= 3[-10 + 4(-1)] + 2$ Simplify inside the ( ) first
$= 3[-10 + (-4)] + 2$ Simplify inside the [ ]
$= 3[-14] + 2$
$= -42 + 2$ Multiply
$= -40$ Add

19b. $-6 + 4[(-2 - 3) + (-5 + 3 \cdot 2)]$
$= -6 + 4[-5 + (-5 + 6)]$ Simplify inside the ( ) first
$= -6 + 4[-5 + 1]$
$= -6 + 4[-4]$ Simplify inside the [ ]
$= -6 + [-16]$ Multiply
$= -22$ Add

19c. $-2\{1 + 3[-2 + 5(-6 + 9)]\}$
$= -2\{1 + 3[-2 + 5(3)]\}$ Simplify inside the ( ) first
$= -2\{1 + 3[-2 + 15]\}$
$= -2\{1 + 3[13]\}$ Simplify inside the [ ]
$= -2\{1 + 39\}$ Simplify inside the { }
$= -2\{40\}$
$= -80$ Multiply

Practice Problem 18 *Simplify.*

    a. $-3[2 + 4(5 - 7)]$ **b.** $-6 + 3[2 + 3(4 - 6) + (-2 - 3)]$

    c. $-4\{2 + 3[-2 + 5(-11 + 5)]\}$

1.7 Exercises

Simplify.

1. $-6 + 2 \cdot 5$
2. $-8 + 30 \div (-5)$
3. $-6 + 6(-3) + 15 \div (-3)$
4. $-6 \cdot 10 \div 2$
5. $-2 - 6 \cdot 5 + (-30) + 6 \div (-2)$
6. $3(-2 - 4)$
7. $(4 + 4)(4 - 4)$
8. $4 + 4(4 - 4)$
9. $-8 + 2(-3 + 5 \cdot 2)$
10. $-5 + (18 + 2 \cdot 3) \div 3$
11. $-18 - 5(-16 + 8 \div 2) + (-6 + 2 \cdot 5)$
12. $3[2 + 6(-2 - 3)]$
13. $-5 + 3[(-5 - 2) + (-7 + 4 \cdot 2)]$
14. $2[-1 + 4\{-2 \cdot (-12 + 3) \cdot 2\} + 5]$
15. $-2 + 3[(-1 + 4 \cdot 2) + (-5)]$
16. $3[-10 + 5(-3 + 1)]$
17. $-11 + 5 \cdot 2 - 55 + 5 - 30 \div 6$
18. $-6 + (-8 + 2 \cdot 2) \div 2 - 4(-8 - 2)$
19. $-60 \div [(-8 + 5) + 4(2 - 4) + 5]$
20. $2\{-3 + 2[-5 + 2(3 + 1)]\}$
21. $\dfrac{-30 - 5}{-5}$
22. $\dfrac{-40 + 5}{-5}$
23. $\dfrac{-10 \cdot 6}{-5 + 10}$
24. $\dfrac{-14 \cdot 5}{2 + 3}$
25. $-10 + 3(2 - 5)$
26. $-15 + 4(8 - 9)$
27. $3(-2 + 5 \cdot 2)$
28. $4(-8 + 2 \cdot 3)$
29. $-8 + (-15 + 2 \cdot 3) \div 3$
30. $-10 + (5 - 4 \cdot 8) \div 3$

Integers

31. $8 - 8 \div 2 + 5 \cdot 3 - 6 \div 2$
32. $-8 - 6 \div 2 + 4 \cdot 2 - 8 \div 4$
33. $-3 + 3(-3 - 3 \cdot 2)$
34. $-4 + 4(-4 - 4 \cdot 2)$
35. $5 - 8[2 - 4(3 + 1 \cdot 4)]$
36. $8 - 10[6 - 7(-3 - 1 \cdot 2)]$
37. $3[-8 + 5(-8 + 2 \cdot 3)] - 10$
38. $2[-7 + 3(-6 + 4 \cdot 2)] - 8$
39. $2\{-4 + 2[-6 + 2(3 - 4)]\}$
40. $3\{-2 + 3[-7 + 2(-3 - 2)]\}$

Fill in the blanks.

41. To simplify an expression involving several operations and no grouping symbols, we should _____ and/or _____ before _____ and/or _____ .

42. If a problem only involves multiplication and division and no grouping symbols, we work from _____ to _____ .

43. To simplify an expression containing several operations and parentheses, we should do all the operations inside the _____ first. However, you must still remember to _____ and _____ before _____ and _____ .

44. In the problem $(6 + 8) \cdot 2$, the parentheses are used to indicate that you should do the operation of _____ before you do the operation of _____ .

45. To simplify an expression involving several operations and nested grouping symbols, you should simplify inside the _____ first.

46. In the problem $2[3 + 5(3 + 2)]$, the parentheses are used as a _____ and also indicate the operation of _____ .

Express each answer as an integer. Show all work.

47. Dan Johnson buys blue jeans for $15 each and resells them for $23 each. If Dan sells 35 pairs of jeans, find the total profit from the sale. (*Hint:* use the fact that the total profit is equal to the resale price minus the original cost times the number of items sold.)

48. In a "going-out-of-business" sale, Cazzie sold three washers at a profit of $30 each, eight dryers at a loss of $10 each, nine suits at a profit of $60 each, and four desks at a loss of $50 each. If profits are represented by a positive integer and losses are represented by a negative integer, find the total profit or loss on these sales.

Simplify each expression.

49. $-432 + 28(-601) - 55$
50. $-107 + 18(-115) + 605 \div (-55)$
51. $-815 + (-770 + 18 \cdot 5) + 2(-55 - 70)$
52. $-755 - 18(-611 + 555) + 810 \div (-30)$

Answers to Practice Problems   16a. 18   b. -76 c. 2   d. -28 17a. -41 b. -6 c. 5   d. 1   18a. 18   b. -33 c. 376

1.8 Algebraic Expressions and Formulas

IDEA 1   A **variable** is a symbol (usually a letter) that represents a number or an unknown quantity. For example, x, y, and z are variables. An **algebraic expression** is a collection of numbers, variables, operation symbols, and/or grouping symbols. For example, $x + 6$, $3y - 9$, and $-5x + 3(x + 2)$ are all algebraic expressions. Note that in the expression $3y - 9$ that $3y$ means 3 times y ($3y = 3 \cdot y$).

An algebraic expression represents a number whenever each variable in the expression is replaced by a number. For example, the algebraic expression $3y - 9$ represents the number -3 when y is replaced by 2.

$$3y - 9 = 3(2) - 9 = 6 - 9 = -3$$
↑
Replace y with 2

This process of finding a value of an algebraic expression is referred to as *evaluating an algebraic expression*.

> **To Evaluate an Algebraic Expression:**
>
> 1. Replace each variable with its given numerical value.
> 2. Simplify the resulting expression.

Example 20 Evaluate each algebraic expression.

    a. $2x - 3y$, $x = 3$ and $y = -2$

    b. $4(x - 7) + 6(z + 4)$, $x = -3$ and $z = -8$

Solution To evaluate an algebraic expression, first replace each letter with its given numerical value. Then, simplify the resulting expression.

20a. $2x - 3y$
$= 2(3) - 3(-2)$ Replace x with 3 and y with -2
$= 6 - (-6)$ Multiply
$= 12$ Subtract

20b. $4(x - 7) + 6(z + 4)$
$= 4(-3 - 7) + 6(-8 + 4)$ Replace x with -3 and y with -8
$= 4(-10) + 6(-4)$ Simplify inside ( )
$= -40 + (-24)$ Multiply
$= -64$ Add

Evaluate each algebraic expression.

    a. $3x - 6$, $x = 1$ **b.** $4x + 3(y - 4)$, $x = 3$ and $y = -2$

IDEA 2

Many problems in mathematics, business, physics, and chemistry require the use of formulas. A mathematical **formula** is a rule that uses letters to express a mathematical relationship between two or more quantities. For example, the formula

$$A = L \cdot W$$

expresses the fact that the area of a rectangle is equal to the length of the rectangle times its width.

Remember that if two or more letters (or a number and a letter) are placed next to each other in a formula, the operation indicated is multiplication. The formula $A = L \cdot W$ is usually written $A = LW$.

The ability to evaluate an algebraic expression will prove to be a very useful skill when evaluating a formula. For example, to find the perimeter (P) of a rectangle

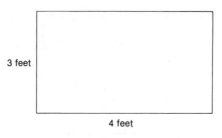

Figure 1-8

(total distance around the figure) whose length (L) is 4 feet and whose width (W) is 3 feet (see Figure 1-8), evaluate the formula $P = 2L + 2W$ when $L = 4$ feet and $W = 3$ feet. In other words,

$$P = 2L + 2W$$
$$= 2(4 \text{ feet}) + 2(3 \text{ feet})$$
$$= 8 \text{ feet} + 6 \text{ feet}$$
$$= 14 \text{ feet}$$

The perimeter of the rectangle is 14 feet.

To Evaluate a Formula:

1. Substitute (replace) the given values for the letters in the formula.
2. Simplify the resulting expression.

Example 21 Evaluate.

a. $E = Q - W$ when $Q = -4$ and $W = 16$

b. $V = S + 4T$ when $S = 6$ and $T = -2$

c. $P = a + 2(c + 3d)$ when $a = -6$, $c = -10$, $d = 5$

Solution To evaluate a formula, replace all letters with their given numerical values and then perform the indicated operations.

21a. $E = Q - W$
$= -4 - 16$
$= -20$

21b. $V = S + 4T$
$= 6 + 4(-2)$
$= 6 + (-8)$
$= -2$

21c. $P = a + 2(c + 3d)$
$= -6 + 2(-10 + 3 \cdot 5)$
$= -6 + 2(-10 + 15)$
$= -6 + 2(5)$
$= -6 + 10$
$= 4$

In solution 21b, when -2 is substituted for T into the formula, the -2 is placed within parentheses to ensure that the operation of multiplication is indicated properly. In general, multiplication of integers a and $-b$ is indicated by the expression $a(-b)$.

1.8 Algebraic Expressions and Formulas

Practice Problem 20 **Evaluate.**

a. $M = a - b + c$ when $a = -6, b = 7, c = -5$

b. $A = b + mx$ when $m = 4, b = -5, x = -3$

c. $B = a + 2(c + 3f)$ when $a = -4, c = -15, f = 4$

Sometimes a formula will contain a bar.

To Evaluate a Formula Containing a Bar:

1. First, substitute the given values for the letters in the formula.
2. Next, simplify the expression above and below the bar separately.
3. Divide the results.

Example 22 Evaluate.

a. $M = \dfrac{a - b}{c - d}$ when $a = 7, b = -5, c = 3, d = 5$

b. $X = \dfrac{md + MD}{m + M}$ when $m = 5, d = -8, M = 2, D = 6$

c. $N = \dfrac{4[x - d]}{d}$ when $x = -6, d = -2$

Solution To evaluate a formula containing a bar, replace all letters with their given numerical values. Next, simplify the expression above and below the bar. Then, divide.

22a. $M = \dfrac{a - b}{c - d}$

$= \dfrac{7 - (-5)}{3 - 5}$

$= \dfrac{12}{-2}$

$= -6$

22b. $X = \dfrac{md + MD}{m + M}$

$= \dfrac{5(-8) + 2(6)}{5 + 2}$

$= \dfrac{-40 + 12}{7}$

$= \dfrac{-28}{7}$

$= -4$

22c. $N = \dfrac{4[x - d]}{d}$

$= \dfrac{4[-6 - (-2)]}{-2}$

$= \dfrac{4[-4]}{-2}$

$= \dfrac{-16}{-2}$

$= 8$

Practice Problem 21 **Evaluate.**

a. $S = \dfrac{a + b}{a - b}$ when $a = -4, b = -2$

b. $X = \dfrac{mb + mc + md}{m + n}$ when $m = 3$, $b = -4$, $c = 2$, $d = -6$, $n = -5$

c. $C = \dfrac{5(F - 32)}{9}$ when $F = -4$

1.8 Exercises

Evaluate.

1. $A = K \div L$ when $K = -60$ and $L = -10$
2. $P = S - C$ when $S = -5$ and $C = -12$
3. $T = L - Bx$ when $L = 10$, $B = -7$, $x = -2$
4. $D = KRt$ when $K = -3$, $R = 5$, $t = -2$
5. $F = c + p$ when $c = -21$ and $p = 57$
6. $B = a - b + c - d$ when $a = -5$, $b = 12$, $c = 13$, $d = 30$

Fill in the blanks with one or more words.

7. A rule that uses letters to express a mathematical relationship between two or more quantities is called a _____.

8. In the formula $y = mx + b$, mx means m _____ x.

9. The first step in finding the value of a formula is to replace the _____ in the formula with their given numerical values.

10. The process of finding a value for an algebraic expression is called _____ an algebraic expression.

Evaluate each expression for $x = -3$, $y = -4$, $z = 5$.

11. $x + yz$
12. $x - yz$
13. $x(y + 2) - 40$
14. $x - 6(y - 2)$
15. $(x - 3)(y - 4)$
16. $(y - 8)(z + 2)$
17. $x - y$
18. $x - z$

Evaluate each expression when $a = 4$ and $b = -2$.

19. $8a + 3b + 5$
20. $4a + 2b + 7$
21. $|b| - a$
22. $-|a| - b$
23. $-6(a - b)$
24. $-8(-a + b)$
25. $2(3a + 4) - 3(2b - 1)$
26. $-4(2a - 1) - 5(3b + 7)$
27. $5a + 8b - 6a + 4b$
28. $-4a + 7b - 6a + 6b$

Evaluate.

29. $-3 + n[2a + (n - 1)d]$ when $a = -15$, $d = 2$, and $n = 10$
30. $-10 + (5 - ab) \div c$ when $a = 4$, $b = 8$, and $c = -3$

The perimeter of a rectangle is given by the formula $P = 2L + 2W$. Find the perimeter of a given rectangle having the following dimensions.

31. $L = 18$ feet and $W = 14$ feet
32. $L = 63$ inches and $W = 19$ inches

The change in internal energy ("E") is expressed by the formula $E = Q - W$, where $Q = $ change in heat and $W = $ work done. Find a value for "E" given the following information:

33. $Q = -10$, $W = -30$
34. $Q = 30$, $W = -10$
35. $Q = 30$, $W = -100$
36. $Q = 0$, $W = 30$
37. $Q = 0$, $W = -30$
38. $Q = -20$, $W = -40$

Given the formula $m = \dfrac{x - y}{a - b}$, *find the value of m given the following information:*

39. $x = -16, y = -2, a = 4, b = 6$
40. $x = -8, y = 2, a = 5, b = 10$
41. $x = 3, y = -12, a = -2, b = -7$

Given the formula $f = \dfrac{p \cdot q}{p + q}$, *find a value for f given the following information:*

42. $p = 12, q = -3$
43. $p = 10, q = -5$
44. $p = -6, q = 10$
45. $p = -8, q = 4$

The formula $°F = 32 + \dfrac{9}{5}°C$ *is used to convert a temperature from Celsius (C) to Fahrenheit (F). Find a value for F given:*

46. $C = -15°$
47. $C = -20°$
48. $C = 5°$
49. $C = 15°$
50. $C = -10°$
51. $C = -30°$
52. $C = 20°$
53. $C = 10°$

Given the formula $M = \dfrac{ax + by + cz}{a + b + c}$, *find the value of M given the following information:*

54. $a = 2, b = 4, c = 6, x = -3, y = -9, z = -7$
55. $a = 10, b = 15, c = 25, x = -5, y = -10, z = 2$

The sum S of the first n terms of an arithmetic progression is given by the formula $S = \dfrac{n[2a + (n - 1)d]}{2}$, *where n = the number of terms, a = the first term, and d = the difference between any two consecutive terms. Find a value for S given the following information:*

56. $a = -15, d = 2, n = 30$
57. $a = -5, d = 3, n = 10$
58. $a = -2, d = -3, n = 14$

Answers to Practice Problems 19a. -3 b. -6 20a. -18 b. -17 c. -10 21a. 3 b. 12 c. -20

Chapter 1 Summary

Important Terms

The numbers 0, 1, 2, 3, 4, 5, ... are called **whole numbers**. [Section 1.1/Idea 1]

The numbers ... $-2, -1, 0, 1, 2, ...$ are called **integers**. [Section 1.1/Idea 1]

A **positive integer** is greater than zero. Positive integers are located to the right of zero on a **number line**. 2, 20, and 101 are positive integers. Positive integers are also called **natural numbers**. [Section 1.1/Idea 1]

A **negative integer** is less than zero. Negative integers are located to the left of zero on a number line. $-6, -20,$ and -101 are negative integers. [Section 1.1/Idea 1]

The **absolute value** of an integer x, written as $|x|$, is the distance between 0 and x on the number line. $|-8| = 8$ and $|8| = 8$. [Section 1.1/Idea 2]

The **absolute value property** says that if x is any positive number or zero, then the absolute value of x and the absolute value of $-x$ are both equal to x. $|x| = x$ and $|-x| = x$. [Section 1.1/Idea 2]

The **commutative law for addition** says

that the order in which two numbers are added does not affect the sum. $a + b = b + a$. [Section 1.2/Idea 3]

The **associative law for addition** says that the order in which three (or more) numbers are added does not affect the sum. $(a + b) + c = a + (b + c)$. [Section 1.2/Idea 3]

The **additive inverse** of an integer x is that integer that can be added to x so that the sum will be zero. The additive inverse of -6 is 6. [Section 1.3/Idea 1]

In the statement $a - b = c$, a is called the **minuend**, b is called the **subtrahend**, and c is called the **difference**. 10 is the minuend, 6 is the subtrahend, and 4 is the difference in $10 - 6 = 4$. [Section 1.3/Idea 2]

The **commutative law for multiplication** says that the order in which two numbers are multiplied does not affect the product. $a \cdot b = b \cdot a$. [Section 1.5/Idea 1]

The **associative law for multiplication** says that the order in which three numbers are multiplied does not affect the product. $(a \cdot b) \cdot c = a \cdot (b \cdot c)$. [Section 1.5/Idea 3]

In the statement $\frac{a}{b} = c$ or $a \div b = c$, a is called the **dividend**, b is called the **divisor**, and c is called the **quotient**. 20 is the dividend, 5 is the divisor, and 4 is the quotient in $\frac{20}{5} = 4$ and $20 \div 5 = 4$. [Section 1.6/Idea 1]

Grouping symbols are pairs of symbols used to indicate that whatever is inside the pairs are to be treated as one quantity or number. Parentheses ( ), brackets [ ], and braces { } are examples of symbols of grouping. [Section 1.7/Idea 2]

A **variable** is a letter that represents a number or unknown quantity. x, y, and z are variables. [Section 1.8/Idea 1]

An **algebraic expression** is a collection of numbers, variables, operation symbols, and/or grouping symbols. $2x + y$ is an algebraic expression. [Section 1.8/Idea 1]

A mathematical **formula** is a rule that uses variables to express a mathematical relationship between two or more quantities. $A = LW$ is a formula where A is the area, L is the length, and W is the width of a rectangle. [Section 1.8/Idea 2]

Important Skills

Adding Integers

To add two or more negative integers, add their absolute values and place a negative sign in front of the sum. [Section 1.2/Idea 1]

To add a positive integer and a negative integer, subtract the smaller absolute value from the larger absolute value. Place the sign of the integer having the larger absolute value in front of this difference. [Section 1.2/Idea 2]

To add three or more integers having different signs, add the negative integers, add the positive integers, and then add the results. [Section 1.2/Idea 4]

Solving Word Problems

To solve word problems involving integers, identify the knowns and unknowns, rewrite the original statement as a mathematical expression, compute an answer, and then check the answer. [Section 1.2/Idea 5]

Subtracting Integers
$a - b = a + (-b)$

To subtract one integer from another, add the additive inverse of the subtrahend to the minuend. [Section 1.3/Idea 2]

Adding and Subtracting Integers

To simplify an expression containing the operations of addition and subtraction, change every operation of subtraction to addition by using the fact that $a - b = a + (-b)$. Then add. [Section 1.4/Idea 1]

Multiplying Integers

To multiply integers, multiply their absolute values. If the original expression contained an even number of negative signs, the product is positive. If the original expression contained an odd number of negative signs, the product is negative. [Section 1.5/Idea 2]

Dividing Integers		To divide one integer by another, divide their absolute values. If the signs are the same, the quotient is positive. If they are different, the quotient is negative. [Section 1.6/Idea 1]
Order of Operations		To simplify an expression containing more than one operation, first simplify within the innermost grouping symbol. Next, repeat step 1 until all grouping symbols have been eliminated. Finally, do all multiplication and/or division before adding and/or subtracting. [Section 1.7/Ideas 1–3]
Evaluating an Algebraic Expression		To evaluate an algebraic expression or a formula, first replace each variable with its given value. Then simplify the resulting expression. [Section 1.8/Ideas 1–2] To evaluate a formula containing a bar, replace all letters with their given numerical values. Next, simplify the expression above and below the bar. Then, divide. [Section 1.8/Idea 2]

Chapter 1 Review Exercises

The answers to these problems—and the sections in which similar problems appear—are given in the Answer Section. It will be helpful to review the sections referred to for the problems you did incorrectly.

Find the following absolute values.

1. $|-20|$
2. $|60|$
3. $|95|$
4. $|0|$

Find the additive inverses of the following integers.

5. -20
6. 60
7. 95
8. 0

Perform the indicated operations and simplify.

9. $-6 + (-9) + (-35)$
10. $4 - (-35)$
11. $-3 + 18 \div (-2) + 40 \cdot 2$
12. $(-8)(5)$
13. $\dfrac{35}{-7}$
14. $\dfrac{-60}{-10}$
15. $-45 + 98$
16. $(-5)(2)(-3)(-4)$
17. $8 + (-16) - (-40) + 35$
18. $35 - 85$
19. $-10 - 6 + 32 - 15$
20. $\dfrac{-11116}{0}$
21. $-5[-8 + 3(-4 - 2) + 9(-8 + 2 \cdot 3)]$

Solve. Show all work.

22. The opening price of a certain stock was listed as 115 (meaning $115). If daily changes in the stock were listed as $+5$, -3, -2, $+10$, -6, $+2$, -5, and $+15$, find the final closing price of the stock.

23. Subtract -15 from 30.

24. In a fire sale, Guerin sold eight coats at a profit of $16 each, three clocks at a loss of $4 each, seven bicycles at a loss of $10 each, and five radios at a profit of $30 each. Find his total profit or loss on these sales, and express the answer as an integer.

25. Howie lost $42 in three weeks. Express his average loss per week as an integer.

26. On October 31, Adrienne has a balance of $500 in her checking account. The chart on the side shows the transactions she made in November. How much was in her account on November 29?

27. Find the value of $M = \dfrac{-4[2(d + 3a)]}{a + d}$ when $a = -2$ and $d = 4$.

28. Find the quotient of -60 and 6.

29. Find the product of -2, 3, -5, and -4.

30. Find the sum of -50 and 80.

31. Find the difference between -2 and 8.

32. Subtract -2 from 8.

33. What is -4 times the sum of -8 and -4.

34. Divide the product of 6 and -8 by 12.

Date	Check Number	Amount	Deposits
Nov. 1	113	$ 50	
Nov. 3	114	75	
Nov. 7	115	100	
Nov. 10			$250
Nov. 15	116	300	
Nov. 21	117	73	
Nov. 23			55
Nov. 23	118	53	
Nov. 26			118
Nov. 28	119	172	

Indicate whether each statement is True or False. If the statement is False, correct it.

35. The absolute value of 6 is -6.

36. The additive inverse of 0 is 0.

37. The absolute value of every integer is a positive integer.

38. The additive inverse of every positive integer is a negative integer.

Perform the indicated operations.

39. $\dfrac{-105}{15}$

40. $\dfrac{-169}{-13}$

41. $\dfrac{-1001}{-7}$

42. $\dfrac{187}{-17}$

43. $|-35| - |50|$

44. $-2 + 3|-6|$

45. $-616 + 15 \cdot 4$

46. $356 + (43)(-191)$

47. $-965 - (-810) + 150 - 699$

48. $5 - (-11) + 6(-3) - 2(-7) + 8$

49. $-10 + 4(-10 + 2 \cdot 8) \div 2 - 4(-8 - 4)$

50. $-8 + 2[-1 + 4\{-2 \cdot (-12 + 6) \cdot 2\} + 5] - 18$

Evaluate.

51. $-4 + x(-x - 3y)$ when $x = 4$ and $y = 2$

52. $(ax + by + cz) \div (a + b + c)$ when $a = 2$, $b = 4$, $c = 6$, $x = -3$, $y = -9$, and $z = -7$

Name: _____

Class: _____

Chapter 1 Test

Find the absolute value of each integer.
1. -85
2. 362

Find the additive inverse of each integer.
3. -6
4. 12

Perform the indicated operation.
5. $-6 + (-10) + (-40)$
6. $55 + (-75)$
7. $-10 + 17 + (-62) + 20 + (-85) + 46$
8. $-8 + 14 + (-18) + (-10) + 16 + (-15) + 70$
9. $20 - 65$
10. $-8 - (-16)$
11. $30 - (-10) + 10 + (-9) - 6$
12. $-6 - 8 - 9 - 2 - 10$
13. $(2)(-2)(-3)(5)(-2)$
14. $(-8)(-10)$
15. $-7(22)$
16. $\dfrac{802}{-2}$
17. $\dfrac{-39}{-13}$
18. $\dfrac{-1001}{0}$
19. $-8 + (-18 + 2 \cdot 4) \div 2$
20. $-8 + 4(-2 - 3 \cdot 2) + (-8 + 2 \cdot 5) \div 2$

Evaluate.
21. $a - 3(c + 4d)$ when $a = 1$, $c = -4$, and $d = -2$
22. $P = 2(L + W)$ when $L = 3$ feet and $W = 6$ feet

Solve.
23. Five people lost a total of $500.00 in a business deal. Represent the average amount of each person's loss as an integer.

1. _____
2. _____
3. _____
4. _____
5. _____
6. _____
7. _____
8. _____
9. _____
10. _____
11. _____
12. _____
13. _____
14. _____
15. _____
16. _____
17. _____
18. _____
19. _____
20. _____
21. _____
22. _____
23. _____

24. One day the temperature climbed to a high of 15° below zero and fell to a low of 35° below zero. Find the difference between the high and low temperatures.

25. Guerin's Cookie Company has $300.00 in cash. In October, his records indicated that he withdrew $50 for personal use, deposited $556, paid $118 for supplies, and that he collected $700. If all of the above transactions were in cash, find Guerin's cash balance.

24. _____

25. _____

2 Fractions and Mixed Numbers

Objectives The objectives for this chapter are listed below along with sample problems for each objective. By the end of this chapter you should be able to find the solutions to the given problems.

1. Find the prime factorization of a number *(Section 2.1/Idea 3)*.
   a. 24   b. 77   c. 297

2. Reduce a fraction to its lowest terms *(Section 2.2/Idea 3)*.
   a. $\dfrac{56}{64}$ b. $\dfrac{273}{182}$

3. Change a mixed number to an improper fraction *(Section 2.3/Idea 1)*.
   a. $3\dfrac{1}{3}$ b. $9\dfrac{5}{8}$

4. Change an improper fraction to a mixed number *(Section 2.3/Idea 2)*.
   a. $\dfrac{15}{2}$ b. $\dfrac{97}{8}$

5. Multiply fractions and mixed numbers *(Section 2.4/Ideas 1–2)*.
   a. $\dfrac{9}{4} \cdot \dfrac{14}{15}$ b. $45 \cdot 2\dfrac{1}{30}$

6. Divide fractions and mixed numbers *(Section 2.5/Ideas 2–3)*.
   a. $\dfrac{27}{16} \div \dfrac{18}{12}$ b. $7\dfrac{3}{7} \div 2\dfrac{1}{30}$

7. Add fractions *(Section 2.6/Ideas 1–5)*.
   a. $\dfrac{5}{9} + \dfrac{1}{66}$ b. $\dfrac{3}{4} + \dfrac{1}{22} + \dfrac{5}{33}$

8. Add mixed numbers *(Section 2.7/Ideas 1–2)*.
   a. $5\dfrac{3}{7} + 6\dfrac{6}{7}$ b. $12\dfrac{3}{8} + 11\dfrac{5}{12} + 5\dfrac{1}{6}$

9. Subtract fractions and mixed numbers *(Section 2.8/Ideas 1–4)*.
   a. $\dfrac{7}{8} - \dfrac{3}{10}$ b. $4\dfrac{9}{14} - 1\dfrac{11}{21}$

10. Add, subtract, multiply, or divide signed fractions *(Section 2.9/Ideas 2–5)*.
    a. $\dfrac{5}{6} - \dfrac{3}{10}$ b. $\dfrac{40}{60} \div \left(-\dfrac{88}{15}\right)$ c. $-\dfrac{2}{3} + \dfrac{9}{10} \cdot \dfrac{8}{27}$

2.1 Finding Prime Factorizations

IDEA 1   A number written in the form $\frac{a}{b}$ with b not equal to zero is called a **fraction**. The number above the fraction bar is the **numerator**, and the number below the bar is the **denominator**. For example, in the fraction $\frac{2}{3}$, 2 is the numerator and 3 is the denominator. The number $\frac{2}{3}$ is also called a **proper fraction** since its numerator is less than its denominator. However, if the numerator of a fraction is greater than or equal to its denominator, the fraction is called an **improper fraction**. Examples of improper fractions include $\frac{3}{2}$, $\frac{5}{3}$, and $\frac{6}{6}$.

Practice Problem 1   *Tell whether each number is a proper or an improper fraction and identify the numerator and denominator.*

     a. $\frac{3}{7}$ **b.** $\frac{5}{3}$ **c.** $\frac{7}{7}$ **d.** $\frac{11}{13}$

In Section 1.6, fractions were used to indicate the operation of division. However, fractions are used more frequently to represent a part of the whole. For example, the rectangle in Figure 2-1 is divided into three equal parts, of which two parts are shaded. That is, $\frac{2}{3}$ (two-thirds) of the rectangle is shaded. Similarly, if you did three out of four problems on a quiz, you completed $\frac{3}{4}$ (three-fourths) of the quiz.

Figure 2-1

Practice Problem 2   *What part of each figure is shaded?*

a. 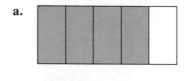   **b.**

c.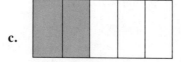

Before considering how to reduce, add, subtract, multiply, and divide fractions, we will discuss the concepts of *factors, prime numbers*, and *prime factorizations*. These concepts will help us to work effectively with fractions.

IDEA 2   If two or more natural numbers (1, 2, 3, etc.) can be multiplied together to form a product K, each of these numbers is called a **factor** or **divisor** of K. For example, 2 and 3 are the factors of 6 since $2 \cdot 3 = 6$. Other factors of 6 are 1 and 6, since $1 \cdot 6 = 6$.

Since 1, 2, 3, and 6 are factors (divisors) of 6, each of these numbers will divide 6 evenly; that is, the division results in a remainder of zero. We can also say that 6 is *divisible* by 1, 2, 3, and 6.

A natural number greater than 1 whose only divisors are 1 and itself is called a **prime number.** The primes less than 50 are 2, 3, 5, 7, 11, 13, 17, 19, 23, 29, 31, 37, 41, 43, and 47. Natural numbers greater than 1 that have more than two divisors are called **composite numbers.** For example, 4, 6, 8, 9, and 10 are composite numbers. The number 1 is neither prime nor composite.

Example 1 Tell whether each number is prime or composite.

  a. 32   **b.** 61   **c.** 85

Solution A prime number has only two divisors or factors—1 and itself. A composite number has more than two divisors (factors).

  1a. 32 is a composite number since 1, 2, 4, 8, 16, and 32 are divisors of 32.

  1b. 61 is a prime number since its only divisors are 1 and 61.

  1c. 85 is a composite number since 1, 5, 17, and 85 are divisors of 85.

Practice Problem 3 *Tell whether each number is a prime or composite.*

  a. 52   **b.** 60   **c.** 71

IDEA 3

All numbers can be factored. That is, they can be written as a product of two natural numbers. If all the factors in this product are prime numbers, this product is the **prime factorization** of the given number. To find the prime factorization of a number, (1) express the number as a product of two numbers; and (2) rewrite all composite factors as a product of primes by repeating step 1.

Example 2 Find the prime factorization of each number.

  a. 30   **b.** 12   **c.** 11

Solution To find the prime factorization of a number, write the number as a product of two natural numbers and then write each factor as a product of primes.

  2a. $30 = 2 \cdot 15$ or $30 = 6 \cdot 5$
           $= 2 \cdot 3 \cdot 5$ $= 2 \cdot 3 \cdot 5$

  2b. $12 = 2 \cdot 6$ or $12 = 4 \cdot 3$
           $= 2 \cdot 2 \cdot 3$ $= 2 \cdot 2 \cdot 3$

  2c. 11 is prime and its only factorization is the trivial factorization $11 = 1 \cdot 11$.

The above example illustrates that no matter what factorization of a number you begin with, you will obtain the same prime factors. It is customary to write the prime factorizations of a number in numerical order (smallest factor to largest factor).

Practice Problem 4 *Find the prime factorization of each number.*

  a. 20   **b.** 19   **c.** 50   **d.** 45

Sometimes it is difficult to find the prime factorization of a large number by first writing the number as a product of two numbers. When this occurs, it will be helpful to use the following *tests of divisibility* to determine whether a number is divisible by 2, 3, or 5 without having to do the division.

44   Fractions and Mixed Numbers

Tests of Divisibility

> 1. A natural number whose last digit is an even number (that is, 0, 2, 4, 6, or 8) is divisible by 2.
>
> 2. If the sum of the digits of a natural number is divisible by 3, the number is divisible by 3.
>
> 3. A natural number whose last digit is 0 or 5 is divisible by 5.

Example 3 Tell whether each number is divisible by 2, 3, and/or 5.

    a. 36   **b.** 861   **c.** 520

Solution

3a. 36 is divisible by 2 since its last digit, 6, is an even number.
36 is divisible by 3 since $3 + 6 = 9$ is divisible by 3.
36 is not divisible by 5 since its last digit is neither 0 nor 5.

3b. 861 is not divisible by 2 since its last digit is not an even number.
861 is divisible by 3 since $8 + 6 + 1 = 15$ is divisible by 3.
861 is not divisible by 5 since its last digit is neither 0 nor 5.

3c. 520 is divisible by 2 since its last digit, 0, is an even number.
520 is not divisible by 3 since $5 + 2 + 0 = 7$ is not divisible by 3.
520 is divisible by 5 since its last digit is 0.

NOTE: Tests of divisibility for the prime numbers 7, 11, 13, . . . were not included because the tests are usually longer than the actual trial division.

> **To Find the Prime Factorization of a Number:**
>
> 1. Determine if the number is divisible by 2, 3, or 5. If it is, then divide. If it is not, then try the other primes (7, 11, 13, 17, etc.) in numerical order. Any time a divisor multiplied by itself is larger than the number you wish to factor, stop trying to find a divisor since this indicates that the number is prime.
>
> 2. Repeat step 1 until your quotient is a prime number.
>
> 3. Write the original number as a product of its prime divisors and the final quotient (a prime number).
>
> 4. Check to see if the product written in step 3 is equal to the original number.

NOTE: To divide one number into another, write

$$2\overline{)100} \qquad \text{or} \qquad \text{divisor}\overline{)\text{dividend}}$$
$$\overline{50} \qquad\qquad\qquad\quad \phantom{\text{divisor})}\overline{\text{quotient}}$$

Practice Problem 5 *Tell whether each number is divisible by 2, 3, and/or 5.*

    a. 24   **b.** 111   **c.** 120   **d.** 169

We can use the tests of divisibility to help us factor. For example, to find the prime factorization of 255, first note that 255 is divisible by 5 (the last digit of 255 is 5) and the quotient is 51. Next, we know that 51 is divisible by 3 (the sum of the digits 5 + 1 = 6 is divisible by 3) and the quotient is 17. The factoring process stops here since the final quotient, 17, is a prime number. The procedure for finding the prime factorization of a number can be conveniently written as:

The prime factorization is the product of the prime divisors and the final quotient (a prime number).

$$\begin{array}{r|l} 5 & 255 \\ 3 & 51 \\ \hline & 17 \end{array}$$

The prime factorization of 255 is $5 \cdot 3 \cdot 17$. This procedure is summarized in the box on page 44.

Example 4 Find the prime factorization of each number.

    a. 36   **b.** 297   **c.** 222   **d.** 187   **e.** 47

Solution To find the prime factorization of a number, repeatedly divide by prime numbers until you obtain a prime number as a quotient. Next, write the original number as a product of its prime divisors and the final quotient.

4a.
$$\begin{array}{r|l} 2 & 36 \\ 2 & 18 \\ 3 & 9 \\ \hline & 3 \end{array}$$

$36 = 2 \cdot 2 \cdot 3 \cdot 3$

4b.
$$\begin{array}{r|l} 3 & 297 \\ 3 & 99 \\ 3 & 33 \\ \hline & 11 \end{array}$$

$297 = 3 \cdot 3 \cdot 3 \cdot 11$

4c.
$$\begin{array}{r|l} 2 & 222 \\ 3 & 111 \\ \hline & 37 \end{array}$$

$222 = 2 \cdot 3 \cdot 37$

4d. By our tests of divisibility, 187 is not divisible by 2, 3, or 5. Now, try 7.

$$\begin{array}{r|l} 7 & 187 \\ \hline & 26 \end{array}$$ remainder is 5

Now, try 11, since $11 \cdot 11$ is less than 187.

$$\begin{array}{r|l} 11 & 187 \\ \hline & 17 \end{array}$$ $187 = 11 \cdot 17$

4e. By our tests of divisibility, 47 is not divisible by 2, 3, 5, or 7. Also, you can stop trying to find a divisor larger than 7 since the product of 7 and itself ($7 \cdot 7 = 49$) is larger than the number you are trying to factor (47). Therefore, 47 is prime.

! Always check to make sure that the final quotient is a prime number before you stop the factoring process.

Practice Problem 6 *Find the prime factorization of each number.*

    a. 60   **b.** 105   **c.** 89   **d.** 143   **e.** 522

2.1 Exercises

Identify the numerator and denominator.

1. $\dfrac{3}{14}$ **2.** $\dfrac{6}{8}$ **3.** $\dfrac{10}{9}$ **4.** $\dfrac{15}{2}$

46 Fractions and Mixed Numbers

What part is shaded?

5.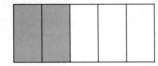

6.

7.

8.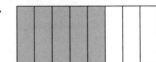

Determine if each number is divisible by 2, 3, and 5.

9. 21   10. 120   11. 201   12. 2754
13. 1755   14. 4901   15. 685,470   16. 16,842

Write the prime factorization of each number.

17. 14   18. 61   19. 60   20. 64   21. 39   22. 72
23. 42   24. 99   25. 77   26. 90   27. 88   28. 66
29. 81   30. 57   31. 25   32. 129   33. 165   34. 198
35. 135   36. 121   37. 115   38. 231   39. 462   40. 234
41. 1001   42. 1599   43. 1598   44. 323   45. 1200   46. 1500
47. 848   48. 101   49. 297   50. 169   51. 2431   52. 1463

Fill in the blanks.

53. The only divisors of 89 are 1 and 89. Therefore, 89 is a _____ number.

54. Since 3 · 5 = 15, we know that 3 and 5 are _____ or _____ of 15. We also know that _____ and _____ will divide _____ evenly.

55. To find the prime factorization of a number, repeatedly divide by _____ numbers until the _____ is a prime number. Next, write the original number as a _____ of its _____ divisors and the final _____.

Solve.

56. Pam completed three questions on a test containing five problems. What fractional part of the test did she complete?

57. In a mathematics class there are 13 men and 17 women. What fractional part of the class is women?

58. Reggie must move five cars. If he moves four of the cars, what fractional part of the job must be completed?

Answers to Practice Problems **1a.** proper fraction (3 = numerator; 7 = denominator) **b.** improper fraction (5 = numerator; 3 = denominator) **c.** improper fraction (7 = numerator; 7 = denominator) **d.** proper fraction (11 = numerator; 13 = denominator) **2a.** $\frac{4}{5}$ **b.** $\frac{1}{2}$ **c.** $\frac{2}{5}$ **3a.** composite **b.** composite **c.** prime **4a.** 2·2·5 **b.** 19 (prime) **c.** 2·5·5 **d.** 3·3·5 **5a.** 2 and 3 **b.** 3 **c.** 2, 3, and 5 **d.** not divisible by 2, 3, or 5 **6a.** 2·2·3·5 **b.** 3·5·7 **c.** 89 (prime) **d.** 11·13 **e.** 2·3·3·29

2.2 Reducing Fractions

IDEA 1 More than one fraction can be used to represent the same part of a unit. We say that such fractions are **equal** or **equivalent**. Figure 2-2 shows that $\frac{1}{2}$ and $\frac{2}{4}$ represent the same part of a whole unit. Therefore, $\frac{1}{2}$ and $\frac{2}{4}$ are equal or equivalent.

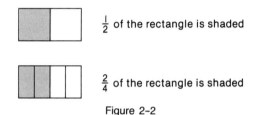

$\frac{1}{2}$ of the rectangle is shaded

$\frac{2}{4}$ of the rectangle is shaded

Figure 2-2

Actually there are many fractions that are equivalent to the fraction $\frac{1}{2}$. For example, Figure 2-3 shows four line segments (one unit in length) that are divided into two, four, six, and eight equal parts, respectively. From this diagram we can see that $\frac{2}{4}$, $\frac{3}{6}$, and $\frac{4}{8}$ are all different ways of representing $\frac{1}{2}$ of the line segment. Therefore, $\frac{1}{2}$, $\frac{2}{4}$, $\frac{3}{6}$, and $\frac{4}{8}$ are equivalent fractions since they are different names for the same number. This equivalence is written as

$$\frac{1}{2} = \frac{2}{4} = \frac{3}{6} = \frac{4}{8}$$

We can also see from Figure 2-3 that $\frac{1}{4} = \frac{2}{8}$, $\frac{6}{8} = \frac{3}{4}$, and $\frac{2}{2} = \frac{4}{4} = \frac{6}{6} = \frac{8}{8} = 1$.

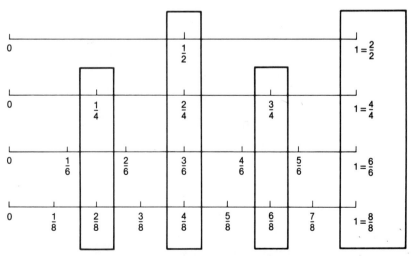

Figure 2-3

If we carefully examine the equivalent fractions $\frac{1}{4}$ and $\frac{2}{8}$, we can see that

$$\frac{1}{4} = \frac{1 \cdot 2}{4 \cdot 2} = \frac{2}{8} \quad \text{and} \quad \frac{2}{8} = \frac{2 \div 2}{8 \div 2} = \frac{1}{4}$$

These relationships suggest the following rule, enabling us to write equivalent fractions for any given fraction.

To Find an Equivalent Fraction:

Multiply or divide both the numerator and denominator of the given fraction by the same nonzero number.

or

$$\frac{a}{b} = \frac{a \cdot c}{b \cdot c} \quad \text{and} \quad \frac{a}{b} = \frac{a \div c}{b \div c}$$

where $\frac{a}{b}$ is any fraction and c does not equal 0.

Example 5 Find three equivalent fractions for each fraction given.

a. $\frac{1}{7}$ b. $\frac{3}{9}$ c. $\frac{4}{5}$

Solution To find an equivalent fraction, multiply or divide the numerator and denominator of the given fraction by the same number.

5a. $\frac{1}{7} = \frac{1 \cdot 3}{7 \cdot 3} = \frac{3}{21}$ 5b. $\frac{3}{9} = \frac{3 \div 3}{9 \div 3} = \frac{1}{3}$

$\frac{1}{7} = \frac{1 \cdot 5}{7 \cdot 5} = \frac{5}{35}$ $\frac{3}{9} = \frac{3 \cdot 2}{9 \cdot 2} = \frac{6}{18}$

$\frac{1}{7} = \frac{1 \cdot 10}{7 \cdot 10} = \frac{7}{70}$ $\frac{3}{9} = \frac{3 \cdot 6}{9 \cdot 6} = \frac{18}{54}$

5c. $\frac{4}{5} = \frac{4 \cdot 3}{5 \cdot 3} = \frac{12}{15}$

$\frac{4}{5} = \frac{4 \cdot 5}{5 \cdot 5} = \frac{20}{25}$

$\frac{4}{5} = \frac{4 \cdot 8}{5 \cdot 8} = \frac{32}{40}$

IDEA 2

To determine if two fractions are equivalent use the fact that $\frac{a}{b}$ and $\frac{c}{d}$ (b and d are not zero) are equivalent if and only if $a \cdot d = b \cdot c$. We say the **cross-products** are equal. For example,

$$\frac{1}{2} = \frac{2}{4} \quad \text{since } 1 \cdot 4 = 2 \cdot 2$$

Similarly,

$$\frac{4}{6} = \frac{2}{3} \quad \text{since } 4 \cdot 3 = 6 \cdot 2$$

> **To Determine If Two Fractions Are Equivalent:**
>
> 1. Multiply the numerator of the first fraction by the denominator of the second fraction.
> 2. Multiply the denominator of the first fraction by the numerator of the second fraction.
> 3. If these two products are equal, the two fractions are equivalent. That is, fractions are equivalent when their cross-products are equal,
>
> or
>
> $$\frac{a}{b} = \frac{c}{d} \text{ when } a \cdot d = b \cdot c$$
>
> where b is not equal to zero, and d is not equal to zero.

Example 6 Determine if the fractions are equivalent.

a. $\frac{1}{2}, \frac{3}{6}$ b. $\frac{2}{9}, \frac{4}{10}$ c. $\frac{12}{16}, \frac{3}{4}$

Solution To determine if two fractions are equivalent, compare their cross-products.

6a. $1 \cdot 6 = 2 \cdot 3$ so $\frac{1}{2} = \frac{3}{6}$ **6b.** $2 \cdot 10 \neq 9 \cdot 4$ so $\frac{2}{9} \neq \frac{4}{10}$

6c. $12 \cdot 4 = 16 \cdot 3$ so $\frac{12}{16} = \frac{3}{4}$

Practice Problem 7 *Determine if the fractions are equivalent.*

a. $\frac{1}{8}, \frac{2}{16}$ b. $\frac{2}{9}, \frac{9}{18}$ c. $\frac{3}{7}, \frac{12}{28}$ d. $\frac{50}{100}, \frac{1}{3}$

IDEA 3

We know that $\frac{1}{2} = \frac{2}{4}$. We can also say that $\frac{1}{2}$ is the simplest form of $\frac{2}{4}$ since 1 is the only number that divides both the numerator and denominator of the fraction $\frac{1}{2}$ evenly. In general, a fraction is said to be **reduced to its lowest terms** when 1 is the only number that divides the numerator and denominator evenly.

To reduce a fraction to its lowest terms, divide both the numerator and denominator by their largest common divisor. The **largest common divisor** of two or more numbers is the largest number that divides each number evenly. In the fraction $\frac{4}{6}$, for example, the largest common divisor of 4 and 6 is 2. Thus,

$$\frac{4}{6} = \frac{4 \div 2}{6 \div 2} = \frac{2}{3}$$

Sometimes it will be difficult to recognize the largest common divisor of a fraction's numerator and denominator. In such a case, use your skills of prime factorization. For example, to reduce

50 *Fractions and Mixed Numbers*

$$\frac{42}{105}$$

first write the prime factorization of the numerator and denominator

$$\frac{42}{105} = \frac{2 \cdot 3 \cdot 7}{3 \cdot 5 \cdot 7}$$

Next, note that the largest common divisor of 42 and 105 is 3 · 7 or 21 since 3 and 7 are divisors of both 42 and 105. Now, divide 42 and 105 by their largest common divisor, 3 · 7. To indicate that you have divided 42 and 105 by 3 · 7, draw slanted lines through 3 and 7. That is,

$$\frac{42}{105} = \frac{2 \cdot \overset{1}{\cancel{3}} \cdot \overset{1}{\cancel{7}}}{\underset{1}{\cancel{3}} \cdot 5 \cdot \underset{1}{\cancel{7}}} = \frac{2}{5}$$

The fraction $\frac{2}{5}$ is equivalent to $\frac{42}{105}$ and it is reduced to its lowest terms since the only common factor of 2 and 5 is 1. Also, as long as you understand that the slanted lines drawn through 3 and 7 indicate that you have divided both 42 and 105 by 3 and 7, you will not need to write a 1 above or below the factors 3 and 7.

This procedure is summarized below.

To Reduce a Fraction:

1. Write the prime factorization of the numerator and denominator.

2. Divide out every factor that appears in both the numerator and denominator. If all the factors divide out in the numerator or denominator, the corresponding numerator or denominator is 1.

3. Multiply the remaining factors in the numerator and multiply the remaining factors in the denominator to obtain a new fraction that is equivalent to the original fraction.

Example 7 Reduce each fraction to its lowest terms.

a. $\frac{42}{36}$ b. $\frac{9}{27}$ c. $\frac{231}{297}$ d. $\frac{25}{49}$ e. $\frac{819}{1599}$

Solution To reduce a fraction to its lowest terms, write the prime factorization of the numerator and denominator and then divide out all of their common factors. Next, find the product of the factors remaining in the numerator and place it over the product of the factors remaining in the denominator.

7a. $\frac{42}{36} = \frac{\cancel{2} \cdot \cancel{3} \cdot 7}{\cancel{2} \cdot 2 \cdot \cancel{3} \cdot 3}$ 7b. $\frac{9}{27} = \frac{\cancel{3} \cdot \cancel{3}}{\cancel{3} \cdot \cancel{3} \cdot 3}$

$\quad\quad = \frac{7}{6}$ $\quad\quad = \frac{1}{3}$

7c. $\frac{231}{297} = \frac{\cancel{3} \cdot 7 \cdot \cancel{11}}{\cancel{3} \cdot 3 \cdot 3 \cdot \cancel{11}}$ $\begin{array}{r}3\,\underline{|231}\\7\,\underline{|77}\\\underline{|11}\end{array}$ $\begin{array}{r}3\,\underline{|297}\\3\,\underline{|99}\\3\,\underline{|33}\\\underline{|11}\end{array}$

$\quad\quad = \frac{7}{9}$

7d. $\dfrac{25}{49} = \dfrac{5 \cdot 5}{7 \cdot 7}$

$= \dfrac{25}{49}$

7e. $\dfrac{819}{1599} = \dfrac{\cancel{3} \cdot 3 \cdot 7 \cdot \cancel{13}}{\cancel{3} \cdot \cancel{13} \cdot 41}$

$= \dfrac{21}{41}$

```
3|819      3|1599
3|273     13|533
7|91       |41
 |13
```

Example 8 Reduce each fraction to its lowest terms.

a. $\dfrac{50}{60}$ b. $\dfrac{36}{72}$

Solution If you can recognize the largest common divisor of the numerator and denominator of a fraction, you do not need to find the prime factorization of these numbers.

8a. $\dfrac{50}{60} = \dfrac{\cancel{10} \cdot 5}{\cancel{10} \cdot 6}$ 8b. $\dfrac{36}{72} = \dfrac{\cancel{36} \cdot 1}{\cancel{36} \cdot 2}$

$= \dfrac{5}{6}$ $= \dfrac{1}{2}$

Practice Problem 8 *Reduce each fraction to its lowest terms.*

a. $\dfrac{16}{24}$ b. $\dfrac{70}{14}$ c. $\dfrac{30}{77}$ d. $\dfrac{65}{169}$ e. $\dfrac{117}{153}$

2.2 Exercises

Determine if the following pairs of fractions are equivalent.

1. $\dfrac{1}{2}, \dfrac{3}{4}$ 2. $\dfrac{3}{5}, \dfrac{30}{50}$ 3. $\dfrac{42}{36}, \dfrac{7}{6}$ 4. $\dfrac{231}{297}, \dfrac{7}{9}$

Reduce each fraction to its lowest terms.

5. $\dfrac{6}{9}$ 6. $\dfrac{10}{15}$ 7. $\dfrac{12}{20}$ 8. $\dfrac{54}{81}$ 9. $\dfrac{36}{48}$ 10. $\dfrac{54}{72}$

11. $\dfrac{16}{24}$ 12. $\dfrac{9}{25}$ 13. $\dfrac{10}{90}$ 14. $\dfrac{35}{91}$ 15. $\dfrac{20}{28}$ 16. $\dfrac{30}{42}$

17. $\dfrac{26}{169}$ 18. $\dfrac{33}{22}$ 19. $\dfrac{77}{121}$ 20. $\dfrac{120}{15}$ 21. $\dfrac{55}{66}$ 22. $\dfrac{52}{78}$

23. $\dfrac{48}{120}$ 24. $\dfrac{26}{39}$ 25. $\dfrac{64}{110}$ 26. $\dfrac{152}{312}$ 27. $\dfrac{66}{210}$ 28. $\dfrac{81}{405}$

29. $\dfrac{215}{258}$ 30. $\dfrac{189}{437}$ 31. $\dfrac{121}{169}$ 32. $\dfrac{53}{159}$ 33. $\dfrac{133}{161}$ 34. $\dfrac{187}{209}$

35. $\dfrac{156}{222}$ 36. $\dfrac{126}{444}$ 37. $\dfrac{115}{135}$ 38. $\dfrac{330}{460}$ 39. $\dfrac{600}{900}$ 40. $\dfrac{400}{800}$

41. $\dfrac{266}{462}$ 42. $\dfrac{1001}{770}$ 43. $\dfrac{126}{132}$ 44. $\dfrac{192}{352}$ 45. $\dfrac{138}{805}$ 46. $\dfrac{132}{220}$

47. $\dfrac{1547}{4641}$ 48. $\dfrac{2310}{980}$ 49. $\dfrac{1575}{4158}$ 50. $\dfrac{252}{3025}$ 51. $\dfrac{3100}{4300}$ 52. $\dfrac{1700}{1900}$

Fill in the blanks.

53. Two fractions are equivalent or equal if their _____ are equal.

54. A fraction is reduced to its lowest terms if the only common divisor of the _____ and _____ is _____.

55. To reduce a fraction to its lowest terms, write the _____ factorization of the _____ and _____ and then divide out all their _____ factors. Next, find the product of the factors remaining in the _____ and place it over the product of the factors remaining in the _____.

Solve and express each answer in its simplest form.

56. Rosalind made a $120 profit by selling Tupperware. If she deposits $105 in her savings account and spends the rest, what fractional part of her profits is she saving?

57. In problem 56, what fractional part of her profits has she spent?

58. Reggie, Laura, and Richard went into business together. If Reggie invested $500, Laura invested $1500, and Richard invested $1000, what fractional part of the business does each person own?

59. Joe cut a pie into six equal pieces and served three pieces. What fractional part of the pie was left?

60. Ed bought a dozen eggs and used six of them in a cake. What fractional part of the dozen eggs did he use?

61. Jane got 60 hits out of 140 times at bat. What fractional part of her times at bat did she get a hit?

62. Ron got 70 hits in 200 times at bat. What fractional part of his times at bat did he get a hit?

63. Joe drank two cans of juice from a pack of six cans. What fractional part of the six-pack did he drink? What fractional part was left?

64. One day, 10 students were absent from a class of 40 students. What fractional part of the class was absent? What fractional part of the class was present?

Answers to Practice Problems   **7a.** yes   **b.** no   **c.** yes   **d.** no   **8a.** $\frac{2}{3}$ **b.** 5   **c.** $\frac{30}{77}$ **d.** $\frac{5}{13}$ **e.** $\frac{13}{17}$

2.3 Improper Fractions and Mixed Numbers

IDEA 1   A **mixed number** is the sum of a whole number and a proper fraction. Mixed numbers are written without a plus sign. For example, $4\frac{3}{5}$, which is read as "four and three-fifths," is equal to $4 + \frac{3}{5}$. Similarly, the mixed number $7\frac{2}{3}$ (read as "seven and two-thirds") is equal to $7 + \frac{2}{3}$.

Mixed numbers can be written as improper fractions, and improper fractions can be expressed as mixed numbers. Examine the rectangles in Figure 2–4. The shaded area can be expressed as the mixed number $1\frac{2}{3}$ since one rectangle plus two-thirds of

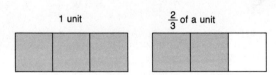

Figure 2–4

another rectangle is shaded. This same shaded area can be represented by the improper fraction $\frac{5}{3}$ since each rectangle is divided into three equal parts and five of these parts are shaded. Thus, $1\frac{2}{3} = \frac{5}{3}$ and $\frac{5}{3} = 1\frac{2}{3}$. Another way to show that $1\frac{2}{3} = \frac{5}{3}$ is by considering that

$1\frac{2}{3} = 1 + \frac{2}{3} =$ ⬤ + ◔ $= 3 \text{ thirds} + 2 \text{ thirds} = 5 \text{ thirds} = \frac{5}{3}$

1 unit = 3 thirds

There is a shortcut for this procedure.

To Change a Mixed Number to an Improper Fraction:

1. Multiply the denominator of the fraction by the whole number.
2. Add the numerator to the product obtained in step 1.
3. Place this sum over the original denominator.

Example 9 Change each mixed number to an improper fraction.

  a. $3\frac{2}{5}$ **b.** $5\frac{3}{7}$ **c.** $15\frac{11}{13}$ **d.** $9\frac{8}{11}$

Solution To change a mixed number to an improper fraction, first multiply the denominator by the whole number. Next, add the numerator. Finally, place this sum over the original denominator.

9a. $3\frac{2}{5} = \frac{5 \cdot 3 + 2}{5}$
$= \frac{15 + 2}{5}$
$= \frac{17}{5}$

9b. $5\frac{3}{7} = \frac{7 \cdot 5 + 3}{7}$
$= \frac{35 + 3}{7}$
$= \frac{38}{7}$

9c. $15\frac{11}{13} = \frac{13 \cdot 15 + 11}{13}$
$= \frac{195 + 11}{13}$
$= \frac{206}{13}$

9d. $9\frac{8}{11} = \frac{11 \cdot 9 + 8}{11}$
$= \frac{99 + 8}{11}$
$= \frac{107}{11}$

Practice Problem 9 *Change each mixed number to an improper fraction.*

  a. $8\frac{3}{5}$ **b.** $6\frac{2}{9}$ **c.** $17\frac{13}{14}$

Fractions and Mixed Numbers

IDEA 2

To change an improper fraction to a mixed number use the fact that the fraction $\frac{x}{y}$ means $x \div y$. For example,

$$\frac{5}{3} = 5 \div 3 = 3\overline{)5} = 1\frac{2}{3}$$

To Change an Improper Fraction to a Mixed Number:

1. Divide the denominator into the numerator.
2. The quotient (Q) is the whole number part of the mixed number. The remainder (R) is the numerator of the fractional part, and the original denominator (d) is the denominator.

In the problem above, $\frac{5}{3} = Q\frac{R}{d} = 1\frac{2}{3}$.

Example 10 Change each fraction to a mixed number.

a. $\frac{29}{3}$ b. $\frac{9}{2}$ c. $\frac{135}{13}$ d. $\frac{302}{8}$

Solution To change an improper fraction to a mixed number, divide the denominator into the numerator. The quotient is the whole number part of the mixed number, and the fractional part is the remainder over the original denominator.

10a. $\frac{29}{3} = 9\frac{2}{3}$

10b. $\frac{9}{2} = 4\frac{1}{2}$

10c. $\frac{135}{13} = 10\frac{5}{13}$

10d. $\frac{302}{8} = 37\frac{6}{8} = 37\frac{3}{4}$

! Remember to always reduce the fractional part of a mixed number to its lowest terms (see Solution 10d).

Practice Problem 10 *Change each fraction to a mixed number.*

a. $\frac{8}{3}$ b. $\frac{17}{4}$ c. $\frac{156}{17}$ d. $\frac{66}{18}$

2.3 Exercises

Change each mixed number to an improper fraction.

1. $2\frac{1}{2}$ 2. $3\frac{3}{4}$ 3. $5\frac{7}{8}$ 4. $6\frac{1}{9}$

5. $10\frac{2}{9}$ 6. $7\frac{3}{11}$ 7. $9\frac{2}{5}$ 8. $15\frac{2}{3}$

9. $12\frac{3}{11}$ 10. $11\frac{11}{13}$ 11. $14\frac{3}{16}$ 12. $21\frac{3}{4}$

13. $35\frac{1}{9}$ 14. $72\frac{13}{19}$ 15. $73\frac{5}{9}$ 16. $27\frac{3}{7}$

17. $5\frac{1}{17}$ 18. $100\frac{18}{29}$ 19. $13\frac{3}{13}$ 20. $262\frac{3}{16}$

Change each fraction to a mixed number.

21. $\frac{15}{2}$ 22. $\frac{16}{7}$ 23. $\frac{60}{7}$ 24. $\frac{19}{4}$

25. $\frac{35}{3}$ 26. $\frac{33}{4}$ 27. $\frac{62}{9}$ 28. $\frac{75}{8}$

29. $\frac{42}{8}$ 30. $\frac{28}{6}$ 31. $\frac{34}{16}$ 32. $\frac{34}{10}$

33. $\frac{180}{50}$ 34. $\frac{106}{13}$ 35. $\frac{102}{18}$ 36. $\frac{120}{48}$

37. $\frac{152}{11}$ 38. $\frac{135}{115}$ 39. $\frac{169}{65}$ 40. $\frac{148}{111}$

Fill in the blanks.

41. A _____ is the sum of a whole number and a proper fraction.

42. To change a mixed number to an improper fraction, multiply the denominator times the _____, add the _____, and place this sum over the original _____.

43. To change an improper fraction to a mixed number, divide the _____ into the _____. The _____ is the whole number part of the mixed number and the fractional part is written as the _____ over the original denominator.

Solve and show all work.

44. The Davis Catering Company must serve 57 pieces of pie at a luncheon. If each pie is divided into five equal pieces, how many pies are needed?

45. Write $6 + \frac{4}{5}$ as an improper fraction.

Answers to Practice Problems 9a. $\frac{43}{5}$ b. $\frac{56}{9}$ c. $\frac{251}{14}$ 10a. $2\frac{2}{3}$ b. $4\frac{1}{4}$ c. $9\frac{3}{17}$ d. $3\frac{2}{3}$

2.4 Multiplying Fractions and Mixed Numbers

IDEA 1 To multiply fractions, multiply numerators, multiply denominators, and reduce (if necessary) the resulting fraction. For example,

$$\frac{1}{2} \cdot \frac{2}{6} = \frac{1 \cdot 2}{2 \cdot 6} = \frac{2}{12} = \frac{1}{6}$$

We can illustrate that $\frac{1}{2} \cdot \frac{2}{6} = \frac{1}{6}$ by noting that $\frac{1}{2} \cdot \frac{2}{6}$ means $\frac{1}{2}$ of $\frac{2}{6}$. In Figure 2-5, two-

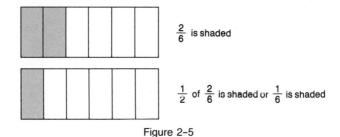

Figure 2-5

sixths $\left(\frac{2}{6}\right)$ of the first rectangle is shaded. Note that half $\left(\frac{1}{2}\right)$ of the shaded area is equal to one-sixth $\left(\frac{1}{6}\right)$ of the rectangle.

As another example, let us consider the number lines in Figure 2-6. Each line is divided into four equal parts.

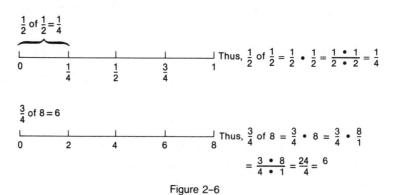

Figure 2-6

Notice that to multiply a whole N times a fraction, we first write N as $\frac{N}{1}$.

Usually it will be inconvenient to multiply fractions and then reduce the result. To avoid this, factor all numerators and denominators and divide out all common factors before actually multiplying. For example,

$$\frac{9}{10} \cdot \frac{4}{6} = \frac{9 \cdot 4}{10 \cdot 6} = \frac{\cancel{3} \cdot 3 \cdot \cancel{2} \cdot 2}{\cancel{2} \cdot 5 \cdot \cancel{2} \cdot \cancel{3}} = \frac{3}{5}$$

↑ Write the numerators and denominator as a product.

↑ Factor and divide out all common factors.

↑ Multiply the remaining factors.

To Multiply Fractions:

1. Write the product of the numerators over the product of the denominators.

2. Reduce the resulting fraction to its lowest terms,

or

$$\frac{a}{b} \cdot \frac{c}{d} = \frac{a \cdot c}{b \cdot d}$$

2.4 Multiplying Fractions and Mixed Numbers

Example 11 Multiply and reduce.

    a. $\dfrac{4}{9} \cdot \dfrac{15}{14}$ **b.** $\dfrac{1}{6} \cdot \dfrac{25}{7}$ **c.** $\dfrac{24}{42} \cdot \dfrac{28}{86}$

    d. $\dfrac{14}{24} \cdot \dfrac{33}{35} \cdot \dfrac{4}{77}$ **e.** $\dfrac{222}{297} \cdot \dfrac{231}{182}$ **f.** $6 \cdot \dfrac{9}{30}$

Solution To multiply fractions, write the factors of the numerators over the factors of the denominators. Reduce the resulting fraction.

11a. $\dfrac{4}{9} \cdot \dfrac{15}{14} = \dfrac{4 \cdot 15}{9 \cdot 14}$

$\phantom{\dfrac{4}{9} \cdot \dfrac{15}{14}} = \dfrac{\boxed{2} \cdot 2 \cdot \boxed{3} \cdot 5}{\boxed{3} \cdot 3 \cdot \boxed{2} \cdot 7}$ Factor and divide out all common factors

$\phantom{\dfrac{4}{9} \cdot \dfrac{15}{14}} = \dfrac{10}{21}$ Multiply

11b. $\dfrac{1}{6} \cdot \dfrac{25}{7} = \dfrac{1 \cdot 25}{6 \cdot 7}$

$\phantom{\dfrac{1}{6} \cdot \dfrac{25}{7}} = \dfrac{1 \cdot 5 \cdot 5}{2 \cdot 3 \cdot 7}$ No common factors

$\phantom{\dfrac{1}{6} \cdot \dfrac{25}{7}} = \dfrac{25}{42}$ Multiply

11c. $\dfrac{24}{42} \cdot \dfrac{28}{86} = \dfrac{24 \cdot 28}{42 \cdot 86}$

$\phantom{\dfrac{24}{42} \cdot \dfrac{28}{86}} = \dfrac{\cancel{2} \cdot \cancel{2} \cdot 2 \cdot \cancel{3} \cdot 2 \cdot 2 \cdot \cancel{7}}{\cancel{2} \cdot \cancel{3} \cdot \cancel{7} \cdot \cancel{2} \cdot 43}$

$\phantom{\dfrac{24}{42} \cdot \dfrac{28}{86}} = \dfrac{8}{43}$

11d. $\dfrac{14}{24} \cdot \dfrac{33}{35} \cdot \dfrac{4}{77} = \dfrac{14 \cdot 33 \cdot 4}{24 \cdot 35 \cdot 77}$

$\phantom{\dfrac{14}{24} \cdot \dfrac{33}{35} \cdot \dfrac{4}{77}} = \dfrac{\cancel{2} \cdot \cancel{7} \cdot \cancel{3} \cdot \cancel{11} \cdot \cancel{2} \cdot \cancel{2}}{\cancel{2} \cdot \cancel{2} \cdot \cancel{2} \cdot \cancel{3} \cdot 5 \cdot \cancel{7} \cdot 7 \cdot \cancel{11}}$

$\phantom{\dfrac{14}{24} \cdot \dfrac{33}{35} \cdot \dfrac{4}{77}} = \dfrac{1}{35}$ Since we divide out all the factors of the numerator, the numerator becomes 1.

11e. $\dfrac{222}{297} \cdot \dfrac{231}{182} = \dfrac{222 \cdot 231}{297 \cdot 182}$

$\phantom{\dfrac{222}{297} \cdot \dfrac{231}{182}} = \dfrac{\cancel{2} \cdot \cancel{3} \cdot 37 \cdot \cancel{3} \cdot \cancel{7} \cdot \cancel{11}}{\cancel{3} \cdot \cancel{3} \cdot 3 \cdot \cancel{11} \cdot \cancel{2} \cdot \cancel{7} \cdot 13}$

$\phantom{\dfrac{222}{297} \cdot \dfrac{231}{182}} = \dfrac{37}{39}$

11f. $\boxed{6} \cdot \dfrac{9}{30} = \dfrac{\boxed{6}}{\boxed{1}} \cdot \dfrac{9}{30}$ Write 6 as a fraction $\left(\dfrac{6}{1}\right)$

$\phantom{6 \cdot \dfrac{9}{30}} = \dfrac{6 \cdot 9}{1 \cdot 30}$

$\phantom{6 \cdot \dfrac{9}{30}} = \dfrac{\cancel{2} \cdot \cancel{3} \cdot 3 \cdot 3}{1 \cdot \cancel{2} \cdot \cancel{3} \cdot 5}$

$\phantom{6 \cdot \dfrac{9}{30}} = \dfrac{9}{5}$

Example 12 Multiply and reduce.

    a. $\dfrac{4}{9} \cdot \dfrac{15}{14}$ **b.** $\dfrac{4}{5} \cdot \dfrac{10}{3} \cdot \dfrac{9}{16}$

58 Fractions and Mixed Numbers

Solution   If you can recognize the largest common divisor of numbers in the numerator and denominator, you will not need to factor.

12a. $\dfrac{4}{9} \cdot \dfrac{15}{14} = \dfrac{\overset{2}{\cancel{4}} \cdot \overset{5}{\cancel{15}}}{\underset{3}{\cancel{9}} \cdot \underset{7}{\cancel{14}}}$ Divide 4 and 14 by 2
Divide 9 and 15 by 3

$= \dfrac{10}{21}$

12b. $\dfrac{4}{5} \cdot \dfrac{10}{3} \cdot \dfrac{9}{16} = \dfrac{\overset{1}{\underset{1}{\cancel{4}}} \cdot \overset{\overset{1}{2}}{\underset{1}{\cancel{10}}} \cdot \overset{3}{\underset{2}{\cancel{9}}}}{\underset{1}{\cancel{5}} \cdot \underset{1}{\cancel{3}} \cdot \underset{4}{\cancel{16}}}$ Divide 4 and 16 by 4
Divide 5 and 10 by 5
Divide 3 and 9 by 3
Divide 2 and 4 by 2

$= \dfrac{3}{2}$

❗ When using the techniques show in Example 12, check to make sure you have reduced your final answer to its lowest terms.

Practice Problem 11   **Multiply and reduce.**

a. $\dfrac{8}{12} \cdot \dfrac{10}{15}$ b. $6 \cdot \dfrac{7}{15}$

c. $\dfrac{14}{20} \cdot \dfrac{66}{42} \cdot \dfrac{2}{55}$ d. $\dfrac{117}{77} \cdot \dfrac{143}{273}$

IDEA 2

Sometimes the factors in a product will be fractions, whole numbers, and mixed numbers. To find this product, first change all mixed numbers to improper fractions and then multiply the resulting fractions. For example,

$4\dfrac{1}{6} \cdot \dfrac{27}{10} = \dfrac{25}{6} \cdot \dfrac{27}{10} = \dfrac{25 \cdot 27}{6 \cdot 10} = \dfrac{\cancel{5} \cdot 5 \cdot 3 \cdot \cancel{3} \cdot 3}{2 \cdot \cancel{3} \cdot 2 \cdot \cancel{5}} = \dfrac{45}{4}$ or $11\dfrac{1}{4}$

> **To Multiply Mixed Numbers:**
>
> 1. Change all mixed numbers to improper fractions.
> 2. Multiply the resulting fractions.
> 3. Write the answer as a mixed number or an improper fraction in reduced form.

Example 13   Multiply.

a. $\dfrac{6}{35} \cdot 2\dfrac{1}{10}$ b. $1\dfrac{22}{43} \cdot 3\dfrac{12}{39} \cdot 1\dfrac{1}{10}$ c. $60 \cdot \dfrac{25}{42} \cdot 2\dfrac{1}{10}$

Solution   To multiply mixed numbers, change all mixed numbers to improper fractions and multiply. Express the final answer as a mixed number or an improper fraction.

13a. $\dfrac{6}{35} \cdot 2\dfrac{1}{10} = \dfrac{6}{35} \cdot \dfrac{21}{10}$ Change a mixed number to a fraction

$= \dfrac{\overset{3}{\cancel{6}} \cdot \overset{3}{\cancel{21}}}{\underset{5}{\cancel{35}} \cdot \underset{5}{\cancel{10}}}$ Divide 6 and 10 by 2
Divide 21 and 35 by 7

$= \dfrac{9}{25}$

13b. $1\dfrac{22}{43} \cdot 3\dfrac{12}{39} \cdot 1\dfrac{1}{10} = \dfrac{65}{43} \cdot \dfrac{129}{39} \cdot \dfrac{11}{10}$

$= \dfrac{65 \cdot 129 \cdot 11}{43 \cdot 39 \cdot 10}$

$= \dfrac{\cancel{5} \cdot \cancel{13} \cdot \cancel{3} \cdot \cancel{43} \cdot 11}{\cancel{43} \cdot \cancel{3} \cdot \cancel{13} \cdot 2 \cdot \cancel{5}}$

$= \dfrac{11}{2}$

$= 5\dfrac{1}{2}$

13c. $60 \cdot \dfrac{25}{42} \cdot 2\dfrac{1}{10} = \dfrac{60}{1} \cdot \dfrac{25}{42} \cdot \dfrac{21}{10}$

$= \dfrac{60 \cdot 25 \cdot 21}{1 \cdot 42 \cdot 10}$

$= \dfrac{\cancel{2} \cdot 2 \cdot \cancel{3} \cdot \cancel{5} \cdot 5 \cdot 5 \cdot 3 \cdot \cancel{7}}{1 \cdot \cancel{2} \cdot \cancel{3} \cdot \cancel{7} \cdot 2 \cdot \cancel{5}}$

$= \dfrac{75}{1}$

$= 75$

Practice Problem 12 **Multiply and express each answer as a mixed number.**

     a. $5\dfrac{2}{5} \cdot 1\dfrac{9}{21}$ **b.** $\dfrac{15}{20} \cdot 2\dfrac{4}{11}$ **c.** $99 \cdot 1\dfrac{25}{66}$

Example 14   Solve and show all work.

     a. Ronnie borrows $2500 from his mother. If he agrees to repay $\dfrac{2}{5}$ of the money in three weeks, how much money should his mother receive at that time?

     b. If Glenn's car can travel $16\dfrac{1}{4}$ miles on one gallon of gas, how far will his car travel on $6\dfrac{4}{5}$ gallons of gas?

Solution   **14a.** *Think:* Find $\dfrac{2}{5}$ of $2500.

$\dfrac{2}{5} \cdot 2500 = \dfrac{2}{5} \cdot \dfrac{2500}{1}$

$= \dfrac{2 \cdot \overset{500}{\cancel{2500}}}{\underset{1}{\cancel{5}} \cdot 1}$ Divide 5 and 2500 by 5

$= \dfrac{1000}{1}$

$= 1000$

60 Fractions and Mixed Numbers

Ronnie's mother should receive $1000.

14b. To find the distance traveled, use the formula:

Miles traveled = (miles per gallon)(gallons of gas used)

$$= \left(16\frac{1}{4}\right)\left(6\frac{4}{5}\right)$$

$$= \frac{65}{4} \cdot \frac{34}{5}$$

$$= \frac{\overset{13}{\cancel{65}} \cdot \overset{17}{\cancel{34}}}{\underset{2}{\cancel{4}} \cdot \underset{1}{\cancel{5}}} \quad \begin{array}{l}\text{Divide 5 and 65 by 5}\\ \text{Divide 4 and 34 by 2}\end{array}$$

$$= \frac{221}{2}$$

$$= 110\frac{1}{2}$$

Glenn's car will travel $110\frac{1}{2}$ miles.

Practice Problem 13 **Solve and show all work.**

a. A real estate broker told Carver that his house is worth $2\frac{1}{2}$ times its original cost. If the original cost was $24,000, how much is the house worth now?

b. If $\frac{3}{4}$ of the 24 people in a social club paid their dues and $\frac{2}{3}$ of these people were men, how many men paid their dues?

2.4 Exercises

Multiply and reduce to lowest terms.

1. $\frac{6}{25} \cdot \frac{5}{8}$
2. $\frac{6}{34} \cdot \frac{51}{56}$
3. $\frac{4}{10} \cdot \frac{6}{8}$
4. $5 \cdot \frac{14}{20}$
5. $\frac{1}{6} \cdot \frac{5}{7}$
6. $10 \cdot \frac{6}{4}$
7. $\frac{45}{44} \cdot \frac{11}{24}$
8. $6 \cdot \frac{4}{9}$
9. $\frac{36}{99} \cdot \frac{25}{40} \cdot \frac{42}{28}$
10. $\frac{86}{65} \cdot \frac{39}{129} \cdot \frac{12}{69}$
11. $135 \cdot \frac{30}{105}$
12. $\frac{14}{60} \cdot \frac{66}{77}$
13. $\frac{65}{56} \cdot \frac{48}{39} \cdot \frac{15}{18}$
14. $\frac{48}{70} \cdot \frac{80}{84} \cdot \frac{154}{192}$
15. $81 \cdot \frac{14}{54}$
16. $\frac{234}{222} \cdot \frac{152}{312}$
17. $\frac{75}{1001} \cdot \frac{154}{125}$
18. $\frac{42}{70} \cdot \frac{25}{35} \cdot \frac{14}{6}$
19. $\frac{21}{36} \cdot \frac{27}{42} \cdot \frac{45}{63} \cdot \frac{21}{10}$
20. $\frac{96}{117} \cdot \frac{48}{192} \cdot \frac{104}{114} \cdot 7$
21. $\frac{7}{8} \cdot \frac{8}{7}$
22. $\frac{3}{5} \cdot \frac{5}{3}$
23. $\frac{16}{27} \cdot \frac{18}{24}$
24. $\frac{14}{35} \cdot \frac{21}{10}$
25. $15 \cdot \frac{4}{35}$
26. $12 \cdot \frac{9}{16}$
27. $\frac{56}{52} \cdot \frac{55}{77} \cdot \frac{26}{20}$
28. $\frac{65}{20} \cdot \frac{60}{91} \cdot \frac{75}{2}$
29. $\frac{8}{69} \cdot \frac{23}{4} \cdot 6$
30. $\frac{20}{91} \cdot \frac{77}{15} \cdot 39$

31. $\dfrac{33}{25} \cdot \dfrac{9}{10} \cdot \dfrac{75}{20}$ 32. $\dfrac{72}{39} \cdot \dfrac{24}{96} \cdot \dfrac{52}{32}$ 33. $\dfrac{32}{80} \cdot \dfrac{25}{28}$

34. $\dfrac{36}{70} \cdot \dfrac{35}{44}$ 35. $\dfrac{54}{143} \cdot \dfrac{13}{270}$ 36. $\dfrac{16}{39} \cdot \dfrac{26}{240}$

37. $\dfrac{2}{11} \cdot 3300$ 38. $\dfrac{3}{13} \cdot 3900$ 39. $\dfrac{33}{105} \cdot \dfrac{6}{1155} \cdot \dfrac{22}{23}$

40. $\dfrac{8}{9} \cdot \dfrac{25}{81} \cdot \dfrac{243}{1000}$

Multiply and express each answer as a mixed number.

41. $\left(1\dfrac{1}{3}\right)\left(1\dfrac{1}{2}\right)$ 42. $\left(8\dfrac{1}{2}\right)\left(2\dfrac{2}{3}\right)$ 43. $\left(3\dfrac{3}{5}\right)(15)$

44. $\left(7\dfrac{1}{5}\right)(25)$ 45. $\left(7\dfrac{1}{2}\right)\left(4\dfrac{3}{4}\right)$ 46. $\left(2\dfrac{1}{7}\right)\left(2\dfrac{4}{5}\right)$

47. $(6)\left(2\dfrac{1}{12}\right)$ 48. $(18)\left(4\dfrac{1}{12}\right)$ 49. $\left(\dfrac{3}{7}\right)\left(3\dfrac{2}{9}\right)$

50. $\left(\dfrac{5}{8}\right)\left(2\dfrac{1}{2}\right)$ 51. $\left(1\dfrac{2}{7}\right)\left(2\dfrac{1}{3}\right)\left(2\dfrac{1}{6}\right)$ 52. $\left(2\dfrac{2}{3}\right)\left(2\dfrac{1}{4}\right)\left(1\dfrac{1}{2}\right)$

53. $\left(10\dfrac{2}{3}\right)\left(\dfrac{5}{8}\right)$ 54. $\left(8\dfrac{3}{4}\right)\left(\dfrac{6}{25}\right)$ 55. $\left(2\dfrac{1}{7}\right)\left(\dfrac{33}{36}\right)\left(4\dfrac{2}{3}\right)$

56. $\left(2\dfrac{5}{8}\right)\left(\dfrac{7}{9}\right)\left(2\dfrac{2}{5}\right)$ 57. $\left(\dfrac{6}{13}\right)\left(2\dfrac{1}{6}\right)$ 58. $\left(\dfrac{7}{15}\right)\left(2\dfrac{1}{7}\right)$

59. $1\dfrac{1}{5} \cdot 1\dfrac{3}{8}$ 60. $2\dfrac{1}{17} \cdot 5\dfrac{1}{10}$ 61. $4\dfrac{1}{6} \cdot 1\dfrac{2}{5}$

62. $14 \cdot 2\dfrac{1}{2}$ 63. $5\dfrac{2}{15} \cdot \dfrac{20}{91}$ 64. $\dfrac{14}{22} \cdot 3\dfrac{3}{10}$

65. $\dfrac{77}{164} \cdot 2\dfrac{25}{49}$ 66. $1\dfrac{2}{3} \cdot \dfrac{48}{70}$ 67. $45 \cdot 2\dfrac{1}{30}$

68. $2\dfrac{2}{5} \cdot 2\dfrac{3}{16}$ 69. $\dfrac{42}{54} \cdot 1\dfrac{5}{7} \cdot 5$ 70. $15\dfrac{7}{15} \cdot 4\dfrac{19}{24}$

71. $5\dfrac{2}{5} \cdot 1\dfrac{9}{21} \cdot \dfrac{14}{81} \cdot 6$ 72. $\dfrac{124}{35} \cdot 1\dfrac{23}{42} \cdot \dfrac{98}{37} \cdot 2\dfrac{3}{4} \cdot \dfrac{3}{155}$ 73. $5\dfrac{5}{9} \cdot 1\dfrac{11}{25} \cdot 1\dfrac{27}{64} \cdot \dfrac{110}{1001} \cdot 3$

Fill in the blanks.

74. To multiply two or more fractions, write the factors of the _____ over the factors of the _____ and _____ the resulting fraction.

75. To multiply a whole number *N* times a fraction, write *N* as _____ and then use the rules for multiplying fractions.

76. To multiply using mixed numbers, change the _____ to _____. Next, use the rules for _____ fractions and express the final answer as a _____ number or _____ fraction.

Solve and show all work.

77. If ADK stock costs $\$12\dfrac{1}{2}$ per share, find the cost of 45 shares.

78. Twenty out of every 45 high school basketball players play college basketball. One out of every 2000 college basketball players play professional basketball. What fractional part of high school basketball players play professional basketball?

79. A warehouse will hold $\dfrac{2}{9}$ of a ton of sugar. If the warehouse is $\dfrac{3}{4}$ full, how much sugar is in the warehouse?

Fractions and Mixed Numbers

80. Cazzie earns $60 a day. If he works only $\frac{7}{12}$ of the day, how much money will he earn?

81. Joyce's car can travel $15\frac{1}{3}$ miles on a gallon of gas. How far can the car travel on $12\frac{3}{4}$ gallons of gas?

82. Terry and Keith bought a painting. Terry invested $2000, and Keith invested $2400. Five years later, they sold the painting for $13,090. How much money should each man receive if the $13,090 is divided fairly? (*Hint:* Whatever fractional part of the money each man originally invested, he should receive that same fractional part back from the sale of the painting.)

83. If two-thirds of the 450 students at a school are men, how many students at the school are men?

84. If three-fifths of the graduating class of 175 seniors are women, how many women are graduating?

85. Find the product of $\frac{123}{77}$ and $\frac{82}{49}$.

86. Find the product of $\frac{21}{27}$ and $\frac{198}{15}$.

87. Multiply $\frac{4}{46}$ by $\frac{69}{12}$.

88. Multiply $\frac{14}{9}$ by $\frac{81}{21}$.

89. Multiply $2\frac{2}{5}$ and 240.

90. Multiply $15\frac{1}{2}$ and 420.

91. Pete's car can travel $16\frac{1}{2}$ miles on one gallon of gas. How far can the car travel on $9\frac{1}{3}$ gallons of gas?

92. Bill's car can travel $17\frac{1}{2}$ miles on one gallon of gas. How far can the car travel on $7\frac{3}{5}$ gallons of gas?

93. A company bought 1000 shares of stock at $25\frac{1}{8}$ per share. Find the total cost of this purchase.

94. Lisa bought 960 shares of a stock at $17\frac{3}{8}$ per share. Find the total cost of this purchase.

95. A truck can hold $\frac{2}{3}$ of a ton of salt. If the truck is $\frac{1}{10}$ full, how much salt is in the truck?

96. A storage room can hold $\frac{3}{4}$ of a ton of flour. If the room is $\frac{1}{6}$ full, how much flour is in the room?

Answers to Practice Problems 11a. $\frac{4}{9}$ b. $\frac{14}{5}$ c. $\frac{1}{25}$ d. $\frac{39}{49}$ 12a. $7\frac{5}{7}$ b. $1\frac{17}{22}$ c. $136\frac{1}{2}$ 13a. $60,000 b. 12

2.5 Dividing Fractions and Mixed Numbers

IDEA 1 Before we discuss division of fractions, let us consider the following special products.

$$\frac{2}{3} \cdot \frac{3}{2} = \frac{6}{6} = 1 \qquad 4 \cdot \frac{1}{4} = \frac{4}{1} \cdot \frac{1}{4} = \frac{4}{4} = 1$$

In each case the product of the two numbers is one. Whenever the product of two numbers is 1, the numbers are said to be **reciprocals** of each other. To find the reciprocal of a nonzero number, interchange the numerator and denominator. For example, the reciprocal of $\frac{3}{4}$ is $\frac{4}{3}$.

Below are three numbers and their reciprocals.

2.5 Dividing Fractions and Mixed Numbers

Number	Reciprocal	Reason
$\frac{1}{8}$	8	$\frac{1}{8} \cdot 8 = \frac{1}{8} \cdot \frac{8}{1} = \frac{8}{8} = 1$
18	$\frac{1}{18}$	$18 \cdot \frac{1}{18} = \frac{18}{1} \cdot \frac{1}{18} = \frac{18}{18} = 1$
0	Does not exist	There is no number that we can multiply 0 by to obtain 1.

IDEA 2 The concept of reciprocals can be used to rewrite a division problem as an equivalent multiplication problem. For example,

Division Problem   *Equivalent Multiplication Problem*

$8 \div 4 = 2$ $8 \cdot \frac{1}{4} = 2$

8 divided by 4 is 2   8 times the reciprocal of 4 is 2

The above example implies that dividing a number x by another number y is the same as multiplying x by the reciprocal of y $\left(x \div y = x \cdot \frac{1}{y}\right)$.

Let us now show that this technique also works for fractions by showing that $\frac{2}{3} \div \frac{7}{5} = \frac{2}{3} \cdot \frac{5}{7}$.

$\frac{2}{3} \div \frac{7}{5} = \dfrac{\frac{2}{3}}{\frac{7}{5}}$ Rewrite the division problem as a fraction

$= \dfrac{\frac{2}{3} \cdot \frac{5}{7}}{\frac{7}{5} \cdot \frac{5}{7}}$ Multiply both the numerator and denominator by $\frac{5}{7}$ (the reciprocal of $\frac{7}{5}$)

$= \dfrac{\frac{2}{3} \cdot \frac{5}{7}}{1}$ Any number times its reciprocal is 1

$= \frac{2}{3} \cdot \frac{5}{7}$ Any number divided by 1 equals that number

Therefore,

$$\frac{2}{3} \div \frac{7}{5} = \frac{2}{3} \cdot \frac{5}{7} = \frac{10}{21}$$

The above result suggests the following rule.

To Divide One Fraction by Another:

1. Keep the dividend the same.
2. Multiply the dividend by the reciprocal of the divisor (interchange the numerator and denominator of the divisor),

or

$$\frac{a}{b} \div \frac{c}{d} = \frac{a}{b} \cdot \frac{d}{c} = \frac{a \cdot d}{b \cdot c}$$

Example 15   Divide and reduce.

a. $\dfrac{9}{16} \div \dfrac{15}{8}$ b. $\dfrac{25}{28} \div 15$

c. $\dfrac{42}{120} \div \dfrac{77}{16}$ d. $\dfrac{10}{1001} \div \dfrac{35}{1309}$

Solution   To divide fractions, multiply the dividend by the reciprocal of the divisor (invert the divisor).

15a. $\dfrac{9}{16} \div \dfrac{15}{8} = \dfrac{9}{16} \cdot \dfrac{8}{15}$ (Reciprocal)

$= \dfrac{\overset{3}{\cancel{9}} \cdot \overset{1}{\cancel{8}}}{\underset{2}{\cancel{16}} \cdot \underset{5}{\cancel{15}}}$ Divide 8 and 16 by 2
Divide 9 and 15 by 3

$= \dfrac{3}{10}$

15b. $\dfrac{25}{28} \div 15 = \dfrac{25}{28} \cdot \dfrac{1}{15}$ Since $15 = \dfrac{15}{1}$, its reciprocal is $\dfrac{1}{15}$

$= \dfrac{\overset{5}{\cancel{25}} \cdot 1}{28 \cdot \underset{3}{\cancel{15}}}$ Divide 25 and 15 by 5

$= \dfrac{5}{84}$

15c. $\dfrac{42}{120} \div \dfrac{77}{16} = \dfrac{42}{120} \cdot \dfrac{16}{77}$

$= \dfrac{42 \cdot 16}{120 \cdot 77}$

$= \dfrac{\cancel{2} \cdot \cancel{3} \cdot \cancel{7} \cdot \cancel{2} \cdot \cancel{2} \cdot 2 \cdot 2}{\cancel{2} \cdot \cancel{2} \cdot \cancel{2} \cdot \cancel{3} \cdot 5 \cdot \cancel{7} \cdot 11}$

$= \dfrac{4}{55}$

2.5 Dividing Fractions and Mixed Numbers

15d. $\dfrac{10}{1001} \div \dfrac{35}{1309} = \dfrac{10}{1001} \cdot \dfrac{1309}{35}$

$= \dfrac{10 \cdot 1309}{1001 \cdot 35}$

$= \dfrac{2 \cdot \cancel{5} \cdot \cancel{7} \cdot \cancel{11} \cdot 17}{\cancel{7} \cdot \cancel{11} \cdot 13 \cdot \cancel{5} \cdot 7}$

$= \dfrac{34}{91}$

Practice Problem 14 *Divide and reduce.*

a. $\dfrac{14}{17} \div \dfrac{28}{51}$ b. $50 \div \dfrac{15}{39}$ c. $\dfrac{85}{90} \div \dfrac{105}{84}$

IDEA 3

Sometimes the divisor and/or the dividend in a division will be a mixed number. When this occurs, change all mixed numbers to improper fractions and then do the resulting division problem. For example,

$$2\dfrac{3}{8} \div 3\dfrac{1}{2} = \dfrac{19}{8} \div \dfrac{7}{2} = \dfrac{19}{8} \cdot \dfrac{2}{7} = \dfrac{19 \cdot 2}{8 \cdot 7} = \dfrac{19 \cdot \cancel{2}}{\cancel{2} \cdot 2 \cdot 2 \cdot 7} = \dfrac{19}{28}$$

↑ Change all mixed numbers to improper fractions
↑ Multiply the dividend by the reciprocal of the divisor
↑ Reduce

To Divide Mixed Numbers:

1. Change all mixed numbers to improper fractions.
2. Divide the resulting fractions.
3. Write the answer as a mixed number or an improper fraction in reduced form.

Example 16 Divide and reduce.

a. $3\dfrac{1}{3} \div 5\dfrac{1}{2}$ b. $45 \div 3\dfrac{3}{5}$ c. $5\dfrac{11}{15} \div 12\dfrac{9}{10}$

Solution To divide using mixed numbers, first change all mixed numbers to improper fractions and then divide.

16a. $3\dfrac{1}{3} \div 5\dfrac{1}{2} = \dfrac{10}{3} \div \dfrac{11}{2}$ Express mixed numbers as improper fractions

$= \dfrac{10}{3} \cdot \dfrac{2}{11}$

$= \dfrac{10 \cdot 2}{3 \cdot 11}$ No common factors

$= \dfrac{20}{33}$

16b. $45 \div 3\frac{3}{5} = \frac{45}{1} \div \frac{18}{5}$

$= \frac{45}{1} \cdot \frac{5}{18}$

$= \frac{\overset{5}{\cancel{45}} \cdot 5}{1 \cdot \underset{2}{\cancel{18}}}$ Divide 45 and 18 by 9

$= \frac{25}{2}$

$= 12\frac{1}{2}$

16c. $5\frac{11}{15} \div 12\frac{9}{10} = \frac{86}{15} \div \frac{129}{10}$

$= \frac{86}{15} \cdot \frac{10}{129}$

$= \frac{86 \cdot 10}{15 \cdot 129}$

$= \frac{2 \cdot 43 \cdot 2 \cdot \cancel{5}}{3 \cdot \cancel{5} \cdot 3 \cdot \cancel{43}}$

$= \frac{4}{9}$

Practice Problem 15 *Divide and reduce.*

a. $1\frac{2}{9} \div \frac{45}{66}$ **b.** $18 \div 8\frac{2}{11}$ **c.** $13\frac{7}{9} \div 1\frac{17}{27}$

Example 17 Carver's car traveled 325 miles on $12\frac{1}{2}$ gallons of gas. How many miles per gallon (mpg) did his car get?

Solution Car's mpg = miles traveled ÷ gallons of gas used

$= 325 \div 12\frac{1}{2}$

$= \frac{325}{1} \div \frac{25}{2}$

$= \frac{325}{1} \cdot \frac{2}{25}$

$= \frac{325 \cdot 2}{1 \cdot 25}$

$= \frac{5 \cdot 5 \cdot 13 \cdot 2}{1 \cdot 5 \cdot 5}$

$= \frac{26}{1}$

$= 26$

Carver's car gets 26 miles per gallon.

Practice Problem 16 *If a car traveled $110\frac{1}{2}$ miles on $6\frac{4}{5}$ gallons of gas, how many miles per gallon did the car get?*

2.5 Exercises

Divide and reduce.

1. $\frac{8}{9} \div \frac{16}{54}$
2. $\frac{15}{4} \div \frac{27}{8}$
3. $\frac{43}{10} \div \frac{86}{25}$
4. $\frac{4}{9} \div 6$
5. $\frac{25}{49} \div \frac{4}{9}$
6. $26 \div \frac{39}{7}$
7. $\frac{6}{23} \div \frac{8}{69}$
8. $\frac{27}{16} \div \frac{18}{24}$
9. $\frac{15}{42} \div \frac{1}{60}$
10. $\frac{123}{77} \div \frac{82}{49}$
11. $\frac{66}{12} \div \frac{44}{45}$
12. $\frac{150}{260} \div \frac{78}{30}$
13. $\frac{75}{24} \div \frac{30}{16}$
14. $\frac{72}{55} \div \frac{108}{33}$
15. $\frac{72}{81} \div \frac{96}{83}$
16. $\frac{429}{165} \div \frac{310}{234}$
17. $\frac{2}{51} \div \frac{1}{119}$
18. $\frac{231}{522} \div \frac{297}{174}$
19. $\frac{4}{9} \div 36$
20. $\frac{5}{6} \div 25$
21. $\frac{35}{16} \div \frac{21}{22}$
22. $\frac{14}{18} \div \frac{42}{15}$
23. $34 \div \frac{17}{14}$
24. $42 \div \frac{21}{13}$
25. $\frac{33}{84} \div \frac{22}{60}$
26. $\frac{56}{15} \div \frac{42}{90}$
27. $\frac{13}{28} \div \frac{26}{56}$
28. $\frac{8}{15} \div \frac{16}{30}$
29. $0 \div \frac{3}{4}$
30. $0 \div \frac{5}{6}$
31. $\frac{3}{4} \div 0$
32. $\frac{5}{6} \div 0$
33. $\frac{36}{10} \div \frac{6}{25}$
34. $\frac{18}{8} \div \frac{9}{12}$
35. $\frac{33}{26} \div \frac{22}{91}$
36. $\frac{18}{35} \div \frac{66}{65}$
37. $2100 \div \frac{7}{3}$
38. $3500 \div \frac{7}{3}$
39. $\frac{123}{92} \div \frac{168}{115}$
40. $\frac{170}{175} \div \frac{340}{50}$
41. $\frac{18}{143} \div \frac{27}{209}$
42. $\frac{12}{429} \div \frac{20}{221}$
43. $1\frac{9}{10} \div 1\frac{11}{14}$
44. $3\frac{1}{7} \div 1\frac{4}{11}$
45. $2\frac{1}{2} \div 1\frac{3}{4}$
46. $5\frac{2}{3} \div 2\frac{2}{3}$
47. $3\frac{1}{3} \div 3\frac{1}{3}$
48. $6\frac{1}{5} \div 6\frac{1}{5}$
49. $2\frac{1}{8} \div \frac{34}{8}$
50. $5\frac{1}{6} \div \frac{62}{12}$
51. $250 \div 2\frac{1}{2}$
52. $49 \div 3\frac{1}{2}$
53. $1\frac{9}{12} \div 4\frac{2}{3}$
54. $2\frac{1}{12} \div 5\frac{5}{9}$
55. $4\frac{5}{7} \div \frac{1}{14}$
56. $7\frac{1}{2} \div \frac{1}{10}$
57. $6\frac{4}{5} \div 1\frac{7}{10}$
58. $8\frac{1}{4} \div 6\frac{7}{8}$
59. $7\frac{1}{9} \div 0$
60. $8\frac{3}{5} \div 0$
61. $6 \div 2\frac{1}{7}$
62. $3\frac{3}{4} \div 5\frac{5}{6}$
63. $6\frac{2}{9} \div 8$
64. $0 \div 7\frac{2}{3}$
65. $\frac{6}{4} \div 1\frac{1}{2}$
66. $3\frac{1}{8} \div \frac{15}{16}$

68 Fractions and Mixed Numbers

67. $5\frac{2}{5} \div \frac{21}{30}$ 68. $10\frac{2}{5} \div \frac{13}{15}$ 69. $2\frac{1}{5} \div \frac{4}{27}$

70. $\frac{56}{72} \div 1\frac{1}{27}$ 71. $\frac{48}{60} \div 1\frac{1}{27}$ 72. $7\frac{3}{7} \div 2\frac{8}{35}$

73. $13\frac{4}{9} \div 6\frac{1}{15}$ 74. $10\frac{5}{9} \div 1\frac{17}{21}$ 75. $12\frac{10}{11} \div 3\frac{13}{33}$

76. $9\frac{5}{8} \div 11\frac{11}{12}$ 77. $7\frac{1}{25} \div 13\frac{1}{5}$ 78. $29\frac{7}{10} \div 23\frac{1}{10}$

Fill in the blanks.

79. To find the _____ of a fraction, interchange the numerator and denominator.

80. To divide one fraction by another fraction, multiply the _____ and the _____ of the divisor.

81. To divide using mixed numbers, change all _____ to _____ and then use the rules for dividing fractions.

Solve and show all work.

82. What is $\frac{99}{120}$ divided by $\frac{66}{132}$?   83. What is $\frac{44}{154}$ divided by $\frac{52}{110}$?

84. Divide $1\frac{1}{65}$ into $1\frac{1}{35}$.   85. Divide $8\frac{2}{5}$ into $5\frac{3}{5}$.

86. Three-fourths of a pound of peanuts must be divided equally among six people. How much will each person receive?

87. Three-fourths of a pound of sugar must be divided equally among nine people. How much will each person receive?

88. Jerry owns 45 acres of land. If he divides this property into smaller lots, each containing $\frac{3}{5}$ of an acre, how many lots will there be?

89. Annette purchased $\frac{3}{4}$ of a pound of meat. If she eats $\frac{1}{8}$ of a pound every meal how many meals will the meat last?

90. A car traveled 180 miles on $11\frac{1}{4}$ gallons of gas. How many miles per gallon did the car get?

91. Bill's car can travel $16\frac{1}{4}$ miles on one gallon of gas. If he drives $110\frac{1}{2}$ miles, how much gas will he use? (*Hint:* gallons of gas used = miles driven ÷ miles per gallon.)

92. If you wanted to travel 42 miles on a bike at a constant rate of $15\frac{3}{4}$ miles per hour, how many hours would it take? (*Hint:* hours traveled = miles traveled ÷ miles per hour.)

93. John has one-half of a tank of gas in his car. If he uses one-tenth of a tank of gas to go to and from school, how many trips can he make to and from school?

94. Bill has three-fifths of a pound of coffee. If he uses one-fifth of a pound of coffee a day, how many days will the coffee last?

95. A car traveled 180 miles on $9\frac{2}{7}$ gallons of gas. How many miles per gallon of gas did the car get?

96. A car traveled 200 miles on $10\frac{2}{5}$ gallons of gas. How many miles per gallon of gas did the car get?

Answers to Practice Problems   14a. $\frac{3}{2}$ b. 130   c. $\frac{34}{45}$ 15a. $1\frac{107}{135}$ b. $2\frac{1}{5}$ c. $8\frac{5}{11}$ 16. $16\frac{1}{4}$

2.6 Adding Fractions

IDEA 1 The sum of $\frac{1}{5}$ and $\frac{3}{5}$ is $\frac{4}{5}$ as shown by Figure 2-7. In other words, 1 fifth + 3 fifths = (1 + 3) fifths = 4 fifths, which can be written as $\frac{1}{5} + \frac{3}{5} = \frac{1+3}{5} = \frac{4}{5}$.

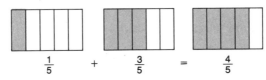

$$\frac{1}{5} \quad + \quad \frac{3}{5} \quad = \quad \frac{4}{5}$$

Figure 2-7

To Add Fractions Having a Common Denominator:

1. Add the numerators and place this sum over the common denominator.
2. If possible, reduce the resulting fraction,

or

$$\frac{a}{b} + \frac{c}{b} = \frac{a+c}{b}$$

Example 18 Add and reduce.

a. $\frac{3}{11} + \frac{4}{11}$ b. $\frac{7}{15} + \frac{2}{15} + \frac{3}{15}$ c. $\frac{11}{117} + \frac{34}{117}$

Solution To add fractions having a common denominator, write the sum of the numerators over the common denominator and reduce the resulting fraction.

18a. $\frac{3}{11} + \frac{4}{11} = \frac{3+4}{11}$

$= \frac{7}{11}$

18b. $\frac{7}{15} + \frac{2}{15} + \frac{3}{15} = \frac{7+2+3}{15}$

$= \frac{12}{15}$ Divide 12 and 15 by 3 to reduce the fraction

$= \frac{4}{5}$

18c. $\frac{11}{117} + \frac{34}{117} = \frac{11+34}{117}$

$= \frac{45}{117}$

$= \frac{\cancel{3} \cdot \cancel{3} \cdot 5}{\cancel{3} \cdot \cancel{3} \cdot 13}$

$= \frac{5}{13}$

Practice Problem 17 *Add and reduce.*

$$\text{a. } \frac{3}{10} + \frac{4}{10} \qquad \text{b. } \frac{5}{18} + \frac{1}{18} + \frac{3}{18} \qquad \text{c. } \frac{32}{105} + \frac{38}{105}$$

IDEA 2 Before considering how to add fractions having different denominators, it will be helpful to review how to write equivalent fractions by multiplying both the numerator and denominator of a fraction by the same nonzero number.

Example 19 Find the missing numerator that makes the fractions equivalent.

$$\text{a. } \frac{1}{2} = \frac{?}{6} \qquad \text{b. } \frac{1}{3} = \frac{?}{6} \qquad \text{c. } \frac{5}{8} = \frac{?}{88}$$

Solution To rewrite a fraction as an equivalent fraction with a larger denominator, divide the original denominator into the new denominator and then multiply the result by the numerator and denominator of the original fraction.

19a. Since $6 \div 2 = 3$, multiply by 3.

$$\frac{1}{2} = \frac{1 \cdot 3}{2 \cdot 3} = \frac{3}{6}$$

19b. Since $6 \div 3 = 2$, multiply by 2.

$$\frac{1}{3} = \frac{1 \cdot 2}{3 \cdot 2} = \frac{2}{6}$$

19c. Since $88 \div 8 = 11$, multiply by 11.

$$\frac{5}{8} = \frac{5 \cdot 11}{8 \cdot 11} = \frac{55}{88}$$

NOTE: Many times you will be able to mentally change a fraction to an equivalent fraction with a larger denominator. Do this by dividing the original denominator into the new denominator and multiplying the result by the numerator and denominator of the original fraction. For example, you can find the missing numerator in the expression $\frac{1}{3} = \frac{?}{6}$ by thinking that 3 divided into 6 is 2, and $2 \cdot 1 = 2$. Thus, $\frac{1}{3} = \frac{2}{6}$. Similarly, you can find the missing numerator in the expression $\frac{1}{2} = \frac{?}{6}$ by thinking that 2 divided into 6 is 3, and $3 \cdot 1 = 3$. Thus, $\frac{1}{2} = \frac{3}{6}$.

Practice Problem 18 *Find the missing numerator that makes the fractions equivalent. Show all work.*

$$\text{a. } \frac{3}{7} = \frac{?}{21} \qquad \text{b. } \frac{6}{13} = \frac{?}{39} \qquad \text{c. } \frac{5}{3} = \frac{?}{18}$$

Practice Problem 19 *Find the missing numerator that makes the fractions equivalent (do this mentally).*

$$\text{a. } \frac{3}{8} = \frac{?}{24} \qquad \text{b. } \frac{5}{9} = \frac{?}{72} \qquad \text{c. } \frac{8}{11} = \frac{?}{77}$$

IDEA 3 Let us now discuss how to add fractions having different denominators. We will begin our discussion by considering how to add $\frac{1}{2}$ and $\frac{1}{3}$. A careful examination of Figure 2-8 will help us understand how to add $\frac{1}{2}$ and $\frac{1}{3}$. Looking at the top line of

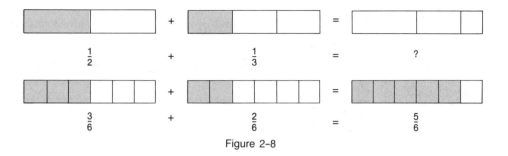

Figure 2-8

Figure 2-8, we cannot visually determine the sum of $\frac{1}{2}$ and $\frac{1}{3}$ since both rectangles are not divided into equally sized parts. However, if we divide each rectangle into 6 equal parts and note that $\frac{1}{2}$ of a rectangle is equivalent to $\frac{3}{6}$ of a rectangle, and that $\frac{1}{3}$ of a rectangle is equivalent to $\frac{2}{6}$ of a rectangle, we will be able to visually add the equal parts of the rectangle. The total is $\frac{5}{6}$ of a rectangle.

IDEA 4

The fractions $\frac{1}{2}$ and $\frac{1}{3}$ can be written with a common denominator of 6, 12, 18, or 24 and then added. However, it is more convenient to use the number 6 since it is the smallest or *lowest common denominator* (LCD) of 2 and 3. The **lowest common denominator** (LCD) of two or more fractions is the smallest number that each denominator divides evenly. The LCD is sometimes referred to as as the **least common multiple** (LCM) of the denominators. Remember that a multiple of a natural number is any number that is the product of the given number and some other natural number. For example, 6 is a multiple of 2 since $2 \cdot 3 = 6$.

By inspection, we know that the LCD of $\frac{1}{2}$ and $\frac{1}{3}$ is 6. However, you will not always be able to find the LCD by inspection. Therefore, to find the LCD of two or more fractions, we factor each denominator and then form the LCD from these factors. For example, the LCD for the fractions $\frac{1}{8}$ and $\frac{3}{10}$ is the smallest number divisible by both 8 and 10. To find this number, first write the prime factorization of 8 and 10.

$$8 = 2 \cdot 2 \cdot 2 \qquad 10 = 2 \cdot 5$$

Next, note that every number divisible by 8 has three factors of 2 and that every number divisible by 10 has one factor of 2 and one factor of 5. Since the LCD of $\frac{1}{8}$ and $\frac{1}{10}$ is the smallest number divisible by both 8 and 10, the LCD is the product of three factors of 2 and one factor of 5.

$$8 = \boxed{2 \cdot 2 \cdot 2} \qquad 10 = 2 \cdot \boxed{5} \qquad \text{LCD} = 2 \cdot 2 \cdot 2 \cdot 5 = 40$$

Since we already have three factors of 2, we do not need to use this factor in the LCD

Factors of 8

Factors of 10

72 Fractions and Mixed Numbers

Finally, the LCD of $\frac{1}{8}$ and $\frac{1}{10}$ is 40. It is the smallest number divisible by both 8 and 10.

The procedure for finding the LCD is summarized below.

To Find the LCD:

1. Write the prime factorization of each denominator.
2. The LCD is the product of the distinct prime factors listed in step 1. Each factor must be used the greatest number of times it occurs in any one factorization.

Example 20 Find the LCD.

a. $\frac{1}{6} + \frac{1}{8}$ **b.** $\frac{3}{26} + \frac{5}{39}$

c. $\frac{1}{21} + \frac{3}{14} + \frac{5}{18}$ **d.** $\frac{3}{44} + \frac{5}{36} + \frac{1}{66}$

Solution To find the LCD, write the prime factorization of each denominator. The LCD is the product of the distinct prime factors of the denominators. Each factor is used the greatest number of times it appears in any one denominator.

20a. $6 = 2 \cdot 3$
$8 = 2 \cdot 2 \cdot 2$
LCD $= 2 \cdot 2 \cdot 2 \cdot 3 = 24$

The factor 2 occurs once in the denominator 6 and three times in the denominator 8, so we put it three times into the LCD. The factor 3 occurs once in the denominator 6, so we put it once into the LCD.

20b. $26 = 2 \cdot 13$
$39 = 3 \cdot 13$
LCD $= 2 \cdot 3 \cdot 13 = 78$

20c. $21 = 3 \cdot 7$
$14 = 2 \cdot 7$
$18 = 2 \cdot 3 \cdot 3$
LCD $= 2 \cdot 3 \cdot 3 \cdot 7 = 126$

20d. $44 = 2 \cdot 2 \cdot 11$
$36 = 2 \cdot 2 \cdot 3 \cdot 3$
$66 = 2 \cdot 3 \cdot 11$
LCD $= 2 \cdot 2 \cdot 3 \cdot 3 \cdot 11 = 396$

Practice Problem 20 *Find the LCD.*

a. $\frac{1}{9} + \frac{1}{6}$ **b.** $\frac{1}{10} + \frac{3}{12} + \frac{4}{15}$ **c.** $\frac{5}{24} + \frac{7}{32}$

IDEA 5 Based on the above information, we can now add fractions having different denominators. For example, to add $\frac{1}{8}$ and $\frac{5}{12}$, first find the LCD. The prime factorization of 8 is $2 \cdot 2 \cdot 2$ and of 12 is $2 \cdot 2 \cdot 3$, so the LCD is $2 \cdot 2 \cdot 2 \cdot 3 = 24$. Next, rewrite $\frac{1}{8}$ and $\frac{5}{12}$ as equivalent fractions having a denominator of 24 (the LCD).

$$\frac{1}{8} + \frac{5}{12} = \underbrace{\frac{1 \cdot 3}{8 \cdot 3} + \frac{5 \cdot 2}{12 \cdot 2}}_{\text{Do mentally}} = \frac{3}{24} + \frac{10}{24}$$

2.6 Adding Fractions   73

Now, add the resulting fractions.

$$\frac{1}{8} + \frac{5}{12} = \frac{3}{24} + \frac{10}{24}$$
$$= \frac{13}{24}$$

The procedure for adding fractions is summarized below.

To Add Fractions Having Different Denominators:

1. Find the LCD.
2. Change each fraction to an equivalent fraction using the LCD as the new denominator.
3. Add the resulting fractions and reduce if possible.

Example 21   Add and reduce.

   a. $\frac{1}{6} + \frac{5}{9}$ b. $\frac{1}{10} + \frac{5}{18} + 2$

   c. $\frac{5}{36} + \frac{7}{66}$ d. $\frac{5}{42} + \frac{3}{28} + \frac{5}{63}$

Solution   To add fractions having different denominators, first find the LCD. Next, express each fraction as an equivalent fraction using the LCD as the new denominator and then add.

21a. $\frac{1}{6} + \frac{5}{9} = \frac{1 \cdot 3}{6 \cdot 3} + \frac{5 \cdot 2}{9 \cdot 2}$ Do this step mentally
$6 = 2 \cdot 3$
$9 = 3 \cdot 3$
$\text{LCD} = 2 \cdot 3 \cdot 3 = 18$

$= \frac{3}{18} + \frac{10}{18}$

$= \frac{13}{18}$

21b. $\frac{1}{10} + \frac{5}{18} + 2 = \frac{1}{10} + \frac{5}{18} + \frac{2}{1}$ $10 = 2 \cdot 5$
$18 = 2 \cdot 3 \cdot 3$
$\text{LCD} = 2 \cdot 3 \cdot 3 \cdot 5 = 90$

$= \frac{9}{90} + \frac{25}{90} + \frac{180}{90}$

$= \frac{214}{90}$ Divide 214 and 90 by 2

$= \frac{107}{45}$

21c. $\frac{5}{36} + \frac{7}{66} = \frac{55}{396} + \frac{42}{396}$ $36 = 2 \cdot 2 \cdot 3 \cdot 3$
$66 = 2 \cdot 3 \cdot 11$
$\text{LCD} = 2 \cdot 2 \cdot 3 \cdot 3 \cdot 11 = 396$

$= \frac{97}{396}$

21d. $\dfrac{5}{42} + \dfrac{3}{28} + \dfrac{5}{63} = \dfrac{30}{252} + \dfrac{27}{252} + \dfrac{20}{252}$ $\quad 42 = 2 \cdot 3 \cdot 7$
$\qquad\qquad\qquad\qquad\qquad\qquad\qquad\qquad\quad 28 = 2 \cdot 2 \cdot 7$
$\qquad\qquad\quad = \dfrac{77}{252}\qquad\qquad\qquad\qquad\quad 63 = 3 \cdot 3 \cdot 7$
$\qquad\qquad\qquad\qquad\qquad\qquad\qquad\quad \text{LCD} = 2 \cdot 2 \cdot 3 \cdot 3 \cdot 7 = 252$
$\qquad\qquad\quad = \dfrac{\cancel{7} \cdot 11}{\cancel{7} \cdot 36}$
$\qquad\qquad\quad = \dfrac{11}{36}$

NOTE: In Example 21, to write $\dfrac{5}{36}$ as $\dfrac{55}{396}$, we divided 36 into 396 and multiplied the result times the numerator and denominator of $\dfrac{5}{36}$. The division can be done either by long division or by thinking of $396 \div 36$ as a fraction that must be reduced. That is,

$$\dfrac{396}{36} = \dfrac{\cancel{2} \cdot \cancel{2} \cdot \cancel{3} \cdot \cancel{3} \cdot 11}{\cancel{2} \cdot \cancel{2} \cdot \cancel{3} \cdot \cancel{3}} = \dfrac{11}{1} = 11$$

The point is that to divide a denominator into the LCD, you can delete all the prime factors that the two numbers have in common, and the product of the factors remaining in the LCD is the quotient. For example, given that

$$36 = 2 \cdot 2 \cdot 3 \cdot 3 \text{ and}$$
$$66 = 2 \cdot 3 \cdot 11,$$
$$\text{LCD} = 2 \cdot 2 \cdot 3 \cdot 3 \cdot 11 = 396$$

If you delete all the prime factors that 36 and 396 have in common, a factor of 11 remains in 396 (LCD). Thus, $396 \div 36 = 11$. Similarly, you can divide 66 into 396 by deleting all their common prime factors (2, 3, and 11) and then noticing that $2 \cdot 3 = 6$ remains in 396. Thus $396 \div 66 = 6$.

IDEA 6

Fractions can also be added by using a vertical format, rather than a horizontal one.

Example 22 Add and reduce.

   a. $\dfrac{1}{6} + \dfrac{1}{8}$ **b.** $\dfrac{5}{12} + \dfrac{2}{21} + \dfrac{1}{14}$

Solution To add fractions having different denominators, first find the LCD. Next, express each fraction as an equivalent fraction using the LCD as the new denominator and then add.

22a. $\begin{aligned}\dfrac{1}{6} &= \dfrac{4}{24}\\ +\dfrac{1}{8} &= \dfrac{3}{24}\\ \hline &\;\dfrac{7}{24}\end{aligned}$ $\quad\begin{aligned}6 &= 2 \cdot 3\\ 8 &= 2 \cdot 2 \cdot 2\\ \text{LCD} &= 2 \cdot 2 \cdot 2 \cdot 3 = 24\end{aligned}$

22b.
$$\frac{5}{12} = \frac{35}{84}$$
$$\frac{2}{21} = \frac{8}{84}$$
$$+\frac{1}{14} = \frac{6}{84}$$
$$\frac{49}{84} = \frac{7}{12}$$

$12 = 2 \cdot 2 \cdot 3$
$21 = 3 \cdot 7$
$14 = 2 \cdot 7$
LCD $= 2 \cdot 2 \cdot 3 \cdot 7 = 84$

Divide 49 and 84 by 7

Practice Problem 21 *Add and reduce.*

a. $\frac{1}{9} + \frac{5}{15}$ b. $\frac{1}{4} + \frac{5}{6} + \frac{3}{20}$ c. $\frac{2}{75} + \frac{7}{30}$

2.6 Exercises

Add and reduce.

1. $\frac{3}{7} + \frac{2}{7}$
2. $\frac{3}{10} + \frac{2}{10}$
3. $\frac{13}{7} + \frac{3}{7}$
4. $\frac{3}{20} + \frac{11}{20}$
5. $\frac{1}{12} + \frac{7}{12}$
6. $\frac{5}{111} + \frac{28}{111}$
7. $\frac{9}{100} + \frac{43}{100} + \frac{13}{100}$
8. $\frac{15}{135} + \frac{20}{135} + \frac{35}{135} + \frac{2}{135}$
9. $\frac{5}{66} + \frac{13}{66} + \frac{15}{66}$
10. $\frac{1}{91} + \frac{3}{91} + \frac{10}{91}$

Find the missing numerator that makes the fractions equivalent.

11. $\frac{3}{8} = \frac{?}{24}$
12. $\frac{5}{7} = \frac{?}{42}$
13. $\frac{1}{30} = \frac{?}{90}$
14. $\frac{3}{35} = \frac{?}{70}$
15. $\frac{5}{17} = \frac{?}{51}$
16. $\frac{7}{3} = \frac{?}{90}$
17. $\frac{1}{15} = \frac{?}{105}$
18. $\frac{13}{18} = \frac{?}{90}$
19. $\frac{7}{36} = \frac{?}{396}$
20. $\frac{3}{37} = \frac{?}{111}$
21. $\frac{1}{18} = \frac{?}{54}$
22. $\frac{1}{16} = \frac{?}{48}$
23. $\frac{5}{53} = \frac{?}{159}$
24. $\frac{7}{49} = \frac{?}{98}$
25. $\frac{5}{6} = \frac{?}{90}$
26. $\frac{4}{15} = \frac{?}{90}$
27. $\frac{7}{30} = \frac{?}{150}$
28. $\frac{13}{50} = \frac{?}{150}$
29. $\frac{3}{14} = \frac{?}{588}$
30. $\frac{5}{12} = \frac{?}{588}$
31. $\frac{2}{49} = \frac{?}{588}$
32. $\frac{7}{42} = \frac{?}{588}$
33. $\frac{5}{12} = \frac{?}{252}$
34. $\frac{8}{21} = \frac{?}{252}$

Find the LCD.

35. $\frac{1}{6} + \frac{2}{9}$
36. $\frac{3}{10} + \frac{7}{15}$
37. $\frac{4}{8} + \frac{3}{10}$
38. $\frac{5}{7} + \frac{3}{14}$
39. $\frac{5}{9} + \frac{7}{10}$
40. $\frac{5}{18} + \frac{7}{12}$

41. $\dfrac{13}{15} + \dfrac{5}{36}$ 42. $\dfrac{1}{24} + \dfrac{5}{36} + \dfrac{1}{40}$ 43. $\dfrac{3}{25} + \dfrac{1}{30} + \dfrac{7}{40}$

44. $\dfrac{1}{4} + \dfrac{5}{6} + \dfrac{2}{15}$ 45. $\dfrac{1}{6} + \dfrac{5}{8} + \dfrac{1}{18}$ 46. $\dfrac{3}{10} + \dfrac{7}{15} + \dfrac{8}{12}$

Add and reduce.

47. $\dfrac{1}{9} + \dfrac{5}{6}$ 48. $\dfrac{3}{4} + \dfrac{5}{6}$ 49. $\dfrac{1}{3} + \dfrac{3}{4}$

50. $\dfrac{5}{12} + \dfrac{1}{9}$ 51. $\dfrac{7}{8} + \dfrac{3}{10}$ 52. $\dfrac{8}{14} + \dfrac{5}{21}$

53. $\dfrac{5}{16} + \dfrac{1}{12}$ 54. $\dfrac{9}{8} + \dfrac{5}{6}$ 55. $\dfrac{4}{15} + \dfrac{5}{12}$

56. $\dfrac{5}{26} + \dfrac{7}{39}$ 57. $\dfrac{3}{10} + \dfrac{14}{15} + \dfrac{7}{12}$ 58. $\dfrac{1}{36} + \dfrac{5}{18} + \dfrac{7}{24}$

59. $\dfrac{5}{24} + \dfrac{7}{36} + \dfrac{1}{48}$ 60. $\dfrac{17}{22} + \dfrac{3}{4} + \dfrac{8}{33}$ 61. $\dfrac{3}{77} + \dfrac{5}{91} + \dfrac{8}{49}$

62. $\dfrac{7}{20} + \dfrac{3}{10} + \dfrac{11}{30}$ 63. $\dfrac{3}{30} + \dfrac{7}{75}$ 64. $\dfrac{3}{48} + \dfrac{7}{54}$

65. $\dfrac{1}{15} + \dfrac{5}{33} + \dfrac{7}{55}$ 66. $\dfrac{5}{36} + \dfrac{7}{66}$ 67. $\dfrac{11}{28} + \dfrac{5}{42}$

68. $\dfrac{11}{18} + \dfrac{7}{24}$ 69. $\dfrac{1}{16} + \dfrac{3}{24}$ 70. $\dfrac{11}{18} + \dfrac{4}{15}$

71. $\dfrac{1}{35} + \dfrac{8}{15}$ 72. $\dfrac{5}{36} + \dfrac{3}{20}$ 73. $\dfrac{9}{14} + \dfrac{5}{42}$

74. $\dfrac{5}{18} + \dfrac{7}{54}$ 75. $\dfrac{17}{40} + \dfrac{3}{16}$ 76. $\dfrac{7}{10} + \dfrac{3}{7}$

77. $\dfrac{8}{13} + \dfrac{3}{5}$ 78. $\dfrac{11}{24} + \dfrac{1}{36}$ 79. $\dfrac{3}{56} + \dfrac{7}{24}$

80. $\dfrac{1}{14} + \dfrac{3}{77}$ 81. $\dfrac{1}{8} + \dfrac{7}{12} + \dfrac{5}{18}$ 82. $\dfrac{5}{6} + \dfrac{9}{10} + \dfrac{11}{24}$

83. $\dfrac{3}{4} + \dfrac{1}{10} + \dfrac{5}{8}$ 84. $\dfrac{5}{6} + \dfrac{9}{8} + \dfrac{1}{12}$ 85. $\dfrac{1}{24} + \dfrac{5}{86} + \dfrac{13}{40}$

86. $\dfrac{3}{25} + \dfrac{7}{30} + \dfrac{1}{40}$ 87. $\dfrac{5}{14} + \dfrac{3}{49} + \dfrac{7}{12}$ 88. $\dfrac{1}{10} + \dfrac{11}{25} + \dfrac{5}{12}$

Fill in the blanks.

89. To add fractions having a common denominator, add the _____ and place this sum over the _____.

90. To add fractions having different denominators, find the _____. Next, express each fraction as an _____ fraction having the _____ as the common denominator and then add.

Solve. Show all work.

91. Find the sum of $\dfrac{3}{10}, \dfrac{7}{8}$, and $\dfrac{5}{6}$.   92. Find the sum of $\dfrac{1}{9}, \dfrac{7}{16}$, and $\dfrac{5}{6}$.

93. Find the sum of $\dfrac{11}{24}$ and $\dfrac{5}{42}$.   94. Find the sum of $\dfrac{5}{36}$ and $\dfrac{7}{48}$.

95. Jim mixed $\dfrac{1}{3}$ pound of jellybeans, $\dfrac{1}{4}$ pound of Spanish peanuts, and $\dfrac{1}{6}$ pound of candy corn. What is the final weight of the mixture?

96. In one week, Linda jogged $\frac{3}{4}$ mile, $\frac{11}{12}$ mile, and $\frac{9}{10}$ mile. How many miles did she jog altogether?

97. On a hunting trip, Joe spent $\frac{1}{4}$ of his money on gas, $\frac{1}{6}$ of his money on food, and $\frac{1}{10}$ of his money on postcards. What fractional part of his money did he spend altogether?

98. Bob drank $\frac{2}{9}$ of his water on Monday, $\frac{1}{3}$ of his water on Tuesday, and $\frac{1}{6}$ of his water on Wednesday. What fractional part of his water did he drink altogether?

99. A recipe calls for $\frac{1}{4}$ teaspoon of thyme, $\frac{1}{3}$ teaspoon of curry powder, and $\frac{1}{8}$ teaspoon of basil. Find the total amount of herbs and spices used in the recipe.

100. Another recipe calls for $\frac{1}{2}$ cup of flour, $\frac{2}{3}$ cup of milk, and $\frac{1}{6}$ cup of sugar. Find the total amount of ingredients used in this recipe.

101. Find the sum of $\frac{6}{1001} + \frac{7}{1001}$.

102. Find the sum of $\frac{7}{715} + \frac{4}{715}$.

103. Bill owns an eighth of a certain company. If he acquires another sixth of the company, what fractional part of the company will he own?

104. Bob owns a tenth of the stock in a certain company. If he acquires another three-eighths of the stock, what fractional part of the company's stock will he own?

105. If three pieces of wood are glued together that are $\frac{1}{4}$ inch thick, $\frac{1}{10}$ inch thick, and $\frac{1}{8}$ inch thick, what is the total thickness of the wood?

Answers to Practice Problems   **17a.** $\frac{7}{10}$ **b.** $\frac{1}{2}$ **c.** $\frac{2}{3}$ **18a.** 9   **b.** 18   **c.** 30   **19a.** 9   **b.** 40   **c.** 56   **20a.** 18   **b.** 60   **c.** 96   **21a.** $\frac{4}{9}$ **b.** $\frac{37}{30}$ **c.** $\frac{13}{50}$

2.7 Adding Mixed Numbers

IDEA 1   To add mixed numbers, we will combine our skills of adding whole numbers and adding fractions. For example,

$$3\frac{1}{6} + 4\frac{1}{4} = \left(3 + \frac{1}{6}\right) + \left(4 + \frac{1}{4}\right)$$ Rewrite the mixed numbers as a sum.

$$= (3 + 4) + \left(\frac{1}{6} + \frac{1}{4}\right)$$ Use the commutative and associative properties of addition.

$$= (3 + 4) + \left(\frac{2}{12} + \frac{3}{12}\right)$$ Rewrite the fractions with an LCD = 12.

$$= 7 + \frac{5}{12}$$ Add whole numbers and add fractions.

$$= 7\frac{5}{12}$$ Write the answer as a mixed number.

Since we obtained the above result by adding the whole number parts (3 + 4) and by adding the fractional parts $\left(\frac{1}{6} + \frac{1}{4}\right)$ of the mixed numbers, we can shorten the

78 *Fractions and Mixed Numbers*

procedure for adding mixed numbers by writing

$$3\frac{1}{6} = 3\frac{2}{12}$$
$$+4\frac{1}{4} = 4\frac{3}{12}$$
$$\overline{7\frac{5}{12}}$$

The procedure for adding mixed numbers is summarized below.

To Add Mixed Numbers:

1. Add the whole number parts.
2. Add the fractional parts (reduce if necessary).
3. If the sum of the fractional parts is an improper fraction, change it to a mixed number and then add the result to the sum of the whole number parts.

Example 23 Add and express each answer as a mixed number.

   a. $3\frac{3}{5} + 5\frac{4}{5}$ **b.** $5\frac{2}{9} + \frac{1}{6}$

   c. $1\frac{1}{8} + 3\frac{7}{12} + 2\frac{5}{18}$ **d.** $16\frac{5}{24} + 23 + 59\frac{3}{32}$

Solution To add using mixed numbers, add the whole number parts and the fractional parts separately. If possible, the final answer should be written as a mixed number.

23a. $\quad 3\frac{3}{5}$
$\quad\quad +5\frac{4}{5}$
$\quad\quad \overline{8\frac{7}{5}} = 8 + 1\frac{2}{5}$ *Remember:* $8\frac{7}{5} = 8 + \frac{7}{5}$
$\quad\quad\quad\quad\;\, = 9\frac{2}{5}$

23b. $\quad 5\frac{2}{9} = 5\frac{4}{18}\quad\quad 9 = 3 \cdot 3$
$\quad\quad\quad\quad\quad\quad\quad\quad\; 6 = 2 \cdot 3$
$\quad\quad +\frac{1}{6} = \frac{3}{18}\quad\quad\; \text{LCD} = 2 \cdot 3 \cdot 3 = 18$
$\quad\quad \overline{5\frac{7}{18}}$

23c. $\quad 1\frac{1}{8} = 1\frac{9}{72}\quad\quad 8 = 2 \cdot 2 \cdot 2$
$\quad\quad\quad\quad\quad\quad\quad\; 12 = 2 \cdot 2 \cdot 3$
$\quad\quad 3\frac{7}{12} = 3\frac{42}{72}\quad\; 18 = 2 \cdot 3 \cdot 3$
$\quad\quad\quad\quad\quad\quad\quad\; \text{LCD} = 2 \cdot 2 \cdot 2 \cdot 3 \cdot 3 = 72$
$\quad\quad +2\frac{5}{18} = 2\frac{20}{72}$
$\quad\quad \overline{6\frac{71}{72}}$

23d.
$$16\frac{5}{24} = 16\frac{20}{96}$$
$$23 = 23$$
$$+59\frac{3}{32} = 59\frac{9}{96}$$
$$\overline{98\frac{29}{96}}$$

$24 = 2 \cdot 2 \cdot 2 \cdot 3$
$32 = 2 \cdot 2 \cdot 2 \cdot 2 \cdot 2$
$LCD = 2 \cdot 2 \cdot 2 \cdot 2 \cdot 2 \cdot 3 = 96$

IDEA 2 Another way to add using mixed numbers is to change all mixed numbers to improper fractions and then add. However, this method can be very tedious since it can lead to computations with very large numbers.

Example 24   Add and express each answer as a mixed number.

a. $3\frac{5}{6} + 4\frac{1}{4}$ b. $2\frac{1}{10} + 5\frac{4}{15}$

Solution   **24a.** $3\frac{5}{6} + 4\frac{1}{4} = \frac{23}{6} + \frac{17}{4}$
$$= \frac{46}{12} + \frac{51}{12}$$
$$= \frac{97}{12}$$
$$= 8\frac{1}{12}$$

$6 = 2 \cdot 3$
$4 = 2 \cdot 2$
$LCD = 2 \cdot 2 \cdot 3 = 12$

24b. $2\frac{1}{10} + 5\frac{4}{15} = \frac{21}{10} + \frac{79}{15}$
$$= \frac{63}{30} + \frac{158}{30}$$
$$= \frac{221}{30}$$
$$= 7\frac{11}{30}$$

$10 = 2 \cdot 5$
$15 = 3 \cdot 5$
$LCD = 2 \cdot 3 \cdot 5 = 30$

Practice Problem 22 *Add and express each answer as a mixed number.*

a. $2\frac{1}{3} + 4\frac{1}{3}$ b. $3\frac{9}{10} + 5\frac{5}{6}$

c. $5\frac{2}{9} + 1\frac{1}{6} + 7\frac{5}{12}$ d. $32\frac{13}{16} + 76\frac{5}{24}$

2.7 Exercises

Add and express each answer as a mixed number.

1. $2\frac{1}{3} + 3\frac{1}{3}$
2. $5\frac{2}{9} + 7\frac{1}{9}$
3. $1\frac{3}{8} + 9$
4. $5\frac{3}{5} + 2\frac{4}{5}$
5. $6\frac{3}{8} + 2\frac{5}{8}$
6. $5\frac{3}{4} + 7\frac{5}{8}$
7. $4\frac{1}{10} + 11\frac{3}{6}$
8. $12\frac{3}{8} + 11\frac{5}{12}$
9. $2\frac{5}{18} + 3\frac{7}{12}$
10. $3\frac{1}{14} + 5\frac{4}{21}$
11. $8\frac{3}{4} + 9\frac{5}{6}$
12. $4\frac{5}{6} + 3\frac{7}{8}$

80 Fractions and Mixed Numbers

13. $6\frac{11}{20} + 7\frac{9}{10}$

14. $5\frac{2}{3} + 6\frac{11}{12}$

15. $7\frac{1}{12} + 8\frac{9}{15}$

16. $4\frac{1}{15} + 3\frac{7}{20}$

17. $3\frac{4}{7} + 5$

18. $4\frac{4}{11} + 5$

19. $\frac{3}{10} + 5\frac{5}{6}$

20. $\frac{3}{8} + 4\frac{1}{6}$

21. $\frac{11}{14} + 6\frac{13}{21} + \frac{1}{7}$

22. $\frac{19}{22} + 7\frac{5}{11} + \frac{1}{33}$

23. $8\frac{3}{8} + 9\frac{1}{12} + 6\frac{8}{9}$

24. $5\frac{1}{6} + 7\frac{9}{10} + 8\frac{7}{15}$

25. $8\frac{3}{16} + 5\frac{1}{24} + \frac{5}{12}$

26. $9\frac{3}{18} + 8\frac{1}{12} + \frac{5}{6}$

27. $15 + 9\frac{5}{24} + \frac{7}{36}$

28. $18 + 8\frac{9}{10} + \frac{11}{30}$

29. $100\frac{7}{20} + 89\frac{1}{30}$

30. $99\frac{7}{12} + 103\frac{1}{18}$

31. $18\frac{7}{30} + 19\frac{13}{15} + 33\frac{19}{20}$

32. $6\frac{7}{10} + 1\frac{2}{15}$

33. $7\frac{7}{9} + 3\frac{4}{15}$

34. $10\frac{3}{20} + \frac{9}{25}$

35. $3\frac{5}{12} + 7\frac{8}{15}$

36. $1\frac{3}{14} + 6\frac{8}{21}$

37. $13\frac{1}{18} + 15\frac{7}{24} + 12\frac{1}{12}$

38. $85\frac{2}{15} + 55\frac{7}{20} + 65\frac{3}{4}$

39. $19\frac{2}{21} + 17\frac{6}{49} + 99\frac{3}{28}$

40. $31\frac{11}{20} + 66 + \frac{7}{12}$

41. $85\frac{3}{10} + 91\frac{1}{14} + 87\frac{6}{35}$

Fill in the blanks.

42. To add using mixed numbers, add the _____ number parts and the _____ parts separately.

Solve and show all work.

43. Frank mixed $4\frac{1}{3}$ pounds of roasted soybeans, $\frac{1}{4}$ pound of peanuts, and $4\frac{1}{6}$ pounds of raisins. What is the total weight of the mixture?

44. Reggie rode his bike for $\frac{5}{6}$ of an hour, jogged for $4\frac{1}{4}$ hours, and played tennis for $1\frac{7}{15}$ of an hour. How many hours did he exercise?

45. In one week, Amy drove $89\frac{3}{10}$ miles; $45\frac{7}{15}$ miles, $75\frac{7}{12}$ miles, and $135\frac{3}{4}$ miles. How many miles did she drive in all?

46. The perimeter of a triangle is the sum of the length of its three sides. If one side of a triangle is $7\frac{5}{9}$ inches, another side is $9\frac{1}{6}$ inches, and the third side is 11 inches, find the perimeter of the triangle.

47. Find the sum of $8\frac{3}{4}$ inches, $9\frac{1}{6}$ inches, and $11\frac{1}{8}$ inches.

48. Find the sum of $9\frac{3}{10}$ miles, $8\frac{7}{15}$ miles, and $12\frac{1}{6}$ miles.

49. Jack weighs $195\frac{3}{4}$ pounds, Ed weighs $187\frac{2}{3}$ pounds, and Paul weighs $187\frac{5}{8}$ pounds. Find the combined weight of the three men.

50. Betty weighs $113\frac{1}{2}$ pounds, Kathy weighs $105\frac{1}{3}$ pounds, and Jane weighs $122\frac{8}{9}$ pounds. Find the combined weight of the three women.

Answers to Practice Problems 22a. $6\frac{2}{3}$ b. $9\frac{11}{15}$ c. $13\frac{29}{36}$ d. $109\frac{1}{48}$

2.8 Subtracting Fractions and Mixed Numbers

IDEA 1 Finding the difference between two fractions is very similar to computing the sum of two fractions.

> **To Subtract Fractions Having a Common Denominator:**
>
> 1. Subtract the numerators and place this difference over the common denominator.
> 2. If possible, reduce the resulting fraction,
>
> or
>
> $$\frac{a}{b} - \frac{c}{b} = \frac{a-c}{b}$$

Example 25 Subtract and reduce.

a. $\frac{7}{15} - \frac{3}{15}$ b. $\frac{9}{10} - \frac{1}{10}$ c. $\frac{25}{91} - \frac{11}{91}$

Solution To subtract one fraction from another fraction having the same denominator, write the difference of the numerators over the common denominator and reduce the resulting fraction.

25a. $\frac{7}{15} - \frac{3}{15} = \frac{7-3}{15}$
$= \frac{4}{15}$

25b. $\frac{9}{10} - \frac{1}{10} = \frac{9-1}{10}$
$= \frac{8}{10}$ Divide 8 and 10 by 2
$= \frac{4}{5}$

25c. $\frac{25}{91} - \frac{11}{91} = \frac{25-11}{91}$
$= \frac{14}{91}$ Divide 14 and 91 by 7
$= \frac{2}{13}$

Practice Problem 23 *Subtract and reduce.*

a. $\frac{3}{5} - \frac{1}{5}$ b. $\frac{6}{14} - \frac{4}{14}$ c. $\frac{43}{115} - \frac{23}{115}$

82   Fractions and Mixed Numbers

IDEA 2   To subtract one fraction from another fraction having a different denominator, use the following rule.

> **To Subtract Fractions Having Different Denominators:**
>
> 1. Find the LCD.
> 2. Change each fraction to an equivalent fraction using the LCD as the new denominator.
> 3. Subtract using the resulting fractions and reduce if possible.

Example 26   Subtract and reduce.

a. $\dfrac{3}{4} - \dfrac{1}{3}$ b. $\dfrac{5}{9} - \dfrac{2}{15}$ c. $\dfrac{19}{2} - 9$ d. $\dfrac{4}{21} - \dfrac{2}{15}$

Solution   To subtract a fraction from another fraction having a different denominator, first find the LCD. Next, express each fraction as an equivalent fraction using the LCD as the new denominator and then subtract.

26a. $\dfrac{3}{4} - \dfrac{1}{3} = \dfrac{3 \cdot 3}{4 \cdot 3} - \dfrac{1 \cdot 4}{3 \cdot 4}$ Do this step mentally
$4 = 2 \cdot 2$
$3 = 3$
$= \dfrac{9}{12} - \dfrac{4}{12}$ LCD $= 2 \cdot 2 \cdot 3 = 12$
$= \dfrac{5}{12}$

26b. $\dfrac{5}{9} - \dfrac{2}{15} = \dfrac{25}{45} - \dfrac{6}{45}$ $9 = 3 \cdot 3$
$15 = 3 \cdot 5$
$= \dfrac{19}{45}$ LCD $= 3 \cdot 3 \cdot 5 = 45$

26c. $\dfrac{19}{2} - 9 = \dfrac{19}{2} - \dfrac{9}{1}$
$= \dfrac{19}{2} - \dfrac{18}{2}$
$= \dfrac{1}{2}$

26d. $\dfrac{4}{21} - \dfrac{2}{15} = \dfrac{20}{105} - \dfrac{14}{105}$ $21 = 3 \cdot 7$
$15 = 3 \cdot 5$
$= \dfrac{6}{105}$ LCD $= 3 \cdot 5 \cdot 7 = 105$
$= \dfrac{2 \cdot 3}{3 \cdot 5 \cdot 7}$
$= \dfrac{2}{35}$

IDEA 3   Fractions can also be subtracted by using a vertical format. For example, subtract $\dfrac{5}{18}$ from $\dfrac{9}{10}$.

$$\frac{9}{10} = \frac{81}{90}$$
$$-\frac{5}{18} = \frac{25}{90}$$
$$\frac{56}{90} = \frac{28}{45}$$

$10 = 2 \cdot 5$
$18 = 2 \cdot 3 \cdot 3$
$LCD = 2 \cdot 3 \cdot 3 \cdot 5 = 90$

Practice Problem 24 *Subtract and reduce.*

a. $\dfrac{3}{4} - \dfrac{3}{5}$ **b.** $\dfrac{5}{8} - \dfrac{3}{6}$ **c.** $\dfrac{11}{24} - \dfrac{1}{20}$

IDEA 4

Mixed numbers can be subtracted in much the same way that they are added.

To Subtract Mixed Numbers:

1. Change each fractional part to an equivalent fraction using the LCD as the new denominator.

2a. When the fraction in the subtrahend (number being subtracted) is *smaller* than the fraction in the minuend, subtract the whole number parts, and then subtract the fractional parts.

2b. When the fraction in the subtrahend is *larger* than the fraction in the minuend, borrow 1 from the whole number part of the minuend and add it to the fractional part. Then subtract the whole number parts and subtract the fractional parts.

3. If possible, express the answer as a mixed number.

Example 27   Subtract and express each answer as a mixed number.

a. $8\dfrac{3}{7} - 5\dfrac{1}{7}$ **b.** $9\dfrac{3}{4} - 7\dfrac{1}{2}$ **c.** $6\dfrac{1}{10} - 4\dfrac{5}{6}$

d. $9\dfrac{1}{12} - 2\dfrac{7}{9}$ **e.** $9 - 3\dfrac{2}{3}$

Solution   To subtract using mixed numbers, subtract the whole number parts and then subtract the fractional parts. Whenever the fraction in the subtrahend is larger than the fraction in the minuend, borrow and then subtract.

27a. $8\dfrac{3}{7}$
         $-5\dfrac{1}{7}$
         $\overline{3\dfrac{2}{7}}$

27b. $9\dfrac{3}{4} = 9\dfrac{3}{4}$
          $-7\dfrac{1}{2} = 7\dfrac{2}{4}$
          $\overline{\phantom{-7\dfrac{1}{2} =}2\dfrac{1}{4}}$

$4 = 2 \cdot 2$
$2 = 2$
$LCD = 2 \cdot 2 = 4$

84 Fractions and Mixed Numbers

27c. $6\dfrac{1}{10} = 6\dfrac{3}{30} = 5\dfrac{33}{30}$

$-4\dfrac{5}{6} = 4\dfrac{25}{30} = 4\dfrac{25}{30}$

$1\dfrac{8}{30} = 1\dfrac{4}{15}$

$10 = 2 \cdot 5$
$6 = 2 \cdot 3$
$\text{LCD} = 2 \cdot 3 \cdot 5 = 30$

Remember: $6\dfrac{3}{30} = 5 + 1\dfrac{3}{30} = 5 + \dfrac{33}{30} = 5\dfrac{33}{30}$

27d. $9\dfrac{1}{12} = 9\dfrac{3}{36} = 8\dfrac{39}{36}$

$-2\dfrac{7}{9} = 2\dfrac{28}{36} = 2\dfrac{28}{36}$

$6\dfrac{11}{36}$

$12 = 2 \cdot 2 \cdot 3$
$9 = 3 \cdot 3$
$\text{LCD} = 2 \cdot 2 \cdot 3 \cdot 3 = 36$

Remember: $9\dfrac{3}{36} = 8 + 1\dfrac{3}{36} = 8 + \dfrac{39}{36} = 8\dfrac{39}{36}$

27e. $9\phantom{\dfrac{2}{3}} = 8\dfrac{3}{3}$

$-3\dfrac{2}{3} = 3\dfrac{2}{3}$

$\phantom{-3\dfrac{2}{3} = \ }5\dfrac{1}{3}$

Remember: $9 = 8 + 1 = 8 + \dfrac{3}{3} = 8\dfrac{3}{3}$

NOTE: Once you understand the concept of borrowing, this technique can be done mentally. For example, you can write $9\dfrac{3}{36}$ as $8\dfrac{39}{36}$ by decreasing the original whole number by 1 and by noting that the new fraction is the sum of the old numerator and denominator over the old denominator. In other words,

$$9\dfrac{3}{36} = (9 - 1)\dfrac{3 + 36}{36} = 8\dfrac{39}{36}$$

IDEA 5 Another way to subtract mixed numbers is to change all mixed numbers to improper fractions and then subtract. When you use this method, you do not have to borrow. However, this method can be very tedious since it can lead to computations with very large numbers.

Example 28 Subtract and express each answer as a mixed number.

a. $6\dfrac{1}{10} - 4\dfrac{5}{6}$ **b.** $9 - 3\dfrac{2}{3}$

Solution **28a.** $6\dfrac{1}{10} - 4\dfrac{5}{6} = \dfrac{61}{10} - \dfrac{29}{6}$

$\phantom{28a.\ 6\dfrac{1}{10} - 4\dfrac{5}{6}\ } = \dfrac{183}{30} - \dfrac{145}{30}$

$\phantom{28a.\ 6\dfrac{1}{10} - 4\dfrac{5}{6}\ } = \dfrac{38}{30}$

$\phantom{28a.\ 6\dfrac{1}{10} - 4\dfrac{5}{6}\ } = 1\dfrac{8}{30}$

$\phantom{28a.\ 6\dfrac{1}{10} - 4\dfrac{5}{6}\ } = 1\dfrac{4}{15}$

Express mixed numbers as fractions
$10 = 2 \cdot 5$
$6 = 2 \cdot 3$
$\text{LCD} = 2 \cdot 3 \cdot 5 = 30$

28b. $9 - 3\frac{2}{3} = \frac{9}{1} - \frac{11}{3}$

$= \frac{27}{3} - \frac{11}{3}$

$= \frac{16}{3}$

$= 5\frac{1}{3}$

Practice Problem 25 *Subtract and express each answer as a mixed number.*

a. $9\frac{5}{7} - 3\frac{2}{7}$ b. $10\frac{7}{8} - 8\frac{1}{6}$

c. $31\frac{7}{15} - 29\frac{11}{20}$ d. $299 - 75\frac{3}{4}$

2.8 Exercises

Subtract and reduce.

1. $\frac{3}{4} - \frac{1}{4}$
2. $\frac{5}{7} - \frac{4}{7}$
3. $\frac{9}{10} - \frac{3}{10}$
4. $\frac{52}{169} - \frac{26}{169}$
5. $\frac{2}{3} - \frac{3}{5}$
6. $\frac{6}{7} - \frac{2}{3}$
7. $\frac{3}{5} - \frac{3}{10}$
8. $\frac{11}{20} - \frac{2}{5}$
9. $\frac{5}{6} - \frac{3}{4}$
10. $\frac{7}{9} - \frac{1}{6}$
11. $\frac{9}{10} - \frac{4}{15}$
12. $\frac{7}{8} - \frac{3}{10}$
13. $\frac{3}{4} - \frac{1}{5}$
14. $\frac{2}{3} - \frac{2}{5}$
15. $\frac{7}{8} - \frac{1}{3}$
16. $\frac{8}{9} - \frac{1}{2}$
17. $\frac{5}{9} - \frac{1}{12}$
18. $\frac{5}{6} - \frac{3}{8}$
19. $\frac{9}{10} - \frac{5}{6}$
20. $\frac{3}{4} - \frac{1}{10}$
21. $\frac{19}{20} - \frac{7}{10}$
22. $\frac{11}{15} - \frac{3}{5}$
23. $\frac{15}{16} - \frac{11}{12}$
24. $\frac{11}{18} - \frac{5}{12}$
25. $\frac{15}{16} - \frac{11}{24}$
26. $\frac{13}{18} - \frac{5}{27}$
27. $\frac{11}{12} - \frac{1}{20}$
28. $\frac{9}{10} - \frac{5}{12}$
29. $\frac{11}{35} - \frac{3}{10}$
30. $\frac{13}{25} - \frac{1}{10}$
31. $\frac{13}{15} - \frac{5}{12}$
32. $\frac{21}{25} - \frac{11}{15}$
33. $\frac{7}{28} - \frac{5}{21}$
34. $\frac{23}{40} - \frac{5}{16}$
35. $\frac{13}{18} - \frac{7}{15}$
36. $\frac{19}{30} - \frac{5}{12}$
37. $\frac{65}{84} - \frac{17}{72}$
38. $\frac{11}{48} - \frac{3}{32}$
39. $\frac{23}{50} - \frac{3}{40}$
40. $\frac{14}{75} - \frac{1}{30}$
41. $\frac{21}{44} - \frac{7}{36}$
42. $\frac{31}{32} - \frac{7}{40}$

Fill in the blanks.

43. $7\frac{3}{5} = 6 + 1\frac{3}{5}$
    $= 6 + \frac{?}{\_\_}$
    $= \underline{\ ?\ }$

44. $9\frac{13}{23} = 8 + 1\frac{13}{23}$
    $= 8 + \frac{?}{\_\_}$
    $= \underline{\ ?\ }$

45. $5\frac{7}{9} = 4 + 1\frac{7}{9}$
    $= 4 + \frac{?}{\_\_}$
    $= \underline{\ ?\ }$

46. $9\frac{5}{11} = 8\frac{?}{11}$

47. $7\frac{3}{10} = 6\frac{?}{10}$

48. $10 = 9\frac{?}{7}$

49. $8\frac{3}{10} = 7\frac{?}{30}$

50. $6\frac{5}{8} = 5\frac{?}{40}$

51. $8 = 7\frac{?}{9}$

52. $6 = 5\frac{?}{6}$

53. $8\frac{9}{13} = 7\frac{?}{65}$

54. $9\frac{11}{15} = 8\frac{?}{60}$

Subtract and express each answer as a mixed number.

55. $8\frac{3}{7} - 5\frac{1}{7}$

56. $9\frac{5}{8} - 6\frac{1}{8}$

57. $5\frac{3}{9} - 2\frac{1}{9}$

58. $6\frac{3}{8} - 4\frac{1}{8}$

59. $8\frac{7}{10} - 6\frac{3}{10}$

60. $5\frac{7}{12} - 2\frac{5}{12}$

61. $8\frac{1}{6} - 2\frac{5}{6}$

62. $9\frac{1}{4} - 6\frac{3}{4}$

63. $8\frac{7}{9} - 5\frac{1}{9}$

64. $5\frac{5}{9} - 3\frac{2}{9}$

65. $9\frac{5}{6} - 7\frac{1}{6}$

66. $7\frac{5}{7} - 4\frac{6}{7}$

67. $9\frac{1}{8} - 6\frac{3}{8}$

68. $11\frac{5}{9} - 5\frac{1}{6}$

69. $4\frac{7}{8} - 1\frac{3}{4}$

70. $9\frac{9}{20} - 7\frac{3}{10}$

71. $2\frac{1}{8} - 1\frac{1}{12}$

72. $6\frac{5}{6} - 2\frac{7}{15}$

73. $6\frac{5}{12} - 3\frac{7}{9}$

74. $13\frac{2}{15} - 9\frac{3}{10}$

75. $5\frac{5}{14} - 2\frac{12}{21}$

76. $9\frac{5}{12} - 2\frac{9}{16}$

77. $41\frac{5}{24} - 13\frac{15}{16}$

78. $6\frac{5}{14} - 3\frac{19}{35}$

79. $11\frac{1}{18} - 9\frac{7}{15}$

80. $55\frac{5}{12} - 30\frac{2}{15}$

81. $73\frac{23}{32} - 15\frac{1}{24}$

82. $15\frac{11}{16} - 9\frac{7}{24}$

83. $8\frac{1}{4} - 6$

84. $9\frac{3}{4} - 7$

85. $8 - 3\frac{2}{3}$

86. $9 - 2\frac{3}{4}$

87. $16 - 7\frac{3}{8}$

88. $5 - 2\frac{1}{5}$

89. $8\frac{3}{14} - 2\frac{13}{18}$

90. $5\frac{1}{12} - 2\frac{13}{27}$

91. $105\frac{1}{36} - 92\frac{19}{20}$

92. $113\frac{1}{32} - 87\frac{11}{20}$

Fill in the blanks.

93. To subtract fractions having a common denominator, subtract the _____ and place this difference over the common _____ .

94. To subtract fractions having different denominators, find the LCD. Next, express each fraction as an _____ fraction having the _____ as the common denominator and then subtract.

95. To subtract using mixed numbers, subtract the _____ number parts and subtract the _____ parts. However, if the fraction in the subtrahend is larger than the fraction in the minuend, then before subtracting you should borrow 1 from the _____ number part of the _____ and add it to the _____ part.

Solve and show all work.

96. Subtract $\dfrac{8}{35}$ from $\dfrac{11}{14}$.

97. Find the difference between $4\dfrac{5}{6}$ and $1\dfrac{1}{4}$.

98. Dan jogged $15\dfrac{3}{4}$ miles and Frank jogged $12\dfrac{1}{6}$ miles. How many more miles did Dan jog than Frank?

99. Cazzie has $\dfrac{7}{8}$ pound of peanuts. If he gives $\dfrac{1}{6}$ pound of his peanuts to Guerin, how many pounds of peanuts does he have left?

100. On Monday morning, the stock of the ABT Clothing Company opened at $115 per share and then dropped $\$5\dfrac{5}{8}$ per share. Find the closing price of ABT stock.

101. Find the difference between 8 and $6\dfrac{1}{2}$.

102. Find the difference between 9 and $6\dfrac{1}{4}$.

103. Subtract $\dfrac{5}{6}$ from $8\dfrac{5}{8}$.

104. Subtract $\dfrac{11}{12}$ from $7\dfrac{3}{10}$.

105. Find the difference between $110\dfrac{5}{6}$ and $75\dfrac{1}{12}$.

106. Find the difference between $103\dfrac{5}{8}$ and $65\dfrac{1}{16}$.

107. What is $\dfrac{13}{66}$ minus $\dfrac{1}{36}$?

108. What is $\dfrac{5}{44}$ minus $\dfrac{1}{20}$?

109. John cut off $15\dfrac{1}{2}$ yards of wire from a piece of wire that was 65 yards. How much wire was left?

110. Betty cut off $\dfrac{1}{8}$ of a pound of fat from a two-pound steak. How much steak was left?

Answers to Practice Problems   23a. $\dfrac{2}{5}$ b. $\dfrac{1}{7}$ c. $\dfrac{4}{23}$ 24a. $\dfrac{3}{20}$ b. $\dfrac{1}{8}$ c. $\dfrac{49}{120}$ 25a. $6\dfrac{3}{7}$ b. $2\dfrac{17}{24}$ c. $1\dfrac{11}{12}$ d. $223\dfrac{1}{4}$

2.9 Signed Fractions

IDEA 1 We will use the term **signed fraction** to mean a positive or negative fraction. The rules, definitions, and procedures used to perform operations on integers and fractions can be extended to signed fractions.

Example 29   Find the additive inverse.

a. $\dfrac{3}{4}$ b. $-\dfrac{6}{5}$

Solution   To find the additive inverse of a number, change its sign.

29a. The additive inverse of $\dfrac{3}{4}$ is $-\dfrac{3}{4}$.

29b. The additive inverse of $-\dfrac{6}{5}$ is $\dfrac{6}{5}$.

88 Fractions and Mixed Numbers

Example 30 Find the absolute value.

    a. $\left|-\dfrac{3}{4}\right|$ **b.** $\left|\dfrac{3}{4}\right|$

Solution If x is a positive number, then $|x| = x$ and $|-x| = x$.

    30a. $\left|-\dfrac{3}{4}\right| = \dfrac{3}{4}$ **30b.** $\left|\dfrac{3}{4}\right| = \dfrac{3}{4}$

Practice Problem 26 **Solve.**

    a. The additive inverse of $\dfrac{9}{13}$ is _____.

    b. The absolute value of $-\dfrac{3}{7}$ is _____.

    c. The additive inverse of $-\dfrac{13}{5}$ is _____.

    d. The absolute value of $-\dfrac{6}{5}$ is _____.

IDEA 2 Signed fractions are added the same way integers are added. It will be helpful to remember that if $\dfrac{a}{b}$ and $\dfrac{c}{b}$ are two fractions, then $\dfrac{a}{b}$ is larger than $\dfrac{c}{b}$ when a is greater than c. For example, $\dfrac{4}{5}$ is larger than $\dfrac{3}{5}$ since 4 is greater than 3.

Example 31 Add and reduce.

    a. $-\dfrac{5}{16} + \left(-\dfrac{7}{16}\right)$ **b.** $-\dfrac{3}{4} + \dfrac{5}{6}$ **c.** $\dfrac{5}{8} + \left(-\dfrac{11}{12}\right)$

Solution To add signed fractions, express each fraction as an equivalent fraction having the LCD as a common denominator. Then add using the rule for addition of integers (see Section 1.2).

    Common sign ↓   Sum of absolute values

31a. $-\dfrac{5}{16} + \left(-\dfrac{7}{16}\right) = -\left(\dfrac{5}{16} + \dfrac{7}{16}\right)$

$= -\dfrac{12}{16}$ Divide 12 and 16 by 4

$= -\dfrac{3}{4}$

31b. $-\dfrac{3}{4} + \dfrac{5}{6} = -\dfrac{9}{12} + \dfrac{10}{12}$ $4 = 2 \cdot 2;\ 6 = 2 \cdot 3;\ \text{LCD} = 2 \cdot 2 \cdot 3 = 12$

$= +\left(\dfrac{10}{12} - \dfrac{9}{12}\right)$ Write a "+" sign. Then write the larger absolute value minus the smaller absolute value

$= \dfrac{1}{12}$ Subtracted

The answer is positive since the sign is determined by the number with the largest absolute value.

2.9 Signed Fractions   89

31c. $\dfrac{5}{8} + \left(-\dfrac{11}{12}\right) = \dfrac{15}{24} + \left(-\dfrac{22}{24}\right)$ $\quad 8 = 2 \cdot 2 \cdot 2$
$\qquad\qquad\qquad\qquad\qquad\qquad\qquad 12 = 2 \cdot 2 \cdot 3$
$\qquad\qquad\qquad = -\left(\dfrac{22}{24} - \dfrac{15}{24}\right)$ $\quad \text{LCD} = 2 \cdot 2 \cdot 2 \cdot 3 = 24$

$\qquad\qquad\qquad = -\dfrac{7}{24}$

Practice Problem 27 **Add and reduce.**

a. $-\dfrac{7}{24} + \left(-\dfrac{11}{24}\right)$ b. $-\dfrac{8}{9} + \dfrac{1}{12}$ c. $\dfrac{7}{8} + \left(-\dfrac{3}{10}\right)$

IDEA 3   Signed fractions are subtracted the same way integers are subtracted.

Example 32   a. $\dfrac{1}{9} - \dfrac{7}{9}$ b. $-\dfrac{5}{6} - \dfrac{1}{10}$ c. $-\dfrac{1}{15} - \left(-\dfrac{5}{12}\right)$

Solution   To subtract one signed fraction from another, keep the minuend the same and add the additive inverse of the subtrahend.

32a. $\dfrac{1}{9} - \dfrac{7}{9} = \dfrac{1}{9} + \left(-\dfrac{7}{9}\right)$

$\qquad\qquad = -\dfrac{6}{9}$

$\qquad\qquad = -\dfrac{2}{3}$

32b. $-\dfrac{5}{6} - \dfrac{1}{10} = -\dfrac{5}{6} + \left(-\dfrac{1}{10}\right)$

$\qquad\qquad\quad = -\dfrac{25}{30} + \left(-\dfrac{3}{30}\right)$ $\quad 6 = 2 \cdot 3$
$\qquad\qquad\qquad\qquad\qquad\qquad\qquad 10 = 2 \cdot 5$
$\qquad\qquad\quad = -\dfrac{28}{30}$ $\qquad\qquad\qquad \text{LCD} = 2 \cdot 3 \cdot 5 = 30$

$\qquad\qquad\quad = -\dfrac{14}{15}$

32c. $-\dfrac{1}{15} - \left(-\dfrac{5}{12}\right) = -\dfrac{1}{15} + \dfrac{5}{12}$

$\qquad\qquad\qquad\quad = -\dfrac{4}{60} + \dfrac{25}{60}$ $\quad 15 = 3 \cdot 5$
$\qquad\qquad\qquad\qquad\qquad\qquad\qquad 12 = 2 \cdot 2 \cdot 3$
$\qquad\qquad\qquad\quad = \dfrac{21}{60}$ $\qquad\qquad \text{LCD} = 2 \cdot 2 \cdot 3 \cdot 5 = 60$

$\qquad\qquad\qquad\quad = \dfrac{7}{20}$

Practice Problem 28 **Subtract and reduce.**

a. $\dfrac{1}{12} - \dfrac{5}{12}$ b. $-\dfrac{4}{15} - \dfrac{7}{20}$ c. $\dfrac{5}{36} - \left(-\dfrac{7}{24}\right)$

IDEA 4   Multiplying signed fractions is similar to multiplying integers.

Example 33   Multiply and reduce.

a. $\left(-\dfrac{6}{8}\right)\left(-\dfrac{12}{15}\right)$ b. $\left(-\dfrac{25}{14}\right)\left(\dfrac{21}{35}\right)$ c. $\left(-\dfrac{14}{24}\right)\left(\dfrac{33}{35}\right)\left(-\dfrac{4}{77}\right)$

90 *Fractions and Mixed Numbers*

Solution To multiply signed fractions, multiply their absolute values and then determine the sign of the product by using the rule for multiplying integers.

2 negative signs

33a. $\left(-\dfrac{6}{8}\right)\left(-\dfrac{12}{15}\right) = +\left(\dfrac{6 \cdot 12}{8 \cdot 15}\right)$ Multiply absolute values of fractions

$= +\left(\dfrac{\cancel{2} \cdot \cancel{3} \cdot \cancel{2} \cdot \cancel{2} \cdot 3}{\cancel{2} \cdot \cancel{2} \cdot \cancel{2} \cdot \cancel{3} \cdot 5}\right)$

$= \dfrac{3}{5}$

The product of an even number of negative numbers is positive.

33b. $\left(-\dfrac{25}{14}\right)\left(\dfrac{21}{35}\right) = -\left(\dfrac{\overset{5}{\cancel{25}} \cdot \overset{3}{\cancel{21}}}{\underset{2}{\cancel{14}} \cdot \underset{7}{\cancel{35}}}\right)$ Divide 25 and 35 by 5
Divide 14 and 21 by 7

$= -\dfrac{15}{14}$

The product of an odd number of negative numbers is a negative number.

33c. $\left(-\dfrac{14}{24}\right)\left(\dfrac{33}{35}\right)\left(-\dfrac{4}{77}\right) = +\left(\dfrac{14 \cdot 33 \cdot 4}{24 \cdot 35 \cdot 77}\right)$

$= +\left(\dfrac{\cancel{2} \cdot \cancel{7} \cdot \cancel{3} \cdot \cancel{11} \cdot \cancel{2} \cdot \cancel{2}}{\cancel{2} \cdot \cancel{2} \cdot \cancel{2} \cdot \cancel{3} \cdot \cancel{5} \cdot \cancel{7} \cdot \cancel{7} \cdot \cancel{11}}\right)$

$= \dfrac{1}{35}$

Practice Problem 29 **Multiply and reduce.**

a. $\left(-\dfrac{4}{9}\right)\left(\dfrac{15}{4}\right)$ **b.** $(-6)\left(-\dfrac{9}{30}\right)$ **c.** $\left(-\dfrac{36}{99}\right)\left(-\dfrac{25}{40}\right)\left(-\dfrac{42}{28}\right)$

IDEA 5

To divide, multiply by the reciprocal of the divisor. To find the reciprocal of a signed fraction, interchange the numerator and denominator. The sign of the reciprocal is the same as the sign of the original fraction.

Example 34 Divide and reduce.

a. $-\dfrac{15}{8} \div \dfrac{10}{12}$ **b.** $-\dfrac{16}{18} \div \left(-\dfrac{24}{22}\right)$ **c.** $\dfrac{25}{28} \div (-15)$

Solution To divide one signed fraction by another, multiply the dividend and the reciprocal of the divisor and then determine the sign of the quotient by using the rules for multiplying integers.

Reciprocal

34a. $-\dfrac{15}{8} \div \boxed{\dfrac{10}{12}} = -\dfrac{15}{8} \cdot \boxed{\dfrac{12}{10}}$

$= -\left(\dfrac{15 \cdot 12}{8 \cdot 10}\right)$ Rule for multiplying signed fractions

$= -\left(\dfrac{3 \cdot \cancel{5} \cdot \cancel{2} \cdot \cancel{2} \cdot 3}{\cancel{2} \cdot \cancel{2} \cdot 2 \cdot 2 \cdot \cancel{5}}\right)$

$= -\dfrac{9}{4}$

The quotient of two numbers having different signs is a negative number.

34b. $-\dfrac{16}{18} \div \left(-\dfrac{24}{22}\right) = -\dfrac{16}{18}\left(-\dfrac{22}{24}\right)$

$= +\left(\dfrac{16 \cdot 22}{18 \cdot 24}\right)$

$= +\left(\dfrac{\cancel{2} \cdot \cancel{2} \cdot \cancel{2} \cdot \cancel{2} \cdot 2 \cdot 11}{\cancel{2} \cdot 3 \cdot 3 \cdot \cancel{2} \cdot \cancel{2} \cdot \cancel{2} \cdot 3}\right)$

$= \dfrac{22}{27}$

The quotient of two numbers having the same sign is a positive number.

34c. $\dfrac{25}{28} \div (-15) = \dfrac{25}{28}\left(-\dfrac{1}{15}\right)$

$= -\left(\dfrac{\overset{5}{\cancel{25}} \cdot 1}{28 \cdot \underset{3}{\cancel{15}}}\right)$ Divide 25 and 15 by 5

$= -\dfrac{5}{84}$

Practice Problem 30 **Divide and reduce.**

     a. $-\dfrac{12}{25} \div \left(-\dfrac{18}{25}\right)$ **b.** $-\dfrac{8}{15} \div \dfrac{12}{10}$ **c.** $\dfrac{20}{30} \div \left(-\dfrac{88}{15}\right)$

IDEA 6   The order of operations rules discussed in Chapter 1 also applies to signed fractions. For example, to simplify the expression $-\dfrac{1}{8} + \dfrac{6}{35} \cdot \dfrac{21}{4}$, multiply $\dfrac{6}{35}$ and $\dfrac{21}{4}$ before doing any addition.

$-\dfrac{1}{8} + \dfrac{6}{35} \cdot \dfrac{21}{4}$

$= -\dfrac{1}{8} + \dfrac{\overset{3}{\cancel{6}} \cdot \overset{3}{\cancel{21}}}{\underset{5}{\cancel{35}} \cdot \underset{2}{\cancel{4}}}$ Multiply first

$= -\dfrac{1}{8} + \dfrac{9}{10}$

$= -\dfrac{5}{40} + \dfrac{36}{40}$ Rewrite equivalent fraction with LCD = 40

$= \dfrac{31}{40}$ Add

Example 35   Simplify $-\dfrac{1}{10} - \dfrac{3}{2}\left(\dfrac{1}{12} - \dfrac{3}{8}\right)$

Solution   To simplify an expression containing more than one operation and only one grouping symbol, first perform the operations inside the grouping symbol. Then, simplify the resulting expression.

$-\dfrac{1}{10} - \dfrac{3}{2}\left(\dfrac{1}{12} - \dfrac{3}{8}\right)$

$= -\dfrac{1}{10} - \dfrac{3}{2}\left(\dfrac{2}{24} - \dfrac{9}{24}\right)$ Simplify with the ( ) first

Fractions and Mixed Numbers

$$= -\frac{1}{10} - \frac{3}{2}\left(-\frac{7}{24}\right)$$

$$= -\frac{1}{10} + \frac{\overset{1}{\cancel{3}} \cdot 7}{2 \cdot \underset{8}{\cancel{24}}} \quad \text{Multiply}$$

$$= -\frac{1}{10} + \frac{7}{16}$$

$$= -\frac{8}{80} + \frac{35}{80} \quad \text{Add}$$

$$= \frac{27}{80}$$

Practice Problem 31 **Simplify each expression.**

a. $-\frac{2}{3} + \frac{5}{8} \cdot \frac{16}{35}$ b. $-\frac{7}{12} - \left(\frac{1}{6} + \frac{5}{12}\right) \div \frac{21}{8}$

2.9 Exercises

Find the additive inverse.

1. $-\frac{3}{5}$
2. $\frac{4}{9}$
3. $-\frac{13}{8}$

Find the value.

4. $\left|-\frac{3}{7}\right|$
5. $\left|\frac{5}{8}\right|$
6. $\left|-\frac{13}{11}\right|$

Add and reduce.

7. $-\frac{2}{9} + \left(-\frac{3}{9}\right)$
8. $-\frac{3}{10} + \frac{1}{10}$
9. $-\frac{7}{10} + \frac{5}{8}$
10. $-\frac{3}{4} + \left(-\frac{1}{3}\right)$
11. $\frac{5}{8} + \left(-\frac{5}{6}\right)$
12. $-\frac{13}{28} + \frac{1}{42}$

Subtract and reduce.

13. $\frac{5}{14} - \frac{1}{14}$
14. $-\frac{5}{12} - \frac{1}{12}$
15. $-\frac{5}{16} - \left(-\frac{7}{12}\right)$
16. $\frac{7}{10} - \left(-\frac{4}{15}\right)$
17. $-\frac{17}{35} - \left(-\frac{5}{14}\right)$
18. $-\frac{31}{32} - \left(-\frac{1}{24}\right)$

Multiply and reduce.

19. $\left(-\frac{6}{8}\right)\left(-\frac{10}{15}\right)$
20. $\left(-\frac{14}{25}\right)\left(\frac{10}{21}\right)$
21. $\left(\frac{72}{35}\right)\left(-\frac{56}{18}\right)$
22. $\left(-\frac{5}{4}\right)\left(\frac{3}{10}\right)\left(-\frac{16}{9}\right)$
23. $\left(-\frac{14}{20}\right)\left(-\frac{66}{42}\right)\left(-\frac{4}{55}\right)$
24. $-8 \cdot \frac{15}{12}$

Find the reciprocal.

25. $-\frac{3}{5}$
26. $\frac{7}{8}$
27. $-\frac{1}{15}$
28. -9

Divide and reduce.

29. $-\frac{8}{14} \div \left(-\frac{10}{21}\right)$
30. $\frac{18}{25} \div \left(-\frac{27}{10}\right)$
31. $\frac{5}{14} \div (-35)$

32. $\dfrac{-15}{56} \div \dfrac{14}{90}$ 33. $\dfrac{12}{25} \div \left(-\dfrac{1}{10}\right)$ 34. $-\dfrac{80}{44} \div \left(-\dfrac{60}{33}\right)$

Do the indicated operation.

35. $-\dfrac{5}{8} + \dfrac{7}{10}$ 36. $\dfrac{20}{21} - \dfrac{28}{16}$ 37. $-\dfrac{36}{81} \div \dfrac{32}{63}$

38. $-\dfrac{7}{66} - \dfrac{5}{36}$ 39. $8 \div \left(-\dfrac{20}{5}\right)$ 40. $\left(-\dfrac{12}{16}\right)\left(\dfrac{18}{15}\right)\left(-\dfrac{20}{3}\right)$

41. $\dfrac{3}{14} - \dfrac{11}{21}$ 42. $\dfrac{9}{20} + \left(-\dfrac{4}{15}\right)$ 43. $-\dfrac{28}{45} \div \left(-\dfrac{42}{15}\right)$

44. $\left(-\dfrac{21}{18}\right)\left(-\dfrac{20}{14}\right)\left(-\dfrac{6}{5}\right)$ 45. $\left(-\dfrac{3}{10}\right) + \left(-\dfrac{5}{12}\right) + \left(-\dfrac{7}{15}\right)$

Solve. Show all work.

46. Find the sum of $-\dfrac{5}{8}$ and $\dfrac{7}{10}$.

47. Find the sum of $-\dfrac{1}{18}$ and $-\dfrac{5}{12}$.

48. Find the difference between $\dfrac{1}{35}$ and $\dfrac{11}{14}$.

49. Find the difference between $-\dfrac{11}{20}$ and $\dfrac{8}{15}$.

50. Find the product of -12 and $\dfrac{7}{10}$.

51. Find the product of -10 and $\dfrac{5}{6}$.

52. Divide $\dfrac{6}{25}$ into $-\dfrac{8}{15}$.

53. Divide $-\dfrac{9}{25}$ into $\dfrac{6}{35}$.

54. Subtract $-\dfrac{5}{12}$ from $\dfrac{7}{18}$.

55. Subtract $-\dfrac{11}{24}$ from $-\dfrac{5}{18}$.

56. Divide -18 by $-\dfrac{1}{2}$.

57. Divide -24 by $\dfrac{1}{3}$.

Simplify each expression.

58. $\dfrac{1}{8} - \dfrac{3}{8} + \dfrac{5}{8}$

59. $\dfrac{1}{10} - \dfrac{7}{10} + \dfrac{3}{10}$

60. $\dfrac{1}{6} - \dfrac{5}{9} + \dfrac{3}{4}$

61. $\dfrac{1}{6} - \dfrac{1}{8} + \dfrac{5}{16}$

62. $-\dfrac{4}{5} + \dfrac{3}{5} \div \dfrac{6}{7}$

63. $-\dfrac{4}{5} + \dfrac{3}{7} \div \dfrac{15}{14}$

64. $-\dfrac{5}{12} - \dfrac{1}{3}\left(-\dfrac{5}{6} + \dfrac{1}{12}\right)$

65. $-\dfrac{7}{18} + \dfrac{5}{6}\left(\dfrac{2}{3} - \dfrac{1}{6}\right)$

66. $-\dfrac{7}{8} - \left(\dfrac{1}{6} + \dfrac{5}{12}\right) \div \dfrac{7}{8}$

67. $-\dfrac{4}{9} + \left(\dfrac{1}{2} - \dfrac{2}{6}\right) \div \dfrac{3}{8}$

68. $2\dfrac{1}{6} - 5\dfrac{1}{8} + 4\dfrac{1}{10}$

69. $3\dfrac{1}{8} - 5\dfrac{1}{10} + 4\dfrac{3}{4}$

70. $-\dfrac{1}{2} - \left(\dfrac{1}{3} + \dfrac{5}{3}\right)\left(\dfrac{4}{7} + \dfrac{1}{14}\right)$

71. $-\dfrac{3}{4} - \left(\dfrac{1}{2} + \dfrac{5}{2}\right)\left(\dfrac{1}{3} + \dfrac{6}{9}\right)$

72. $32 \div (-8) - 5\left(7 - \dfrac{6-2}{5}\right)$

73. $36 \div (-9) - 6\left(4 - \dfrac{9-7}{3}\right)$

74. $-\dfrac{1}{5} + \left\{\dfrac{2}{3} - \left(\dfrac{5}{8} - \dfrac{1}{2} \cdot \dfrac{3}{5}\right)\right\}$

75. $-3\dfrac{1}{8} - 5\dfrac{1}{10} + 4\dfrac{3}{4}$

76. $-8\dfrac{1}{6} - 3\dfrac{3}{4} + 5\dfrac{5}{8}$

77. $-145\dfrac{1}{18} + 150\dfrac{7}{24} - \left(-75\dfrac{1}{12}\right)$

78. $-175\dfrac{3}{10} - \left(-17\dfrac{1}{14}\right) + 87\dfrac{6}{35}$

Answers to Practice Problems **26a.** $-\frac{9}{13}$ **b.** $\frac{3}{7}$ **c.** $\frac{13}{5}$ **d.** $\frac{6}{5}$ **27a.** $-\frac{3}{4}$ **b.** $-\frac{29}{36}$ **c.** $\frac{23}{40}$ **28a.** $-\frac{1}{3}$ **b.** $-\frac{37}{60}$ **c.** $\frac{31}{72}$ **29a.** $-\frac{5}{3}$ **b.** $\frac{9}{5}$ **c.** $-\frac{15}{44}$ **30a.** $\frac{2}{3}$ **b.** $-\frac{4}{9}$ **c.** $-\frac{5}{44}$ **31a.** $-\frac{8}{21}$ **b.** $-\frac{29}{36}$

Chapter 2 Summary

Important Terms

A **fraction** is any number that can be written in the form $\frac{a}{b}$, where b is not equal to zero. The number a is called the **numerator** and the number b is called the **denominator**. $\frac{32}{39}$ and $\frac{79}{13}$ are fractions. [Section 2.1/Idea 1]

A **proper fraction** is a fraction whose numerator is less than its denominator. $\frac{3}{4}$ and $\frac{6}{9}$ are proper fractions. [Section 2.1/Idea 1]

An **improper fraction** is a fraction whose numerator is greater than (or equal to) its denominator. $\frac{10}{9}$ and $\frac{26}{13}$ are improper fractions. [Section 2.1/Idea 1]

Numbers that are multiplied together are called **factors** or **divisors** of the product. 1, 2, 5, and 10 are factors of 10. [Section 2.1/Idea 2]

A **prime number** is a natural number greater than 1 whose only divisors are one and itself. 7 and 29 are prime numbers. [Section 2.1/Idea 2]

A **composite number** is a natural number greater than 1 that has more than two divisors. 12 and 100 are composite numbers. [Section 2.1/Idea 2]

The **prime factorization** of a number is the representation of that number as the product of its prime factors. The prime factorization of 12 is $2 \cdot 2 \cdot 3$. [Section 2.1/Idea 3]

Equivalent fractions are fractions that have the same numerical value. $\frac{3}{4}$ and $\frac{6}{8}$, and $\frac{7}{1}$ and $\frac{14}{2}$ are two pairs of equivalent fractions. [Section 2.2/Idea 1]

A fraction is **reduced to lowest terms** if the only common factor of its numerator and denominator is 1. $\frac{2}{3}$ and $\frac{8}{7}$ are reduced to lowest terms. [Section 2.2/Idea 3]

The **largest common divisor** of two or more numbers is the largest number that divides each number evenly. 4 is the largest common divisor of 12 and 16. [Section 2.2/Idea 3]

A **mixed number** is the sum of a whole number and a proper fraction. $1\frac{7}{10}$ and $14\frac{18}{19}$ are mixed numbers. [Section 2.3/Idea 1]

Any two numbers whose product is 1 are called **reciprocals**. $\frac{1}{3}$ and 3 are reciprocals; $\frac{4}{7}$ and $\frac{7}{4}$ are reciprocals. [Section 2.5/Idea 1]

The **lowest common denominator** (LCD) of two or more fractions is the smallest number that each denominator divides evenly. The LCD is sometimes referred to as the **least common multiple** (LCM) of the denominators. 24 is the LCD of $\frac{1}{6}$ and $\frac{3}{8}$. [Section 2.6/Idea 4]

A **signed fraction** is a positive or negative fraction. $-\frac{1}{2}$ and $+\frac{3}{4}$ are signed fractions. [Section 2.9/Idea 1]

Important Skills

Finding Prime Factorizations
To find the prime factorization of a number, repeatedly divide by prime numbers until the quotient is a prime number. Write the original number as a product of its prime divisors and the final quotient. [Section 2.1/Idea 3]

Finding Equivalent Fractions
$\dfrac{a}{b} = \dfrac{a \cdot c}{b \cdot c}$ and
$\dfrac{a}{b} = \dfrac{a \div c}{b \div c}$

To find a fraction equivalent to a given fraction, multiply or divide both the numerator and denominator by the same nonzero number. [Section 2.1/Idea 1]

Determining If Fractions Are Equivalent
$\dfrac{a}{b} = \dfrac{c}{d}$ when
$a \cdot d = b \cdot c$

To determine if two fractions are equivalent, compare their cross-products. If the cross-products are equal, the fractions are equivalent. [Section 2.2/Idea 2]

Reducing Fractions
To reduce a fraction to its lowest terms, factor the numerator and denominator and then divide out all of their common factors. Finally, multiply the remaining factors in the numerator, and the remaining factors in the denominator. [Section 2.2/Idea 3]

Changing Mixed Numbers To Improper Fractions
To change a mixed number to an improper fraction, multiply the denominator by the whole number, add the numerator, and place this sum over the original denominator. [Section 2.3/Idea 1]

Changing Improper Fractions to Mixed Numbers
To change an improper fraction to a mixed number, divide the denominator into the numerator. The quotient is the whole number part of the mixed number, and the fractional part is the remainder over the original denominator. [Section 2.3/Idea 2]

Multiplying Fractions
$\dfrac{a}{b} \cdot \dfrac{c}{d} = \dfrac{a \cdot c}{b \cdot d}$

To multiply fractions, write the product of the numerators over the product of the denominators. Reduce the resulting fraction. [Section 2.4/Idea 1]

Multiplying Mixed Numbers
To multiply mixed numbers, change all mixed numbers to improper fractions, and then multiply the resulting fractions. [Section 2.4/Idea 2]

Dividing Fractions
$\dfrac{a}{b} \div \dfrac{c}{d} =$
$\dfrac{a}{b} \cdot \dfrac{d}{c} = \dfrac{a \cdot d}{b \cdot c}$

To divide fractions, multiply the dividend by the reciprocal of the divisor. [Section 2.5/Idea 2]

Dividing Mixed Numbers
To divide mixed numbers, change all mixed numbers to improper fractions, and then divide the resulting fractions. [Section 2.5/Idea 3]

Adding Fractions Having a Common Denominator
$\dfrac{a}{b} + \dfrac{c}{b} = \dfrac{a + c}{b}$

To add fractions having a common denominator, write the sum of the numerators over the common denominator and reduce the resulting fraction. [Section 2.6/Idea 1]

Finding the LCD
To find the LCD of two or more fractions, write the prime factorizations of each denominator. The LCD is the product of the distinct prime factors. Each factor must be used the greatest number of times it occurs in any one factorization. [Section 2.6/Idea 4]

Adding Fractions Having Different Denominators	To add fractions having different denominators, first find the LCD. Next, change each fraction to an equivalent fraction using the LCD as the new denominator. Add the resulting fractions and reduce if possible. [Section 2.6/Idea 5]
Adding Mixed Numbers	To add mixed numbers, add the whole number parts and the fractional parts separately. If the sum of the fractional parts is an improper fraction, change it to a mixed number and add it to the whole number parts. [Section 2.7/Idea 1]
Subtracting Fractions Having a Common Denominator $\frac{a}{b} - \frac{c}{b} = \frac{a - c}{b}$	To subtract fractions having a common denominator, subtract the numerators and place this difference over the common denominator, then reduce the resulting fraction if possible. [Section 2.8/Idea 1]
Subtracting Fractions Having Different Denominators	To subtract fractions having different denominators, first find the LCD. Next, change each fraction to an equivalent fraction using the LCD as the new denominator. Subtract the resulting fractions and reduce if possible. [Section 2.8/Idea 2]
Subtracting Mixed Numbers	To subtract mixed numbers, change each fractional part to an equivalent fraction using the LCD as the new denominator. Whenever the fraction in the subtrahend is larger than the fraction in the minuend, borrow one from the whole number part of the minuend and add 1 to its fractional part. Then, subtract the whole number parts and the fractional parts separately. [Section 2.8/Idea 4]
Operating on Signed Fractions	To add, subtract, multiply, and divide signed fractions, do so in the same way that you add, subtract, multiply, and divide integers. [Section 2.9/Idea 2–5] To simplify signed fractions involving more than one operation, use the order of operations rules discussed in Section 1.7. [Section 2.9/Idea 6]

Chapter 2 Review Exercises

Write the prime factorization.

1. 36
2. 121
3. 522
4. 1001

Reduce.

5. $\frac{12}{20}$
6. $\frac{35}{91}$
7. $\frac{21}{49}$
8. $\frac{231}{297}$

Express as a mixed number.

9. $\frac{17}{5}$
10. $\frac{77}{6}$
11. $\frac{121}{9}$

Express as an improper fraction.

12. $3\frac{3}{4}$
13. $7\frac{5}{8}$
14. $82\frac{3}{4}$

Do the indicated operation.

15. $\frac{18}{25} \cdot \frac{35}{27}$
16. $-\frac{5}{12} - \frac{4}{15}$
17. $\frac{5}{6} + \frac{7}{9}$
18. $-\frac{25}{56} \div \frac{5}{12}$
19. $2\frac{2}{5} \cdot 2\frac{3}{16}$
20. $\frac{7}{9} - \frac{4}{15}$

21. $8\frac{1}{12} - 4\frac{7}{9}$ **22.** $\frac{8}{39} \div \frac{36}{26}$ **23.** $5\frac{9}{10} + 6\frac{5}{6}$

24. $10\frac{5}{9} \div 1\frac{17}{21}$ **25.** $9 - \frac{2}{3}$ **26.** $-30 \div \frac{1}{7}$

Solve.

27. If three-fourths of the students in Math 111E passed a test, what fractional part of the class did not pass?

28. Determine what fraction part $3\frac{1}{3}$ is of 15.

29. Kathy owes Betty $1000. If Kathy gives Betty three-fourths of the money that she owes her, how much money will Betty receive?

30. One day APD stock opened at $35 and closed at $31\frac{3}{8}$. How much did the price of this stock drop?

31. A recipe for fruit punch requires $\frac{1}{4}$ cup of lemon juice, $5\frac{3}{8}$ cups of cranberry juice, $\frac{1}{6}$ cup of lime juice, and $3\frac{5}{12}$ cups of raspberry juice. How much juice is required for the punch?

32. Lorie owns 28 acres. If she subdivides this property into lots containing $3\frac{1}{2}$ acres, how many lots will she have?

33. Frank has a piece of wood that is $8\frac{1}{4}$ inches in length. If the board is $2\frac{3}{8}$ too long, what is the length of the piece of board Frank needs.

34. Subtract $\frac{7}{24}$ from $\frac{11}{32}$.

35. Edgar must serve 37 pieces of pie at a business meeting. If each pie is divided into 4 equal pieces, how many pies are needed?

36. On a map 1 inch equals $12\frac{1}{2}$ miles. How many miles is represented by $8\frac{4}{5}$ inches?

37. On her vacation, Jill drove $75\frac{3}{10}$ miles, $85\frac{5}{12}$ miles, and $15\frac{5}{6}$ miles. Find the total number of miles Jill drove.

Perform the indicated operation.

38. $\frac{117}{143} \cdot \frac{8}{153}$ **39.** $72 \cdot 1\frac{25}{27}$ **40.** $-5\frac{1}{16} \cdot 1\frac{13}{27}$

41. $-\frac{28}{45} \div \frac{21}{40}$ **42.** $15\frac{7}{11} \div 2\frac{46}{55}$ **43.** $-\frac{15}{91} + \left(-\frac{20}{91}\right)$

44. $\frac{1}{24} + \frac{5}{36} + \frac{3}{40}$ **45.** $8\frac{7}{8} + 6\frac{11}{12}$ **46.** $6\frac{5}{18} - 3\frac{4}{21}$

47. $\frac{7}{26} - \frac{1}{39}$ **48.** $\left(-\frac{40}{30}\right)\left(-\frac{15}{88}\right)$ **49.** $(-60)\left(2\frac{1}{10}\right) \div \left(-\frac{42}{25}\right)$

50. $-\frac{1}{4} + \frac{8}{15} - \frac{3}{10}$ **51.** $\left(\frac{4}{5} - \frac{3}{8}\right) - \frac{1}{2}\left(\frac{2}{5} - \frac{1}{2}\right) + \frac{5}{16}$

52. $-\frac{3}{8} \div \left(-\frac{5}{12} + \frac{3}{8}\right)$ **53.** $\frac{7}{8} - \left(-\frac{1}{6} + \frac{5}{12}\right) \div \frac{7}{8}$

54. $-\frac{1}{10} + \left(-\frac{1}{8} + \frac{3}{10} \cdot \frac{25}{9}\right) \div \frac{17}{12}$

55. $-\frac{1}{24} + \frac{5}{42} - \left(-\frac{3}{28}\right) + \frac{1}{7} \cdot \frac{1}{9}$

56. $\left(5\frac{1}{5} + 2\frac{2}{3}\right) \div \left(6\frac{7}{10} - 4\frac{1}{2}\right)$

Chapter 2 Test

Name: _____

Class: _____

1. Write the prime factorization of 121.
2. Reduce $\dfrac{12}{20}$ to its lowest terms.
3. Reduce $\dfrac{156}{222}$ to its lowest terms.
4. Express $7\dfrac{3}{11}$ as an improper fraction.
5. Express $35\dfrac{1}{9}$ as an improper fraction.
6. Express $\dfrac{15}{2}$ as a mixed number.
7. Express $\dfrac{180}{50}$ as a mixed number.

Perform the indicated operation.

8. $\dfrac{14}{24} \cdot \dfrac{33}{35} \cdot \dfrac{4}{77}$
9. $60 \cdot \dfrac{25}{42} \cdot 2\dfrac{1}{10}$
10. $\dfrac{6}{23} \div \dfrac{8}{69}$
11. $9\dfrac{5}{8} \div 11\dfrac{11}{22}$
12. $\dfrac{3}{10} + \dfrac{14}{15} + \dfrac{7}{12}$
13. $18\dfrac{7}{30} + 19\dfrac{13}{15} + 33\dfrac{19}{20}$
14. $\dfrac{19}{2} - 9$
15. $\dfrac{4}{21} - \dfrac{2}{15}$
16. $299 - 75\dfrac{3}{4}$

Solve.

17. Find the difference between $\dfrac{1}{35}$ and $\dfrac{11}{14}$.
18. Divide -24 by $\dfrac{1}{3}$.
19. The perimeter of a triangle is the sum of the length of its three sides. If one side of a tri-

1. _____
2. _____
3. _____
4. _____
5. _____
6. _____
7. _____
8. _____
9. _____
10. _____
11. _____
12. _____
13. _____
14. _____
15. _____
16. _____
17. _____
18. _____
19. _____
20. _____
21. _____
22. _____
23. _____

angle is $7\frac{5}{9}$ inches, another side is $9\frac{1}{6}$ inches, and the third side is 11 inches, find the perimeter of the triangle.

20. Dan jogged $15\frac{3}{4}$ miles and Frank jogged $12\frac{1}{6}$ miles. How many more miles did Dan jog than Frank?

24. _____

25. _____

Perform the indicated operation.

21. $-\frac{7}{12} - \left(-\frac{8}{15}\right)$

22. $-\frac{15}{28} \div \left(-\frac{9}{14}\right)$

23. $-1\frac{1}{6} + 2 - \frac{1}{10}$

24. $-\frac{5}{6} + \frac{14}{24} \div \frac{147}{112}$

25. $27 \div (-9) - 5\left(6 - \frac{8-4}{5}\right)$

3 Decimals and Percents

Objectives The objectives for this chapter are listed below along with sample problems for each objective. By the end of this chapter you should be able to find the solutions to the given problems.

1. Write the word name for a decimal *(Section 3.1/Idea 2)*.
   a. 0.03   b. 2.415

2. Add decimals *(Section 3.2/Idea 1)*.
   a. 0.31 + 3.1   b. 5.31 + 8.2 + 39.158

3. Subtract one decimal from another *(Section 3.2/Idea 2)*.
   a. 8.321 − 6.55   b. 11 − 0.328

4. Multiply decimals *(Section 3.3/Idea 1)*.
   a. (0.2)(0.8)   b. (3.89)(7.3)

5. Round decimals *(Section 3.4/Idea 1)*.
   a. 0.323 (hundredth)   b. 7.951 (tenth)

6. Divide one decimal by another *(Section 3.5/Idea 1)*.
   a. 76.13 ÷ 2.3   b. 24.36 ÷ 12

7. Change a fraction to a decimal *(Section 3.6/Idea 1)*.
   a. $\frac{3}{5}$ b. $\frac{8}{3}$

8. Change a decimal to a fraction *(Section 3.6/Idea 2)*.
   a. 0.35   b. 3.125

9. Add, subtract, multiply, and divide signed decimals *(Section 3.7/Idea 1–6)*.
   a. −0.35 − (−0.8)   b. (−3.2)(0.3)(−11)   c. −0.45 + 0.646 ÷ (−0.2)

10. Express a percent as a fraction *(Section 3.8/Idea 2)*.
    a. 65%   b. 0.08%   c. $\frac{3}{8}$%

11. Express a percent as a decimal *(Section 3.8/Idea 3)*.
    a. 65%   b. 0.08%   c. $\frac{3}{8}$%

12. Express a decimal as a percent *(Section 3.9/Idea 1)*.
    a. 0.45   b. 0.0095   c. 3.2

13. Express a fraction as a percent *(Section 3.9/Idea 2)*.
    a. $\frac{3}{4}$ b. $\frac{1}{8}$ c. $\frac{2}{3}$

3.1 Reading and Writing Decimals

IDEA 1   A **decimal** or **decimal fraction** is a number that can be written as a fraction whose denominator is 1, 10, 100, or 1000, and so on. Some examples are:

$$\text{Decimal notation} = \text{Fraction notation}$$
$$.4 = \frac{4}{10}$$
$$.071 = \frac{71}{1000}$$
$$2.47 = \frac{247}{100}$$
$$5 = \frac{5}{1}$$

The "." is a decimal point. In a whole number, the decimal point is located just to the right of the digit in the ones place (for example, 5 = 5.).

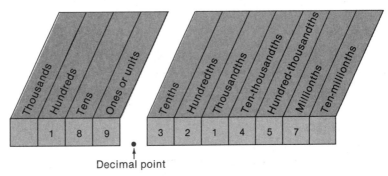

Figure 3-1

The **place value** of a digit is determined by where it is in relation to the decimal point. The digits to the left of the decimal point have place values of 1, 10, 100, 1000, and so on. The digits to the right of the decimal point have place values of $\frac{1}{10}$, $\frac{1}{100}$, $\frac{1}{1000}$, and so on (see Figure 3-1). Remember, each place is $\frac{1}{10}$ as large as the first place to its left. The **value** of each digit, on the other hand, is determined by multiplying the digit by its place value. For example, the place value of the number 2 in Figure 3-1 is hundredths $\left(\frac{1}{100}\right)$, and its value is $2 \cdot \frac{1}{100} = \frac{2}{100}$.

IDEA 2   The concept of place value is useful in helping us to read or write word names for decimals.

Example 1   Write the word name for each decimal.

     a. 0.32   **b.** 0.051   **c.** 2.6   **d.** 17.0031

Solution   **1a.** 0.32 = thirty-two hundredths (Hundredths place)

     1b. 0.051 = fifty-one thousandths

To Write the Word Name for a Decimal:
If there is no number other than zero to the left of the decimal point, omit steps 1 and 2.

1. Write the name for the whole number to the left of the decimal point.
2. Write the word "and" for the decimal point.
3. Write the name for the number to the right of the decimal point as if it were a whole number. Then write the name for the place value of the last digit on the right.

1c. 2.6 = two and six tenths

1d. 17.0031 = seventeen and thirty-one ten-thousandths

Practice Problem 1 Write the word name for each decimal.
a. 0.7   b. 0.06   c. 3.112   d. 9.000051

3.1 Exercises

Identify the place value of the 5.

1. 35.01
2. 17.056
3. 0.513
4. 51.731
5. 0.0051
6. 3.00051
7. 6.00375
8. 7.123045
9. 175124.8

Write the word name for each decimal.

10. 0.03
11. 0.813
12. 0.5
13. 0.1357
14. 0.0404
15. 0.000034
16. 7.15
17. 6.034
18. 17.05
19. 0.031
20. 8.158
21. 6.703
22. 0.051
23. 7.7
24. 17.0101

Express the following as decimals.

25. Eight and seventeen hundredths
26. Two hundred five thousandths
27. Forty-six ten-thousandths
28. Six hundred and seven hundred-thousandths
29. Ten and five millionths
30. Ninety-four hundredths
31. Sixteen and six thousandths
32. Nineteen and eight ten-thousandths
33. Three thousand and three thousandths
34. Eight hundred and eight hundredths
35. Twenty thousand nine and fifty-six hundred-thousandths
36. Two million four hundred and forty-two millionths

Fill in the blanks.

37. A _____ is a number that can be written as a fraction whose denominator is 1, 10, 100, 1000, and so on.

38. Given the number 3.125, the place value of the 5 is _____ and its value is _____.

39. To write a word name for a decimal, write the name for the _____ number to the left of the decimal point. Write the word "_____" for the decimal point. Finally, write the name for the number to the right of the decimal point as if it were a _____ number and then write the _____ for the last digit on the right.

Answers to Practice Problems **1a.** seven tenths **b.** six hundredths **c.** three and one hundred twelve thousandths **d.** nine and fifty-one millionths

3.2 Adding and Subtracting Decimals

IDEA 1 Let us now consider how to add decimals. For example, we could add 4.3 and 3.6 by first changing each decimal to a mixed number.

$$\begin{aligned} 4.3 &= 4\frac{3}{10} \\ +3.6 &= 3\frac{6}{10} \\ \hline & 7\frac{9}{10} \text{ or } 7.9 \end{aligned}$$

However, notice that this same result can be obtained by just adding the digits having the same place value and then writing the decimal point in the sum directly below the decimal points in the addends.

$$\begin{array}{rl} 4.3 & \text{Addend} \\ +3.6 & +\text{Addend} \\ \hline 7.9 & \text{Sum} \end{array}$$

The above statement suggests the following procedure for adding decimals.

> **To Add Decimals:**
>
> 1. Write each number so that the decimal points are lined up.
> 2. Add the numbers as if they were whole numbers.
> 3. Place the decimal point in the answer (sum) so it is lined up with the other decimal points.

It may be helpful to write zeros to the right of the digits following the decimal point to ensure that we only add digits having the same place value. Remember, writing in extra zeros on the right will not change the value of the original number (for example 0.32 = 0.320).

Example 2 Add.

    a. $0.32 + 0.6 + 0.315$ **b.** $7.3 + 0.16 + 5.1$

    c. $17.16 + 5.14 + 9.1345$ **d.** $8.37 + 91 + 5.2$

Solution To add decimals, line up the decimal points. Add as if you were adding whole numbers and then line up the decimal point in the answer with the other decimal points.

```
2a.    0.320     Line up the "."    2b.    7.30
       0.600                               0.16
     + 0.315                             + 5.10
       1.235                              12.56

2c.   17.1600                       2d.    8.37
       5.1400                              91.00
     + 9.1345                            +  5.20
      31.4345                             104.57
```

Practice Problem 2 *Add.*

    a. $5.3 + 6.11$ **b.** $0.005 + 0.07 + 0.9$

    c. $17.8 + 5.81 + 1.857$ **d.** $99.5 + 165.35 + 0.316$

IDEA 2

The procedure for subtracting one decimal from another can also be developed by converting the given decimals to mixed numbers. For example,

$$\begin{aligned} 8.9 &= 8\frac{9}{10} \\ -2.6 &= 2\frac{6}{10} \\ \hline &\ 6\frac{3}{10} \quad \text{or } 6.3 \end{aligned}$$

Again, notice that the same result could be obtained by simply subtracting the digits having the same place value and then writing the decimal point in the difference directly below the decimal points in the minuend and the subtrahend.

$$\begin{array}{rl} 8.9 & \text{Minuend} \\ -2.6 & -\text{Subtrahend} \\ \hline 6.3 & \text{Difference} \end{array}$$

The above statements suggest the following rule for subtracting one decimal from another.

To Subtract Decimals:

1. Write the subtrahend beneath the minuend and line up the decimal points.

2. Subtract the numbers as if they were whole numbers. (If necessary, write in zeros.)

3. Place the decimal point in the answer (difference) so it is lined up with the other decimal points.

Example 3 Subtract.

    a. $14.97 - 10.54$ **b.** $0.6323 - 0.429$

    c. $16.57 - 8.193$ **d.** $91 - 17.34$

Solution To subtract decimals, line up the decimal points in the minuend and the subtrahend. Subtract as if you were subtracting whole numbers. Line up the decimal point in the answer with the other decimal points.

3a.
```
  14.97     Line up the "."
- 10.54
  -----
   4.43
```
3b.
```
   0.6̶3̷23
 - 0.4 2 9
   -------
   0.2 0 33
```

3c.
```
   16.5̶7̶0
 -  8.1 9 3
   -------
    8.3 7 7
```
3d.
```
   9̶1̶.0̶ 0
 - 1 7.3 4
   -------
   7 3.6 6
```

Practice Problem 3 **Subtract.**

    a. $5.9 - 3.7$ **b.** $0.64315 - 0.5324$

    c. $35.6 - 29.54$ **d.** $18 - 0.008$

Example 4 Solve and show all work.

    a. Pam bought a shirt for $35.82, a tie for $15.85, and a belt for $18. How much money did she spend?

    b. Subtract 15.87 from 39.1.

Solution **4a.** Find the sum of the three items.

```
    35.82
    15.85
  + 18.00
  -------
    69.67    Pam spent $69.67
```

4b. The subtrahend is 15.87, and the minuend is 39.1.

```
   3̶9̶.1̶ 0
 - 1 5.8 7
   -------
   2 3.2 3
```

Practice Problem 4 **Solve and show all work.**

    a. Matt ran 5 miles on Monday, 6.3 miles on Wednesday, 8.35 miles on Friday, and 10.8 miles on Sunday. What was the total distance he ran over the seven-day period?

    b. Ernie bought a book for $8.43 and paid for it with a $20 bill. How much change should he receive?

3.2 Exercises

Add.

1. $0.31 + 0.64$ **2.** $7.3 + 5.6$ **3.** $0.5 + 0.8$

4. $0.3 + 0.15$ **5.** $9.22 + 8.5$ **6.** $7.321 + 0.52$

106 Decimals and Percents

7. 17 + 18.34
8. 9.7 + 8.135
9. 0.71 + 3.578
10. 17.31 + 18.09
11. 5.3 + 2.389
12. 9.51 + 99.531
13. 8.35 + 7.3 + 9.584
14. 8.35 + 0.115 + 16.9 + 13
15. 35.13 + 75.195 + 0.3378
16. 0.8 + 0.95 + 8.192 + 0.7
17. 3.6 + 5.18 + 6 + 0.1891
18. 1.5 + 8.31 + 35.111
19. 8.5 + 3.17 + 18.501 + 98
20. 16.19 + 0.8 + 17.987

Subtract.

21. 15.3 − 8.1
22. 18.5 − 17.3
23. 5.37 − 3.25
24. 0.63 − 0.34
25. 19.1 − 19
26. 0.37 − 0.2
27. 8.7 − 4.623
28. 4.87 − 3.9
29. 84.07 − 83.97
30. 4.006 − 3.94
31. 73.16 − 10.8
32. 96 − 35.1
33. 85 − 11.08
34. 2.07 − 0.007
35. 57.88 − 41
36. 15 − 3.9
37. 97.815 − 17.5
38. 89.58 − 8.731

Perform the indicated operation.

39. 0.51 + 0.8
40. 0.61 + 0.7
41. 3.22 + 8.1 + 9.531
42. 4.54 + 7.1 + 6.351
43. 0.81 − 0.62
44. 0.93 − 0.75
45. 8 − 3.64
46. 6 − 2.52
47. 1.035 + 3.57 + 4
48. 3.125 + 4.63 + 5
49. 6.3 + 8.2 + 9.35 + 0.81
50. 7.2 + 9.3 + 7.45 + 0.63
51. 8357 + 631
52. 9349 − 852
53. 552 + 85 + 1347
54. 675 − 23 + 2356
55. 574.2 − 234.71
56. 837.5 − 236.53
57. 1234.45 − 39.4
58. 2365.83 − 79.6
59. 0.0085 − 0.0063
60. 0.0099 − 0.0078
61. 9.19 + 0.3 + 5.1 + 8.2
62. 6.15 + 0.5 + 6.3 + 8.7
63. 2.395 + 5.74 + 4.3 + 0.963
64. 1.385 + 7.54 + 3.4 + 0.693
65. 0.08 + 1.5834 + 3.463
66. 0.09 + 2.8543 + 4.346
67. 20.5 − 0.5
68. 30.7 − 0.7
69. 33 + 18.7 + 99.35
70. 44 + 19.8 + 73.67
71. 2335.78 − 235.9
72. 3226.87 − 325.9
73. 6.2 + 8.2 + 9.2 + 0.21
74. 2.6 + 8.6 + 5.6 + 0.37
75. 234 − 17.3
76. 335 − 19.7
77. 3374.52 + 831.57
78. 4437.85 − 733.96
79. 87.0468 − 72.3007
80. 78.4068 − 27.7003
81. 175.02 − 139.2
82. 185.03 − 149.3

Fill in the blanks.

83. To add decimals, line up the _____. Add as if you were adding _____ numbers and then line up the decimal point in the answer with the other _____.

84. To subtract using decimals, line up the _____ in the _____ and the subtrahend. Subtract as if you were subtracting _____ numbers and then line up the decimal point in the answer with the other _____.

85. When adding or subtracting decimals, remember that you can only add or subtract digits having the same _____.

Solve and show all work.

86. Subtract 15.78 from 33.2.

87. Find the difference between 193.87 and 79.354.

88. During a vacation, Buford's records showed gasoline purchases of 19.75 gallons, 15.4 gallons, 16 gallons, and 13.85 gallons. How many gallons of gas did he buy during his vacation?

89. The perimeter of a triangle is equal to the sum of the length of its sides. Find the perimeter of a triangle if the first side measures 8.75 inches, the second side measures 9.6 inches, and the third side measures 10.375 inches.

90. Rochelle has $331.64 in her savings account. If she withdraws $58.97, what is the new balance in her account?

91. The original price of a shirt is $23.73. If you receive a $4.82 discount, how much does the shirt cost?

92. One week a jogger ran the following distances: 15.3 miles, 18.75 miles, 19 miles, 21.5 miles, 25.375 miles, and 30.25 miles. Find the total distance the jogger ran.

95. Subtract 92.4 from 235.

94. Subtract 16.45 from 335.

95. Find the sum of $3.25, $5.04, and $17.63.

96. Find the sum of $2.35, $4.05, and $71.36.

97. Find the difference between 85.17 and 58.09.

98. Find the difference between 58.71 and 18.04.

99. Find the total of $19.45, $72.53, $18, $55, and $153.45.

100. Find the total of $13.54, $27.35, $81, $75, and $213.17.

101. Find the difference between $131.57 and $19.99.

102. Find the difference between $268.76 and $17.68.

103. Joe has $575.88 in his checking account. If he made deposits of $93.64 and $33, what is his new balance?

104. Susan earns $75 per week from one job and $37.18 per week from another. Find her total weekly salary.

105. Jane earns $85 per week babysitting and $48.50 per week tutoring students in English. Find her total weekly salary.

Answers to Practice Problems 2a. 11.41   b. 0.975   c. 25.467   d. 265.166   3a. 2.2   b. 0.11075   c. 6.06   d. 17.992   4a. 30.45 miles   b. $11.57

3.3 Multiplying Decimals

IDEA 1 To determine how to multiply decimals, we can convert the decimals to fractions, multiply the fractions, convert the answer to a decimal, and then compare the answer to the original problem. In other words,

$$(0.23)(0.5) = \left(\frac{23}{100}\right)\left(\frac{5}{10}\right) \quad \text{Convert decimals to fractions}$$
$$= \frac{115}{1000} \quad \text{Multiply}$$
$$= 0.115 \quad \text{Convert fraction to a decimal}$$

108 Decimals and Percents

The number 0.115 has 3 decimal places or 3 digits to the right of the decimal point. In general, the number of *decimal places* in a decimal is the number of digits written to the right of the decimal point.

Before writing a rule for multiplying decimals, let us examine several other products.

Example 5 Change each decimal to a fraction and multiply.

    a. (0.3)(3)   **b.** (0.3)(0.3)   **c.** (0.3)(0.03)

Solution

5a. $(0.3)(3) = \frac{3}{10} \cdot \frac{3}{1} = \frac{9}{10} = 0.9$ 0.3 (1 place)
or   3 (0 places)
0.9 (1 place)

5b. $(0.3)(0.3) = \frac{3}{10} \cdot \frac{3}{10} = \frac{9}{100} = 0.09$ 0.3 (1 place)
or 0.3 (1 place)
0.09 (2 places)

5c. $(0.3)(0.03) = \frac{3}{10} \cdot \frac{3}{100} = \frac{9}{1000} = 0.009$ or   0.03 (2 places)
0.3 (1 place)
0.009 (3 places)

If we examine these examples carefully, we can see that each product was determined by multiplying 3 and 3 and then determining the number of decimal places in the product. The number of decimal places in the product is equal to the sum of the number of decimal places in the numbers being multiplied.

The above discussion is summarized below.

> **To Multiply Decimals:**
>
> 1. Multiply the numbers as if they were whole numbers.
>
> 2. Place the decimal point in the product so that the number of decimal places in the product equals the sum of the number of decimal places in the factors. If necessary, write in zeros.

Example 6 Multiply.

    a. (2.3)(5.3)   **b.** (131)(1.02)

    c. (9.6)(0.0004)   **d.** (42.3)(3.5)(0.23)

Solution To multiply decimals, multiply as if you were multiplying whole numbers. Place the decimal point in the product.

6a.   2.3   1 decimal place
       5.3   1 decimal place
       6 9
    11 5
   12.1 9   1 + 1 = 2 decimal places

6b.   1 31   0 decimal places
      1.02   2 decimal places
      2 62
    131 0
   133.62   0 + 2 = 2 decimal places

6c.	9.6	1 decimal place			
	0.000 4	4 decimal places			
	0.0038 4	1 + 4 = 5 decimal places. Write two zeros to the left of the 3 since the product must have 5 decimal places.			
6d.	4 2.3	1 decimal place	1 48.05	2 decimal places	
	3.5	1 decimal place	0.23	2 decimal places	
	21 1 5		4 44 15		
	126 9		29 61 0		
	148.0 5	1 + 1 = 2 decimal places	34.05 15	2 + 2 = 4 decimal places	

Practice Problem 5 **Multiply.**

a. (7.5)(0.33)   **b.** (17.5)(35)

c. (0.031)(0.007)   **d.** (3.2)(15)(0.075)

Example 7 Solve and show all work.

Barbara was reimbursed 19 cents per mile for mileage for business purposes. If she drove 105 miles, how much money (in dollars) should she be reimbursed?

Solution Recall that 19 cents = $0.19. Thus,
Reimbursement = (miles driven)(reimbursement per mile)
= (105)($0.19)
= $19.95

1 05
0.19
9 45
10 5
19.95

Barbara should receive $19.95.

Practice Problem 6 **Solve and show all work.**

Shirley earns $6.75 an hour. How much will she earn if she works 39 hours?

3.3 Exercises

Place a decimal point correctly in each product. If necessary, write in zeros.

1. (1.74)(0.33) = 5742   **2.** (17.4)(3.3) = 5742   **3.** (17.4)(0.33) = 5742

4. (0.174)(0.33) = 5742   **5.** (1.74)(0.033) = 5742   **6.** (0.174)(33) = 5742

Multiply.

7. 3.2
       0.3

8. 0.65
       0.2

9. 3.4
       0.32

10. 35
        0.21

11. 0.32
        0.005

12. 0.2
        0.3

13. 15.3
        5.2

14. 35.4
        1.05

15. 0.421
        0.24

16. 77.2
        0.015

17. 69.5
        0.34

18. 145.3
        0.15

19. 0.0529
        0.0031

20. 89.98
        7.09

21. 19.7
        85.1

Multiply.

22. (4.6)(869)   **23.** (13.75)(9.7)   **24.** (0.35)(0.271)

110   Decimals and Percents

25. (22.5)(29.8)
26. (2.51)(0.0031)
27. (0.56)(4.12)
28. (25.32)(35.2)
29. (2.1049)(0.513)
30. (2.3)(0.35)(8.5)
31. (15.1)(17.2)(19.3)
32. (0.15)(1.2)(0.009)
33. (17.5)(0.008)(99)
34. (2.35)(0.4)
35. (6.25)(0.6)
36. (3.15)(3.15)
37. (4.62)(4.62)
38. (41.1067)(0.003)
39. (52.1245)(0.005)
40. (3.51)(0.008)

Fill in the blanks.

41. The number of decimal places in a decimal is the number of _____ written to the _____ of the decimal point.

42. To multiply decimals, multiply the numbers as if they were _____ numbers. Place the _____ in the product. The number of decimal places in the product must equal the _____ of the number of decimal places in the _____.

43. When multiplying decimals, you may need to write _____ in the product to make sure that you have enough decimal places.

Solve and show all work.

44. John pays a monthly payment of $175.58 for a car. If it takes 36 months to pay for the car, what is the total cost of the car?

45. A coach buys 19 pairs of socks at $2.25 per pair. What is the total cost of the socks?

46. If you earn $7.57 an hour, how much should you earn for a 40-hour week?

47. Rick buys 19 gallons of gas at $1.48 per gallon. What is the total cost of the gas?

48. What is the cost (in dollars) for having 317 pages xeroxed at nine cents per page?

49. A jogger ran 35.8 miles per day for 7 days. How many miles did he run?

50. Bill traveled 500.9 miles per day for 13 days. How many miles did he travel?

51. Find the product of 2.3, 4.5, and 0.06.
52. Find the product of 3.2, 5.4, and 0.08.

53. Multiply 0.894 by 0.32.
54. Multiply 0.498 by 0.23.

55. Joel earns $835.50 per month. If Rich earns twice as much as Joel, how much does Rich earn per month?

56. Dan ran 11.8 miles on Sunday. If Betty ran twice as far as Dan did, how far did Betty run?

57. Find the product of ninety-three thousand and three and six tenths.

58. Find the product of eighty-five thousand and eleven and three tenths.

59. James earns $10.50 per hour. If he works 17 hours, how much will he earn?

60. Gilberto earns $9.75 per hour. If he works 19 hours, how much will he earn?

61. Betty drove 35.7 miles per day for three days. How many miles did she drive?

62. Rich drove 53.7 miles per day for five days. How many miles did he drive?

Answers to Practice Problems   **5a.** 2.475   **b.** 612.5   **c.** 0.000217   **d.** 3.6   **6.** $263.25

3.4 Rounding Decimals

IDEA 1   When you want to find an approximate value for a given number, use the concept of **rounding.** We speak of "rounding to the nearest tenth," for example, when we want the decimal to have just one decimal place. Similarly "round to the nearest hundredth'' means we want the decimal to have just two decimal places. We use the symbol ≐ to mean "approximately equal to."

Example 8   Round each number to the nearest tenth.

   a. 3.72   **b.** 5.68   **c.** 6.75

Solution   **8a.** Since 3.72 is closer to 3.7 than to 3.8 (see Figure 3-2), round *down* to 3.7.

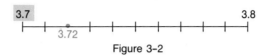

Figure 3-2

8b. Since 5.68 is closer to 5.7 than to 5.6 (see Figure 3-3), round *up* to 5.7.

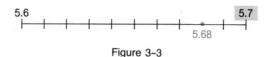

Figure 3-3

8c. Any time a number is halfway between its two closest approximations, round up. For example, 6.75 rounded to the nearest tenth is 6.8 (see Figure 3-4).

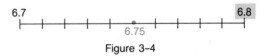

Figure 3-4

Example 8 suggests a rule for rounding a decimal to a given place value.

> **To Round a Decimal:**
>
> 1. Identify the first digit to the right of the given place value.
>
> 2. If this digit is *less than* 5 (0, 1, 2, 3, or 4), round *down* by changing all digits to the right of the given place value to zero.
>
>   If this digit is *5 or greater* (5, 6, 7, 8, or 9), round *up* by adding 1 to the digit in the given place value and changing all digits to the right of it to zero.
>
> 3. Omit any zeros that appear to both the right of the decimal point and to the right of the digit in the given place value.

Omitting zeros that are both to the right of the decimal point and to the right of the last digit of a decimal will not change its value (for example, 0.50 = 0.5).

Example 9   **a.** Round 7.8691 to the nearest hundredth.

   b. Round 10.867542 to the nearest ten-thousandth.

   c. Round 1751.79 to the nearest hundred.

   d. Round 15.987 to the nearest tenth.

Solution To round a decimal to a given place value, identify the digit to the right of the given place value. If it is 5 or greater, round up. If it is less than 5, round down.

┌─ Hundredths place   ┌─ Ten-thousandths place

9a. 7.8**6**91 ≐ 7.8700   **9b.** 10.867**5**42 ≐ 10.867500
              ≐ 7.87   ≐ 10.8675

9c. 1751.79 ≐ 1800.00   **9d.** 15.987 ≐ 16.000
              ≐ 1800   ≐ 16.0

Practice Problem 7   *Solve.*

a. Round 6.761 to the nearest tenth.

b. Round 0.7658 to nearest thousandth.

c. Round 145.77 to the nearest ten.

d. Round 35.7976 to the nearest hundredth.

Example 10   a. Round 0.578 to one decimal place.

b. Round 3.6519 to two decimal places.

Solution To round a decimal to a given number of places, use the same rule for rounding decimals to a given place value.

10a. 0.578 ≐ 0.600   **10b.** 3.6519 ≐ 3.6500
             ≐ 0.6   ≐ 3.65

Practice Problem 8   *Solve.*

a. Round 13.7591 to three decimal places.

b. Round 0.854371 to four decimal places.

3.4 Exercises

Round to the nearest tenth.

1. 0.17	2. 0.25	3. 0.31
4. 6.178	5. 18.117	6. 9.35
7. 0.91	8. 0.97	9. 15.7516

Round to the nearest hundredth.

10. 0.788	11. 0.854	12. 0.735
13. 0.1666	14. 7.658	15. 1.379
16. 5.795	17. 2.7543	18. 7.5582

Round to the nearest thousandth.

19. 0.1456	20. 0.7852	21. 0.6785
22. 7.8519	23. 5.6784	24. 1.77777
25. 95.3333	26. 8.9595	27. 9.99999

Round **1335.3745898** *to the nearest:*

28. tenth
29. thousandth
30. ten
31. hundred
32. hundredth
33. millionth
34. thousand
35. ten-thousandth
36. one

Fill in the blanks.

37. The symbol ÷ means "is _____."

38. To round a decimal to a given place value, identify the _____ to the _____ of the given place value. If it is less than _____, round down. If it is _____ or greater, round up.

39. Rounding a decimal to two decimal places is the same as rounding to the nearest _____.

Round each decimal to the indicated place.

40. 155.77 (hundreds)
41. 75.7781 (two decimal places)
42. 165,781 (thousands)
43. 0.73852 (three decimal places)
44. 1785.77 (tens)
45. 4.963 (one decimal place)

Answers to Practice Problems 7a. 6.8 b. 0.766 c. 150 d. 35.80 8a. 13.759 b. 0.8544

3.5 Dividing Decimals

IDEA 1 If you multiply the divisor and the dividend in a division problem by the same nonzero number, the value of the quotient does not change. This concept will be useful in helping us to develop a rule for dividing decimals. For example,

$$7.2 \div 2.4 = \frac{7.2}{2.4}$$ Write the division as a fraction

$$= \frac{7.2(10)}{2.4(10)}$$ Multiply numerator and denominator by 10

$$= \frac{72}{24}$$

$$= 72 \div 24$$ Write the fraction as a division problem

$$= 3$$ Divide

Before writing a rule for dividing decimals, let us consider two other quotients.

Example 11 Change each decimal to a fraction and divide.

  a. 0.84 ÷ 0.04   **b.** 8 ÷ 0.4

Solution   **11a.** $0.84 \div 0.04 = \frac{0.84}{0.04} = \frac{0.84(100)}{0.04(100)} = \frac{84}{4} = 84 \div 4 = 4\overline{)84} = 21$

$$\begin{array}{r} 21 \\ 4\overline{)84} \\ \underline{8} \\ 4 \\ \underline{4} \\ 0 \end{array}$$

  11b. $8 \div 0.4 = \frac{8}{0.4} = \frac{8(10)}{0.4(10)} = \frac{80}{4} = 80 \div 4 = 4\overline{)80} = 20$

$$\begin{array}{r} 20 \\ 4\overline{)80} \\ \underline{8} \\ 0 \end{array}$$

114 Decimals and Percents

In Example 11, after writing the division problem as a fraction we multiplied the numerator and denominator by the smallest possible number that would make the divisor a whole number. Note that the same result could be obtained by moving the decimal point in the divisor and the dividend the same number of places to the right. For example, in Solution 11a, we have that

$$0.84 \div 0.04 = 84 \div 4 \quad \text{or} \quad 0.04\overline{)0.84} = 4\overline{)84.}^{21.}$$

Similarly, in Solution 11b, we have that

$$8 \div 0.4 = 80 \div 4 \quad \text{or} \quad 0.4\overline{)8} = 4\overline{)80.}^{20.}$$

Based on the above statements, we can see that to divide decimals we move the decimal point in the divisor to make it a whole number. Next, we move the decimal point in the dividend the same number of places and we align the decimal points in the dividend and the quotient. Finally, we divide as if we were dividing whole numbers.

To Divide One Decimal by Another:

1. Rewrite the problem in the long division format.
2. Move the decimal points in the divisor and the dividend the same number of places, until the divisor is a whole number.
3. Divide as if dividing whole numbers.
4. Place the decimal point in the quotient so that it is directly above the decimal point in the dividend.

Example 12   Divide.

   a. $9.28 \div 3.2$ **b.** $38.76 \div 12$ **c.** $6.936 \div 1.20$ **d.** $0.0350 \div 2.5$

Solution   To divide using decimals, rewrite the problem in the long division format. Move the decimal points to the right the same number of places until the divisor is a whole number. Line up the decimal points in the quotient and the dividend and then divide.

12a. $3.2\overline{)9.28} = 32\overline{)92.8}$

$$\begin{array}{r} 2.9 \\ 32\overline{)92.8} \\ \underline{64} \\ 28\,8 \\ \underline{28\,8} \\ 0 \end{array}$$

Move the decimal points one place to the right

Therefore, $9.28 \div 3.2 = 2.9$

12b. $12\overline{)38.76}$

$$\begin{array}{r} 3.23 \\ 12\overline{)38.76} \\ \underline{36} \\ 2\,7 \\ \underline{2\,4} \\ 36 \\ \underline{36} \\ 0 \end{array}$$

Since the divisor is a whole number do not move the decimal point

3.5 Dividing Decimals

$$
\textbf{12c.}\ 1.20\overline{)6.936} = 12\overline{)69.36}
$$

$$
\begin{array}{r}
5.78 \\
12\overline{)69.36} \\
\underline{60} \\
9\,3 \\
\underline{8\,4} \\
96 \\
\underline{96} \\
0
\end{array}
$$

We only needed to move the decimal point 1 place to obtain a whole number. That is, $12 = 12.0$

$$
\textbf{12d.}\ 2.5\overline{)0.0350} = 25\overline{)0.350}
$$

$$
\begin{array}{r}
0.014 \\
25\overline{)0.350} \\
\underline{25} \\
100 \\
\underline{100} \\
0
\end{array}
$$

Practice Problem 9 *Solve.*

a. $8.80 \div 3.2$ **b.** $0.01302 \div 0.21$ **c.** $86.4 \div 72$

Example 13 Solve.

a. $0.48 \div 0.032$ **b.** $1 \div 8$

Solution Sometimes extra zeros must be written in the dividend.

$$
\textbf{13a.}\ 0.032\overline{)0.48} = 32\overline{)480.}
$$

$$
\begin{array}{r}
15. \\
32\overline{)480.} \\
\underline{32} \\
160 \\
\underline{160} \\
0
\end{array}
$$

Write in a zero as a place holder

$$
\textbf{13b.}\ 8\overline{)1} = 8\overline{)1.000}
$$

$$
\begin{array}{r}
0.125 \\
8\overline{)1.000} \\
\underline{8} \\
20 \\
\underline{16} \\
40 \\
\underline{40} \\
0
\end{array}
$$

Insert zeros to complete the division

Example 14 Divide: $0.314 \div 0.17$

Solution When you do not obtain a remainder of zero after carrying division out 3 places, round the quotient to the nearest hundredth.

$$
0.17\overline{)0.314} = 17\overline{)31.400} = 1.85
$$

$$
\begin{array}{r}
1.847 \\
17\overline{)31.400} \\
\underline{17} \\
14\,4 \\
\underline{13\,6} \\
80 \\
\underline{68} \\
120 \\
\underline{119} \\
1
\end{array}
$$

To round to the nearest hundredth, always carry the division out to the thousandths place in the quotient.

Decimals and Percents

Practice Problem 10 **Divide. If you do not obtain a remainder of zero, round the quotient to the nearest hundredth.**

    a. $4.03 \div 0.032$ **b.** $3 \div 6$ **c.** $0.961 \div 7.5$

Example 15   Solve and show all work.

Marcia borrowed $846.48. If she pays the loan in 24 equal monthly payments, how much is each payment?

Solution   Payments = amount borrowed ÷ number of payments
= $846.48 ÷ 24
= $35.27

$$\begin{array}{r} 35.27 \\ 24\overline{)846.48} \\ \underline{72} \\ 126 \\ \underline{120} \\ 6\ 4 \\ \underline{4\ 8} \\ 1\ 68 \\ \underline{1\ 68} \\ 0 \end{array}$$

Practice Problem 11 **Solve and show all work.**

Larry pays $34.73 a month toward his loan. How many months will it take him to pay a balance of $382.03?

3.5 Exercises

Divide. If you do not obtain a remainder of zero, round the quotient to the nearest hundredth.

1. $431.6 \div 13$
2. $50.04 \div 18$
3. $156.619 \div 23$
4. $19.95 \div 105$
5. $148.05 \div 42.3$
6. $0.0384 \div 9.6$
7. $1.219 \div 0.23$
8. $57.22 \div 3.3$
9. $5.722 \div 1.74$
10. $3.486 \div 0.42$
11. $0.0735 \div 17.5$
12. $9.368 \div 0.04$
13. $0.0035 \div 7$
14. $65 \div 0.013$
15. $11 \div 55$
16. $9.875 \div 2.3$
17. $7.647 \div 1.4$
18. $35.3 \div 0.06$
19. $9 \div 5$
20. $13.75 \div 3.2$
21. $0.1571 \div 4$
22. $0.8215 \div 5$
23. $282.94 \div 4.3$
24. $224.74 \div 3.9$
25. $57.8 \div 0.01$
26. $65.7 \div 0.01$
27. $125 \div 0.001$
28. $375 \div 0.001$
29. $1.78 \div 10$
30. $3.87 \div 10$
31. $11.74 \div 0.33$
32. $15.66 \div 0.44$
33. $1.566 \div 35.6$
34. $1.174 \div 35.6$
35. $0.054 \div 0.11$
36. $0.06 \div 0.15$
37. $15 \div 55$
38. $20 \div 44$
39. $287.738 \div 6.38$
40. $18.8048 \div 3.68$

Fill in the blanks.

41. If you multiply the divisor and the _____ in a division problem by the same nonzero number, the value of the _____ does not change.

42. To divide using decimals, move the decimal point until the _____ is a whole number. Move the decimal point in the _____ the same number of places. Place the decimal point in the _____ directly above the decimal point in the _____ and then divide.

Solve and show all work.

43. Three men paid a total of $10.26 for lunch. If the cost of the meal is divided equally, how much is each man's share?
44. Joe worked 32 hours and earned $138.56. How much was he paid per hour?
45. Betty earns $8.35 per hour. If her salary one week was $177, how many hours did she work?
46. When Jerry drove 400 miles, he used 12.7 gallons of gas. How many miles per gallon of gas did he get?
47. Adrienne's car gets 23.8 miles per gallon of gas. How many gallons of gas will she need if she drives 202.3 miles?
48. John's odometer read 38,769.1 miles at the beginning of his trip. When the trip ended, it read 39,365.8 miles. If he used 31.5 gallons of gas, how many miles per gallon of gas did he get? Round the answer to the nearest tenth.
49. Divide 0.36 into 1.836.
50. Divide 3.21 into 18.939.
51. Divide 189.55 by 22.3.
52. Divide 593.9 by 11.4.
53. Divide four and five tenths into sixteen and fifty-six hundredths.
54. Divide nine and eight tenths by fourteen thousandths.
55. If 24 cartons of orange juice cost $51.60, find the cost of one carton.
56. If 48 cans of cola cost $21.60, find the cost of one can.
57. Linda's car gets 16.5 miles per gallon of gas. How many gallons of gas will she need if she drives 336.6 miles?
58. Bob's car gets 25.4 miles per gallon of gas. How many gallons of gas will he need if he drives 266.7 miles?
59. If Jane pays $52.26 per month on her charge account, how long will it take her to pay a balance of $1254.24?
60. If John pays $48.17 per month on his charge account, how long will it take him to pay a balance of $1734.12?

Answers to Practice Problems   **9a.** 2.75   **b.** 0.062   **c.** 1.2   **10a.** 125.94   **b.** 0.5   **c.** 0.13   **11.** 11

3.6 Converting Fractions and Decimals

IDEA 1   In mathematics, you will sometimes need to change a fraction to a decimal. Since a fraction denotes the operation of division, to express a fraction as a decimal, just divide the denominator into the numerator. For example,

$$\frac{1}{2} = 1 \div 2 = 0.5 \qquad 2\overline{)1.0} \quad \begin{array}{r} 0.5 \\ \underline{10} \\ 0 \end{array}$$

The rule for expressing a fraction as a decimal is summarized below.

To Express a Fraction as a Decimal:

1. Divide the denominator into the numerator.
2. If you do not obtain a remainder of zero after carrying the division out three places, round the answer (quotient) to the nearest hundredth.

Example 16   Express each fraction as a decimal. If necessary, round the answer to the nearest hundredth.

   a. $\dfrac{3}{4}$ b. $\dfrac{7}{8}$ c. $\dfrac{7}{6}$ d. $\dfrac{5}{19}$

Solution   To change a fraction to a decimal, divide the numerator by the denominator.

16a. $\dfrac{3}{4} = 0.75$

$$\begin{array}{r} 0.75 \\ 4\overline{)3.00} \\ \underline{2\ 8} \\ 20 \\ \underline{20} \\ 0 \end{array}$$

16b. $\dfrac{7}{8} = 0.875$

$$\begin{array}{r} 0.875 \\ 8\overline{)7.000} \\ \underline{6\ 4} \\ 60 \\ \underline{56} \\ 40 \\ \underline{40} \\ 0 \end{array}$$

16c. $\dfrac{7}{6} \doteq 1.17$

$$\begin{array}{r} 1.166 \\ 6\overline{)7.000} \\ \underline{6} \\ 1\ 0 \\ \underline{6} \\ 40 \\ \underline{36} \\ 40 \\ \underline{36} \\ 4 \end{array}$$

16d. $\dfrac{5}{19} \doteq 0.26$

$$\begin{array}{r} 0.263 \\ 19\overline{)5.000} \\ \underline{3\ 8} \\ 1\ 20 \\ \underline{1\ 14} \\ 60 \\ \underline{57} \\ 3 \end{array}$$

Since we obtained a repeating remainder (4) in Solution 16c, the quotient of 7 and 6 is a repeating decimal. That is, $\dfrac{7}{6} = 1.1\overline{6}$. The bar indicates the digit that repeats.

Practice Problem 12   *Express each fraction as a decimal. If necessary, round the answer to the nearest hundredth.*

   a. $\dfrac{2}{5}$ b. $\dfrac{8}{5}$ c. $\dfrac{23}{33}$ d. $\dfrac{2}{3}$

Example 17   Solve and show all work.

Cazzie got 28 hits in 85 times at bat. What is his batting average? Round to the nearest thousandth.

Solution   Batting average = $\dfrac{\text{hits}}{\text{times at bat}}$

= $\dfrac{28}{85}$

= 0.329

$$\begin{array}{r} 0.3294 \\ 85\overline{)28.0000} \\ \underline{25\ 5} \\ 2\ 50 \\ \underline{1\ 70} \\ 800 \\ \underline{765} \\ 350 \\ \underline{340} \\ 10 \end{array}$$

Cazzie's batting average is 0.329.

Practice Problem 13   *Solve and show all work.*

Penny got 16 hits in 49 times at bat. What is her batting average? Round to the nearest thousandth.

IDEA 2

Sometimes it may be necessary to change a decimal to a reduced fraction. To convert a decimal to a fraction, just rewrite the decimal as a fraction and reduce it to its lowest terms. For example,

3.6 Converting Fractions and Decimals

$$0.5 \text{ (5 tenths)} = \frac{5}{10} = \frac{1}{2}$$

The rule for expressing decimals as fractions is summarized below.

To Express a Decimal as a Fraction:

1. Identify the place value of the last digit to the right of the decimal point.
2. Multiply this place value times the decimal number *without the decimal point*.
3. Reduce the resulting fraction, if possible.

Example 18 Express each decimal as a fraction.

    a. 0.71   **b.** 0.075   **c.** 13.5   **d.** 3.0007   **e.** 249

Solution To express a decimal as a fraction, multiply the decimal without a decimal point by the place value of the last digit to the right of the decimal point.

18a. $0.71 = 71\left(\frac{1}{100}\right)$ The place value of 1 in .71 is $\frac{1}{100}$

$$= \frac{71}{100}$$

18b. $0.075 = \frac{75}{1000}$ **18c.** $13.5 = \frac{135}{10}$

$$= \frac{3}{40} \qquad\qquad\qquad = \frac{27}{2}$$

18d. $3.0007 = \frac{30007}{10{,}000}$ **18e.** $249 = \frac{249}{1}$

NOTE: Some fractions and decimals are used so often that it may be helpful to memorize the fraction-decimal equivalents shown below.

TABLE OF EQUIVALENTS

Fraction	$\frac{1}{2}$	$\frac{1}{4}$	$\frac{1}{3}$	$\frac{2}{3}$	$\frac{1}{5}$	$\frac{2}{5}$	$\frac{3}{5}$	$\frac{3}{4}$	$\frac{1}{8}$
Decimal	0.5	0.25	$0.\overline{3}$	$0.\overline{6}$	0.2	0.4	0.6	0.75	0.125

Practice Problem 14   *Express each decimal as a fraction.*

    a. 0.15   **b.** 3.62   **c.** 0.0013   **d.** 32

IDEA 3   In some problems containing both fractions and decimals, to find the solution it may be necessary to change the fractions to decimals or the decimals to fractions.

Example 19   Solve. Express each answer both as a fraction and a decimal.

    a. $0.21 + \frac{3}{5}$ **b.** $\frac{9}{32} \div 0.125$

120 Decimals and Percents

Solution To simplify an expression containing fractions and decimals, express all fractions as decimals or all decimals as fractions (whichever is more convenient) and then do the indicated operations.

19a. Method 1
$$0.21 + \frac{3}{5}$$
$$= 0.21 + 0.6$$
$$= 0.81 \text{ or } \frac{81}{100}$$

Method 2
$$0.21 + \frac{3}{5}$$
$$= \frac{21}{100} + \frac{60}{100}$$
$$= \frac{81}{100} \text{ or } 0.81$$

19b. Method 1
$$\frac{9}{32} \div 0.125 = \frac{9}{32} \div \frac{125}{1000}$$
$$= \frac{9}{32} \cdot \frac{1000}{125}$$
$$= \frac{9 \cdot 1000}{32 \cdot 125}$$
$$= \frac{3 \cdot 3 \cdot \cancel{2} \cdot \cancel{5} \cdot \cancel{2} \cdot \cancel{5} \cdot \cancel{2} \cdot \cancel{5}}{\cancel{2} \cdot \cancel{2} \cdot \cancel{2} \cdot 2 \cdot 2 \cdot \cancel{5} \cdot \cancel{5} \cdot \cancel{5}}$$
$$= \frac{9}{4} \text{ or } 2.25$$

Method 2
If we change $\frac{9}{32}$ to a decimal we will obtain a repeating decimal and only an approximate value for $\frac{9}{32} \div 0.125$. Whenever this situation occurs, express all decimals as fractions and then perform the indicated operation.

Practice Problem 15 Solve. Express each answer both as a fraction and a decimal.

a. $3\frac{3}{4} + 2.1$ **b.** $\left(\frac{90}{70}\right)(1.05)$

3.6 Exercises

Express each fraction as a decimal. If necessary, round the answer to the nearest hundredth.

1. $\frac{3}{5}$ 2. $\frac{1}{2}$ 3. $\frac{1}{4}$ 4. $\frac{1}{5}$ 5. $\frac{9}{8}$

6. $\frac{23}{8}$ 7. $\frac{7}{25}$ 8. $\frac{19}{20}$ 9. $\frac{1}{3}$ 10. $\frac{7}{11}$

11. $\frac{5}{7}$ 12. $\frac{11}{18}$ 13. $\frac{17}{40}$ 14. $\frac{3}{16}$ 15. $\frac{31}{50}$

16. $\frac{27}{20}$ 17. $\frac{9}{10}$ 18. $\frac{5}{16}$ 19. $\frac{32}{125}$ 20. $\frac{3}{8}$

21. $\frac{3}{4}$ 22. $\frac{4}{5}$ 23. $\frac{7}{2}$ 24. $\frac{17}{4}$ 25. $\frac{1}{8}$

26. $\frac{5}{8}$ 27. $\frac{1}{16}$ 28. $\frac{1}{13}$ 29. $\frac{1}{100}$ 30. $\frac{1}{1000}$

31. $\frac{7}{400}$ 32. $\frac{9}{400}$ 33. $\frac{31}{5}$ 34. $\frac{2}{5}$ 35. $\frac{3}{16}$

36. $\frac{11}{15}$ 37. $\frac{7}{6}$ 38. $\frac{11}{9}$ 39. $\frac{7}{500}$ 40. $\frac{19}{200}$

Express each decimal as a reduced fraction.

41. 0.3   42. 0.15   43. 0.031   44. 0.5   45. 0.05

3.6 Exercises 121

46. 1.8	47. 2.5	48. 3.011	49. 8	50. 34
51. 5.12	52. 1.01	53. 6.1103	54. 0.80	55. 165
56. 0.0005	57. 0.125	58. 0.0625	59. 0.3125	60. 0.8125

Express each decimal as a reduced fraction.

61. 0.45	62. 0.85	63. 0.111	64. 0.513	65. 13
66. 17	67. 0.30	68. 0.90	69. 3.5	70. 6.8
71. 0.225	72. 0.155	73. 7.01	74. 8.001	75. 0.36
76. 0.7	77. 0.1500	78. 0.8000	79. 246.8	80. 348.5

Fill in the blanks.

81. To change a fraction to a decimal, divide the _____ into the _____. If you do not obtain a remainder of zero, _____ the answer to the desired place value.

82. If you divide one number into another and you obtain a _____ remainder, this implies that your quotient is a repeating decimal.

83. To express a decimal as a fraction, _____ the decimal without a decimal point by the _____ of the last digit to the right of the decimal point. Reduce the resulting fraction.

Solve. Express each answer both as a fraction and a decimal.

84. $1.25 - \dfrac{4}{5}$ 85. $\dfrac{7}{8} + 3.25$ 86. $(0.15)\left(\dfrac{52}{100}\right)$

87. $\dfrac{3}{16} \div 0.015$ 88. $\left(\dfrac{75}{22}\right)(0.33)$ 89. $\dfrac{11}{5} - \dfrac{2}{3}(1.8)$

Solve and show all work.

90. Dick got 19 hits in 60 times at bat. What is his batting average? Round to the nearest thousandth.

91. Barbara paid $15 for 35 notebooks. What was the cost of each notebook? Round to the nearest cent (hundredth).

92. John earned $45 for working 11 hours. How much did he earn each hour? Round to the nearest cent.

93. Pam must drill a 0.5625-inch hole through a table. What is the fractional size of the drill she must use?

94. Joe mixed three-fourths of a quart of orange juice and 0.6 of a quart of pineapple juice together to make a pineapple-orange drink. How many quarts was the final mixture?

95. Express $3\dfrac{1}{4}$ as a decimal.   96. Express $6\dfrac{1}{5}$ as a decimal.

97. Find the sum of $\dfrac{3}{4}$ and 6.75.   98. Find the sum of $\dfrac{4}{5}$ and 3.85.

99. Find the product of 0.17 and $\dfrac{1}{100}$.   100. Find the product of 0.001 and $\dfrac{7}{1000}$.

101. Lynn must drill a 0.125-inch hole. What is the fractional size of the drill she must use?

102. John must drill a 0.625-inch hole. What is the fractional size of the drill he must use?

Answers to Practice Problems   12a. 0.4   b. 1.6   c. 0.70   d. 0.67   13. 0.327   14a. $\dfrac{3}{20}$ b. $\dfrac{181}{50}$ c. $\dfrac{13}{10,000}$ d. $\dfrac{32}{1}$ 15a. 5.85 or $\dfrac{117}{20}$ b. 1.35 or $\dfrac{27}{20}$

3.7 Signed Decimals

IDEA 1 We will use the term **signed decimal** to mean a positive or a negative decimal. The rules, definitions, and procedures used to perform operations on integers and decimals also can be applied to signed decimals.

Example 20 Find the additive inverse.

 a. 3.25   **b.** -0.834

Solution To find the additive inverse of a number, change its sign.

20a. The additive inverse of 3.25 is -3.25.

20b. The additive inverse of -0.834 is 0.834.

Example 21 Find the absolute value.

 a. $|-6.8|$ **b.** $|0.326|$

Solution If x is a positive number, then $|x| = x$ and $|-x| = x$.

21a. $|-6.8| = 6.8$ **21b.** $|0.326| = 0.326$

Practice Problem 16 **Fill in the blanks.**

a. The additive inverse of 0.85 is _____.

b. $|-0.73| =$ _____.

c. The additive inverse of -3.17 is _____.

d. $|3.92| =$ _____.

IDEA 2 Signed decimals are added the same way integers are added. It will be helpful to remember that to compare two decimals, first write in extra zeros so that each number has the same number of decimal places. Then, compare the numbers as if they were whole numbers. For example, 0.3 is larger than 0.19 since

$$0.3 = 0.30$$
$$0.19 = 0.19$$

and 30 is larger than 19.

Example 22 Add.

 a. $-0.35 + (-0.7)$ **b.** $-0.653 + 0.71$ **c.** $-6.85 + 3.98$

Solution To add signed decimals, express each number with the same number of decimal places. Then add using the rule for adding integers.

22a. $-0.35 + (-0.7) = -0.35 + (-0.70)$
$= -(0.35 + 0.70)$
$= -1.05$

```
0.35
0.70
────
1.05
```

22b. $-0.653 + 0.71 = -0.653 + 0.710$
$= +(0.710 - 0.653)$
$= 0.057$

```
0.710
0.653
─────
0.057
```

The answer is positive since the sign is determined by the number having the largest absolute value.

3.7 Signed Decimals   123

$$22c.\ -6.85 + 3.98 = -(6.85 - 3.98)$$
$$= -2.87$$

$$\begin{array}{r} 0.85 \\ \underline{3.98} \\ 2.87 \end{array}$$

Practice Problem 17 **Add.**

    a. $-0.39 + 0.8$ **b.** $-3.37 + (-8.32)$ **c.** $0.651 + (-0.72)$

IDEA 3 Signed decimals are subtracted the same way integers are subtracted.

Example 23 Subtract.

    a. $3.84 - 7.95$ **b.** $-0.8 - 0.51$ **c.** $-0.953 - (-8)$

Solution To subtract using signed decimals, keep the same minuend and add the additive inverse of the subtrahend.

$$23a.\ 3.84 - 7.95 = 3.84 + (-7.95)$$
$$= -4.11$$

$$\begin{array}{r} 7.95 \\ \underline{3.84} \\ 4.11 \end{array}$$

$$23b.\ -0.8 - 0.51 = -0.8 + (-0.51)$$
$$= -1.31$$

$$\begin{array}{r} 0.80 \\ \underline{0.51} \\ 1.31 \end{array}$$

$$23c.\ -0.953 - (-8) = -0.953 + 8$$
$$= 7.047$$

$$\begin{array}{r} 8.000 \\ \underline{.953} \\ 7.047 \end{array}$$

Practice Problem 18 **Subtract.**

    a. $-0.36 - 0.8$ **b.** $-0.85 - (-0.111)$ **c.** $-3.65 - (-8.3)$

IDEA 4 Multiplying signed decimals is similar to multiplying integers.

Example 24 Multiply.

    a. $(-0.2)(-0.3)$ **b.** $(-3.2)(3.1)$ **c.** $(-0.09)(0.7)(-2)$

Solution To multiply signed decimals, multiply their absolute values and determine the sign of the product.

$$24a.\ (-0.2)(-0.3) = +[(0.2)(0.3)]$$
$$= 0.06$$

The product of an even number of negative signs is a positive number.

$$24b.\ (-3.2)(3.1) = -[(3.2)(3.1)]$$
$$= -9.92$$

$$\begin{array}{r} 3.2 \\ \underline{3.1} \\ 3\ 2 \\ \underline{96} \\ 9.92 \end{array}$$

The product of an odd number of negative signs is a negative number.

$$24c.\ (-0.09)(0.7)(-2) = +[(0.09)(0.7)(2)]$$
$$= 0.126$$

Practice Problem 19 **Multiply.**

    a. $(-2.3)(0.2)$ **b.** $(-31.2)(-0.31)$ **c.** $(-0.3)(-0.3)(-0.3)$

IDEA 5 Dividing signed decimals is similar to dividing integers.

Example 25 Divide.

a. $\dfrac{-0.88}{-0.2}$ b. $\dfrac{35}{-0.7}$ c. $\dfrac{-3.754}{2.3}$

Solution To divide using signed decimals, rewrite the problem with a divisor that is a whole number. Then use the rule for dividing integers.

25a. $\dfrac{-0.88}{-0.2} = \dfrac{-8.8}{-2}$ The quotient of two numbers having the same sign is positive.
$= 4.4$

25b. $\dfrac{35}{-0.7} = \dfrac{350}{-7}$ The quotient of two numbers having different signs is negative.
$= -50$

25c. $\dfrac{-3.754}{2.3} = \dfrac{-37.54}{23}$
$\doteq -1.63$

$$\begin{array}{r} 1.632 \\ 23\overline{)37.540} \\ \underline{23} \\ 14\,5 \\ \underline{13\,8} \\ 74 \\ \underline{69} \\ 50 \\ \underline{46} \\ 4 \end{array}$$

Practice Problem 20 *Divide.*

a. $\dfrac{-3.2}{0.8}$ b. $\dfrac{-1}{-.1}$ c. $\dfrac{4.824}{-1.2}$

IDEA 6

The order of operations rules discussed in Chapter 1 also apply for signed decimals. For example, to simplify the expression

$$-0.55 + 0.75(0.4)$$

multiply 0.75 and 0.4 before doing any addition.

$-0.55 + 0.75(0.4)$
$= -0.55 + 0.3$ Multiply first
$= -0.25$ Add

Example 26 Simplify.

$$-6.14 + 0.25(3.45 - 8) + 8.75 \div 0.5$$

Solution To simplify an expression containing more than one operation and one pair of grouping symbols, first perform all operations inside the grouping symbols. Then, simplify the resulting expression.

$-6.14 + 0.25\,(3.45 - 8) + 8.75 \div 0.5$
$= -6.14 + 0.25(-4.55) + 8.75 \div 0.5$ Simplify inside ()
$= -6.14 + (-1.1375) + 17.5$ Multiply and divide
$= -7.2775 + 17.5$ Add negative integers
$= 10.2225$ Add

Practice Problem 21 *Simplify.*

a. $-5.45 + 2.3(-5.1)$ b. $-8 + 3.2(-0.5 - 8.5) + 3.8$

3.7 Exercises

Find the additive inverse.

1. -3.21
2. 0.53
3. -8.231

Find the absolute value.

4. $|-0.5|$
5. $|0.85|$
6. $|-7.28|$

Tell which is the larger number.

7. 0.96, 0.8
8. 0.82, 0.9
9. 2.06, 2.008
10. 0.95, 0.099
11. 8.5, 8.58
12. 4.006, 4.04
13. 0.521, 0.52
14. 0.99, 1
15. 5.08, 5.7
16. 0.85, 0.6

Add.

17. $-0.6 + (-0.31)$
18. $-0.009 + 0.05$
19. $-10.91 + 18$
20. $-3.21 + (-6.03)$
21. $-0.631 + 0.7$
22. $-8.31 + (-0.795)$

Subtract.

23. $-0.61 - 0.32$
24. $-0.85 - (-0.91)$
25. $5.2 - (-8.34)$
26. $-4.51 - (-8)$
27. $0.231 - 0.45$
28. $-3.25 - 2.051$

Multiply.

29. $(-0.3)(0.5)$
30. $(-3.3)(5)$
31. $(-0.4)(-0.2)(-0.3)$
32. $(-0.31)(-3.8)$
33. $(-6.1)(0.3)$
34. $(-0.013)(-13)$

Divide.

35. $\dfrac{8.88}{-0.4}$
36. $\dfrac{-0.12}{-0.06}$
37. $\dfrac{-90}{.9}$
38. $\dfrac{-0.7}{-0.8}$
39. $\dfrac{-1.302}{0.3}$
40. $\dfrac{-16.016}{-1.6}$

Perform the indicated operation.

41. $-0.35 + 0.621$
42. $5.32 - 7$
43. $3.2 - (-8.11)$
44. $-0.561 + 0.87$
45. $-0.3333 \div 0.11$
46. $(-0.5)(-0.2)(-10)$
47. $(9.1)(-0.2)$
48. $-5.8 - (-0.65)$
49. $-8.08 \div (-0.4)$
50. $(-3.2)(0.005)(-5)$
51. $0.35 + (-0.44)$
52. $0.65 + (-0.88)$
53. $-6.2 + (-3.24)$
54. $-8.4 + (-4.33)$
55. $-0.45 + 6$
56. $-0.35 + 9$
57. $-0.031 + 0.04$
58. $-0.041 + 0.05$
59. $-0.25 - 0.83$
60. $-0.35 - 0.76$
61. $-0.85 - (-0.9)$
62. $-0.35 - (-0.5)$
63. $3.25 - (-0.75)$
64. $4.36 - (-0.64)$
65. $89.5 - 98.35$
66. $78.6 - 97.45$
67. $(-0.3)(-0.7)$

126 Decimals and Percents

68. $(-0.4)(-0.9)$
69. $(-3.6)(4)$
70. $(-5.4)(6)$
71. $(-0.2)(0.3)(-0.2)$
72. $(-0.3)(0.2)(-0.3)$
73. $(-0.1)(-0.4)(0.3)(-2)$
74. $(-0.2)(-0.5)(0.3)(-4)$
75. $-0.66 \div 0.11$
76. $-0.77 \div 0.7$
77. $-88.8 \div (-2)$
78. $-99.9 \div (-3)$
79. $14 \div (-0.7)$
80. $65 \div (-0.5)$
81. $1.645 \div (-0.35)$
82. $4.355 \div (-0.65)$

Solve. Show all work.

83. Find the sum of -3.25, 6.8, -7.891, and 0.3658.
84. Find the sum of -4.35, 6.7, -3.158, and 0.4769.
85. Subtract 8.51 from -13.6.
86. Subtract 6.39 from -8.7.
87. Find the product of -0.1, -0.2, -0.4, and -0.15.
88. Find the product of -0.2, -0.3, -0.1, and -0.21.
89. Divide -593.94 by 11.4.
90. Divide -390.05 by 14.5.
91. Find the difference between 8.51 and -13.6.
92. Find the difference between 6.39 and -8.7.
93. Divide -1.5 into 0.051.
94. Divide -3.2 into 0.144.

Simplify each expression.

95. $-6.3 + (2.8)(-5)$
96. $-8.1 + (3.5)(-8)$
97. $-6.45 + 8.7 - (-8.65) - 8$
98. $-7.35 + 3.8 - (-8.75) - 9$
99. $-4.38 + 6.84 \div (-0.4)$
100. $-6.75 + 4.42 \div (-0.2)$
101. $-8.2 + 4(-3.1 - 3.1)$
102. $-6.4 + 8(-2.3 - 2.3)$
103. $-8.75 + 6(-8.2 + 8 \div 0.4)$
104. $-6.85 + 3(-6.5 + 6 \div 0.3)$
105. $-4.02 [0.8 + 100(-0.105 - 0.004)] - 8.2$
106. $-15 + 100 [-0.9 + 3(-0.08 + 0.98)]$
107. $-6 + 4 [(-2.1 - 4.5) + (-8.6 + 6.88 \div 0.04)]$
108. $-8 + 5 [(-6.1 - 4.9) + (-8.6 + 4.44 \div 0.12)]$

Answers to Practice Problems   **16a.** -0.85 **b.** 0.73 **c.** 3.17 **d.** 3.92 **17a.** 0.41 **b.** -11.69 **c.** -0.069 **18a.** -1.16 **b.** -0.739 **c.** 4.65 **19a.** -0.46 **b.** 9.672 **c.** -0.027 **20a.** -4 **b.** 10 **c.** -4.02 **21a.** -17.18 **b.** -33

3.8 The Meaning of Percent

IDEA 1 The word **percent** means *per hundred*, and the percent symbol (%) means *hundredths*. For example, 9% means 9 hundredths $\left(0.09 \text{ or } \frac{9}{100}\right)$. Some other examples are:

$$7\% = 7 \text{ hundredths} \qquad 7\% = 7 \text{ hundredths}$$
$$= \frac{7}{100} \qquad\qquad = 0.07$$

$$13\% = 13 \text{ hundredths} \qquad 13\% = 13 \text{ hundredths}$$
$$= \frac{13}{100} \qquad\qquad = 0.13$$

3.8 The Meaning of Percent

IDEA 2   When P is any number, we can see from the above examples that

$$P\% \text{ means } \frac{P}{100} \quad \text{or} \quad P\left(\frac{1}{100}\right) \quad \text{or} \quad P(0.01)$$

We can use the fact that $P\%$ means $P\left(\frac{1}{100}\right)$ to express a percent as a fraction. For example,

$$7\% \text{ means } 7\left(\frac{1}{100}\right) = \frac{7}{100}$$

To Express a Percent as a Fraction:

1. Drop the percent symbol.
2. Multiply this number by $\frac{1}{100}$, and reduce the resulting fraction.

or

$$P\% = P\left(\frac{1}{100}\right)$$

Example 27   Express each percent as a reduced fraction.

   a. 27%   **b.** $\frac{3}{4}\%$ **c.** 315%   **d.** 80%   **e.** $66\frac{2}{3}\%$ **f.** 16.4%

Solution   To express a percent as a reduced fraction, use the formula $P\% = P\left(\frac{1}{100}\right)$. Reduce the answer to its lowest terms.

27a. $27\% = 27\left(\frac{1}{100}\right)$
$= \frac{27}{100}$

27b. $\frac{3}{4}\% = \frac{3}{4}\left(\frac{1}{100}\right)$
$= \frac{3}{400}$

27c. $315\% = 315\left(\frac{1}{100}\right)$
$= \frac{315}{100}$
$= \frac{63}{20}$

27d. $80\% = 80\left(\frac{1}{100}\right)$
$= \frac{80}{100}$
$= \frac{4}{5}$

Convert the decimal to a fraction

27e. $66\frac{2}{3}\% = \left(66\frac{2}{3}\right)\left(\frac{1}{100}\right)$
$= \frac{200}{3} \cdot \frac{1}{100}$
$= \frac{\overset{2}{\cancel{200}} \cdot 1}{3 \cdot \underset{1}{\cancel{100}}}$
$= \frac{2}{3}$

27f. $16.4\% = 16\frac{2}{5}\%$ $.4 = \frac{4}{10} = \frac{2}{5}$
$= \frac{82}{5} \cdot \frac{1}{100}$
$= \frac{\overset{41}{\cancel{82}} \cdot 1}{5 \cdot \underset{50}{\cancel{100}}}$
$= \frac{41}{250}$

Practice Problem 22 **Express each percent as a fraction.**

    a. 35%   **b.** 110%   **c.** $\frac{1}{2}$%   **d.** 3.25%

IDEA 3

Expressing percents as decimals is done in much the same way that we converted percents to fractions. That is, since $P\%$ also means $P(0.01)$, we have that 7% means $7(0.01) = 0.07$.

To Express a Percent as a Decimal:

1. Drop the percent symbol.
2. Multiply this number by 0.01,

        or

$$P\% = P(0.01)$$

Example 28   Express each percent as a decimal.

    a. 27%   **b.** 80%   **c.** 100%   **d.** 0.05%   **e.** $2\frac{3}{4}$%   **f.** $\frac{3}{5}$%

Solution   To express as a decimal, use the formula $P\% = P(0.01)$.

28a.   27% = 27(0.01)   **28b.**   80% = 80(0.01)
           = 0.27   = 0.80 or 0.8

28c.   100% = 100(0.01)   **28d.**   0.05% = 0.05(0.01)
            = 1   = 0.0005

Convert the fraction to a decimal

28e. $2\frac{3}{4}$% = 2.75% $\frac{3}{4} = .75$ **28f.** $\frac{3}{5}$% = 0.6%
           = 2.75 (0.01)   = 0.6(0.01)
           = 0.0275   = 0.006

Practice Problem 23 **Express each percent as a decimal.**

    a. 33%   **b.** 0.08%   **c.** $3\frac{1}{4}$%   **d.** $\frac{4}{5}$%

Based on the results in Example 28 the rule for changing a percent to a decimal can be restated as follows.

To Express a Percent as a Decimal:

1. Drop the percent symbol.
2. Move the decimal point two places to the left.

Example 29 Express each percent as a decimal.

a. 71% **b.** 1.05% **c.** 130% **d.** $66\frac{2}{3}\%$

Solution To express a percent as a decimal, move the decimal point two places to the left and remove the %.

29a. 71% = 71.%
       = 0.71

29b. 1.05% = 01.05%
       = 0.0105

Convert the fraction to a decimal

29c. 130% = 1 30 %
       = 1.30

29d. $66\frac{2}{3}\%$ = 66.$\overline{6}$%
       = 0.66$\overline{6}$
       ≐ 0.67

Practice Problem 24 *Express each percent as a decimal (see Example 4).*

a. 15% **b.** 0.05% **c.** 105% **d.** $3\frac{3}{4}\%$

3.8 Exercises

Express each percent as a fraction.

1. 13%   2. 15%   3. 31%   4. 12%   5. 55%
6. 331%   7. 2%   8. 1.06%   9. $6\frac{1}{3}\%$ 10. $\frac{1}{2}\%$
11. 100%   12. $\frac{1}{3}\%$ 13. $37\frac{1}{2}\%$ 14. 0.02%   15. 1.5%
16. $\frac{7}{8}\%$ 17. 150%   18. $8\frac{2}{3}\%$ 19. 0.004%   20. $4\frac{1}{16}\%$
21. 17%   22. 19%   23. 18%   24. 16%   25. 115%
26. 102%   27. 1.04% 28. 1.08% 29. $8\frac{1}{3}\%$ 30. $6\frac{2}{3}\%$
31. $\frac{3}{4}\%$ 32. $\frac{1}{4}\%$ 33. 500%   34. 200%   35. $55\frac{1}{8}\%$
36. $65\frac{5}{8}\%$ 37. 0.006%   38. 0.008%   39. 65%   40. 75%

Express each percent as a decimal.

41. 13%   42. 15%   43. 31%   44. 12%   45. 55%
46. 331%   47. 2%   48. 1.06% 49. $6\frac{1}{4}\%$ 50. $\frac{1}{2}\%$
51. 100%   52. 0.8%   53. $\frac{1}{8}\%$ 54. 851%   55. 250%
56. 3.5%   57. $10\frac{3}{4}\%$ 58. 0.51% 59. $\frac{17}{5}\%$ 60. 2.011%
61. 17%   62. 19%   63. 16%   64. 18%   65. 115%

66. 102%   67. 1.04%   68. 1.08%   69. $8\frac{1}{3}$%   70. $6\frac{2}{3}$%

71. $\frac{3}{4}$%   72. $\frac{1}{4}$%   73. 500%   74. 200%   75. $55\frac{1}{8}$%

76. $65\frac{5}{8}$%   77. 0.006%   78. 0.008%   79. 65%   80. 75%

Fill in the blanks.

81. The word _____ means per hundred.

82. To express a percent as a fraction, replace the percent symbol with _____. Next, multiply this number by the number preceding the _____ and _____ the resulting fraction.

83. To express a percent as a decimal, replace the percent symbol with _____. Next, multiply this number by the number preceding the _____.

84. To express a percent as a decimal, move the decimal point _____ places to the _____ and remove the _____.

83. If 23% of the students at a certain college jog, this means that _____ out of every _____ students jog.

Solve.

86. John saved 25% of his annual income. What fractional part of his income did he save?

87. If 22.5% of the students at a community college attended a basketball game, what fractional part of the student body attended the game?

88. The Second National Bank pays $5\frac{1}{4}$% interest on a savings account. Express this interest rate as a decimal.

89. Jim spends $12\frac{1}{2}$% of his time each day reading. Express the amount of time he spends reading as a decimal.

90. The price of a coat is discounted $33\frac{1}{3}$%. What fractional part of the original price must a customer pay?

91. Express 131% as a fraction.   92. Express 231% as a fraction.

93. Change $\frac{3}{8}$% to a decimal.   94. Change $\frac{11}{40}$% to a decimal.

95. Rich saved $37\frac{1}{8}$% of his annual income. What fractional part of his income did he save?

Answers to Practice Problems 22a. $\frac{7}{20}$ b. $\frac{11}{10}$ c. $\frac{1}{200}$ d. $\frac{13}{400}$ 23a. 0.33 b. 0.0008 c. 0.0325 d. 0.008 24a. 0.15 b. 0.0005 c. 1.05 d. 0.0375

3.9 Writing Decimals and Fractions as Percents

IDEA 1 In Example 28c we showed that 100% = 1. We also know that multiplying a number by 1 does not change the value of that number. Using these two facts, we can convert decimals to percents. For example,

$$0.18 = 0.18(1) = 0.18(100\%) = 18\%$$

Example 30   Express each decimal as a percent.

a. 0.83   b. 0.389   c. 0.5

Solution   To express a decimal as a percent, write the decimal and 100% as a product and then multiply.

30a. $0.83 = 0.83(1)$
$= 0.83(100\%)$
$= 83\%$

30b. $0.389 = 0.389(1)$
$= 0.389(100\%)$
$= 38.9\%$

30c. $0.5 = 0.5(1)$
$= 0.5(100\%)$
$= 50\%$

The results in Example 30 indicate that when we multiply decimals by 100 the decimal point moves 2 places to the right and it suggests the following rule.

To Express a Decimal as a Percent:

1. Move the decimal point two places to the right.
2. Write a percent symbol.

Example 31   Express each decimal as a percent.

a. 0.91   b. 0.7   c. 0.145   d. 5.3361

Solution   To express a decimal as a percent, move the decimal point two places to the right and write a percent symbol.

31a. $0.91 = 0.91$
$= 91\%$

31b. $0.7 = 0.70$
$= 70\%$

31c. $0.145 = 0.14\,5$
$= 14.5\%$

31d. $5.3361 = 5.33\,61$
$= 533.61\%$

Practice Problem 25   *Express each decimal as a percent.*

a. 0.13   b. 0.8   c. 0.115   d. 2.1374

IDEA 2   You have learned how to change decimals to percents and fractions to decimals. By combining these two procedures, you can change fractions to percents. For example,

$$\frac{1}{4} = 4\overline{)1} = 0.25 = 25\%$$

To Express a Fraction as a Percent:

1. Change the fraction to a decimal.
2. Express the decimal as a percent.

Some fractions can be expressed as percents without first changing the fraction to a decimal. For example, $\frac{17}{100} = 17\%$ by using the definition of percent. Similarly,

$$\frac{4}{5} = \frac{4 \cdot 20}{5 \cdot 20} = \frac{80}{100} = 80\%$$

Example 32 Express each fraction as a percent.

a. $\frac{3}{4}$ b. $\frac{1}{5}$ c. $\frac{3}{500}$ d. $\frac{5}{2}$ e. $\frac{9}{25}$

Solution To express a fraction as a percent, first change the fraction to a decimal and then express the decimal as a percent.

32a. $\frac{3}{4} = 0.75$ $\begin{array}{r} 0.75 \\ 4\overline{)3.00} \\ \underline{2\,8} \\ 20 \\ \underline{20} \end{array}$ 32b. $\frac{1}{5} = 0.2$ $\begin{array}{r} 0.2 \\ 5\overline{)1.0} \\ \underline{1\,0} \end{array}$
$= 75\%$ $= 20\%$

32c. $\frac{3}{500} = 0.006$ $\begin{array}{r} 0.006 \\ 500\overline{)3.000} \\ \underline{3\,000} \end{array}$ 32d. $\frac{5}{2} = 2.5$ $\begin{array}{r} 2.5 \\ 2\overline{)5.0} \\ \underline{4} \\ 1\,0 \\ \underline{1\,0} \end{array}$
$= 0.6\%$ $= 250\%$

32e. $\frac{9}{25} = \frac{9 \cdot 4}{25 \cdot 4} = \frac{36}{100}$
$= 36\%$

Example 33 Express each fraction to the nearest tenth of a percent.

a. $\frac{1}{3}$ b. $\frac{7}{320}$

Solution To express a fraction to the nearest tenth of a percent, first change the fraction to a decimal rounded to the nearest thousandth. Then, express the decimal as a percent.

33a. $\frac{1}{3} \doteq 0.333$ $\begin{array}{r} 0.3333 \\ 3\overline{)1.0000} \\ \underline{9} \\ 10 \\ \underline{9} \\ 10 \\ \underline{9} \\ 10 \\ \underline{9} \end{array}$ 33b. $\frac{7}{320} \doteq 0.022$ $\begin{array}{r} 0.0218 \\ 320\overline{)7.0000} \\ \underline{6\,40} \\ 600 \\ \underline{320} \\ 2800 \\ \underline{2560} \end{array}$
$= 33.3\%$ $= 2.2\%$
Thus, $\frac{1}{3} \doteq 33.3\%$ Thus, $\frac{7}{320} \doteq 2.2\%$

Practice Problem 26 *Express each fraction to the nearest tenth of a percent.*

a. $\frac{3}{7}$ b. $\frac{2}{3}$ c. $\frac{5}{6}$

Frequently in business applications you will need to find the *exact* percent equivalent of a fraction rather than an approximate value.

Example 34 Express $\frac{1}{3}$ as a percent.

Solution The fraction $\frac{1}{3}$ is usually represented by a repeating decimal ($0.3\overline{3}$). Therefore, to find its exact percent equivalent, we will follow two steps.

Step 1. Express the fraction as a decimal that contains a fraction.

$$\frac{1}{3} = 0.33\frac{1}{3} \qquad 3\overline{)1.00}$$

Carry out the division two places and then write the remainder over the divisor.

Step 2. Next, express the decimal $\left(0.33\frac{1}{3}\right)$ as a percent $\left(33\frac{1}{3}\%\right)$. Thus,

$$\frac{1}{3} = 0.33\frac{1}{3} = 33\frac{1}{3}\%$$

Practice Problem 27 **Express each fraction as a percent (see Example 34)**

a. $\frac{3}{7}$ b. $\frac{2}{3}$ c. $\frac{5}{6}$

3.9 Exercises

Express each decimal as a percent.

1. 0.65	2. 0.33	3. 2.3	4. 32	5. 0.006
6. 5	7. 0.81	8. 0.00768	9. 0.75	10. 0.17
11. 0.6	12. 0.8	13. 0.4	14. 0.1115	15. 51.2
16. 0.8842	17. 0.007	18. 3.05	19. $0.33\frac{1}{3}$	20. $0.66\frac{2}{3}$
21. 0.13	22. 0.38	23. 8	24. 9	25. 0.008
26. 0.009	27. 0.5	28. 0.9	29. 0.111	30. 0.222
31. 8.1	32. 1.8	33. 0.0045	34. 0.0065	35. 0.6
36. 0.7	37. $0.55\frac{1}{2}$	38. $0.65\frac{1}{2}$	39. $0.8\frac{1}{2}$	40. $0.9\frac{3}{4}$

Express each fraction as a percent.

41. $\frac{3}{5}$	42. $\frac{1}{2}$	43. $\frac{1}{4}$	44. $\frac{1}{5}$	45. $\frac{9}{8}$
46. $\frac{23}{50}$	47. $\frac{7}{25}$	48. $\frac{19}{20}$	49. $\frac{1}{3}$	50. $\frac{7}{11}$
51. $\frac{5}{7}$	52. $\frac{11}{18}$	53. $\frac{17}{40}$	54. $\frac{3}{16}$	55. $\frac{31}{50}$
56. $\frac{27}{20}$	57. $\frac{9}{10}$	58. $\frac{5}{8}$	59. $\frac{3}{4}$	60. $\frac{1}{16}$
61. $\frac{1}{10}$	62. $\frac{11}{20}$	63. $\frac{2}{3}$	64. $\frac{3}{8}$	65. $\frac{11}{40}$
66. $\frac{22}{50}$	67. $\frac{7}{10}$	68. $\frac{3}{5}$	69. $\frac{15}{10}$	70. $\frac{8}{5}$

134   Decimals and Percents

71. $\dfrac{7}{1000}$ 72. $\dfrac{3}{10,000}$ 73. $\dfrac{5}{1}$ 74. $\dfrac{8}{1}$ 75. $\dfrac{1}{40}$

76. $\dfrac{1}{20}$ 77. $\dfrac{5}{16}$ 78. $\dfrac{7}{16}$ 79. $\dfrac{11}{10}$ 80. $\dfrac{23}{20}$

Fill in the blanks.

81. To express a decimal as a percent, move the _____ two places to the _____ and write a percent symbol.

82. To express a fraction as a percent, change the _____ to a _____ and then express the _____ as a _____.

83. The fraction $\dfrac{2}{3}$ is represented by a repeating decimal. Thus, to find its exact percent equivalent, first express the _____ as a _____ which contains a fraction. Next, express the _____ as a _____.

Solve.

84. A clothing store advertised that all items would be reduced by $\dfrac{1}{4}$ of their original cost. Write this discount as a percent.

85. Three-fifths of the employees at a supermarket smoke. What percent of the employees smoke?

86. A student answered 33 out of 40 questions on an exam correctly. What percent of the questions did he answer incorrectly?

87. When Jim purchased a car, he was told that the interest rate was 0.1455. Express this rate as a percent.

88. A certain type of medicine is 0.057 alcohol. What percent of the medicine is alcohol?

89. A survey indicated that 0.155 of all students at a college exercise daily. What percent of the students exercise daily?

90. A survey indicated that 0.205 of all employees at a certain company jog. What percent of the employees jog?

91. Express $\dfrac{8}{7}$ as a percent.   92. Express $\dfrac{10}{7}$ as a percent.

93. Express $0.52\dfrac{1}{8}$ as a percent.   94. Express $0.78\dfrac{3}{8}$ as a percent.

95. A medicine is 0.187 alcohol. What percent of the medicine is alcohol?

Answers to Practice Problems   25a. 13%   b. 80%   c. 11.5%   d. 213.74%   26a. 42.9%   b. 66.7%   c. 83.3%   27a. $42\dfrac{6}{7}\%$ b. $66\dfrac{2}{3}\%$ c. $83\dfrac{1}{3}\%$

Chapter 3 Summary

Important Terms

A **decimal** or **decimal fraction** is a number that can be written as a fraction with a denominator of 1, 10, 100, 1000, and so on. The numbers 3, 6.5, and −0.48 are decimals. [Section 3.1/Idea 1]

[Section 3.4/Idea 1]

A **signed decimal** is a positive or a negative decimal. −3.75 and +0.017 are signed decimals. [Section 3.7/Idea 1]

Chapter 3 Summary

Rounding is a procedure used to find an approximate value for a given number. 0.35 is the approximate value for 0.348 when rounded to the nearest hundredth.

Percent means per hundred. 8 percent (8%) means 8 hundredths $\left(0.08 \text{ or } \frac{8}{100}\right)$. [Section 3.8/Idea 1].

Important Skills

Writing Word Names for Decimals
To write the word name for a decimal, write the name for the whole number to the left of the decimal point. Next, write "and" for the decimal point. Finally, write the number to the right of the decimal point as if it were a whole number, and then write the name for the place value of the last digit on the right. [Section 3.1/Idea 2]

Adding Decimals
To add decimals, write each number so that the decimal points are lined up. Add as if adding whole numbers, and then place the decimal point in the sum so it is lined up with the other decimal points. [Section 3.2/Idea 1]

Subtracting Decimals
To subtract one decimal from another, write the subtrahend beneath the minuend so that the decimal points are lined up. Subtract as if subtracting whole numbers, and then place the decimal point in the difference so it is lined up with the other decimal points. [Section 3.2/Idea 2]

Multiplying Decimals
To multiply decimals, multiply as if multiplying whole numbers. Place the decimal point in the product so that the number of decimal places in the product equals the sum of the number of decimal places in the factors. [Section 3.3/Idea 1]

Rounding Decimals
To round a decimal to a given place value, identify the first digit to the right of the given place value. If it is 5 or greater, round up by adding 1 to the digit in the given place value and changing all digits to the right of it to zero. If it is less than 5, round down by changing all digits to the right of the given place value to zero. [Section 3.4/Idea 1]

Dividing Decimals
To divide one decimal by another, first rewrite the problem in the long division format. Next, move the points in the divisor and the dividend the same number of places to the right until the divisor is a whole number. Divide as if dividing whole numbers. Finally, place the decimal point in the quotient so it is directly above the decimal point in the dividend. [Section 3.5/Idea 1]

Expressing Fractions as Decimals
To express a fraction as a decimal, divide the denominator into the numerator. [Section 3.6/Idea 1]

Expressing Decimals as Fractions
To express a decimal as a fraction, identify the place value of the last digit to the right of the decimal point. Multiply this place value times the decimal number without the decimal point. Reduce the resulting fraction, if possible. [Section 3.6/Idea 2]

Operating on Signed Decimals
To add, subtract, multiply, or divide signed decimals, do so in the same way you add, subtract, multiply, or divide integers. [Section 3.7/Ideas 2–5]

To simplify signed decimals involving more than one operation, use the order of operations rules discussed in Section 1.7. [Section 3.7/Idea 6]

Expressing Percents as Fractions
$P\% = P\left(\frac{1}{100}\right)$

To express a percent as a fraction, drop the percent symbol and multiply by $\frac{1}{100}$. Reduce the resulting fraction to lowest terms. [Section 3.8/Idea 2]

Expressing Percents as Decimals
$P\% = P(0.01)$

To express a percent as a decimal, drop the percent symbol, and move the decimal point two places to the left. [Section 3.8/Idea 3]

136   *Decimals and Percents*

Expressing Decimals as Percents	To express a decimal as a percent, move the decimal point two places to the right and write a percent symbol. [Section 3.9/Idea 1]
Expressing Fractions as Percents	To express a fraction as a percent, first change the fraction to a decimal and then express the decimal as a percent. [Section 3.9/Idea 2]

Chapter 3 Review Exercises

Write the word name.

1. 5.009
2. 0.2318
3. 0.35

Complete the following tables by filling in the equivalent fraction or decimal.

	Fraction	Decimal
4.	$\frac{61}{20}$	
5.		0.31
6.	$\frac{1}{3}$	

	Fraction	Decimal
7.	$\frac{11}{19}$	
8.		0.65
9.		0.125

Round 1751.85928 to the nearest:

10. thousandth
11. hundred

Perform the indicated operation.

12. 6.54 + 8.3
13. 9.325 − 8.41
14. (7.89)(0.0005)
15. 86.16 ÷ 0.19
16. −8.79 + (−3.125)
17. 66 ÷ (−0.6)
18. (−8.3)(1.02)
19. (0.25)(35.7)
20. 0.897 − 0.2345
21. 0.358 + 0.21
22. 0.85 − 0.9
23. 6.58 + 3.2 + 8.101

Solve.

24. If gas costs $1.45 per gallon, how much will 10.4 gallons of gas cost?

25. Kay saved $369 in one year. Find her average monthly savings.

26. Tom went shopping with $90 in cash. He bought a record for $8.50, a shirt for $28.75, a sweater for $42, and he spent $4.23 for lunch. If Tom paid cash for all items, how much cash does he have left?

27. Vanessa earns $8.80 per hour. If she worked $35\frac{3}{4}$ hours last week, how much did she earn?

Perform the indicated operation.

28. −6.85 + (2.4)(−5)
29. −6.2 − 8.4 ÷ 0.14
30. $-\frac{1}{2} + 0.75\left(-\frac{2}{5}\right)$
31. −10 + 5[−2.4 + 3(−0.84 + 1 ÷ 0.1)]
32. 18.357 − 101.68 + 147.6345 + (−66.345) − (−1007)
33. Find the reciprocal of −2.5.

Complete the following tables by filling in the equivalent fraction, decimal, or percent.

	Decimal	Fraction	Percent
34.	0.5		
35.		$\frac{1}{5}$	
36.			25%
37.	0.61		
38.		$\frac{1}{4}$	
39.			80%
40.	0.125		
41.		$\frac{1}{3}$	

	Decimal	Fraction	Percent
42.			$66\frac{2}{3}\%$
43.	0.6		
44.		$\frac{3}{4}$	
45.			37.5%
46.	5		
47.			$\frac{3}{5}\%$
48.		$\frac{7}{500}$	

Name: _____

Class: _____

Chapter 3 Test

1. Write the word name for 3.005.
2. 18.14 + 6.24 + 9.1365
3. 92 − 0.0006
4. 643.89 − 99.483
5. (18.5)(0.24)
6. Round 8.7634 to the nearest hundredth.
7. Round 145.36 to the nearest ten.
8. 45.9 ÷ 2.5
9. −13.33 ÷ 3.4
10. Express $\frac{1}{4}$ as a decimal.
11. Express $\frac{2}{3}$ as a decimal.
12. Express 0.036 as a reduced fraction.
13. −0.3 − (−0.45)
14. $(-2.4)(-2)\left(-\frac{2}{5}\right)$
15. Betty has $431.87 in a checking account. If she withdraws $37.89, what is the new balance in her account?
16. If Lorie pays $52.57 per month on her charge account, how long will it take her to pay a balance of $1892.52.
17. If Liz buys 19.8 gallons of gas at 95 cents per gallon, what is the total cost of the gas?
18. Express 35% as a fraction.
19. Express 1.05% as a fraction.
20. Express 35% as a decimal.
21. Express $\frac{3}{8}$% as a decimal.
22. Express 0.7 as a percent.

1. _____
2. _____
3. _____
4. _____
5. _____
6. _____
7. _____
8. _____
9. _____
10. _____
11. _____
12. _____
13. _____
14. _____
15. _____
16. _____
17. _____
18. _____
19. _____
20. _____
21. _____
22. _____

23. Express 0.918 as a percent.

24. Express $\frac{3}{4}$ as a percent.

25. Express $\frac{1}{9}$ as a percent.

23. _____

24. _____

25. _____

4 Exponential Expressions

Objectives The objectives for this chapter are listed below along with sample problems. By the end of this chapter you should be able to find the solutions to the given problems.

1. Find the value of a number written in exponential form *(Section 4.1/Idea 2)*.
   a. 2^3 b. 2^{-3} c. $(-2)^3$

2. Multiply exponential expressions having the same base *(Section 4.2/Idea 1)*.
   a. $7^{10} \cdot 7^{40}$ b. $x^{-18} \cdot x^{30}$

3. Divide exponential expressions having the same base *(Section 4.2/Idea 2)*.
   a. $\dfrac{7^5}{7^{-8}}$ b. $\dfrac{x^{-9}}{x^9}$

4. Raise an exponential expression to a power *(Section 4.3/Idea 1)*.
   a. $(x^2)^4$ b. $(7^{-3})^5$

5. Raise a product to a power *(Section 4.3/Idea 2)*.
   a. $(3 \cdot 6)^{11}$ b. $(-3x)^4$

6. Raise a fraction to a power *(Section 4.3/Idea 3)*.
   a. $\left(\dfrac{2}{3}\right)^3$ b. $\left(\dfrac{x}{-5}\right)^3$

7. Multiply a decimal by a power of ten *(Section 4.4/Idea 1)*.
   a. 2.35×10^3 b. 62.8×10^{-4}

8. Express a decimal in scientific notation *(Section 4.4/Idea 2)*.
   a. 68,000   b. 0.000068

4.1 The Meaning of Exponential Expressions

IDEA 1 The number 3^4 (read as "three to the fourth power") is said to be written in **exponential form,** and 3^4 means that 3 is to be used as a factor four times ($3^4 = 3 \cdot 3 \cdot 3 \cdot 3$). The 3 is called the **base** and the 4 is called the **exponent.** In general, an exponent shows how many times the base is to be used as a factor. When a number does not have an exponent, assume that the exponent is 1. For example,

$$6 = 6^1$$

Example 1   Explain the meaning of each exponential expression. Tell which is the base and which is the exponent.

   a. 3^7 **b.** 2^3 **c.** -4^2 **d.** $(-4)^2$

Solution   To explain the meaning of an exponential expression, remember that an exponent tells you how many times the base is to be used as a factor.

   1a. 3^7 means $3 \cdot 3 \cdot 3 \cdot 3 \cdot 3 \cdot 3 \cdot 3$. The base is 3 and the exponent is 7.

   1b. 2^3 means $2 \cdot 2 \cdot 2$. The base is 2 and the exponent is 3. Read 2^3 as "two to the third power" or "two cubed."

   1c. -4^2 means $-(4 \cdot 4)$. The base is 4 and the exponent is 2. Read -4^2 as "the negative of four to the second power" or "the negative of four squared."

   1d. $(-4)^2$ means $(-4)(-4)$. The base is -4 and the exponent is 2. Read $(-4)^2$ as "negative four to the second power" or "negative four squared."

!   The base of an exponent is the number immediately to the left and below the exponent. That is,

   -4^2 The base is the number 4.
   $(-4)^2$ The base is the number -4, which is inside the parentheses.

Practice Problem 1   *Explain the meaning of each expression. Identify the base and the exponent.*

   a. 7^5 **b.** 3^2 **c.** -2^6 **d.** $(-2)^6$

Example 2   Write each product in exponential form.

   a. $3 \cdot 3$ **b.** $(-5)(-5)(-5)$ **c.** $7 \cdot 7 \cdot 7 \cdot 7 \cdot 7 = 7$

Solution   To write a product in exponential form, remember that an exponent indicates the number of times the base is used as a factor.

   2a. $3 \cdot 3 = 3^2$ **2b.** $(-5)(-5)(-5) = (-5)^3$ **2c.** $7 \cdot 7 \cdot 7 \cdot 7 \cdot 7 \cdot 7 = 7^6$

Practice Problem 2   *Write each product in exponential form.*

   a. $8 \cdot 8 \cdot 8 \cdot 8$ **b.** $(-2)(-2)(-2)$ **c.** $10 \cdot 10 \cdot 10 \cdot 10 \cdot 10$

IDEA 2   Sometimes in mathematics you will need to find the value of a number written in exponential form.

Exponential Expressions

To Find the Value of an Exponential Expression:

If a is any number and n is a positive integer

1. To find the value of a^n, use a as a factor n times.
2. To find the value of a^{-n}, write a fraction whose numerator is 1 and whose denominator is a^n.
3. The value of a^0 is 1, if a is not equal to zero.

Or

$$a^n = a \cdot a \cdot a \cdots a \quad (n \text{ factors of } a)$$
$$a^{-n} = \frac{1}{a^n}$$
$$a^0 = 1$$

Example 3 Find the value of each expression.

    a. 5^2 **b.** 4^3 **c.** $(-3)^4$ **d.** -3^4 **e.** $(0.2)^4$

Solution To find the value of an exponential expression with a positive exponent n, use the base as a factor n times.

  3a. $5^2 = 5 \cdot 5$ **3b.** $4^3 = 4 \cdot 4 \cdot 4$ **3c.** $(-3)^4 = (-3)(-3)(-3)(-3)$
       $= 25$ $= 64$ $= 81$

  3d. $-3^4 = -(3 \cdot 3 \cdot 3 \cdot 3)$ **3e.** $(0.2)^4 = (0.2)(0.2)(0.2)(0.2)$
        $= -81$ $= 0.0016$

Practice Problem 3 *Find the value of each expression.*

    a. 6^2 **b.** 2^5 **c.** -5^3 **d.** $(-5)^3$

Example 4 Find the value of each expression.

    a. 3^0 **b.** $(-5)^0$ **c.** -5^0 **d.** $\left(\frac{2}{3}\right)^0$

Solution To raise any number (except zero) to the zero power, remember that $a^0 = 1$.

  4a. $3^0 = 1$ **4b.** $(-5)^0 = 1$ **4c.** $-5^0 = -(5^0) = -1$ **4d.** $\left(\frac{2}{3}\right)^0 = 1$

Example 5 Find the value of each expression.

    a. 5^{-2} **b.** 6^{-3} **c.** $(-3)^{-4}$ **d.** -3^{-4} **e.** $(-4)^{-3}$

Solution To raise any number (except zero) to a negative power, remember $a^{-n} = \frac{1}{a^n}$.

  5a. $5^{-2} = \frac{1}{5^2}$ **5b.** $6^{-3} = \frac{1}{6^3}$ **5c.** $(-3)^{-4} = \frac{1}{(-3)^4}$
          $= \frac{1}{25}$ $= \frac{1}{216}$ $= \frac{1}{81}$

5d. $-3^{-4} = -\dfrac{1}{3^4}$
$= -\dfrac{1}{81}$

5e. $(-4)^{-3} = \dfrac{1}{(-4)^3}$
$= \dfrac{1}{-64}$
$= -\dfrac{1}{64}$

Practice Problem 4 Find the value of each expression.

a. 7^0 **b.** 3^{-3} **c.** $(-8)^0$ **d.** $(-2)^{-5}$ **e.** -2^{-5}

4.1 Exercises

Explain the meaning of each expression.

1. 3^5 2. 13^2 3. 5^7 4. 6^4 5. 7^6

Write each product in exponential notation.

6. $3 \cdot 3 \cdot 3 \cdot 3$ 7. $9 \cdot 9$ 8. $2 \cdot 2 \cdot 2$
9. $6 \cdot 6 \cdot 6 \cdot 6 \cdot 6$ 10. $7 \cdot 7 \cdot 7 \cdot 7 \cdot 7$ 11. $(.5)(.5)$
12. $(-13)(-13)$ 13. $4 \cdot 4 \cdot 4 \cdot 4 \cdot 4 \cdot 4$

Find the value of each exponential expression.

14. 2^3 15. 3^2 16. 5^3 17. 6^2 18. 4^3
19. 7^{-2} 20. 9^{-3} 21. 10^{-3} 22. 4^{-3} 23. 2^{-4}
24. 5^0 25. 9^0 26. $(-6)^0$ 27. -6^0 28. 35^0
29. $(0.3)^2$ 30. $(0.2)^3$ 31. $(1.1)^3$ 32. $(-0.5)^2$ 33. $(-3)^4$
34. $(-5)^3$ 35. -3^4 36. 3^{-4} 37. $(-6)^{-2}$ 38. -6^2
39. $(-7)^{-3}$ 40. -19^0 41. -2^{-5} 42. 2^{-5} 43. $(0.01)^5$
44. $(-0.4)^3$

Fill in the blanks.

45. An exponent indicates how many times the _____ is to be used as a factor.

46. Any number (except zero) raised to the zero power is equal to _____.

47. If you raise a number (except zero) to a negative power, the result is a _____ in which the numerator is _____ and the denominator is that number raised to the corresponding _____ power.

Answers to Practice Problems **1a.** $7 \cdot 7 \cdot 7 \cdot 7 \cdot 7$, base = 7, exponent = 5 **b.** $3 \cdot 3$, base = 3, exponent = 2 **c.** $-(2 \cdot 2 \cdot 2 \cdot 2 \cdot 2 \cdot 2)$, base = 2, exponent = 6 **d.** $(-2)(-2)(-2)(-2)(-2)(-2)$, base = -2, exponent = 6 **2a.** 8^4 **b.** $(-2)^3$ **c.** 10^5 **3a.** 36 **b.** 32 **c.** -125 **d.** -125 **4a.** 1 **b.** $\dfrac{1}{27}$ **c.** 1 **d.** $-\dfrac{1}{32}$ **e.** $-\dfrac{1}{32}$

4.2 Multiplying and Dividing Exponential Expressions

IDEA 1 We can use the definition of a^n, where n is any positive integer, to develop a rule for multiplying exponential expressions. For example, consider the product $5^2 \cdot 5^4$.

Exponential Expressions

$$5^2 \cdot 5^4 = (5 \cdot 5)(5 \cdot 5 \cdot 5 \cdot 5)$$
$$= 5 \cdot 5 \cdot 5 \cdot 5 \cdot 5 \cdot 5$$
$$= 5^6$$

A careful analysis of the above example suggests that the same result could be obtained by keeping the same base and adding the exponents. That is,

$$5^2 \cdot 5^4 = 5^{2+4} = 5^6$$

To Multiply Exponential Expressions Having the Same Base:

1. Keep the same base.
2. Add the exponents.

Or

$$a^m \cdot a^n = a^{m+n}$$

where m and n are integers.

Example 6 Multiply and express each answer with a positive exponent.

a. $5^2 \cdot 5^8$ b. $2^{-6} \cdot 2^{16}$ c. $x^{-5} \cdot x^{-8}$ d. $7 \cdot 7^3 \cdot 7^9$

Solution To multiply exponential expressions having the same base, use the rule $a^m \cdot a^n = a^{m+n}$.

6a. $5^2 \cdot 5^8 = 5^{2+8}$
           $= 5^{10}$

6b. $2^{-6} \cdot 2^{16} = 2^{-6+16}$
             $= 2^{10}$

6c. $x^{-5} \cdot x^{-8} = x^{-5+(-8)}$
              $= x^{-13}$
              $= \dfrac{1}{x^{13}}$

6d. $7 \cdot 7^3 \cdot 7^9 = 7^{1+3+9}$
              $= 7^{13}$ Note: $7 = 7^1$

Practice Problem 5 *Multiply and express each answer with a positive exponent.*

a. $3^2 \cdot 3^5$ b. $x^{-5} \cdot x^{-66}$ c. $7^{-15} \cdot 7^{10}$
d. $b^{-6} \cdot b^{10}$ e. $y^5 \cdot y^9 \cdot y^{15} \cdot y$

IDEA 2

We can develop the rule for dividing exponential expressions in much the same way that we developed the rule for multiplying them. For example,

$$\frac{5^6}{5^2} = \frac{\cancel{5} \cdot \cancel{5} \cdot 5 \cdot 5 \cdot 5 \cdot 5}{\cancel{5} \cdot \cancel{5}} = 5 \cdot 5 \cdot 5 \cdot 5 = 5^4$$

A careful analysis of the above example suggests that the same result could be obtained by keeping the same base and subtracting the exponent in the denominator from the exponent in the numerator. That is,

$$\frac{5^6}{5^2} = 5^{6-2} = 5^4$$

4.2 Multiplying and Dividing Exponential Expressions

> **To Divide Exponential Expressions Having the Same Base:**
>
> 1. Keep the same base.
> 2. Subtract the exponents.
>
> Or
>
> $$\frac{a^m}{a^n} = a^{m-n}$$
>
> where m and n are integers and a is not equal to zero.

Example 7 Divide and express each answer with a positive exponent.

a. $\dfrac{7^8}{7^2}$ b. $\dfrac{7^2}{7^8}$ c. $\dfrac{x^{13}}{x^{-2}}$ d. $\dfrac{b^{-19}}{b}$

Solution To divide using exponential expressions having the same base, use the rule

$$\frac{a^m}{a^n} = a^{m-n}$$

7a. $\dfrac{7^8}{7^2} = 7^{8-2}$ 7b. $\dfrac{7^2}{7^8} = 7^{2-8}$ 7c. $\dfrac{x^{13}}{x^{-2}} = x^{13-(-2)}$
$= 7^6$ $= 7^{-6}$ $= x^{15}$
                           $= \dfrac{1}{7^6}$

7d. $\dfrac{b^{-19}}{b} = b^{-19-1}$
$= b^{-20}$
$= \dfrac{1}{b^{20}}$

Practice Problem 6 *Divide and express each answer with a positive exponent.*

a. $\dfrac{3^{11}}{3^3}$ b. $\dfrac{5^7}{5^7}$ c. $\dfrac{x^{-13}}{x^{-21}}$ d. $\dfrac{a^{-7}}{a}$

IDEA 3 To find the sum or difference of exponential expressions, or the product or quotient of exponential expressions having different bases, first find the value of each expression.

> **To Add or Subtract Exponential Expressions, or Multiply or Divide Exponential Expressions Having Different Bases:**
>
> 1. Find the value of each exponential expression.
> 2. Perform the indicated operation.

Exponential Expressions

Example 8 Find the value of each expression.

a. $\dfrac{3^3}{4^2}$ b. $-3^2 \cdot 2^5$

Solution   8a. $\dfrac{3^3}{4^2} = \dfrac{3 \cdot 3 \cdot 3}{4 \cdot 4}$ 8b. $-3^2 \cdot 2^5 = -9 \cdot 32$
$= -288$
$= \dfrac{27}{16}$

Practice Problem 7 Find the value of each expression.

a. $5^2 - 5^3$ b. $-2^3 \cdot 3^2$ c. $3^2 + 2^3$

4.2 Exercises

Multiply. Express each answer with a positive exponent.

1. $3^2 \cdot 3^9$
2. $5^6 \cdot 5^8$
3. $8^7 \cdot 8^8$
4. $3^5 \cdot 3^8$
5. $4^3 \cdot 4^{-10}$
6. $7^8 \cdot 7^{-2}$
7. $2^{-5} \cdot 2^{-6}$
8. $9 \cdot 9^{-5}$
9. $n^{10} \cdot n^{-10}$
10. $y^{-1} \cdot y^{-8}$
11. $m \cdot m^6$
12. $p^0 \cdot p^9$
13. $6^{-10} \cdot 6^3$
14. $2^2 \cdot 2^5 \cdot 2^3$
15. $x^4 \cdot x^5 \cdot x^{10}$

Divide. Express each answer with a positive exponent.

16. $\dfrac{5^{15}}{5^6}$
17. $\dfrac{5^6}{5^{15}}$
18. $\dfrac{7^{18}}{7^2}$
19. $\dfrac{3^5}{3^{11}}$
20. $\dfrac{x^{11}}{x^2}$
21. $\dfrac{3^{-3}}{3^5}$
22. $\dfrac{2^{-8}}{2}$
23. $\dfrac{a^2}{a^{-19}}$
24. $\dfrac{n^{-8}}{n^{-2}}$
25. $\dfrac{m^{-51}}{m^{-10}}$
26. $\dfrac{y^{10}}{y^{35}}$
27. $\dfrac{x^{-14}}{x^2}$
28. $\dfrac{t^5}{t^5}$
29. $\dfrac{7^9}{7^{15}}$
30. $\dfrac{x^{-9}}{x^{-10}}$
31. $\dfrac{n^{-5}}{n^{-7}}$
32. $\dfrac{n^{-8}}{n}$
33. $\dfrac{x^{-5}}{x^{-5}}$

Simplify.

34. $\dfrac{x^{-8}}{x^{-8}}$
35. $2^5 \cdot 2^{-8}$
36. $5^6 \cdot 5^{-9}$
37. $\dfrac{2^{-3}}{2^{-8}}$
38. $\dfrac{3^{-5}}{3^{-15}}$
39. $\dfrac{x^8}{x^{15}}$
40. $\dfrac{n^{15}}{n^{21}}$
41. $n^{-8} \cdot n^{-12}$
42. $x^{-6} \cdot x^{-10}$
43. $\dfrac{7^5}{7^{-10}}$
44. $\dfrac{5^7}{5^{-18}}$
45. $\dfrac{x^{-8}}{x^{-12}}$
46. $\dfrac{b^{-3}}{b^{-9}}$
47. $x^{-6} \cdot x^{10}$
48. $x^{-7} \cdot x^{15}$
49. $\dfrac{x^{-65}}{x^{-10}}$
50. $\dfrac{n^{-75}}{n^{-3}}$

Fill in the blanks.

51. To multiply exponential expressions having the same base, the _____ remains the same and we _____ the exponents.

52. To divide using exponential expressions having the same base, the _____ remains the same and we _____ the exponents. Always write the exponent in the _____ minus the exponent in the _____.

53. To find the product or quotient of exponential expressions having different bases, compute the value of each _____ and then do the indicated _____.

Find the value of each expression.

54. $3^2 \cdot 2^3$
55. $7^0 \cdot 7^2$
56. $3^2 \cdot 2^4$
57. $2^{-3} + 2^{-3}$
58. $\dfrac{3^4}{5^3}$
59. $-8^0 + 8^2$
60. $\dfrac{-3^4}{(-2)^6}$
61. $3^{-2} \cdot 6^2$
62. $-3^2 + (-3)^2$

Answers to Practice Problems   **5a.** 3^7 **b.** $\dfrac{1}{x^{71}}$ **c.** $\dfrac{1}{7^5}$ **d.** b^4 **e.** y^{30} **6a.** 3^8 **b.** 1 **c.** x^8 **d.** $\dfrac{1}{a^8}$
7a. -100 **b.** -72 **c.** 17

4.3 Raising Products, Quotients, and Powers to Higher Powers

IDEA 1   The next topic we will develop involves raising an exponential expression to a power. For example,

$$(2^3)^2 = 2^3 \cdot 2^3 = 2^{3+3} = 2^6$$

Notice that the same result could be obtained by keeping the same base and multiplying the exponents. That is,

$$(2^3)^2 = 2^{3 \cdot 2} = 2^6$$

> **To Raise an Exponential Expression to a Power:**
>
> 1. Keep the same base.
> 2. Multiply the exponents.
>
> Or
> $$(a^m)^n = a^{m \cdot n}$$
>
> where m and n are integers.

Example 9   Simplify and express each answer with a positive exponent.

     a. $(5^2)^9$ **b.** $(3^2)^{-8}$ **c.** $(x^{-5})^{-3}$ **d.** $(x^{-2})^3$

Solution   To raise an exponential expression to a power, use the rule

148 *Exponential Expressions*

$$(a^m)^n = a^{m \cdot n}$$

9a. $(5^2)^9 = 5^{2 \cdot 9}$
$ = 5^{18}$

9b. $(3^2)^{-8} = 3^{2(-8)}$
$\phantom{(3^2)^{-8}} = 3^{-16}$
$\phantom{(3^2)^{-8}} = \dfrac{1}{3^{16}}$

9c. $(x^{-5})^{-3} = x^{(-5)(-3)}$
$\phantom{(x^{-5})^{-3}} = x^{15}$

9d. $(x^{-2})^3 = x^{-2 \cdot 3}$
$\phantom{(x^{-2})^3} = x^{-6}$
$\phantom{(x^{-2})^3} = \dfrac{1}{x^6}$

Practice Problem 8 **Simplify and express each answer with a positive exponent.**

a. $(x^{11})^2$ **b.** $(8^5)^{-2}$ **c.** $(x^{-7})^5$ **d.** $(13^{-2})^{-7}$

IDEA 2

Let us now develop a rule for raising a product to a power. For example,

$$(2 \cdot 5)^3 = (2 \cdot 5)(2 \cdot 5)(2 \cdot 5)$$
$$= 2 \cdot 5 \cdot 2 \cdot 5 \cdot 2 \cdot 5$$
$$= 2 \cdot 2 \cdot 2 \cdot 5 \cdot 5 \cdot 5$$
$$= 2^3 \cdot 5^3$$

In the above example, notice that the same result could be obtained by raising each factor inside the parentheses (2 and 5) to the third power. That is, $(2 \cdot 5)^3 = 2^3 \cdot 5^3$.

To Raise a Product to a Power:

1. Raise each factor to that power.
2. Simplify the resulting expression.

Or

$$(a \cdot b)^n = a^n \cdot b^n$$

where *n* is any integer.

Example 10 Simplify.

a. $(3 \cdot 8)^9$ **b.** $(4x)^2$ **c.** $(-3xyz)^3$

Solution   To raise a product to a power, use the formula $(a \cdot b)^n = a^n \cdot b^n$. Also, if variables (or a number and a variable) appear next to each other, the operation indicated is multiplication. For example, $4x = 4 \cdot x$.

10a. $(3 \cdot 8)^9 = 3^9 \cdot 8^9$

10b. $(4x)^2 = 4^2 x^2$
$ = 16x^2$

10c. $(-3xyz)^3 = (-3)^3 x^3 y^3 z^3$
$ = -27x^3 y^3 z^3$

Practice Problem 9 **Simplify.**

a. $(3 \cdot 7)^{15}$ **b.** $(-4x)^3$ **c.** $(3xy)^4$

IDEA 3

The last rule for exponential expressions that we will develop involves raising a fraction to a power. Consider the following example.

$$\left(\frac{3}{4}\right)^3 = \frac{3}{4} \cdot \frac{3}{4} \cdot \frac{3}{4}$$
$$= \frac{3 \cdot 3 \cdot 3}{4 \cdot 4 \cdot 4}$$
$$= \frac{3^3}{4^3}$$

In the above example notice that the same result could be obtained by raising the numerator and denominator of the fraction to the third power. That is,

$$\left(\frac{3}{4}\right)^3 = \frac{3^3}{4^3} = \frac{27}{64}$$

To Raise a Fraction to a Power:

1. Raise the numerator and denominator to that power.
2. Simplify the resulting expression.

Or

$$\left(\frac{a}{b}\right)^n = \frac{a^n}{b^n}, \quad b \neq 0$$

where n is any integer.

Example 11 Simplify.

a. $\left(\frac{2}{3}\right)^3$ b. $\left(\frac{x}{5}\right)^9$ c. $\left(\frac{-3}{x}\right)^3$

Solution   To raise a fraction to a power, use the formula $\left(\frac{a}{b}\right)^n = \frac{a^n}{b^n}$.

11a. $\left(\frac{2}{3}\right)^3 = \frac{2^3}{3^3}$ **11b.** $\left(\frac{x}{5}\right)^9 = \frac{x^9}{5^9}$ **11c.** $\left(\frac{-3}{x}\right)^3 = \frac{(-3)^3}{x^3}$
$\qquad\qquad\quad = \frac{8}{27}$ $\qquad\qquad\qquad\qquad\qquad\qquad\qquad\qquad\quad = \frac{-27}{x^3}$
$\qquad\qquad\qquad\qquad\qquad\qquad\qquad\qquad\qquad\qquad\qquad\qquad\quad = -\frac{27}{x^3}$

Practice Problem 10 *Simplify.*

a. $\left(\frac{3}{5}\right)^2$ b. $\left(\frac{1}{n}\right)^{35}$ c. $\left(\frac{m}{-5}\right)^3$

IDEA 4

To simplify some expressions, we will need to use more than one rule for exponential expressions.

Example 12 Simplify.

a. $(-3x^3)^2$ b. $\dfrac{5^{10} \cdot 5^5 \cdot 5^9}{5^{25}}$ c. $\left(\dfrac{5}{x}\right)^{-3}$

Solution

12a. $(-3x^3)^2 = (-3)^2(x^3)^2$
$= 9x^6$

Rule
$(a \cdot b)^n = a^n \cdot b^n$
$(a^n)^m = a^{n \cdot m}$

12b. $\dfrac{5^{10} \cdot 5^5 \cdot 5^9}{5^{25}} = \dfrac{5^{24}}{5^{25}}$
$= 5^{-1}$
$= \dfrac{1}{5}$

12c. $\left(\dfrac{5}{x}\right)^{-3} = \dfrac{1}{\left(\dfrac{5}{x}\right)^3} = \dfrac{1}{\dfrac{125}{x^3}}$
$= 1 \cdot \dfrac{x^3}{125}$
$= \dfrac{x^3}{125}$

Practice Problem 11 **Simplify.**

a. $\left(\dfrac{2}{5}\right)^{-3}$ b. $(-4x^2y^5)^3$ c. $\left(\dfrac{x^2}{7}\right)^2$

4.3 Exercises

Simplify. The final answer should not contain negative exponents.

1. $(2^3)^9$
2. $(x^5)^{12}$
3. $(y^4)^2$
4. $(7^{15})^2$
5. $(3^{-2})^9$
6. $(x^{-2})^{-4}$
7. $(5^2)^{-8}$
8. $(n^5)^{-1}$
9. $(x^3)^{-9}$
10. $(n^a)^b$
11. $(x^{-a})^{-b}$
12. $(9^{11})^{10}$
13. $(x^2)^{-6}$
14. $(x)^{-9}$
15. $(x^{-2})^{-3}$
16. $(x^{-6})^{-8}$
17. $(2^5)^{16}$
18. $(3^8)^{15}$
19. $(2^{-2})^{-2}$
20. $(3^{-1})^{-2}$

Simplify.

21. $(3 \cdot 5)^7$
22. $(4x)^3$
23. $(5n)^{10}$
24. $(2 \cdot 7)^8$
25. $(-3p)^3$
26. $(-2y)^4$
27. $(4xy)^2$
28. $(13xyz)^2$
29. $(abcd)^8$
30. $(-5xp)^{10}$
31. $(-7x)^2$
32. $(xyz)^n$
33. $\left(\dfrac{2}{3}\right)^4$
34. $\left(\dfrac{x}{5}\right)^2$
35. $\left(\dfrac{3}{n}\right)^4$
36. $\left(\dfrac{a}{b}\right)^3$
37. $\left(\dfrac{-x}{2}\right)^3$
38. $\left(\dfrac{n}{-7}\right)^2$
39. $\left(\dfrac{b}{5}\right)^8$
40. $\left(\dfrac{n}{-3}\right)^4$
41. $\left(\dfrac{x}{y}\right)^n$
42. $\left(\dfrac{1}{x}\right)^5$
43. $\left(\dfrac{n}{-5}\right)^3$
44. $\left(\dfrac{x}{a}\right)^{15}$
45. $(2x)^5$
46. $(3x)^4$
47. $(-2x)^3$
48. $(-5n)^3$
49. $(xyz)^9$
50. $(xyz)^{11}$
51. $(7x)^2$
52. $(-9x)^2$
53. $\left(\dfrac{1}{6}\right)^2$
54. $\left(\dfrac{1}{8}\right)^2$
55. $\left(\dfrac{-n}{3}\right)^4$
56. $\left(\dfrac{-x}{2}\right)^4$
57. $\left(\dfrac{x}{-7}\right)^3$
58. $\left(\dfrac{n}{-8}\right)^3$
59. $\left(\dfrac{1}{5}\right)^{-2}$
60. $\left(\dfrac{1}{10}\right)^{-2}$

Fill in the blanks.

61. To raise a product to a power, raise each _____ to that _____.

62. To raise an exponential expression to a power, the _____ remains the same and we _____ the exponents.

63. To raise a fraction to a power, raise the _____ and the _____ to that _____.

Simplify. The final answer should not contain negative exponents.

64. $(3x^2)^3$
65. $(-2x^2y^3)^4$
66. $(5x^{-2})^3$
67. $(3^a)^{-b}$
68. $\left(\dfrac{2}{3}\right)^{-3}$
69. $\dfrac{(m^2)^4}{(m^3)^5}$
70. $\left(\dfrac{x^2}{-2}\right)^5$
71. $(a^2b^{-4})^n$
72. $\dfrac{a^5 \cdot a^{19} \cdot a^{15}}{(a^{12})^3}$
73. $\dfrac{(x^{-5})^2}{(x^{-3})^{-4}}$
74. $\dfrac{(n^7)^{-3}}{(n^{-4})^{-2}}$
75. $(xy^{-2}z^4)^{-3}$
76. $(a^3b^{-6}c^{-1})$
77. $\left(\dfrac{8x^{-4}}{x^3}\right)^{-2}$
78. $\left(\dfrac{-6x^4}{x^{-2}}\right)^{-2}$
79. $\left(\dfrac{4a^{-2}b}{c^{-2}}\right)^3$
80. $\left(\dfrac{2xy^{-4}}{z^8}\right)^{-4}$
81. $\left(\dfrac{4x^{-2}y^3}{z^{-3}}\right)^{-3}$

Answers to Practice Problems 8a. x^{22} b. $\dfrac{1}{8^{10}}$ c. $\dfrac{1}{x^{35}}$ d. 13^{14} 9a. $3^{15} \cdot 7^{15}$ b. $-64x^3$ c. $81x^4y^4$ 10a. $\dfrac{9}{25}$ b. $\dfrac{1}{n^{35}}$ c. $-\dfrac{m^3}{125}$ 11a. $\dfrac{125}{8}$ b. $-64x^6y^{15}$ c. $\dfrac{x^4}{49}$

4.4 Scientific Notation

IDEA 1 Numbers occurring in science are sometimes very large or very small. Because computations with these types of numbers can be difficult, scientists find it very useful to express these numbers in scientific notation, which involves writing a number using powers of ten. Before considering how to write numbers in scientific notation, it will be helpful to discuss how to multiply a number by a power of 10. First study the series of statements below.

$$3 \cdot 10^3 = 3 \cdot 1000 = 3000$$
$$3 \cdot 10^2 = 3 \cdot 100 = 300$$
$$3 \cdot 10^1 = 3 \cdot 10 = 30$$
$$3 \cdot 10^0 = 3 \cdot 1 = 3$$
$$3 \cdot 10^{-1} = 3 \cdot \dfrac{1}{10} = \dfrac{3}{10} = 0.3$$
$$3 \cdot 10^{-2} = 3 \cdot \dfrac{1}{100} = \dfrac{3}{100} = 0.03$$
$$3 \cdot 10^{-3} = 3 \cdot \dfrac{1}{1000} = \dfrac{3}{1000} = 0.003$$

In general, these statements suggest that multiplying a decimal by a power of ten only changes the position of the decimal point.

152 Exponential Expressions

> **To Multiply a Decimal by a Power of Ten:**
>
> 1. If we multiply by 10^n, where n is a positive integer, the decimal point moves n places to the right.
>
> 2. If we multiply by 10^{-n}, where n is a positive integer, the decimal point moves n places to the left.

Example 13   Find each product mentally.

   a. 6.37×10^4 **b.** 0.0035×10^5

   c. 5.8×10^{-3} **d.** 6.37×10^{-4}

Solution   If n is any positive integer, multiplying a number by 10^n moves the decimal point n places to the right. Multiplying a number by 10^{-n} moves the decimal point n places to the left.

   13a. $6.37 \times 10^4 = 63700$ **13b.** $0.0035 \times 10^5 = 350$

   13c. $5.8 \times 10^{-3} = 0.0058$ **13d.** $6.37 \times 10^{-4} = 0.000637$

Practice Problem 12   *Find each product mentally.*

   a. 5.01×10^3 **b.** 6.345×10^{-2}

   c. 7.385×10^6 **d.** 0.365×10^{-5}

IDEA 2

The products in Examples 13 a, c, and d are written in scientific notation. In other words, a number is in **scientific notation** when it is written as a number that is at least 1 but less than 10 (one nonzero digit is to the left of the decimal point) times a power of ten.

Until now, we have used "·" to mean multiplication. In scientific notation, however, products are expressed using the "×" symbol. For example,

$$\begin{array}{rl} \textit{Decimal notation} & \textit{Scientific notation} \\ 365 = & 3.65 \times 10^2 \\ 0.0365 = & 3.65 \times 10^{-2} \\ 485{,}000{,}000 = & 4.85 \times 10^8 \\ 0.00000485 = & 4.85 \times 10^{-6} \end{array}$$

By analyzing the above results we can see that when numbers greater than 10 are converted to scientific notation, the exponent is a positive integer that indicates the number of places the decimal point was moved to obtain a number that is at least 1 but less than 10.

$$365 = \underset{\substack{\uparrow \\ \text{number} \\ \text{between} \\ 1 \text{ and } 10}}{3.65} \times \underset{\substack{\uparrow \\ 10^n}}{10^2}$$

The 2 indicates the number of places the decimal point in 365 was moved to obtain 3.65.

4.4 Scientific Notation

Similarly, when positive numbers less than 1 are converted to scientific notation, the exponent is a negative integer that indicates the number of places the decimal point was moved to obtain a number that is at least 1 but less than 10.

$$0.0365 = 3.65 \times 10^{-2}$$

The 2 indicates the number of places the decimal point in 0.0365 was moved to obtain 3.65.

number between 1 and 10 × 10^{-n}

Let us now summarize the above statements.

To Express a Decimal in Scientific Notation:

1a. If the decimal is 10 or more, it is written as

$$M \times 10^n$$

1b. If the decimal is less than 1, it is written as

$$M \times 10^{-n}$$

In either case M is at least 1 but less than 10, and n is the number of places the decimal point was moved to make M at least 1 but less than 10.

2. Check to be sure that $M \times 10^n$ or $M \times 10^{-n}$ is equal to the original number.

Example 14 Express each number in scientific notation.

a. 855 b. 0.0095

c. 775,000 d. 0.00003801

Solution To write a decimal in scientific notation, move the decimal point until there is only one nonzero digit to the left of the decimal point. Next, multiply this new number M times the appropriate power of 10.

14a. $855 = 8.55 \times 10^n$
$= 8.55 \times 10^2$

14b. $0.0095 = 9.5 \times 10^{-n}$
$= 9.5 \times 10^{-3}$

14c. $775,000 = 7.75 \times 10^n$
$= 7.75 \times 10^5$

14d. $0.00003801 = 3.801 \times 10^{-n}$
$= 3.801 \times 10^{-5}$

Practice Problem 13 **Express each number in scientific notation.**

a. 363 b. 0.0363 c. 17,600,000 d. 0.00000329

IDEA 3 Scientific notation is useful in multiplying or dividing very large or small numbers.

Example 15 Do the indicated operation.

a. (630,000)(0.000003)   **b.** $\dfrac{36{,}000{,}000}{0.0009}$

Solution To multiply (or to divide) very large or small numbers, first express the numbers in scientific notation. Then multiply (divide) the decimals and multiply (divide) the exponential expressions. Express the answer without exponents.

15a. $(630{,}000)(0.000003)$
$= (6.3 \times 10^5)(3 \times 10^{-6})$
$= (6.3 \times 3)(10^5 \times 10^{-6})$
$= 18.9 \times 10^{-1}$
$= 1.89$

15b. $\dfrac{36{,}000{,}000}{0.0009}$
$= \dfrac{3.6 \times 10^7}{9 \times 10^{-4}}$
$= \dfrac{3.6}{9} \times \dfrac{10^7}{10^{-4}}$
$= 0.4 \times 10^{11}$
$= 40{,}000{,}000{,}000$

Practice Problem 14 **Do the indicated operation.**

a. $(75{,}000{,}000)(0.0005)$ **b.** $\dfrac{666{,}600{,}000}{0.011}$

4.4 Exercises

Find each product mentally.

1. 3.72×10^2
2. 3.72×10^{-2}
3. 8.57×10^{-8}
4. 6.7×10^{11}
5. 8.75×10^{-1}
6. 3.25×10^7
7. 0.3×10^6
8. 3.07×10^{-5}
9. 7×10^{-8}
10. -7.35×10^2
11. -6.77×10^{-4}
12. 76.8×10^{10}
13. 8.1×10^4
14. 1.8×10^4
15. 5.35×10^{-4}
16. 7.5×10^{-5}
17. -3.12×10^6
18. -5.61×10^7
19. 8.1×10^{11}
20. 1.8×10^{13}

Express each number in scientific notation.

21. 275
22. 8571
23. 0.000275
24. 0.031
25. 31.8
26. 0.002017
27. 11,000
28. 0.000011
29. 750,000,000
30. 1750
31. 0.0005
32. 7,000,000,000
33. 0.00000619
34. 0.817
35. 378,000,000,000,000,000
36. 875,000,000
37. 625
38. 893
39. 0.0875
40. 0.0832
41. 35,000,000,000
42. 83,000,000
43. 0.000000051
44. 0.0000000032

Fill in the blanks.

45. If you multiply a decimal by 10^n, where n is a positive integer, the decimal point moves _____ places to the _____ .

46. If you multiply a decimal by 10^{-n}, where n is a positive integer, the decimal point moves _____ places to the _____ .

47. When we write a number greater than 10 in scientific notation, we do so in the form $M \times 10^n$. M is at least _____ but less than _____, and n is the number of places the _____ was moved to make M at least _____ but less than _____.

48. When we write a number less than 1 in scientific notation, we do so in the form $M \times 10^{-n}$. M is at least _____ but less than _____, and n is the number of places the _____ was moved to make M at least _____ but less than _____.

Perform the indicated operation.

49. $(0.0044)(7,000)$

50. $(0.000055)(0.00081)$

51. $(32,000)(0.005)(200,000)$

52. $\dfrac{900,000}{0.003}$

53. $\dfrac{0.000016}{40,000}$

54. $\dfrac{0.000091}{0.13}$

55. $(32,000)(2,000)(30,000)$

56. $(24,000)(5,000)(1,000)$

57. $\dfrac{9,000,000}{0.0003}$

58. $\dfrac{16,000,000}{0.004}$

59. $\dfrac{(35,000)(0.005)}{25,000,000}$

60. $\dfrac{(45,000,000)(0.00009)}{0.0405}$

61. $(-0.0000005)(0.00031)$

62. $(-0.0000061)(0.005)$

Solve.

63. The distance from the earth to the sun is 93,000,000 miles. Express this number in scientific notation.

64. The average diameter of the sun is about 865,000 miles. Express this number in scientific notation.

65. Under certain conditions, hydrogen gas weighs 0.00005190 ounces per cubic inch. Express this number in scientific notation.

66. A fly's wing weighs approximately 0.0000044 pounds. Express this number in scientific notation.

67. A cosmic year is approximately 225,000,000 years long. Express this number in scientific notation.

68. Express 6,590,000,000,000,000,000,000 tons—which is the approximate weight of the earth—in scientific notation.

Answers to Practice Problems **12a.** 5010 **b.** 0.06345 **c.** 7,385,000 **d.** 0.00000365 **13a.** 3.63×10^2 **b.** 3.63×10^{-2} **c.** 1.76×10^7 **d.** 3.29×10^{-6} **14a.** 37,500 **b.** 60,600,000,000

Chapter 4 Summary

Important Terms

An **exponential expression** is a number written in the form a^n. a is the **base** and n is the **exponent**. 3^4 is an exponential expression with a base of 3 and an exponent of 4. [Section 4.1/Idea 1]

A number is written in **scientific notation** when it is expressed as a number that is at least 1 but less than 10 times a power of 10. 3.34×10^3 and 6.78×10^{-5} are written in scientific notation. [Section 4.4/Ideas 1–2]

Important Skills

Finding the Value of Exponential Expressions
$a^n = a \cdot a \cdot \cdots \cdot a$
(n factors of a)

To find the value of a^n, use a as a factor n times. [Section 4.1/Idea 2]

$a^{-n} = \dfrac{1}{a^n}$

$a^0 = 1$

To find the value of a^{-n}, write a fraction whose numerator is 1 and whose denominator is a^n. [Section 4.1/Idea 2]

The value of a^0 is 1, if a is not equal to zero. [Section 4.1/Idea 2]

Multiplying Exponential Expressions Having the Same Base

$a^m \cdot a^n = a^{m+n}$

To multiply exponential expressions having the same base, keep the same base and add the exponents. [Section 4.2/Idea 1]

Dividing Exponential Expressions Having the Same Base

$\dfrac{a^m}{a^n} = a^{m-n}$

To divide exponential expressions having the same base, keep the same base and subtract the exponents. [Section 4.2/Idea 2]

Adding or Subtracting Exponential Expressions, or Multiplying and Dividing Exponential Expressions Having Different Bases

To add or subtract exponential expressions or multiply or divide exponential expressions having different bases, find the value of each expression. Then, perform the indicated operation. [Section 4.2/Idea 3]

Raising Powers to a Power

$(a^m)^n = a^{m \cdot n}$

To raise an exponential expression to a power, keep the same base and multiply the exponents. [Section 4.3/Idea 1]

Raising Products to a Power

$(a \cdot b)^n = a^n \cdot b^n$

To raise a product to a power, raise each factor to that power. [Section 4.3/Idea 2]

Raising Fractions to a Power

$\left(\dfrac{a}{b}\right)^n = \dfrac{a^n}{b^n}, \quad b \neq 0$

To raise a fraction to a power, raise the numerator and denominator to that power. [Section 4.3/Idea 3]

Multiplying Decimals by a Power of 10

To multiply a decimal by 10^n, where n is a positive integer, move the decimal point n places to the right. To multiply a decimal by 10^{-n}, where n is a positive integer, move the decimal point n places to the left. [Section 4.4/Idea 1]

Expressing Decimals in Scientific Notation

To express a decimal greater than 10 in scientific notation, write $M \times 10^n$. To express a positive decimal less than 1 in scientific notation write $M \times 10^{-n}$, where M is at least 1 but less than 10 and n is the number of places the decimal point was moved to obtain M. [Section 4.4/Idea 2]

Chapter 4 Review Exercises

Find the value of each exponential expression.

1. 3^4
2. 3^{-4}
3. $(-3)^4$
4. -3^4

Perform the indicated operations and simplify.

5. $(x^{15})(x^{-35})$
6. $\left(\dfrac{11}{-4}\right)^2$
7. $\dfrac{x^{-9}}{x^{-12}}$
8. $(x^3)^{10}$
9. $(-2p)^3$
10. $5^{18} \cdot 5^{35}$
11. $2^{-3} + 2^4$
12. $\dfrac{9}{9^{25}}$
13. $\left(\dfrac{3}{4}\right)^2$
14. $(2^{-3})^{-5}$
15. $(2x)^{15}$
16. $\dfrac{2^3}{3^2}$
17. $(3x^2)^4$
18. $\dfrac{x^5}{x^{-15}}$
19. $\dfrac{(5^2)^3(5^3)^{-5}}{(5^4)^5}$

Find each product mentally.

20. 3.47×10^8
21. 4.06×10^{-7}
22. -2.47×10^3
23. -86.2×10^{-4}

Express each number in scientific notation.

24. 15,000,000
25. 0.000000312
26. 0.000325
27. 870,000,000,000

Solve.

28. The Gross National Product (GNP) in 1972 for the United States was $1,155,200,000,000. Express the GNP for 1972 in scientific notation.

Perform the indicated operation. Express the answer without exponents.

29. $(6 \times 10^{53})(3.1 \times 10^{-50})$
30. $\dfrac{(82,000,000)(0.0003)}{0.02}$
31. $\dfrac{(9000)(800,000)}{(0.0000006)(0.0012)}$

Simplify.

32. $\dfrac{(x^{-2})^3(x^4)}{x^8}$
33. $\left(\dfrac{x^2}{4}\right)^3$
34. $\left(\dfrac{x^3 \cdot x^2}{x^{-8}}\right)^{-2}$
35. $(3^{-2} + 3^{-2})^{-2}$
36. $\left(\dfrac{3x^{-2}y}{z^{-4}}\right)^{-2}$
37. $(-2x^4y^2)^3$
38. $\left(\dfrac{x^{-6}y^3}{x^5y^{-1}}\right)^{-2}$
39. $\left(\dfrac{2}{7}\right)^{-2}$
40. $(x^{-4}y^2)^{-5}$

Name: _____

Class: _____

Chapter 4 Test

Find the value of each expression.
1. 9^2
2. 5^{-3}
3. $(-0.3)^3$
4. -5^0
5. $2^{-2} + 2^{-3}$

Perform the indicated operation.
6. $6^{-8} \cdot 6^6$
7. $x^{-5} \cdot x^{-7} \cdot x^{-10}$
8. $\dfrac{n^{35}}{n^{-7}}$
9. $\dfrac{5^{-30}}{5}$
10. $\left(\dfrac{m^{10}}{m^{38}}\right)^{-2}$
11. $(x^{-3})^5$
12. $(x^{-8})^{-7}$
13. $(-0.2x)^4$
14. $(abc)^8$
15. $\left(\dfrac{x}{4}\right)^3$
16. $\left(\dfrac{n}{-7}\right)^3$
17. $(-5x^2y^3)^3$
18. 2.34×10^5
19. Express 0.000246 in scientific notation.
20. $\dfrac{(36{,}000{,}000)(0.003)}{27{,}000}$

1. _____

2. _____

3. _____

4. _____

5. _____

6. _____

7. _____

8. _____

9. _____

10. _____

11. _____

12. _____

13. _____

14. _____

15. _____

16. _____

17. _____

18. _____

19. _____

20. _____

5 Polynomials

Objectives The objectives for this chapter are listed below along with sample problems for each objective. By the end of this chapter you should be able to find the solutions to the given problems.

1. Identify the like terms of a polynomial (Section 5.1/Idea 2).
   a. $-7x - 7x^2 - 8x + 2$ b. $3x^2 + 3y^2$

2. Combine the like terms of a polynomial (Section 5.1/Idea 2).
   a. $-8x - 9x$ b. $-8x + x + 7$

3. Evaluate a polynomial (Section 5.2/Idea 1).
   a. $6x^2 - 5x$ when $x = -3$ b. $x^3 + 2x^2 + 2$ when $x = 0.2$

4. Add polynomials (Section 5.3/Idea 1).
   a. $(5x - 6) + (x - 7)$ b. $(3y + z - 6x) + (7x - 3y - z)$

5. Subtract polynomials (Section 5.4/Idea 1).
   a. $(4x - 7) - (3x - 8)$ b. $(3.4x^2 - 0.7x - 0.3) - (1.5x + 7x^2 + 1)$

6. Multiply monomials (Section 5.5/Idea 1).
   a. $(-3x^4)(-7x^5y)$ b. $\left(-\frac{6}{34}x^5\right)\left(\frac{51}{56}x\right)$

7. Multiply a monomial and a polynomial (Section 5.5/Idea 2).
   a. $3xy^2(5x^2y + 3xy - 2)$ b. $-0.3x^3(0.2x^2 - 2.1x + 10)$

8. Multiply polynomials (Section 5.5/Idea 3).
   a. $(2x - 3)(x^2 + 5x - 1)$ b. $(x - y)(x^2 + xy + y^2)$

9. Multiply binomials (Section 5.6/Ideas 1–3).
   a. $(2x - 5)(3x + 2)$ b. $(2x + 3)(2x - 3)$ c. $(2x - 1)^2$

10. Divide one monomial by another (Section 5.7/Idea 1).
    a. $\dfrac{-77x^5y}{11x^2z}$ b. $\dfrac{35x^2}{-0.5x^5}$

11. Divide a polynomial by a monomial (Section 5.7/Idea 2).
    a. $\dfrac{10x^6 - 12x^4 + 8x}{2x^3}$ b. $\dfrac{9x^5y^2 - 18x^4y}{15x^4y^2}$

12. Divide one polynomial by another (Section 5.8/Idea 1).
    a. $(12x^8 + x^4 - 6) \div (3 + 4x^4)$ b. $(n^3 - 27) \div (n - 3)$

5.1 Combining Like Terms

IDEA 1 Expressions like $2x$, $-5xy$, n^3, and 3 are called **monomials**. That is, a monomial is a constant (number), a variable, or the product of a constant and one or more variables. In a monomial, only positive integers or zero may appear as exponents, and a variable must not appear in the denominator of the expression.

A monomial or the sum of two or more monomials is called a **polynomial**. For example, $6x^2$ and $3n^2 + 5m - 6$ are polynomials, whereas $3x^{-2} + 5x$ is not a polynomial. In addition, we say that $3x^2 + 7x - 9$ is a **polynomial in x**, and $6x^2 - 10y^2$ is a **polynomial in x and y**.

When a polynomial involves the operation of addition, the parts being added are called **terms**. The numerical factor (number part) of a term is called a **numerical coefficient**.

Example 1 List each term of the polynomial and identify its numerical coefficient.

  a. $6x^2 - 7x + 9$ **b.** $5x^3 - 6x^2 - x + 8$

Solution To identify the terms and numerical coefficients in a polynomial, rewrite the polynomial using only the operation of addition.

1a. $6x^2 - 7x + 9 = 6x^2 + (-7x) + 9$

The terms are $6x^2$, $-7x$, and 9. The numerical coefficients are 6, -7, and 9, respectively.

1b. $5x^3 - 6x^2 - x + 8 = 5x^3 + (-6x^2) + (-x) + 8$

The terms are $5x^3$, $-6x^2$, $-x$, and 8. The numerical coefficients are 5, -6, -1, and 8, respectively. The numerical coefficient of $-x$ is -1 since $-x = -1 \cdot x$.

We can also identify terms and numerical coefficients by remembering that if a "+" or "−" sign appears between two monomials, the "+" or "−" is the sign of the term and the operation is understood to be addition. For example, the terms of the polynomial $5x^2 - 3x$ (think: $5x^2$ plus $-3x$) are $5x^2$ and $-3x$.

Practice Problem 1 *List the terms of each polynomial and identify their numerical coefficients.*

  a. $16x^2 + 7x - 6$ **b.** $13p^3 - p^2 - 8p + 7$
  c. $15x - 9$ **d.** $7x^2 - 9xy + y^2$

Polynomials can be classified by the number of terms they contain. That is, **monomials, binomials,** and **trinomials** are polynomials containing one, two, and three terms, respectively.

Example 2 Tell whether each polynomial is a monomial, binomial, or trinomial.

  a. $6x^2 + 9x - 7$ **b.** $3x^2 - 5$ **c.** $3xy^4$

Solution To classify a polynomial, count the terms.

2a. Since $6x^2 + 9x - 7$ contains three terms, it is a trinomial.

2b. Since $3x^2 - 5$ contains two terms, it is a binomial.

2c. Since $3xy^4$ contains one term, it is a monomial.

Practice Problem 2 *Tell whether each polynomial is a monomial, binomial, or trinomial.*

    a. $-7x^3yz$ **b.** $7x^2 - 9x - 2$ **c.** $6x^2 - 6y^2$

IDEA 2

Two or more terms of a polynomial that contain the same variable(s) with the same exponent(s) are called **like** or **similar terms**. For example, x^2 and $6x^2$ are like terms since they have the same variable factor x^2.

Example 3 List the like terms.

    a. $6x^3 - 8x^2y - 7x^3 + 9x^2y$ **b.** $x^2 + 9 - 16x^2 - 8 + 6x^2$

    c. $6x^2 + 6y^2$

Solution To identify the like terms of a polynomial, remember that like terms contain the same variable(s) raised to the same power.

3a. $6x^3 - 8x^2y - 7x^3 + 9x^2y = 6x^3 + (-8x^2y) + (-7x^3) + 9x^2y$
Like terms: $6x^3, -7x^3$ and $-8x^2y, 9x^2y$

3b. $x^2 + 9 - 16x^2 - 8 + 6x^2 = x^2 + 9 + (-16x^2) + (-8) + 6x^2$
Like terms: $x^2, -16x^2, 6x^2$ and $9, -8$

3c. $6x^2 + 6y^2$
Like terms: none

NOTE: All constant terms are like terms. For example, 5 and 3 are like terms.

Practice Problem 3 *List the like terms.*

    a. $6x^3 - 7x^2 + 9x^3$ **b.** $8xy^2 + 8x^2y - 9xy^2$

    c. $3y^2 - 6xy + 9x^2$ **d.** $5n^2 - 9m^2 + 8m^2 + 9$

Sometimes we can simplify a polynomial by combining or collecting like terms. The rule for collecting like terms is an application of the *distributive law*.

Distributive Law

> Factors are distributed over sums,
>
> or
>
> $a \cdot (b + c) = a \cdot b + a \cdot c$ and $(b + c) \cdot a = b \cdot a + c \cdot a$
>
> where a, b, and c are any numbers.

> **To Combine Like Terms:**
>
> 1. Identify the like terms.
> 2. Find the sum of their numerical coefficients; then multiply this sum times the common variable factor(s) of the like terms.

Example 4   Combine the like terms.

a. $6x + 3x^2 + 2x$ b. $-7x^2 + 9x^2 + 8x^2$

c. $xy - 8xy$ d. $-8n^3 - 15n^3$

e. $0.3x^3 + 0.21x^3$ f. $-\frac{1}{18}y^5 + \frac{7}{10}y^5$

Solution   To combine like terms, find the sum of their numerical coefficients and then multiply this sum times the common variable factor(s) of the like terms.

4a. $6x + 3x^2 + 2x = (6 + 2)x + 3x^2$ Use the distributive law
$ = 8x + 3x^2$ Add

4b. $-7x^2 + 9x^2 + 8x^2 = (-7 + 9 + 8)x^2$
$ = 10x^2$

4c. $xy - 8xy = 1xy - 8xy$ Think: $1xy$ plus $-8xy$
$ = (1 - 8)xy$
$ = -7xy$

4d. $-8n^3 - 15n^3 = (-8 - 15)n^3$
$ = -23n^3$

4e. $0.3x^3 + 0.21x^3 = (0.3 + 0.21)x^3$
$ = 0.51x^3$

4f. $-\frac{1}{18}y^5 + \frac{7}{10}y^5 = \left(-\frac{1}{18} + \frac{7}{10}\right)y^5$

$\phantom{-\frac{1}{18}y^5 + \frac{7}{10}y^5} = \left(-\frac{5}{90} + \frac{63}{90}\right)y^5$ $18 = 2 \cdot 3 \cdot 3$
$10 = 2 \cdot 5$
LCD $= 2 \cdot 3 \cdot 3 \cdot 5 = 90$

$\phantom{-\frac{1}{18}y^5 + \frac{7}{10}y^5} = \frac{58}{90}y^5$

$\phantom{-\frac{1}{18}y^5 + \frac{7}{10}y^5} = \frac{29}{45}y^5$

In Solutions 4c and 4d, the sum of the coefficients is written without a plus sign. That is, $1 + (-8)$ is written $1 - 8$ and $-8 + (-15)$ is written as $-8 - 15$. Remember, if a and b are positive numbers,

$$a + (-b) = a - b \quad \text{and} \quad -a + (-b) = -a - b.$$

Practice Problem 4   **Combine the like terms.**

a. $-6x^5 + 4x^5$ b. $6xy + 9x^2y - 9xy$

c. $\frac{5}{14}p^5 - \frac{3}{4}p - \frac{1}{10}p^5$ d. $-5.3x^3 - 3.32x^3$

IDEA 3   To complete our introduction to polynomials, we discuss the concept of degree. The **degree of a term** containing only one variable is indicated by the exponent of the variable. For example, the degree of the term $8x^3$ is 3. If the term contains more than one variable, the degree of the term is the sum of the exponents of the variables. For example, the degree of the term $-6x^2y$ is 3 since the exponents are 2 and 1, and $2 + 1 = 3$.

Example 5 List each term and identify its degree.

    a. $6x^2 - 7x + 8$ **b.** $6xy^7 - 3x^4y^3z^3$

Solution   **5a.** $6x^2 - 7x + 8$

Term	Degree	
$6x^2$	2	
$-7x$	1	$-7x = -7x^1$
8	0	$8 = 8x^0$

5b. $6xy^7 - 3x^4y^3z^3$

Term	Degree	
$6xy^7$	8	$1 + 7 = 8$
$-3x^4y^3z^3$	10	$4 + 3 + 3 = 10$

The **degree of a polynomial** is defined as the degree of its highest term. Thus the degree of the polynomial $6x^2 - 7x + 8$ is 2. Zero (0) is not assigned a degree. Finally, a polynomial is said to be written in **descending order** when (a) the variables in each term are in alphabetical order, and (b) the terms are arranged in decreasing powers of the first variable. For example, $6x^2 - 7x + 8$ and $x^2 - xy + 2y^2$ are polynomials written in descending order.

Practice Problem 5 List each term and identify its degree. Also, identify the degree of the polynomial.

    a. $7x^2 - 6xy + 8y^2$ **b.** $8p^4 - 6p^3 + 5p + 9$

5.1 Exercises

Tell whether each expression is a polynomial.

1. $6x^2 - 7x + 9$
2. $5x^3y - 6xy^{-2}$
3. $\dfrac{5}{x - 6}$
4. $-5x^2$

List each term of the polynomial and identify its numerical coefficient.

5. $7x^3 - 7x^2 + 8$
6. $8x^3 + 7xy^2 - x^2y + y^3$
7. $-3x^2 - 7x + 5$
8. $15p^5 - 8p^4 + 3p^3 - p + 2$

Tell whether each polynomial is a monomial, binomial, or trinomial.

9. $7x^3 - 7x^2 + 8$
10. $6x^2 - 9$
11. $3x^3y$
12. $7x^3y - 7xy^3$
13. $-6x^5$
14. $x^2 + xy + y^2$

List the like terms.

15. $7x^3 - 6x^2 + 3x^3$
16. $6y^2 - 6xy + 9y^2$
17. $3x^3 - 7x^2 + 5x^3 - x^2$
18. $6x^3 + 6x^2 + 6x$
19. $3xy^3 - 6xy^3 + xy^3$
20. $-5p^4 - 8p^3 + p^4 + 7p^3$
21. $7x^2 + 9 - 6x^2 + 5x - 2$
22. $3xy^3 - 4x^3y - 2xy^3$
23. $5x^3 + 3x^2 - x^3 + 3x + 5 + x^2 - 9$

Combine the like terms.

24. $5x + 3x$
25. $6y^2 - 3y^2$
26. $-5xy - 7xy$
27. $7x^5 - 9x^5$
28. $-9px + 2px$
29. $-7x + x$
30. $0.5n^2 + 1.71n^2$
31. $6x^3y + 6xy^3$
32. $\dfrac{3}{8}x^3 + \dfrac{5}{12}x^3$
33. $6x^3 + 7x + 3x^3$
34. $7n^2 - 18n^2 + 5n$
35. $-7xy + 8y + 9xy$

36. $-3t^4 - 7t^3 - 6t^4$
37. $6x^2 + 9x^2 + 3x^2$
38. $-7y + 9y + 18y$
39. $6xy - 9xy + xy$
40. $-3y^2 - 8y^2 - 2y^2$
41. $0.3x + 0.51x + x^2$
42. $0.3x^2 - 5.71x^2 + 8$
43. $0.5x^3 + 3x^2 - x^3$
44. $-\frac{7}{18}x^2 + \frac{1}{12}x^2 + \frac{3}{4}x$
45. $\frac{5}{24}x^2 - \frac{1}{18}x^2 + \frac{1}{3}x$
46. $-\frac{11}{36}y^2 + \frac{13}{36}y - \frac{1}{66}y^2$
47. $6xy - 9xy$
48. $9xy - 12xy$
49. $0.3x - 0.51x$
50. $0.6y + 0.61y$
51. $\frac{1}{6}x^2y - \frac{1}{8}x^2y$
52. $\frac{3}{8}xy^2 - \frac{3}{10}xy^2$
53. $-9x - 9y + 9x$
54. $-3n - 3n + 3n$
55. $\frac{3}{16}x^2 - \frac{1}{8}y^2 + \frac{5}{24}x^2$
56. $\frac{5}{18}n + \frac{1}{9}m + \frac{7}{36}n$
57. $3.5a - 3.58a$
58. $7.6ay - 7.68ay$
59. $-3.6b + 7.8b$
60. $-6.3b + 9.2b$
61. $\frac{1}{10}x^3 + \frac{3}{15}x^3 + \frac{1}{18}x^3$
62. $\frac{5}{12}x^4 + \frac{3}{18}x^4 + \frac{5}{24}x^4$
63. $0.3x + \frac{1}{5}x$
64. $0.1x + \frac{3}{5}x$
65. $x - \frac{3}{4}x$
66. $n - \frac{5}{6}n$
67. $-5.6y - \frac{1}{2}y$
68. $-3.2x - \frac{3}{10}x$

Determine the degree of each term and the degree of the polynomial.

69. $6x^3 - 7x^2 + x + 9$
70. $3x^3 - 7x^2y + 2x^2y^2$

Fill in the blanks.

71. A _____ is a constant, a variable, or the product of a constant and one or more variables.
72. The sum of two or more monomials is called a _____.
73. When a polynomial involves only the operation of addition, the parts being added are called _____.
74. The numerical part of a term is called its _____.
75. Two or more terms that contain the same variable(s) with the same exponent(s) are called _____ or _____ terms.
76. To combine like terms, find the sum of their _____ and then place the common variable factor(s) to the right of this sum.

Answers to Practice Problems

1a.
Term	Coefficient
$16x^2$	16
$7x$	7
-6	-6

b.
Term	Coefficient
$13p^3$	13
$-p^2$	-1
$-8p$	-8
7	7

c.
Term	Coefficient
$15x$	15
-9	-9

d.
Term	Coefficient
$7x^2$	7
$-9xy$	-9
y^2	1

2a. monomial b. trinomial c. binomial 3a. $6x^3$ and $9x^3$ b. $8xy^2$ and $-9xy^2$ c. none d. $-9m^2$ and $8m^2$ 4a. $-2x^5$ b. $-3xy + 9x^2y$ c. $\frac{9}{35}p^5 - \frac{3}{4}p$ d. $-8.62x^3$

5a.
Term	Degree
$7x^2$	2
$-6xy$	2
$8y^2$	2
Polynomial's degree = 2	

b.
Term	Degree
$8p^4$	4
$-6p^3$	3
$5p$	1
9	0
Polynomial's degree = 4	

5.2 Evaluating Polynomials

IDEA 1 Frequently in mathematics you must find the value of a polynomial for a specific value of a variable. This process is called **evaluating a polynomial** and you can do it by following the steps described below.

> **To Evaluate a Polynomial:**
>
> 1. Replace the variable with its given value.
> 2. Find a value for all exponential expressions.
> 3. Simplify the resulting expression.

Example 6 Evaluate each polynomial for the given value.

a. $6x^2 - 5x$, $x = -2$ **b.** $-x^3 - 4x^2 + 3$, $x = -2$

c. $x^3 + 3x^2 + 3$, $x = 0.3$ **d.** $6x^3 - 5x^2 + 2$, $x = \frac{2}{3}$

Solution To evaluate a polynomial, replace the variable with its given value. Next, find a value for all exponentials. Finally, do all multiplication and then do all addition and/or subtraction.

6a. $6x^2 - 5x = 6(-2)^2 - 5(-2)$
$= 6(4) - 5(-2)$
$= 24 + 10$ *Think:* $6(4) + (-5)(-2)$
$= 34$

6b. $-x^3 - 4x^2 + 3 = -(-2)^3 - 4(-2)^2 + 3$
$= -(-8) - 4(4) + 3$
$= 8 - 16 + 3$ *Think:* $8 + (-16) + 3$
$= -5$

6c. $x^3 + 3x^2 + 3 = (0.3)^3 + 3(0.3)^2 + 3$
$= 0.027 + 3(0.09) + 3$
$= 0.027 + 0.27 + 3$
$= 3.297$

6d. $6x^3 - 5x^2 + 2 = 6\left(\frac{2}{3}\right)^3 - 5\left(\frac{2}{3}\right)^2 + 2$
$= 6\left(\frac{8}{27}\right) - 5\left(\frac{4}{9}\right) + 2$
$= \frac{\overset{2}{\cancel{6}} \cdot 8}{\underset{9}{\cancel{27}}} - \frac{5 \cdot 4}{9} + 2$
$= \frac{16}{9} - \frac{20}{9} + \frac{2}{1}$
$= \frac{16}{9} - \frac{20}{9} + \frac{18}{9}$ *Recall:* $\frac{2}{1} = \frac{2 \cdot 9}{1 \cdot 9} = \frac{18}{9}$
$= \frac{16 - 20 + 18}{9}$
$= \frac{14}{9}$

Practice Problem 6 *Evaluate each polynomial for the given value.*

$\quad$ **a.** $-6x^3 - 3x + 3, x = -1$ $\quad$ **b.** $3x^3 + 2x^2 - 2, x = -0.2$

$\quad$ **c.** $6x^3 - 3x^2 + 1, x = \dfrac{3}{2}$

Example 7 *Evaluate each polynomial for the given value.*

$\quad$ **a.** $2x^2 - 3x - 4.1, x = -0.2$

$\quad$ **b.** $\dfrac{25}{36}x^3 - \dfrac{15}{8}x^2 + \dfrac{1}{18}, x = \dfrac{2}{5}$

Solution

7a. $2x^2 - 3x - 4.1 = 2(-0.2)^2 - 3(0.2) - 4.1$
$\qquad\qquad\qquad\quad = 2(0.04) - 3(-0.2) - 4.1$
$\qquad\qquad\qquad\quad = 0.08 + 0.6 - 4.1$
$\qquad\qquad\qquad\quad = -3.42$

$\quad$ $\begin{array}{rr} 0.08 & -4.10 \\ +0.60 & +0.68 \\ \hline 0.68 & -3.42 \end{array}$

7b. $\dfrac{25}{36}x^3 - \dfrac{15}{8}x^2 + \dfrac{1}{18} = \dfrac{25}{36}\left(\dfrac{2}{5}\right)^3 - \dfrac{15}{8}\left(\dfrac{2}{5}\right)^2 + \dfrac{1}{18}$

$\qquad\qquad\qquad\qquad\quad = \dfrac{25}{36} \cdot \dfrac{8}{125} - \dfrac{15}{8} \cdot \dfrac{4}{25} + \dfrac{1}{18}$

$\qquad\qquad\qquad\qquad\quad = \dfrac{\overset{1}{\cancel{25}} \cdot \overset{2}{\cancel{8}}}{\underset{9}{\cancel{36}} \cdot \underset{5}{\cancel{125}}} - \dfrac{\overset{3}{\cancel{15}} \cdot \overset{1}{\cancel{4}}}{\underset{2}{\cancel{8}} \cdot \underset{5}{\cancel{25}}} + \dfrac{1}{18}$

$\qquad\qquad\qquad\qquad\quad = \dfrac{2}{45} - \dfrac{3}{10} + \dfrac{1}{18}$ $\quad\begin{array}{l} 45 = 3 \cdot 3 \cdot 5 \\ 10 = 2 \cdot 5 \\ 18 = 2 \cdot 3 \cdot 3 \\ \text{LCD} = 2 \cdot 3 \cdot 3 \cdot 5 = 90 \end{array}$

$\qquad\qquad\qquad\qquad\quad = \dfrac{4}{90} - \dfrac{27}{90} + \dfrac{5}{90}$

$\qquad\qquad\qquad\qquad\quad = \dfrac{4 - 27 + 5}{90}$

$\qquad\qquad\qquad\qquad\quad = -\dfrac{18}{90}$

$\qquad\qquad\qquad\qquad\quad = -\dfrac{1}{5}$

5.2 Exercises

Evaluate each polynomial for $x = 2$.

1. $3x^2 - 5x$ $\qquad\qquad\qquad$ **2.** $x^3 - 9x^2$ $\qquad\qquad\qquad$ **3.** $-4x - 7$

Evaluate each polynomial for $x = -3$.

4. $x^3 - 4x^2$ $\qquad\qquad\qquad$ **5.** $6x^2 - 7x + 1$ $\qquad\qquad$ **6.** $-3x^4 + 5$

Evaluate each polynomial for $x = -2$.

7. $6x^3 - 7x^2 + 5$ $\qquad\qquad$ **8.** $3x^2 - 5x + 2$ $\qquad\qquad$ **9.** $5x^3 - 7x - 1$

10. $-3x^3 - 2x^2 - x$

Evaluate each polynomial for $x = 0.3$.

11. $x^2 + 5x$ $\qquad\qquad\qquad$ **12.** $x^3 - 2x^2$ $\qquad\qquad\qquad$ **13.** $3x^2 - 4$

168   Polynomials

Evaluate each polynomial for $x = -0.1$.

14. $7x^3 + 6x^2 + 3x$

15. $3x^2 - 2x + 3.7$

Evaluate each polynomial for $x = \frac{2}{3}$.

16. $6x^2 + 2x$

17. $x^3 - x^2$

18. $12x^3 + 3$

Evaluate each polynomial for $x = -\frac{3}{4}$.

19. $6x^2 - x + 2$

20. $\frac{16}{15}x^3 - \frac{1}{2}x + \frac{1}{10}$

Evaluate each polynomial for the given value.

21. $5x^4 - 6x^3 + 7x^2 - 5x, \ x = -1$

22. $-3x^3 + 4x^2 - 5, \ x = -3$

23. $0.5x^2 - 0.4x + 8.2, \ x = 0.2$

24. $3x^3 - 6x^2 + 4, \ x = \frac{2}{3}$

25. $\frac{14}{15}x^2 - \frac{1}{9}x + \frac{1}{15}, \ x = -\frac{3}{7}$

26. $5x^3 - 0.5x^2 + 3.2x, \ x = -0.2$

Fill in the blanks.

27. When we find the value of a polynomial for a specific value of the variable, this process is called _____ the polynomial.

28. To evaluate a polynomial, replace the _____ with its given value. Next, find a value for all _____. Finally, do all _____ and then do the _____ and/or subtraction.

Solve. Show all work.

29. Suppose that the cost (in dollars) of manufacturing p kitchen tables can be approximated by the polynomial $5p^2 - 400p + 15{,}500$. Find the cost of manufacturing 20 tables by evaluating the polynomial when $p = 20$.

30. The amount of profit (in dollars) for selling 2000 toys at x dollars per toy can be approximated by the polynomial $-400x^2 + 6800x - 12{,}000$. Find the earned profits if all 2000 toys are sold at $10 per toy.

31. The cost in dollars of manufacturing x units of a perfume is $x^3 - 18x^2 + 500x - 50$. Find the cost of manufacturing 20 units of the perfume by evaluating the polynomial when $x = 20$.

32. The consumer demand for a certain brand of blue jeans is $-300p + 16{,}000$ pairs per month, with the price being p dollars per pair. Find the consumer demand for the jeans when the price is $45 per pair.

33. Evaluate $0.005x^2 - 0.35x + 15$ when $x = 50$.

34. Evaluate $0.005x^2 - 0.35x + 15$ when $x = 70$.

35. The cost in dollars of manufacturing x units of cologne is $x^3 - 18x^2 + 500x - 50$. Find the cost of manufacturing 50 units by evaluating the polynomial when $x = 50$.

36. Evaluate the polynomial in problem 35 for $x = 80$ to find the cost of manufacturing 80 units of cologne.

37. Evaluate the polynomial $0.4x^2 - 4.5x + 2$ for $x = -\frac{1}{2}$.

38. Evaluate the polynomial $0.6x^2 - 3.8 + 1$ for $x = -\frac{1}{4}$.

39. Evaluate $\frac{6}{25}x^3 - \frac{3}{10}x^2 + \frac{7}{15}$ for $x = -\frac{5}{3}$.

40. Evaluate $\frac{8}{21}x^2 - \frac{4}{35}x - \frac{7}{10}$ for $x = -\frac{7}{2}$.

Answers to Practice Problems   6a. 12   b. -1.944 c. $\frac{29}{2}$ 7a. 0.53   b. $\frac{31}{72}$

5.3 Adding Polynomials

IDEA 1   Adding polynomials is similar to collecting like terms.

> **To Add Polynomials:**
> 1. Combine the like terms.
> 2. If necessary, rearrange the terms in descending order.

The commutative and associative laws for addition allow us to rearrange the terms of a polynomial.

Example 8   Add.

a. $(5x^2 - 7x + 2) + (x^2 + 9x - 7)$

b. $(7x^2 - 3x - 1) + (3x^3 - 9x^2 + 13x + 7)$

c. $(-5y^2 + 3xy + 7x^2) + (7y^2 - 4xy + 3x^2)$

Solution   To add polynomials, combine the like terms and then arrange the resulting terms in descending order.

8a. $(5x^2 - 7x + 2) + (x^2 + 9x - 7)$
$= (5 + 1)x^2 + (-7 + 9)x + (2 - 7)$ Combine like terms
$= 6x^2 + 2x + (-5)$
$= 6x^2 + 2x - 5$ *Think*: $2x + (-5) = 2x - 5$

8b. $(7x^2 - 3x - 1) + (3x^3 - 9x^2 + 13x + 7)$
$= (7 - 9)x^2 + (-3 + 13)x + (-1 + 7) + 3x^3$
$= -2x^2 + 10x + 6 + 3x^3$
$= 3x^3 - 2x^2 + 10x + 6$ Rearrange the terms

8c. $(-5y^2 + 3xy + 7x^2) + (7y^2 - 4xy + 3x^2)$
$= (-5 + 7)y^2 + (3 - 4)xy + (7 + 3)x^2$
$= 2y^2 + (-1xy) + 10x^2$
$= 10x^2 - xy + 2y^2$

Practice Problem 8   *Add*.

a. $(5x^2 + 7x - 3) + (-6x^2 - 9x - 5)$

b. $(7x^2 + 4x^3 + 9) + (-x^2 - 7x - 11)$

c. $(-5y + 7z - 6x) + (8x - 3y + 6z)$

Example 9   Add.

a. $(0.6x^2 - 6.3x + 0.08) + (0.3x^2 + 2.35x - 3)$

b. $\left(\dfrac{3}{5}x^2 + \dfrac{1}{8}x - \dfrac{7}{10}\right) + \left(\dfrac{1}{5}x^2 + \dfrac{3}{10}x + 2\right)$

Solution

9a. $(0.6x^2 - 6.3x + 0.08) + (0.3x^2 + 2.35x - 3)$
$= (0.6 + 0.3)x^2 + (-6.3 + 2.35)x + (0.08 - 3)$
$= 0.9x^2 + (-3.95x) + (-2.92)$
$= 0.9x^2 - 3.95x - 2.92$

9b. $\left(\frac{3}{5}x^2 + \frac{1}{8}x - \frac{7}{10}\right) + \left(\frac{1}{5}x^2 + \frac{3}{10}x + 2\right)$
$= \left(\frac{3}{5} + \frac{1}{5}\right)x^2 + \left(\frac{1}{8} + \frac{3}{10}\right)x + \left(-\frac{7}{10} + 2\right)$
$= \frac{4}{5}x^2 + \left(\frac{5}{40} + \frac{12}{40}\right)x + \left(-\frac{7}{10} + \frac{20}{10}\right)$ If possible, add the fractions mentally.
$= \frac{4}{5}x^2 + \frac{17}{40}x + \frac{13}{10}$

Practice Problem 9 Add.

a. $(0.31x^3 - 0.5x^2 + 5) + (3.7x^3 - 0.3x^2 - 4.08)$

b. $\left(\frac{5}{14}x^2 + \frac{3}{8}x - 5\right) + \left(\frac{1}{14}x^3 + \frac{1}{10}x^2 - \frac{6}{8}x\right)$

IDEA 2 Polynomials can also be added by using a vertical or column format. This format will be useful when we discuss multiplication of polynomials.

Example 10 Add $(6x^3 - 7x^2 + 5) + (5x^2 - 9x - 7)$.

Solution To add polynomials by using a vertical format, arrange each polynomial in descending order. Next, write the like terms in the same column and then add.

$$\begin{array}{r} 6x^3 - 7x^2 + 5 \\ 5x^2 - 9x - 7 \\ \hline 6x^3 - 2x^2 - 9x - 2 \end{array}$$

Practice Problem 10 Add.

a. $(3x + 5) + (6x - 7)$ b. $(x^5 + 5x - 8) + (x^3 - 7x + 5)$

5.3 Exercises

Arrange each polynomial in descending order.

1. $6x + 5x^2 - 7$
2. $8y - 7z + 8x$
3. $n - 5 + 6n^3 - 7n^2$
4. $x^2y^2 - 9x^3 - 6y^2 + 7xy^2$
5. $6x + 7x^3 - 9x^5 + 7x^4 - 8x^6 + 5$

Add.

6. $(3x + 2) + (5x + 7)$
7. $(3n + 5) + (2n - 7)$
8. $(x - 5) + (2x - 5)$
9. $(-5x + 3) + (2x - 8)$
10. $(7x^2 - 9) + (-6x^2 + 3)$
11. $(3y + 7x) + (-7y - 8x)$
12. $(x^2 - 7x + 1) + (x^2 + 9x + 3)$
13. $(n^3 + 5n^2 - 7) + (-4n^3 - 15n^2 - 9)$

14. $(6x^2 + 7xy + 2y^2) + (x^2 - 4xy - 5y^2)$
15. $(3x - 7y + 8z) + (-5x - 8y + 2z)$
16. $(3x^2 - 7x + 5) + (3x^3 - 7x^2 + 5x - 2)$
17. $(4s^3 - 3s^2 - 2) + (5s^2 + 7s + 2)$
18. $(7xy - 8x^2 + 3y^2) + (xy - 13y^2 + 5x^2)$
19. $(-8ab + 5a^2b + 9b^2) + (2a^2b - 10b^2 + 10ab)$
20. $(5x^3 - 7x^2y + 9xy^2 - y^3) + (3x^2 + 9x^2y - 7xy^2 + y^3)$
21. $(8x^4 + 7x^2 - 8x - 5) + (-3x^3 - 18x^2 - 8x + 2)$
22. $(-8x^2 - 7x - 1) + (-7x^2 - 7x + 5)$
23. $(-3x^3 - 7x^2 + 8x - 9) + (5x^2 + 6x^2 - 5x + 9)$
24. $(3x + 5) + (-7x + 9) + (8x - 5)$
25. $(3c^2 - 4ac + 5a^2) + (c^2 - 8ac) + (5c^2 - 8a^2)$
26. $(0.3x + 2.1) + (0.31x + 3.2)$
27. $(3.5x + 0.5) + (1.8x - 2)$
28. $(0.8x^2 + 0.3x - 0.5) + (0.3x^2 - 0.51x + 3.5)$
29. $(3x^2 - 0.5xy - 3.4y^2) + (0.3x^2 + 4.8xy - 0.41y^2)$
30. $(x^2 + 0.8x - 0.35) + (x^2 - 0.17x + 4)$
31. $(0.19x^2 - 3x + 0.8) + (8x - x^2 - 3.7)$
32. $\left(\frac{3}{5}x + \frac{1}{3}\right) + \left(\frac{1}{5}x + \frac{1}{3}\right)$
33. $\left(\frac{4}{7}x - 9\right) + \left(\frac{15}{21}x + \frac{3}{8}\right)$
34. $\left(x^2 + \frac{5}{12}x - \frac{3}{10}\right) + \left(6x^2 + \frac{1}{18}x + \frac{1}{10}\right)$
35. $\left(\frac{5}{36}x^2 - \frac{1}{18}x - \frac{5}{11}\right) + \left(\frac{7}{66}x^2 + \frac{4}{15}x + 3\right)$
36. $\left(\frac{7}{9}x^3 - \frac{3}{4}x^2 + \frac{4}{9}\right) + \left(\frac{1}{4}x^2 + \frac{3}{8}x + \frac{5}{6}\right)$
37. $\left(\frac{3}{10}x^2 - \frac{7}{10}x + \frac{3}{22}\right) + \left(\frac{1}{10}x^2 + \frac{7}{15}x + \frac{1}{33}\right)$

Fill in the blanks.

38. When the terms of a polynomial are arranged in decreasing powers of a variable, the terms of the polynomial are said to be in _____ order.

39. To write a polynomial containing two or more variables in descending order, we first write the variables of each term in _____ order. Next, arrange the terms in decreasing powers of the _____ variable.

40. To add polynomials, combine the _____ terms and then arrange the terms in _____ order.

Solve.

41. Find the sum of $(2x^2 - 5x + 7)$ and $(3x^2 - 8x + 8)$.

42. Find the sum of $(3x^2 - 6x + 8)$ and $(4x^2 + 8x - 10)$.

43. Add $\left(\frac{3}{4}x^2 - \frac{3}{10}\right)$ and $\left(\frac{5}{6}x^2 - \frac{1}{8}\right)$.

44. Add $\left(\frac{1}{6}x^2 - \frac{7}{15}\right)$ and $\left(\frac{5}{8}x^2 - \frac{1}{6}\right)$.

45. Find the sum of $(8.5x - 1.4y)$ and $(-7.05x + 4.58y)$.

46. Find the sum of $(5.8x - 4.1y)$ and $(-3.06x + 7.85y)$.

47. Add $(3x - 5y + 6z)$ and $(-5.1x + 5y - z)$.

48. Add $(4x - 6y + 7z)$ and $(-6.2x + 6y - z)$.

Answers to Practice Problems   **8a.** $-x^2 - 2x - 8$ **b.** $4x^3 + 6x^2 - 7x - 2$ **c.** $2x - 8y + 13z$ **9a.** $4.01x^3 - 0.8x^2 + 0.92$ **b.** $\frac{1}{14}x^3 + \frac{16}{35}x^2 - \frac{3}{8}x - 5$ **10a.** $9x - 2$ **b.** $x^5 + x^3 - 2x - 3$

5.4 Subtracting Polynomials

IDEA 1   Before stating a rule for subtracting one polynomial from another, we must define the additive inverse of a polynomial. The **additive inverse of a polynomial** is the polynomial that results when you replace the coefficient in each term with its additive inverse. For example, the additive inverse of $2x - 5$ is $-2x + 5$. Similarly, the additive inverse of $x^2 - 7x + 4$ is $-x^2 + 7x - 4$.

Polynomials

> **To Subtract Polynomials:**
>
> 1. Keep the minuend (first polynomial) the same.
> 2. Add the additive inverse of the subtrahend (second polynomial) to the minuend.

Example 11 Subtract.

  a. $(6x^2 - 7x + 5) - (4x^2 - 9x + 8)$

  b. $(5x^3 - 7x^2 + 5) - (8 + x^2 - 6x^3 - 8x)$

Solution To subtract using polynomials, add the additive inverse of the subtrahend.

 11a. $(6x^2 - 7x + 5) - (4x^2 - 9x + 8)$
  $= (6x^2 - 7x + 5) + \boxed{(-4x^2 + 9x - 8)}$ Additive inverse of subtrahend
  $= 2x^2 + 2x - 3$ Combine like terms

 11b. $(5x^3 - 7x^2 + 5) - (8 + x^2 - 6x^3 - 8x)$
  $= (5x^3 - 7x^2 + 5) + (-8 - x^2 + 6x^3 + 8x)$
  $= 11x^3 - 8x^2 - 3 + 8x$
  $= 11x^3 - 8x^2 + 8x - 3$ Rearrange the terms

Practice Problem 11 **Subtract.**

  a. $(6x^2 - 8x + 5) - (x^2 + 7x + 6)$

  b. $(-7y^2 - 5xy - 9x^2) - (5y^2 - 5xy - 10x^2)$

Example 12 Subtract.

  a. $(3.5x^2 - 0.71x - 0.3) - (8x^2 + 0.3x - 0.56)$

  b. $\left(\dfrac{3}{5}x^2 - \dfrac{5}{6}x - \dfrac{1}{3}\right) - \left(\dfrac{1}{5}x^2 - \dfrac{7}{10}x - \dfrac{3}{9}\right)$

Solution **12a.** $(3.5x^2 - 0.71x - 0.3) - (8x^2 + 0.3x - 0.56)$
  $= (3.5x^2 - 0.71x - 0.3) + (-8x^2 - 0.3x + 0.56)$
  $= (3.5 - 8)x^2 + (-0.71 - 0.3)x + (-0.3 + 0.56)$
  $= -4.5x^2 - 1.01x + 0.26$

 12b. $\left(\dfrac{3}{5}x^2 - \dfrac{5}{6}x - \dfrac{1}{3}\right) - \left(\dfrac{1}{5}x^2 - \dfrac{7}{10}x - \dfrac{3}{9}\right)$
  $= \left(\dfrac{3}{5}x^2 - \dfrac{5}{6}x - \dfrac{1}{3}\right) + \left(-\dfrac{1}{5}x^2 + \dfrac{7}{10}x + \dfrac{3}{9}\right)$
  $= \left(\dfrac{3}{5} - \dfrac{1}{5}\right)x^2 + \left(-\dfrac{5}{6} + \dfrac{7}{10}\right)x + \left(-\dfrac{1}{3} + \dfrac{3}{9}\right)$
  $= \dfrac{2}{5}x^2 + \left(-\dfrac{25}{30} + \dfrac{21}{30}\right)x + \left(-\dfrac{3}{9} + \dfrac{3}{9}\right)$
  $= \dfrac{2}{5}x^2 - \dfrac{4}{30}x + 0$
  $= \dfrac{2}{5}x^2 - \dfrac{2}{15}x$

Practice Problem 12 **Subtract.**

  a. $(3.2x^2 - 0.75) - (1.8x^2 - 0.35)$ **b.** $\left(\dfrac{3}{4}x^2 - \dfrac{1}{18}x + 3\right) - \left(\dfrac{5}{4}x^2 + \dfrac{4}{15}x - \dfrac{1}{2}\right)$

IDEA 2   Subtraction can also be done by using a vertical or column format. This format will be useful when we discuss division of polynomials.

Example 13   Subtract.

     a. $(6x - 5) - (8x + 3)$ **b.** Subtract $(4 - 7x + x^2)$ from $(3x^2 - 10x - 9)$

Solution   To subtract polynomials by using a vertical format, first arrange the terms in descending order. Next, write the subtrahend directly below the minuend and align the like terms. Finally, change the sign of each term in the subtrahend and then add.

$$\begin{array}{ll} \textbf{13a.} & 6x - 5 \\ & \ominus \\ & \ominus 8x + 3 \\ \hline & -2x - 8 \end{array} \qquad \begin{array}{l} \textbf{13b.} \quad 3x^2 - 10x - 9 \\ \phantom{\textbf{13b.} \quad} \oplus \ominus \\ \phantom{\textbf{13b.} \quad} \ominus x^2 - 7x + 4 \\ \hline \phantom{\textbf{13b.} \quad} 2x^2 - 3x - 13 \end{array}$$

It may be helpful to draw a circle around the new signs.

Practice Problem 13   **Subtract.**

     a. $(6a - b) - (3a + 4b)$
     b. Subtract $(3n^2 - 7n + 4)$ from $(-n^2 - 8n + 7)$

5.4 Exercises

Find the additive inverse.

1. $2x - 6$
2. $-3x^3 + 5x$
3. $-4x - 5$
4. $6x^3 - 7x^2 + 5x$
5. $x - 3y + 4z$

Subtract.

6. $(3x - 7) - (2x + 7)$
7. $(9x^2 - 6) - (3x^2 - 4)$
8. $(7x - 3y) - (4x - 5y)$
9. $(6a + 9) - (2a - 3)$
10. $(x + 5) - (2x + 8)$
11. $(6x - 7) - (9x - 9)$
12. $(7x^3 - 9x + 3) - (5x^3 + 9x - 9)$
13. $(3x^2 - 8x + 1) - (5x^2 - 9x + 5)$
14. $(6x - 3y + 5) - (2x - 3y - 8)$
15. $(3y^2 - 7xy + 2x^2) - (2y^2 - 7xy - y^2)$
16. $(a^2 - 5a + 2) - (a^2 - 9a - 7)$
17. $(7n^2 - n + 2) - (5 + n - 3n^2)$
18. $(x^3 - 7x^2 + 5) - (3x^2 - 5x + 7)$
19. $(8y^2 + 7y - 8) - (10y^2 + 9y - 10)$
20. $(3 + 5b - 6b^2) - (7 + 3b - 9b^2)$
21. $(6x^3 - 7x^2 + 3) - (3x^3 + 5x^2 - 7)$
22. $(5x^2 - 9x + 5) - (7x^3 + 5x^2 - 9x + 11)$
23. $(xy^2 + 7xy + 3) - (-xy^2 + 9xy - 7)$
24. Subtract $(11a - 9b + 7c)$ from $(9a - 7b + 5c)$
25. Subtract $(7a - 3b + 8c)$ from $(7c - 13b + 5a)$
26. $(0.3x^2 - 7) - (0.2x^2 - 9)$
27. $(4.1x + 0.7) - (3.2x + 3)$
28. $(0.2x^2 - 0.71x - 2) - (0.37x^2 + 3.1x - 5)$
29. Subtract $(7.3x^2 + 0.3x - 0.35)$ from $(3.7x^2 - 0.5x + 0.5)$
30. $(0.3x - 4.1y + 3.1z) - (0.5x + 3.15y + 5z)$
31. $\left(\frac{3}{5}x - \frac{1}{3}\right) - \left(\frac{1}{5}x + \frac{1}{3}\right)$
32. $\left(\frac{5}{9}x^2 - \frac{2}{7}\right) - \left(\frac{2}{9}x^2 + 2\right)$
33. $\left(\frac{3}{8}a^2 - \frac{2}{5}a + \frac{1}{10}\right) - \left(\frac{1}{6}a^2 - \frac{4}{5}a + 2\right)$
34. $\left(\frac{3}{7}x^2 - \frac{7}{9}x + \frac{3}{8}\right) - \left(-\frac{4}{7}x^2 - \frac{1}{15}x + \frac{1}{8}\right)$
35. $\left(\frac{5}{14}x^2 - \frac{1}{12}x + \frac{1}{6}\right) - \left(\frac{1}{10}x^2 - \frac{3}{10}x + \frac{1}{9}\right)$

174 Polynomials

Fill in the blanks.

36. To find the additive inverse of a polynomial, change the _____ of each _____ of the polynomial.

37. To subtract one polynomial from another, add the _____ of the _____ to the _____.

38. To subtract one polynomial from another by using a vertical format, arrange the _____ in descending order. Next, write the subtrahend directly below the _____ and align the _____ terms. Finally, change the _____ of each term in the _____ and then add.

Subtract.

39. $5x + 2$
$3x + 1$

40. $7x + 8$
$5x + 2$

41. $-4x - 3$
$-3x + 3$

41. $-6x - 5$
$-2x + 5$

43. $4.4x^2 - 2.8y + 4.7$
$8.2x^2 + 4.2y - 0.09$

44. $5.5x^2 - 3.9y + 5.8$
$9.1x^2 + 4.5y + 0.08$

Solve.

45. Subtract $(7x^2 + 2x - 4)$ from $(7x^3 - 3x^2 + 7)$.

46. Subtract $(8x^2 + 3x - 4)$ from $(8x^3 - 6x^2 + 9)$.

47. Find the difference between $\left(\frac{1}{2}x^2 - \frac{3}{8}x + 4\right)$ and $\left(\frac{3}{2}x^2 - \frac{1}{10}x + \frac{1}{5}\right)$.

48. Find the difference between $\left(\frac{1}{4}x^2 - \frac{1}{12}x + 5\right)$ and $\left(\frac{5}{4}x^2 - \frac{5}{8}x + \frac{1}{4}\right)$.

Answers to Practice Problems   **11a.** $5x^2 - 15x - 1$ **b.** $x^2 - 12y^2$ **12a.** $1.4x^2 - 0.4$ **b.** $-\frac{1}{2}x^2 - \frac{29}{90}x + \frac{7}{2}$
13a. $3a - 5b$ **b.** $-4n^2 - n + 3$

5.5 Multiplying Polynomials

IDEA 1 We can find the product of two monomials by using the commutative law, the associative law, and the multiplication rule for exponential expressions. For example,

$$(-6x^2)(5x^4) = -6 \cdot 5 \cdot x^2 \cdot x^4$$
$$= -30 \cdot x^6$$
$$= -30x^6$$

From this example, we can see that the commutative and associative laws for multiplication allow us to rearrange our factors so that we can multiply all numerical coefficients together and multiply all exponential expressions together.

> **To Multiply Monomials:**
>
> 1. Multiply the numerical coefficients.
>
> 2. Multiply the exponential expressions.
>
> 3. Multiply the results in steps 1 and 2.

5.5 Multiplying Polynomials

Example 14 Multiply.

   a. $(-9x^5)(4x^3)$ **b.** $(-7xy)(8x^3)(-2y)$

   c. $\left(\dfrac{35}{9}n^5\right)\left(\dfrac{15}{14}n^4\right)$ **d.** $(0.3x^3y^2)(0.2x^2y^4)$

Solution To multiply two or more monomials, multiply the numerical coefficients and then use the multiplication rule for exponential expressions.

14a. $(-9x^5)(4x^3) = -36x^{5+3}$
$= -36x^8$

14b. $(-7xy)(8x^3)(-2y) = 112x^{1+3}y^{1+1}$
$= 112x^4y^2$

14c. $\left(\dfrac{35}{9}n^5\right)\left(\dfrac{15}{14}n^4\right) = \dfrac{\overset{5}{\cancel{35}} \cdot \overset{5}{\cancel{15}}}{\underset{3}{\cancel{9}} \cdot \underset{2}{\cancel{14}}} n^{5+4}$
$= \dfrac{25}{6}n^9$

14d. $(0.3x^3y^2)(0.2x^2y^4) = 0.06x^{3+2}y^{2+4}$
$= 0.06x^5y^6$

Practice Problem 14 *Multiply*.

   a. $(5x^7)(-8x^{10})$ **b.** $(6xy^2)(-7x)$

   c. $(3.1m^4)(0.2m^5)$ **d.** $\left(\dfrac{8}{12}x^5\right)\left(\dfrac{10}{15}x^3y\right)$

IDEA 2 To find the product of a monomial and a polynomial, we will use the distributive law, which states that $a(b + c) = a \cdot b + a \cdot c$. For example,

$$2x(3x + 5) = 2x \cdot 3x + 2x \cdot 5$$

Notice from the above example that to multiply the sum of two or more terms by a monomial, multiply each term inside the parentheses by the monomial.
This procedure is summarized below.

> **To Multiply a Polynomial by a Monomial:**
>
> 1. Multiply each term of the polynomial by the monomial.
> 2. Express the products in step 1 as a sum and then simplify.

Example 15 Multiply.

   a. $5x^2(3x + 5)$ **b.** $3x^4(x^2 - 7x + 5)$

   c. $-5n^2(3n^2 - 7n + 2)$ **d.** $-1.5x^7y(0.3xy - 4)$

Solution To multiply a monomial times a polynomial, use the distributive law.

15a. $5x^2(3x + 5) = 5x^2(3x) + 5x^2(5)$ Distributive law
$= 15x^3 + 25x^2$ Multiply monomials

15b. $3x^4(x^2 - 7x + 5) = 3x^4(x^2) + 3x^4(-7x) + 3x^4(5)$
$= 3x^6 - 21x^5 + 15x^4$

15c. $-5n^2(3n^2 - 7n + 2) = (-5n^2)(3n^2) + (-5n^2)(-7n) + (-5n^2)(2)$
$= -15n^4 + 35n^3 - 10n^2$

15d. $-1.5x^7y(0.3xy - 4) = (-1.5x^7y)(0.3xy) + (-1.5x^7y)(-4)$
$= -0.45x^8y^2 + 6x^7y$

Practice Problem 15 **Multiply.**

a. $5x^3(7x^5 - 8x^2 + 2)$ b. $-3x^2(x^2 - 7x + 2)$

c. $\frac{4}{9}x\left(\frac{5}{7}x^3 - \frac{6}{14}x^2 + \frac{1}{6}\right)$

IDEA 3

The distributive law can also be used to find the product of any two polynomials. For example, to find the product of the polynomials $x + 2$ and $x^2 - 4x + 3$, we will think of $x + 2$ as a single factor and distribute it over $x^2 - 4x + 3$.

$$(x + 2)(x^2 - 4x + 3) = x(x^2 - 4x + 3) + 2(x^2 - 4x + 3)$$

Now use the distributive law to find $x(x^2 - 4x + 3)$ and $2(x^2 - 4x + 3)$ and then combine like terms.

$$(x + 2)(x^2 - 4x + 3) = x(x^2 - 4x + 3) + 2(x^2 - 4x + 3)$$
$$= x^3 - 4x^2 + 3x + 2x^2 - 8x + 6$$
$$= x^3 - 2x^2 - 5x + 6$$

A careful analysis of the above statements suggests that to multiply two polynomials, we multiply each term of the first polynomial by each term of the second polynomial and simplify the resulting expression.

This procedure is summarized below.

To Multiply Polynomials:

1. Multiply each term of the first polynomial by each term of the second polynomial.

2. Express the products in step 1 as a sum and then simplify.

Example 16 **Multiply.**

a. $(x + 3)(x + 4)$ b. $(2x - 5y)(3x + 4y)$

c. $(5x - 3)(x^2 - 7x + 3)$

Solution To multiply one polynomial times another, use the distributive law. Remember that *each term* of the first polynomial must be multiplied by *each term* of the second polynomial.

16a. $(x + 3)(x + 4) = x(x + 4) + 3(x + 4)$ Distributive law
$= x^2 + 4x + 3x + 12$ Multiply
$= x^2 + 7x + 12$ Combine like terms

16b. $(2x - 5y)(3x + 4y) = 2x(3x + 4y) + (-5y)(3x + 4y)$
$= 6x^2 + 8xy - 15xy - 20y^2$
$= 6x^2 - 7xy - 20y^2$

16c. $(5x - 3)(x^2 - 7x + 3) = 5x(x^2 - 7x + 3) + (-3)(x^2 - 7x + 3)$
$= 5x^3 - 35x^2 + 15x - 3x^2 + 21x - 9$
$= 5x^3 - 38x^2 + 36x - 9$

Practice Problem 16 **Multiply.**

    a. $(2x - 7)(3x - 5)$ b. $(4x - 7)(3x^2 + 2x - 5)$

    c. $(2x^2 + y^2)(x^4 - 3x^2y^2 - y^4)$

IDEA 4

We can multiply two polynomials by using a vertical or column format.

Example 17   Multiply.

    a. $(3x + 5)(6x^2 - x + 1)$ b. $(3x^2 - 5x + 1)(2x^2 + 3x - 5)$

Solution   To multiply two polynomials by using a vertical format, multiply every term in the polynomial at the top by every term in the polynomial at the bottom. Then, combine like terms.

17a. $6x^2 - x + 1$
          $3x + 5$
       $\overline{18x^3 - 3x^2 + 3x}$ ⟶ Think: $3x(6x^2 - x + 1)$
         $+ 30x^2 - 5x + 5$ ⟶ $5(6x^2 - x + 1)$
       $\overline{18x^3 + 27x^2 - 2x + 5}$

17b. $3x^2 - 5x + 1$
         $2x^2 + 3x - 5$
       $\overline{6x^4 - 10x^3 + 2x^2}$
            $+ 9x^3 - 15x^2 + 3x$
                   $- 15x^2 + 25x - 5$
       $\overline{6x^4 - x^3 - 28x^2 + 28x - 5}$

Practice Problem 17 **Multiply.**

    a. $(3x - 2)(x^2 - 7x + 5)$ b. $(2x + y - 2)(3x - y + 2)$

5.5 Exercises

Multiply.

1. $(-3x^5)(5x^4)$
2. $(-6x)(7x)$
3. $(5n^5)(-3n^4)$
4. $(5m^2)(-8m^3)$
5. $(-9xy)(5x^3y)$
6. $(-3xy)(-5x)$
7. $(-8x)(-7x^4)$
8. $(5ab)(-7a^3)$
9. $(-3x)(-3y)$
10. $(-9ab^2)(7a^3)$
11. $(2x)(-3x^2)(-5x^3)$
12. $(-5x^2)(-2x^3)(-3y)$
13. $(3.1x^3)(0.3x^5)$
14. $(-0.2x)(-0.4x^5)$
15. $(-5n^5)(2.4n^7)$

16. $(-0.2x^4)(0.2x^3)(-0.2x)$
17. $(-3.7n^3)(0.11n^4)$
18. $\left(\frac{2}{3}x\right)\left(\frac{4}{7}x^3\right)$
19. $\left(\frac{6}{25}x^4\right)\left(-\frac{5}{8}x^3\right)$
20. $\left(-\frac{6}{34}n\right)\left(-\frac{51}{56}n\right)$
21. $(81x^5)\left(-\frac{14}{54}x^2\right)$
22. $\left(-\frac{14}{60}a^2b\right)\left(\frac{66}{77}a^3\right)$
23. $2x(x+5)$
24. $7x^2(5x+4)$
25. $7x^2(x^3 - 7x^2 + 2)$
26. $3xy(7x^2y + 5xy^2 - 3)$
27. $-3x^2(5x^2 - 4x + 2)$
28. $-7np^2(3np - 5np^2 + 3p^4)$
29. $4n^5(3n^2 - 7n - 8)$
30. $-5x(-7x^2 - 7xy + 3y^2)$
31. $0.2x^4(0.3x^2 - 3.1x)$
32. $2.2n^4(0.3n^4 + 5)$
33. $-0.5x(3.1x^2 - 0.2x + 2.1)$
34. $2.1p^4(3.1p^3 - 5.2p^2)$
35. $\frac{4}{9}x^2\left(\frac{5}{9}x - \frac{15}{14}\right)$
36. $-\frac{9}{10}n\left(\frac{1}{2}n^2 - \frac{15}{21}\right)$
37. $(x+3)(x+5)$
38. $(3x^2 + 5)(3x^2 + 5)$
39. $(7x - 2y)(3x - 5y)$
40. $(2x - 3)(3x - 2)$
41. $(x - 2)(x^2 + 5x + 2)$
42. $(a - b)(a^2 + 2ab + b^2)$
43. $(4n - 3)(6n^2 - 7n + 4)$
44. $(3a + 2)(5a^2 - a + 2)$
45. $(7n^2 - 5n + 1)(n^2 + 4n - 7)$
46. $\left(2x - \frac{1}{3}\right)\left(5x - \frac{3}{4}\right)$
47. $(x^2 - 5x + 2)(x^2 + 3x + 1)$
48. $(x + 0.5)(x^2 - 0.2)$

Fill in the blanks.

49. To multiply two or more monomials, multiply the _____. Next, find the product of the _____ and place this value to the right of the _____.

50. To multiply a monomial times a polynomial we use the _____ law.

51. To multiply one polynomial times another, write the products obtained when each term of the first polynomial is _____ by the second polynomial. Next, express these products as a _____, multiply, and then combine the _____ terms.

Perform the indicated operation.

52. $(3xy)(-6x^2y)$
53. $(4xy)(-8x^3y)$
54. $(-6x)(-7y)$
55. $(-8x)(-9y)$
56. $(0.2x^2)(-0.4x^2)$
57. $(0.3x^3)(-0.2x^5)$
58. $\left(-\frac{8}{21}x^4\right)\left(\frac{35}{44}xy\right)$
59. $\left(-\frac{6}{25}x^5\right)\left(\frac{30}{33}xy^2\right)$
60. $3x^2(5x^2 - 7x + 3)$
61. $4y^3(6y^2 - 8y + 5)$
62. $-4x(6x^2 - 7xy + y^2)$
63. $-8x(2x^2 - 6xy + y^2)$
64. $5xy(6x^2y + 3xy - 6)$
65. $6xy(5x^2y + 4xy - 6)$
66. $-6xy^2(-4xy - 5xy^2 - 6y^2)$
67. $-2xy^2(-5xy - 6xy^2 - 3y^2)$
68. $0.3x^2(0.2x^2 - 4.1)$
69. $0.2x^3(0.3x^2 - 3.1)$
70. $2.1b^2(5b^2 - 0.4b + 0.05)$
71. $4.1c^2(6c^2 - 0.3c + 0.06)$
72. $\frac{6}{15}x^3\left(\frac{10}{12}x^2 - 30\right)$
73. $\frac{4}{10}x^4\left(\frac{6}{8}x^2 - 20\right)$
74. $\frac{1}{5}x^2y^2\left(\frac{3}{8}xy - \frac{5}{7}xy^2\right)$
75. $\frac{3}{7}a^2b^2\left(\frac{1}{4}ab - \frac{7}{8}ab^2\right)$
76. $(x - y)(x + y)$
77. $(x + 4)(x - 4)$
78. $(x + y)(x + y)$
79. $(x + 4)(x + 4)$
80. $(x + 3)(x + 3)$
81. $(x + 5)(x + 5)$
82. $(4x - 1)(3x^2 - 5x - 2)$
83. $(3x - 1)(4x^2 - 6x - 3)$
84. $(3x + 0.5)(4x - 0.2)$
85. $(4x + 0.3)(2x - 0.5)$
86. $(x + 2.1)(0.5x^2 - 3.1x + 4)$
87. $(x + 3.1)(0.6x^2 - 4.2x + 5)$
88. $(x^2 + x + 1)(x^2 - x + 1)$
89. $(x^4 + x^2 + 2)(x^4 - x^2 + 1)$
90. $(3x + 2y - 4z)(x - 2y + 3z)$
91. $(2x + 3y - z)(x - 3y + 5z)$
92. $(x - y)(x^2 + xy + y^2)$
93. $(x + y)(x^2 - xy + y^2)$

Solve.

94. Find the product of $(x - 7)$ and $(4x^2 + 3x - 6)$.
95. Find the product of $(x - 2)$ and $(3x^2 + 2x - 7)$.

96. Multiply $3x^2 - 6x + 2$ by $-6x$.

97. Multiply $4x^2 - 5x + 1$ by $-3x$.

98. Find the product of $-0.7x^3$ and $-\frac{1}{2}x$.

99. Find the product of $-0.6x^3$ and $-\frac{3}{5}x$.

Answers to Practice Problems   **14a.** $-40x^{17}$ **b.** $-42x^2y^2$ **c.** $0.62m^9$ **d.** $\frac{4}{9}x^8y$ **15a.** $35x^8 - 40x^5 + 10x^3$ **b.** $-3x^4 + 21x^3 - 6x^2$ **c.** $\frac{20}{63}x^4 - \frac{4}{21}x^3 + \frac{2}{27}x$ **16a.** $6x^2 - 31x + 35$ **b.** $12x^3 - 13x^2 - 34x + 35$ **c.** $2x^6 - 5x^4y^2 - 5x^2y^4 - y^6$ **17a.** $3x^3 - 23x^2 + 29x - 10$ **b.** $6x^2 + xy - 2x - y^2 + 4y - 4$

5.6 Special Products

IDEA 1   Quite often in mathematics you must find the product of two binomials. In Section 5.5 you did this by using this distributive property. For example,

$$(x^2 + 2)(x + 3) = x^2(x + 3) + 2(x + 3)$$
$$= x^3 + 3x^2 + 2x + 6$$

$$(x - 6)(x + 3) = x(x + 3) + (-6)(x + 3)$$
$$= x^2 + 3x - 6x - 18$$
$$= x^2 - 3x - 18$$

If we carefully analyze the methods used to find the product of any two polynomials, we can develop a shortcut procedure for finding the product of two binomials. To do this let us examine the second example above very carefully.

$$\overset{①\quad②\quad③\quad④}{(x - 6)(x + 3)} = x^2 + 3x - 6x - 18$$
$$= x^2 - 3x - 18$$

① The first term is the product of the first terms of the binomials. $(x - 6)(x + 3)$ $\uparrow x^2 \uparrow$

② The second term is the product of the outer terms of the binomials. $(x - 6)(x + 3)$ $\uparrow \quad 3x \quad \uparrow$

③ The third term is the product of the inner terms of the binomials. $(x - 6)\quad(x + 3)$ $\uparrow -6x \uparrow$

④ The last term is the product of the last terms of the binomial. $(x - 6)(x + 3)$ $\uparrow \ -18 \ \uparrow$

Our final step, of course, is to combine like terms.

A summary of this procedure is given in the box on page 180.

The word FOIL will be useful in helping you to remember how to multiply binomials. That is, F means to multiply the first terms, O means to multiply the outer terms, I means to multiply the inner terms, and L means to multiply the last terms. This procedure is sometimes called the **FOIL method** and the outer (or inner) products are called **cross-products**.

> **To Multiply Two Binomials:**
>
> 1. Multiply the first terms of the binomials.
> 2. Multiply the outside terms of the binomials.
> 3. Multiply the inside terms of the binomials.
> 4. Multiply the last terms of the binomials.
> 5. Express the products as a sum and then simplify.
>
> Or
>
> $$(a + b)(c + d) = a \cdot c + a \cdot d + b \cdot c + b \cdot d$$

Example 18 Multiply.

a. $(x + 3)(x + 5)$ b. $(x - 6)(x + 7)$ c. $\left(x - \frac{1}{2}\right)\left(x + \frac{3}{4}\right)$

d. $(x^2 - 3)(x^2 - 4)$ e. $(6x + 5y)(3x - 7y)$

Solution To multiply two binomials, use the FOIL method.

18a. $(x + 3)(x + 5)$
   First term, Last term, First term, Last term

$$= x \cdot x + 5 \cdot x + 3 \cdot x + 5 \cdot 3$$
$$= x^2 + 5x + 3x + 15$$
$$= x^2 + 8x + 15$$

18b. $(x - 6)(x + 7)$
$$= x^2 + 7x - 6x - 42$$
$$= x^2 + x - 42$$

18c. $\left(x - \frac{1}{2}\right)\left(x + \frac{3}{4}\right)$
$$= x^2 + \frac{3}{4}x - \frac{1}{2}x - \frac{3}{8}$$
$$= x^2 + \frac{1}{4}x - \frac{3}{8}$$

18d. $(x^2 - 3)(x^2 - 4)$
$$= x^4 - 4x^2 - 3x^2 + 12$$
$$= x^4 - 7x^2 + 12$$

18e. $(6x + 5y)(3x - 7y)$
$$= 18x^2 - 42xy + 15xy - 35y^2$$
$$= 18x^2 - 27xy - 35y^2$$

With practice, you should be able to multiply two binomials mentally.

Practice Problem 18 *Multiply using the FOIL method.*

a. $(x + 2)(x + 6)$ b. $(x - 4)(x + 6)$ c. $(x + 0.3)(x - 0.2)$

d. $(3x - 2y)(5x + 7y)$ e. $\left(3x + \frac{1}{3}\right)\left(4x + \frac{2}{3}\right)$ f. $(5x - 2)(5x + 2)$

Practice Problem 19 *Multiply mentally. Just write the answer.*

a. $(x + 3)(x + 4)$ b. $(2x - 3)(3x - 2)$ c. $(x + 8)(x - 3)$

IDEA 2 Two special binomial products occur so often that it will be helpful to be able to recognize these products and to memorize the form of the answer. The first product is the square of a binomial. For example,

$$(x + 3)^2 = (x + 3)(x + 3) \quad \text{Definition of } a^n$$
$$= x^2 + 3x + 3x + 9 \quad \text{FOIL method}$$
$$= x^2 + 6x + 9 \quad \text{Combine like terms}$$

In the above example notice that the square of a binomial is a trinomial in which the first term is the square of the first term of the binomial, the second term is twice the product of the two terms of the binomial, and the last term is the square of the last term of the binomial.

This result is summarized below.

To Square a Binomial:

1. Square the first term of the binomial.
2. Add twice the product of the first and last term of the binomial.
3. Add the square of the last term of the binomial.

Or

$$(a + b)^2 = a^2 + 2ab + b^2$$

The square of a binomial written in the form $a^2 + 2ab + b^2$ is called a **perfect square trinomial**.

Example 19 Square each binomial.

    a. $(x + 5)^2$ **b.** $(x - 5)^2$ **c.** $(4x + 3)^2$

    d. $(3x^3 - 2y)^2$ **e.** $(x + 0.1)^2$ **f.** $\left(3x - \frac{1}{4}\right)^2$

Solution To square a binomial, use the formula $(a + b)^2 = a^2 + 2ab + b^2$.

19a. $(x + 5)^2 = x^2 + 2(x)(5) + 5^2$
$$= x^2 + 10x + 25$$

19b. $(x - 5)^2 = x^2 + 2(x)(-5) + (-5)^2$
$$= x^2 - 10x + 25$$

19c. $(4x + 3)^2 = (4x)^2 + 2(4x)(3) + 3^2$
$$= 16x^2 + 24x + 9 \quad \textit{Think: } (4x)^2 = 4^2 \cdot x^2 = 16x^2$$

19d. $(3x^3 - 2y)^2 = (3x^3)^2 + 2(3x^3)(-2y) + (-2y)^2$
$$= 9x^6 - 12x^3y + 4y^2$$

19e. $(x + 0.1)^2 = x^2 + 2(x)(0.1) + 0.1^2$
$$= x^2 + 0.2x + 0.01$$

19f. $\left(3x - \frac{1}{4}\right)^2 = (3x)^2 + 2(3x)\left(-\frac{1}{4}\right) + \left(-\frac{1}{4}\right)^2$
$$= 9x^2 - \frac{3}{2}x + \frac{1}{16}$$

With practice you should be able to square a binomial mentally.

Practice Problem 20   **Square each binomial.**

a. $(x - 6)^2$ b. $(2x + y)^2$ c. $\left(5x - \dfrac{3}{4}\right)^2$

Practice Problem 21 *Square each binomial mentally. Just write the answer.*

a. $(y + 5)^2$ b. $(2x - 3)^2$ c. $\left(x - \dfrac{1}{2}y\right)^2$

IDEA 3

The second special binomial product that occurs frequently in mathematics involves multiplying the sum and the difference of the same two terms. For example,

$$(x + 3)(x - 3) = x^2 - 3x + 3x - 9 \quad \text{FOIL method}$$
$$= x^2 - 9 \quad \text{Combine like terms}$$

In the above example the final product contains only two terms since the sum of the outer and inner products is zero. Thus, the sum times the difference of the same terms is a binomial which equals the square of the first term minus the square of the second term.

To Multiply a Sum and a Difference of the Same Terms:

1. Square the first term.
2. Write a minus sign and then square the last term.

Or

$$(a + b)(a - b) = a^2 - b^2$$

The product of a sum and difference of the same terms written in the form $a^2 - b^2$ is called the **difference of two squares**.

Example 20 Multiply.

a. $(x + 5)(x - 5)$ b. $(7x + 3)(7x - 3)$ c. $(2x - 5y)(2x + 5y)$

Solution To multiply the sum and the difference of the same terms, use the formula $(a + b)(a - b) = a^2 - b^2$.

20a. $(x + 5)(x - 5) = x^2 - 5^2$
$\qquad\qquad\qquad = x^2 - 25$

20b. $(7x + 3)(7x - 3) = (7x)^2 - 3^2$
$\qquad\qquad\qquad\qquad = 49x^2 - 9$

20c. $(2x - 5y)(2x + 5y) = (2x)^2 - (5y)^2$
$\qquad\qquad\qquad\qquad\quad = 4x^2 - 25y^2$

With practice you should be able to multiply the sum and the difference of the same terms mentally.

Practice Problem 22 *Multiply.*

a. $(x + 7)(x - 7)$ b. $\left(3x + \dfrac{1}{2}\right)\left(3x - \dfrac{1}{2}\right)$ c. $(9x - y)(9x + y)$

Practice Problem 23 *Multiply mentally. Just write the answer.*

a. $(3x + 4)(3x - 4)$ b. $(x - 1)(x + 1)$ c. $(y + 0.3)(y - 0.3)$

5.6 Exercises

Multiply using the FOIL method.

1. $(2x + 3)(3x + 2)$
2. $(x - 5)(x + 7)$
3. $(4x - 1)(x + 5)$
4. $(x - 7)(x - 8)$
5. $(5x + 1)(x - 9)$
6. $\left(x + \frac{1}{2}\right)\left(x - \frac{1}{3}\right)$
7. $(x + 0.2)(x - 0.3)$
8. $(3x + y)(2x - 5y)$
9. $(xz - 5)(2xz + 2)$
10. $(x - 3y)(x + 4y)$
11. $(x - 2y)(x + 3y)$
12. $(3x - 5)(2x - 1)$
13. $(2x - 3)(3x - 1)$
14. $(x + 3)(x + 3)$
15. $(x + 4)(x + 4)$
16. $(2x - 7)(3x - 6)$
17. $(3x - 5)(2x - 5)$
18. $(x + 2)(x - 2)$
19. $(x + 5)(x - 5)$
20. $(x - y)(x + 2y)$
21. $(x - y)(x + 3y)$
22. $(5x + 7)(5x - 7)$
23. $(3x + 4)(3x - 4)$
24. $(3x - 2y)(2x - 3y)$
25. $(2x - 5y)(3x - y)$
26. $(x + 5y)(2x - y)$
27. $(5x - y)(x + 3y)$
28. $(7x + 5y)(5x - 7y)$
29. $(6x + 1)(3x - 2)$
30. $(11x - 2)(11x - 5)$
31. $(x - 0.2)(x - 0.2)$
32. $(x - 0.3)(x - 0.3)$
33. $(x + 0.5y)(x - 0.5y)$
34. $\left(x + \frac{3}{4}\right)\left(x - \frac{2}{3}\right)$
35. $\left(x + \frac{2}{3}\right)\left(x - \frac{5}{2}\right)$
36. $\left(\frac{1}{2}x + \frac{1}{3}y\right)\left(\frac{1}{2}x - \frac{1}{3}y\right)$

Square each binomial (see Example 19).

37. $(x + 2)^2$
38. $(x + 5)^2$
39. $(x - 7)^2$
40. $(x - 2y)^2$
41. $(2x + y)^2$
42. $(8x - 3y)^2$
43. $\left(2x + \frac{1}{4}\right)^2$
44. $(x - 0.5)^2$
45. $(5x + 2y)^2$
46. $(x + 8)^2$
47. $(x + 9)^2$
48. $(x - 4)^2$
49. $(x - 3)^2$
50. $(x + 6)^2$
51. $(x + 7)^2$
52. $(x - 2)^2$
53. $(x - 6)^2$
54. $(x + 2y)^2$
55. $(x + 3y)^2$
56. $(x - 7y)^2$
57. $(x - 5y)^2$
58. $(3x - 5y)^2$
59. $(2x - 3y)^2$
60. $(x + 0.1)^2$
61. $(x + 0.2)^2$
62. $\left(3x + \frac{1}{3}\right)^2$
63. $\left(2x + \frac{1}{2}\right)^2$
64. $(2x + 0.1)^2$
65. $(3x + 0.1)^2$
66. $(3x + 6y)^2$
67. $(2x + 7y)^2$
68. $(x^2 + 11)^2$
69. $(x^2 + 13)^2$

Multiply (see Example 20).

70. $(x + 5)(x - 5)$
71. $(x + 3)(x - 3)$
72. $(2x - 3)(2x + 3)$
73. $(5x + 1)(5x - 1)$
74. $(a + b)(a - b)$
75. $(3x^2 - 10)(3x^2 + 10)$
76. $(xy - 1)(xy + 1)$
77. $(ab + 1)(ab - 1)$
78. $(2x + y)(2x - y)$
79. $\left(x + \frac{3}{4}\right)\left(x - \frac{3}{4}\right)$
80. $\left(x + \frac{1}{3}\right)\left(x - \frac{1}{3}\right)$
81. $\left(\frac{x}{2} + 5\right)\left(\frac{x}{2} - 5\right)$

Fill in the blanks.

82. The word FOIL is useful in helping us to multiply two _____. The F means multiply the _____, O means multiply _____, I means multiply the _____, and L means multiply the _____.

83. To square a binomial, we _____ the first term of the binomial, add _____ the product of the first and last term of the binomial, and then add the _____ of the last term of the binomial.

84. To multiply the sum and the difference of the same two terms, we _____ the first term of the binomial, write a _____ sign, and then _____ the last term.

184   Polynomials

Multiply mentally. Just write the answer.

85. $(4x - 3)(5x - 1)$
86. $(x + 8)(x - 6)$
87. $(x + 9)^2$
88. $(x + y)(x - y)$
89. $(x + 2)(x + 4)$
90. $(x - 2y)(x + 2y)$
91. $(3x - 2)^2$
92. $(3x - 1)(3x - 5)$
93. $(x + 2)(x - 2)$
94. $(x - 10)^2$
95. $\left(x - \frac{3}{4}\right)^2$
96. $(2x + 9y)(3x - y)$
97. $\left(x + \frac{1}{2}\right)\left(x - \frac{1}{2}\right)$
98. $(x + 0.2)^2$
99. $(3x + 5y)(3x - 5y)$
100. $(7x - y)(3x + 2y)$
101. $(x + 5)(2x - 7)$
102. $(3x - 4)(3x + 5)$
103. $(x - 8)^2$
104. $\left(\frac{1}{4}x - y\right)\left(\frac{1}{4}x + y\right)$
105. $(3x^3 - y^2)(2x^3 + y^2)$

Answers to Practice Problems   18a. $x^2 + 8x + 12$ b. $x^2 + 2x - 24$ c. $x^2 + 0.1x - 0.06$ d. $15x^2 + 11xy - 14y^2$ e. $12x^2 + \frac{10}{3}x + \frac{2}{9}$ f. $25x^2 - 4$ 19a. $x^2 + 7x + 12$ b. $6x^2 - 13x + 6$ c. $x^2 + 5x - 24$ 20a. $x^2 - 12x + 36$ b. $4x^2 + 4xy + y^2$ c. $25x^2 - \frac{15}{2}x + \frac{9}{16}$ 21a. $y^2 + 10y + 25$ b. $4x^2 - 12x + 9$ c. $x^2 - xy + \frac{1}{4}y^2$ 22a. $x^2 - 49$ b. $9x^2 - \frac{1}{4}$ c. $81x^2 - y^2$ 23a. $9x^2 - 16$ b. $x^2 - 1$ c. $y^2 - 0.09$

5.7 Dividing by Monomials

IDEA 1   We can find the quotient of two monomials by using the quotient rule for exponential expressions and the fact that since

$$\frac{a}{b} \cdot \frac{c}{d} = \frac{a \cdot c}{b \cdot d} \text{ then } \frac{a \cdot c}{b \cdot d} = \frac{a}{b} \cdot \frac{c}{d}$$

For example,

$$\frac{16x^5}{4x^2} = \frac{16 \cdot x^5}{4 \cdot x^2} = \frac{16}{4} \cdot \frac{x^5}{x^2} = 4 \cdot x^3 = 4x^3$$

The above example suggests that the quotient of two monomials is the quotient of the numerical coefficients times the quotient of the exponential expressions.

> **To Divide One Monomial by Another:**
>
> 1. Write the numerical coefficients as a quotient and write the exponential expressions having the same base as a quotient. Also, write any exponential expressions that appear only in the numerator or denominator as a quotient.
>
> 2. Express the quotients in step 1 as a product and then simplify.

Example 21   Divide.

5.7 Dividing by Monomials

a. $\dfrac{-35x^6}{5x^2}$ b. $\dfrac{42x^5y^2}{77x^3y^5}$ c. $\dfrac{n^5}{2n^4m^2}$ d. $\dfrac{-35x^7y}{0.5xz}$

Solution To divide one monomial by another, write the numerical coefficients as a quotient and write the exponential expressions as a quotient. Next, express these quotients as a product and then simplify.

21a. $\dfrac{-35x^6}{5x^2} = \dfrac{-35}{5} \cdot \dfrac{x^6}{x^2}$

$= -7x^4$ *Think:* $\dfrac{x^6}{x^2} = x^{6-2} = x^4$

21b. $\dfrac{42x^5y^2}{77x^3y^5} = \dfrac{\overset{6}{\cancel{42}}}{\underset{11}{\cancel{77}}} \cdot \dfrac{x^5}{x^3} \cdot \dfrac{y^2}{y^5}$

$= \dfrac{6}{11} \cdot x^2 \cdot \dfrac{1}{y^3}$ *Recall:* $\dfrac{y^2}{y^5} = y^{-3} = \dfrac{1}{y^3}$

$= \dfrac{6x^2}{11y^3}$ *Think:* $\dfrac{6}{11} \cdot \dfrac{x^2}{1} \cdot \dfrac{1}{y^3}$

21c. $\dfrac{n^5}{2n^4m^2} = \dfrac{1}{2} \cdot \dfrac{n^5}{n^4} \cdot \dfrac{1}{m^2}$

$= \dfrac{1}{2} \cdot n \cdot \dfrac{1}{m^2}$

$= \dfrac{n}{2m^2}$

21d. $\dfrac{-35x^7y}{0.5xz} = \dfrac{-35}{0.5} \cdot \dfrac{x^7}{x} \cdot \dfrac{y}{z}$

$= -70 \cdot x^6 \cdot \dfrac{y}{z}$

$= \dfrac{-70x^6y}{z}$

$= -\dfrac{70x^6y}{z}$

Your answer is correct if the quotient times the divisor equals the dividend. That is, $\dfrac{-35x^6}{5x^2} = -7x^4$ since $-7x^4 \cdot 5x^2 = -35x^6$.

Practice Problem 24 *Divide.*

a. $\dfrac{-48x^9}{-12x^2}$ b. $\dfrac{35x^5y^2}{91x^8y}$ c. $\dfrac{35x^3}{0.5xy}$

IDEA 2

To find the quotient of a polynomial and a monomial, we will use the definition for addition of fractions. That is, since

$$\dfrac{a}{b} + \dfrac{c}{b} = \dfrac{a+c}{b} \quad \text{then} \quad \dfrac{a+c}{b} = \dfrac{a}{b} + \dfrac{c}{b}$$

For example,

$$\dfrac{16x^3 + 8x^2 + 4x}{2x} = \dfrac{16x^3}{2x} + \dfrac{8x^2}{2x} + \dfrac{4x}{2x}$$
$$= 8x^2 + 4x + 2$$

186 Polynomials

This example suggests that to divide a polynomial by a monomial we divide each term of the polynomial by the monomial.

> **To Divide a Polynomial by a Monomial:**
>
> 1. Divide each term of the polynomial by the monomial.
> 2. Express the quotients in step 1 as a sum and then simplify.

Example 22 Divide.

a. $\dfrac{8x^4 + 12x^3}{2x^2}$ b. $\dfrac{15x^7 - 30x^6 + 40x^2}{10x^4}$

c. $\dfrac{15x^6 - 10x^4 + 25x^2}{-5x^2}$ d. $\dfrac{35x^2y^5 + 45x}{-5xy}$

Solution To divide a polynomial by a monomial, write each term of the polynomial over the monomial. Express these quotients as a sum and then divide.

22a. $\dfrac{8x^4 + 12x^3}{2x^2} = \dfrac{8x^4}{2x^2} + \dfrac{12x^3}{2x^2}$
$= 4x^2 + 6x$

Check: $\dfrac{8x^4 + 12x^3}{2x^2} = 4x^2 + 6x$ since $2x^2(4x^2 + 6x) = 8x^4 + 12x^3$

22b. $\dfrac{15x^7 - 30x^6 + 40x^2}{10x^4} = \dfrac{15x^7}{10x^4} + \dfrac{-30x^6}{10x^4} + \dfrac{40x^2}{10x^4}$
$= \dfrac{3}{2}x^3 - 3x^2 + \dfrac{4}{x^2}$

22c. $\dfrac{15x^6 - 10x^4 + 25x^2}{-5x^2} = \dfrac{15x^6}{-5x^2} + \dfrac{-10x^4}{-5x^2} + \dfrac{25x^2}{-5x^2}$
$= -3x^4 + 2x^2 - 5$

22d. $\dfrac{35x^2y^5 + 45x}{-5xy} = \dfrac{35x^2y^5}{-5xy} + \dfrac{45x}{-5xy}$
$= -7xy^4 - \dfrac{9}{y}$

Practice Problem 25 Divide.

a. $\dfrac{6x^3 - 4x^2 + 8x}{2x}$ b. $\dfrac{8a^6b^6 - 12a^4b + 18b^6}{-6a^2b^2}$

5.7 Exercises

Divide.

1. $\dfrac{-8x^5}{-2x^2}$ 2. $\dfrac{-16n^8}{4n}$ 3. $\dfrac{-6x^5y^5}{9x^2y^2}$

4. $\dfrac{6.5x}{1.5x^3}$ 5. $\dfrac{-21x^2y}{-3xy^2}$ 6. $\dfrac{35ab^3}{91ab}$

7. $\dfrac{n^6 m}{5n^2}$ 8. $\dfrac{66x^8}{0.3x^5}$ 9. $\dfrac{-15xy^5}{5x^3y^3}$

10. $\dfrac{35a^8}{-7a^9}$ 11. $\dfrac{35x^5}{0.5x^2}$ 12. $\dfrac{14a^3b^2}{4ab^3}$

13. $\dfrac{x^7 y^8}{5x^6 y^2}$ 14. $\dfrac{0.75x^6}{-0.5x}$ 15. $\dfrac{55x^4 y}{65x^2}$

16. $\dfrac{100a^3 b^3 c^3}{-10abc}$ 17. $\dfrac{-51x^3}{-68x}$ 18. $\dfrac{85x^3 y}{-17xy}$

19. $\dfrac{6x^5 + 8x^3}{2x^2}$ 20. $\dfrac{9a^3 - 6a^2}{3a}$ 21. $\dfrac{8x^6 + 6x^4 - 4x^2}{2x}$

22. $\dfrac{10x^6 - 70x^4 - 15x^2}{5x^2}$ 23. $\dfrac{25x^8 - 15x^7}{-5x^3}$ 24. $\dfrac{14x^6 - 21x^5 + 35x^3}{-7x^2}$

25. $\dfrac{9a^3 - 6a^2 b^4}{3a^2 b^2}$ 26. $\dfrac{8x^3 y^5 - 4xy}{-4xy^3}$ 27. $\dfrac{24x^3 - 18x^2 + 10x}{6x^2}$

28. $\dfrac{16n^5 - 48n^3 + 72n^2}{12n^3}$ 29. $\dfrac{9n^5 m^2 - 18n^4 m}{15n^4 m^2}$ 30. $\dfrac{80x^7 y^4 - 70x^3 y^6}{-10b^5}$

31. $\dfrac{45a^4 + 15a^3 b^2 - 96b^4}{-3ab}$ 32. $\dfrac{20b^4 - 30b^2 + 10b}{-10b^5}$ 33. $\dfrac{6x^8 y^6 - 8x^6 y^4 + 4x^4 y^2 + 12x^2 y + 18x}{4x^4 y^4}$

Fill in the blanks.

34. To divide one monomial by another, write the _____ coefficients as a quotient and write the _____ expressions as a quotient. Express these quotients as a _____ and then simplify.

35. To divide a polynomial by a monomial, write the quotients obtained when each term of the _____ is divided by the _____. Express these quotients as a _____ and then simplify.

Perform the indicated operation.

36. $\dfrac{-111x^6}{-37x^2}$ 37. $\dfrac{-155x^8}{-31x^2}$ 38. $\dfrac{222x}{-3x^4}$

39. $\dfrac{333x}{-3x^5}$ 40. $\dfrac{8a^3 b}{-2a^2 b}$ 41. $\dfrac{6x^3 y}{-2x^2 y}$

42. $\dfrac{16x^5 y^2}{24x^5 y}$ 43. $\dfrac{12x^8 y^3}{18x^8 y}$ 44. $\dfrac{-15x^9}{0.3x^6}$

45. $\dfrac{-65x^8}{0.5x^2}$ 46. $\dfrac{-7.5x^6}{-1.5x^8}$ 47. $\dfrac{-9.9x^{11}}{-3.3x^{15}}$

48. $\dfrac{77x^2 y}{121xy^2}$ 49. $\dfrac{85a^2 b}{119ab^2}$ 50. $\dfrac{45x^3 y^3}{-35xy}$

51. $\dfrac{55a^3 b^3}{-65ab}$ 52. $\dfrac{57x^2 y^2}{-76x^3 y}$ 53. $\dfrac{90a^2 b^2}{72a^3 b}$

54. $\dfrac{-8.5x^4}{-50x}$ 55. $\dfrac{-6.5x^8}{-50x}$ 56. $\dfrac{36x^5 - 18x^4 + 12x^3}{6x^2}$

57. $\dfrac{15x^6 - 25x^4 + 35x^3}{5x^2}$ 58. $\dfrac{9x^3 - 6x^2}{3x}$ 59. $\dfrac{10x^5 - 8x^3}{2x}$

60. $\dfrac{14x^5 - 28x^2 + 35x}{-7x}$ 61. $\dfrac{15x^6 - 18x^4 + 21x}{-3x}$ 62. $\dfrac{16x^4 - 8x^2 + 4x}{4x^3}$

63. $\dfrac{20a^4 - 8a^2 + 4a}{4a^3}$ 64. $\dfrac{9x^3 - 6x^2 y^4}{3x^2 y^2}$ 65. $\dfrac{6x^4 - 9x^3 y^5}{3x^3 y^3}$

66. $\dfrac{30x^3 - 18x^2 + 15x}{6x^2}$ 67. $\dfrac{40x^4 - 32x^3 + 12x}{8x^3}$ 68. $\dfrac{9x^5 y^2 - 18x^4 y}{15x^4 y^2}$

188   Polynomials

69. $\dfrac{18x^6y^3 - 24x^5y^2}{15x^5y^3}$

70. $\dfrac{45x^4 + 15x^3y^2 - 96x^3}{-3xy}$

71. $\dfrac{50x^5 + 20x^4y^3 - 55x^4}{-5xy}$

72. $\dfrac{x^{15} + x^{13} + x^{11} - x^9}{-x^3}$

73. $\dfrac{x^{16} + x^{14} + x^{12} - x^{10}}{-x^4}$

74. $\dfrac{-9a - 6b + 15c}{-3}$

75. $\dfrac{-6x - 9y + 15z}{-3}$

76. Divide $20x^4 - 30x^2 + 10x$ by $-10x$.

77. Divide $30x^5 - 40x^3 + 20x^2$ by $-5x^6$.

78. Divide $-77x^8$ into $66x^5$.

79. Divide $-55x^9$ into $44x^2$.

80. Divide $13abc^2$ into $39a^3bc^2 - 65ab^2c^3 - 78a^2b^2c^2$.

81. Divide $15xyz^2$ into $45x^3yz^2 - 60xy^2z^3 - 75x^2y^2z^2$.

Answers to Practice Problems   24a. $4x^7$ b. $\dfrac{5y}{13x^3}$ c. $\dfrac{70x^2}{y}$ 25a. $3x^2 + 2x + 4$ b. $-\dfrac{4}{3}a^4b^4 + \dfrac{2a^2}{b} - \dfrac{3b^4}{a^2}$

5.8 Dividing by Polynomials

IDEA 1   When we must find the quotient of two polynomials in which the divisor is not a monomial, we will use a long division format similar to the one used to divide whole numbers. For example, to divide 594 by 22 we write

$$\begin{array}{r} 27 \\ 22\overline{)594} \\ 44 \\ \hline 154 \\ 154 \\ \hline 0 \end{array}$$ which is a shortcut for $$\begin{array}{r} 25 + 2 \\ 20 + 2\overline{)500 + 90 + 4} \\ 500 + 50 \\ \hline 40 + 4 \\ 40 + 4 \\ \hline 0 \end{array}$$

Notice that in the above example we always divided by 20 and multiplied the result times the complete divisor $20 + 2$. That is, we first divided 20 into 500 to obtain the first term of the quotient (25) and then we divided 20 into 40 to obtain the second term of the quotient (2).

This is basically the same procedure used to find the quotient of two polynomials. For example, to divide $(x^2 + 6x + 8)$ by $(x + 2)$ we write

$$\begin{array}{r} x \\ x + 2\overline{)x^2 + 6x + 8} \end{array}$$ Divide x into x^2

$$\begin{array}{r} x \\ x + 2\overline{)x^2 + 6x + 8} \\ x^2 + 2x \end{array}$$ Multiply x times $x + 2$

$$\begin{array}{r} x \\ x + 2\overline{)x^2 + 6x + 8} \\ x^2 + 2x \\ \hline 4x + 8 \end{array}$$ Subtract and bring down the next term

$$\begin{array}{r} x + 4 \\ x + 2\overline{)x^2 + 6x + 8} \\ x^2 + 2x \\ \hline 4x + 8 \\ 4x + 8 \end{array}$$ Divide x into $4x$ and then multiply the result times $x + 2$

Check: $(x^2 + 6x + 8) \div (x + 2) = x + 4$ since $(x + 4)(x + 2) = x^2 + 6x + 8$.

5.8 Dividing by Polynomials

This procedure is summarized below.

> **To Divide One Polynomial by Another:**
>
> 1. Arrange the polynomials in descending powers of a variable.
> 2. Set up the problem in the long division format. Also, write in a zero as the coefficient of any missing term.
> 3. To obtain the first term of the quotient, divide the first term of the divisor into the first term of the dividend.
> 4. Multiply the value obtained in step 3 times each term in the divisor.
> 5. Place the product obtained in step 4 beneath the dividend (align the like terms). Then, subtract and bring down the next term. This polynomial is your new dividend.
> 6. Repeat steps 3–5 until you obtain a remainder of zero or until the degree of the remainder is less than the degree of the divisor.

 To check your answer, multiply the quotient and the divisor and add the remainder. This result must be equal to the dividend.

Example 23 Divide.

a. $(6x^2 + x - 15) \div (3x + 5)$ **b.** $(a^3 + 1) \div (a + 1)$

b. $(4 - 11x + 5x^2) \div (x - 2)$ **d.** $\dfrac{5xy + 6x^2 - 8y^2}{-2y + 3x}$

Solution To divide a polynomial by a polynomial, arrange each polynomial in descending power of a variable. Next, write the polynomials in a long division format and then divide as if you were dividing using whole numbers.

23a.
$$
\begin{array}{r}
2x - 3 \\
3x + 5 \overline{\smash{)}6x^2 + x - 15} \\
\underline{6x^2 + 10x} \\
- 9x - 15 \\
\underline{- 9x - 15} \\
\end{array}
$$

23b.
$$
\begin{array}{r}
a^2 - a + 1 \\
a + 1 \overline{\smash{)}a^3 + 0a^2 + 0a + 1} \\
\underline{a^3 + a^2} \\
- a^2 + 0a \\
\underline{- a^2 - a} \\
a + 1 \\
\underline{a + 1} \\
\end{array}
$$
⟵ Insert zeros as the coefficients of the missing terms

23c.
$$
\begin{array}{r}
5x - 1 \\
x - 2 \overline{\smash{)}5x^2 - 11x + 4} \\
\underline{5x^2 - 10x} \\
- x + 4 \\
\underline{- x + 2} \\
2 \\
\end{array}
$$
⟵ Arrange the dividend in descending powers of x

The quotient is $5x - 1$ with a remainder of 2. The answer can be written as

$$5x - 1, \text{ R2 or } 5x - 1 + \frac{2}{x - 2} \left(\text{quotient} + \frac{\text{remainder}}{\text{divisor}}\right)$$

23d.
$$\begin{array}{r} 2x + 3y \\ 3x - 2y \overline{\smash{\big)}\, 6x^2 + 5xy - 8y^2} \\ \underline{6x^2 - 4xy} \\ 9xy - 8y^2 \\ \underline{9xy - 6y^2} \\ -2y^2 \end{array}$$ ← Arrange the dividend and divisor in descending powers of x

Thus, $\dfrac{5xy + 6x^2 - 8y^2}{-2y + 3x} = 2x + 3y + \dfrac{-2y^2}{-2y + 3x}$

Practice Problem 26 **Divide.**

a. $(x^2 + 6x + 8) \div (x + 4)$ b. $(x^3 - 27) \div (x - 3)$

c. $\dfrac{12x^3 + 10y^3 - 11x^2y + 17xy^2}{3x - 2y}$

5.8 Exercises

Divide.

1. $(x^2 + 5x + 6) \div (x + 2)$
2. $(x^2 + 3x - 10) \div (x - 2)$
3. $(9x^2 - 9x - 11) \div (3x - 5)$
4. $(8x^2 - 22x + 15) \div (2x - 3)$
5. $(3x^2 - 5x - 2) \div (x - 2)$
6. $(2x^2 - 9x - 18) \div (x - 6)$
7. $(10x^2 - xy - 3y^2) \div (y + 2x)$
8. $(24x^2 - xy - 3y^2) \div (y + 3x)$
9. $\dfrac{15x^3 - x^2y - 8xy^2 - 2y^3}{3x + y}$
10. $\dfrac{12x^3 - 10x^2y + 8xy^2 - 3y^3}{2x - y}$
11. $\dfrac{4x^3 + 27x + 31x^2 + 42}{x + 7}$
12. $\dfrac{4x^3 + 12x + 11x^2 + 12}{x + 2}$
13. $(4x^2 - 9y^2) \div (2x + 3y)$
14. $(16x^2 - 25y^2) \div (4x - 5y)$
15. $(x^2 + 3x - 5) \div (x - 3)$
16. $(x^2 + 5x - 12) \div (x - 6)$
17. $(6x^2 - 13x - 6) \div (2x - 3)$
18. $(15x^2 - 11x - 2) \div (3x - 2)$
19. $\dfrac{6a^3 - 28a + 3a^2 + 15}{2a - 3}$
20. $\dfrac{15x^3 - 8x - 31x^2 - 4}{3x - 2}$
21. $(x^2 - y^2) \div (x - y)$
22. $(a^2 - b^2) \div (a + b)$
23. $(x^3 - y^3) \div (x - y)$
24. $(x^3 + y^3) \div (x + y)$
25. $\dfrac{2a^5 - 3a^4 + 5a^2 - 5a - 5}{2a^2 - a - 1}$
26. $\dfrac{2x^5 + x^3 - 3x^2 - 7x + 5}{x^2 + 2x - 1}$
27. $\dfrac{3x^4 + 5x^2 - 2}{2 + x^2}$
28. $\dfrac{2m^4 - 5m^2 + 3}{-1 + m^2}$
29. $(x^2 - 4) \div (x - 2)$
30. $(10a^2 - ab - 3b^2) \div (b + 2a)$
31. $(a^3 - 8) \div (a - 2)$
32. $(3x^2 + 13x - 10) \div (-x - 5)$
33. $\dfrac{6x^3 - 10x^2 - 2x + 2}{2x + 2}$
34. $\dfrac{25x^2 - 4}{5x + 2}$
35. $\dfrac{7 - 9x + 10x^2}{2x + 1}$
36. $\dfrac{a^3 - 1}{a - 1}$
37. $\dfrac{12x^8 + x^4 - 6}{3 + 4x^4}$
38. $\dfrac{x^3 - x^2 + x - 1}{x + 1}$
39. $\dfrac{9t^2 + 3 + 9t}{3t + 1}$
40. $\dfrac{10a^3 + 8a - 16a^2 + 2}{5a^2 - 3a + 1}$

Fill in the blanks.

41. To divide a polynomial by a polynomial, arrange each polynomial in _____ powers of a variable. Next, write the polynomials in a _____ format and then divide as if you were dividing using _____ numbers.

Solve.

42. Divide $x - 2$ into $3x^2 - 5x - 2$.

43. Divide $2x + 3$ into $2x^2 - x - 6$.

44. Divide $a^2 - b^2$ by $a - b$.

45. Divide $a^2 - b^2$ by $a + b$.

Answers to Practice Problems   **26a.** $x + 2$ **b.** $x^2 + 3x + 9$ **c.** $4x^2 - xy + 5y^2 + \dfrac{20y^3}{3x - 2y}$

Chapter 5 Summary

Important Terms

A **monomial** is a constant, a variable, or the product of a constant and a variable. In a monomial, only positive integers or zero may appear as exponents, 3, x^2, and $3x^2$ are monomials. A **polynomial** is a monomial or the sum of two or more monomials. When x is the variable we say the polynomial is a **polynomial in** x.

The quantities being added in a polynomial are called **terms**. $3x^2$, $-5x$, and 4 are the terms of the polynomial in x $3x^2 - 5x + 4$. The **numerical coefficient** is the numerical factor (number part) of a term. 3 is the numerical coefficient of the term $3x^2$. [Section 5.1/Idea 1]

A **binomial** is a polynomial that contains two terms. A **trinomial** is a polynomial that contains three terms. $2x + 5$ is a binomial and $x^2 + 5x - 9$ is a trinomial. [Section 5.1/Idea 1]

Like terms or **similar terms** are terms having the same variable factors. $-3x^2$ and x^2 are like terms. [Section 5.1/Idea 2]

The **distributive law** says that factors are distributed over sums. $a \cdot (b + c) = a \cdot b + a \cdot c$ and $(b + c) \cdot a = b \cdot a + c \cdot a$. [Section 5.1/Idea 2]

The **degree of a term** containing only one variable is the exponent of the variable. The degree of the term $-8x^5$ is 5. The **degree of a polynomial** is the degree of its highest term. The degree of the polynomial $8x^4 - 6$ is 4. A polynomial is said to be written in **descending order** when (a) the variables in each term are in alphabetical order, and (b) the terms are arranged in decreasing powers of the first variable. $7x^3t^4 + 10xt^6$ is a polynomial in descending order. [Section 5.1/Idea 3]

Evaluating a polynomial is the process of finding the value of a polynomial for a specific value of a variable. If we evaluate the polynomial $6x^2 - 5x$ when $x = -2$, we obtain the value -2. [Section 5.2/Idea 1]

The **additive inverse of a polynomial** is the polynomial that results when you replace the coefficient in each term with its additive inverse. The additive inverse of $3x^2 - 2$ is $-3x^2 + 2$. [Section 5.6/Idea 1]

A **perfect square trinomial** is the square of a binomial, $(a + b)^2$, written in the form $a^2 + 2ab + b^2$. $x^2 + 6x + 9$ is a perfect square trinomial. [Section 5.6/Idea 2]

A **difference of two squares** is the product of the sum and difference of the same terms, $(a + b)(a - b)$, written in the form $a^2 - b^2$. $x^2 - 25$ is a difference of two squares. [Section 5.6/Idea 3]

Important Skills

Combining Like Terms   To combine like terms, find the sum of their numerical coefficients and then multiply this sum times the common variable factor(s) of the like terms. [Section 5.1/Idea 2]

Evaluating Polynomials	To evaluate a polynomial, replace the variable with its given value. Next, find a value for all exponential expressions. Finally, simplify the resulting expression. [Section 5.2/ Idea 1]
Adding Polynomials	To add polynomials, combine the like terms and then arrange the resulting terms in descending order. [Section 5.3/Idea 1]
Subtracting Polynomials	To subtract one polynomial from another, add the additive inverse of the subtrahend to the minuend. [Section 5.4/Idea 1]
Multiplying Monomials	To multiply two or more monomials, multiply the numerical coefficients, multiply the exponential expressions, and then multiply the results. [Section 5.5/Idea 1]
Multiplying Monomials Times Polynomials	To multiply a monomial and a polynomial, multiply each term of the polynomial by the monomial. Express the products as a sum and simplify. [Section 5.5/Idea 2]
Multiplying Polynomials	To multiply one polynomial times another, multiply each term of the first polynomial by each term of the second polynomial. Express the products as a sum and simplify. [Section 5.5/Idea 3]
Multiplying Binomials $(a + b)(c + d) = a \cdot c + a \cdot d + b \cdot c + b \cdot d$	To multiply two binomials, compute the product of the first, outer, inner, and last terms of the binomials and then add the results and simplify. [Section 5.6/Idea 1]
Squaring Binomials $(a + b)^2 = a^2 + 2ab + b^2$	To square a binomial, square the first term, add twice the product of the first and last term of the binomial, and add the square of the last term of the binomial. [Section 5.6/Idea 2]
Multiplying the Sum and the Difference of the Same Terms $(a + b)(a - b) = a^2 - b^2$	To multiply the sum times the difference of the same terms, square the first term, write a minus sign, and then square the last term. [Section 5.6/Idea 3]
Dividing Monomials	To divide one monomial by another, write the numerical coefficients as a quotient and write the exponential expressions as a quotient. Next, express these quotients as a product and then simplify. [Section 5.7/Idea 1]
Dividing Polynomials by Monomials	To divide a polynomial by a monomial, divide each term of the polynomial by the monomial. Express these quotients as a sum and simplify. [Section 5.7/Idea 2]
Dividing Polynomials	To divide a polynomial by a polynomial, arrange each polynomial in descending order. Next, write the polynomials in a long division format and then divide as if you were dividing whole numbers. [Section 5.8/Idea 1]

Chapter 5 Review Exercises

Identify the like terms.

1. $6x^2 - 7x - 16x^2 + 3x$
2. $x^3 - 5x^2y - 3x^2 + x^2y$

Combine the like terms.

3. $5x^2 + 3x - 8x^2$
4. $0.4x + 5x^2 + 0.51x$
5. $\frac{5}{8}x^3y + \frac{7}{12}x^3y$
6. $-\frac{5}{36}x^2 + \frac{1}{36}y^2 - \frac{1}{66}x^2$

Evaluate.

7. $-3x^3 + 4x^2 - 5$ when $x = -2$
8. $3x^3 + 6x^2 + 4$ when $x = \frac{2}{3}$

Perform the indicated operations.

9. $\left(\frac{4}{15}x + \frac{1}{10}\right) + \left(\frac{3}{10}x - \frac{11}{12}\right)$
10. $\frac{-35x^4y^2}{7x^2y^4}$
11. $(-7x^5y^2)(5.2xy^3)$
12. $(7x^2 - 8y^2) - (2x^3 - 9y^2)$
13. $(6x^2 - 4x + 2) + (2x^3 - 8x^2 - 4)$
14. $(x^3 - 8) \div (x - 2)$
15. $-0.5x^2(3x^2 - 0.1x + 4)$
16. $\frac{18a^4 - 24a^3 + 15a}{-6a^2}$
17. $(3x - 2)(x^2 - 8x + 3)$
18. $\frac{(-3x^2y)^3}{36xy^3z}$
19. $(-6y^2 - 5xy + 9x^2) - (5y^2 + 5xy + x^2)$

Solve.

20. An efficiency study indicates that a worker at the Pace Clock Company who arrives at work at 7:00 A.M. should have assembled $-x^3 + 5x^2 + 7x$ clocks x hours later. If Howie arrived at work at 7:00 A.M., how many clocks should he have assembled by 10:00 A.M.?

21. The cost in dollars of manufacturing y units of a product is $y^3 + 18y^2 - 500y - 50$. Find the cost of manufacturing 20 units of this product.

22. Evaluate $0.6x^2 - 3.8x - 1$ for $x = -0.25$.

Perform the indicated operation.

23. $(3x - 2)(3x - 5)$
24. $(2x + 5)^2$
25. $(x + 9)(x - 9)$
26. $(x - 0.2)^2$
27. $(x + 8)(x - 9)$
28. $\left(\frac{1}{4}x - 2\right)\left(\frac{1}{4}x + 2\right)$
29. $(x + 0.3)(x - 0.5)$
30. $\left(x + \frac{1}{3}\right)\left(x - \frac{3}{4}\right)$
31. $\left(2x - \frac{1}{2}\right)^2$
32. $\frac{a^3 + 2a^2 - 8a - 35}{a - 5}$
33. $\frac{15x^3y^2 + 6x^2y + 9xy^3}{18x^2y^4}$
34. $-5n - 4n + (n - 6) - (2n - 7)$
35. $5x(8 - 3x)(5 - 2x)$
36. $-8x^2(5x + 2) + 2x(x - 5)$
37. $(x - 5)^2 + (x + 5)^2$
38. $(0.32x^3 - 0.6x^2 + 5) + (3.7x^3 - 0.4x^2 - 4.08)$
39. $\left(\frac{5}{14}x^2 + \frac{3}{8}x^2 - 5\right) - \left(\frac{1}{14}x^3 + \frac{1}{10}x - \frac{2}{3}\right)$
40. $-1.5xy^6(0.4xy - 1.2x^2y^2 + 10)$
41. $\frac{81}{7}x^5 - \frac{14}{54}x^5$
42. $-0.25x^3 + \frac{3}{8}x^3$

Name: _____

Class: _____

Chapter 5 Test

1. Identify the like terms in the polynomial
   $3x^3 + 3x - 6x^3$

Combine the like terms.

2. $0.3x^3 - 5x - 0.51x^3$
3. $-\dfrac{5}{18}x^2 + \dfrac{2}{3}x + \dfrac{1}{10}x^2$

Evaluate each polynomial for the given value.

4. $-x^4 - 4x^2 + 3,\ x = -2$
5. $-6x^3 - 5x^2 + 2,\ x = -\dfrac{2}{3}$

Perform the indicated operation.

6. $(6x^2 - 5x + 2) + (x^2 + 7x - 7)$
7. $(0.7x^2 - 6.1x + 0.08) - (0.3x^2 + 2.34x - 1)$
8. $\left(x^2 - \dfrac{5}{12}x + \dfrac{3}{10}\right) - \left(6x^2 - \dfrac{5}{8}x + 2\right)$
9. $(0.5xy)(8x^3y^2)$
10. $-\dfrac{8}{21}x^6\left(-\dfrac{35}{44}x^7y^2\right)$
11. $0.4x^2(5x^2 - 3x + 20)$
12. $-5x^2(7x^3 - 2xy + 3y^2)$
13. $(2x - 3)(3x + 2)$
14. $(3x - 2)(5x^2 + 3x - 7)$
15. $\dfrac{-45x^8}{5x^2}$
16. $\dfrac{-6.5x^2}{0.05x^5}$
17. $\dfrac{28x^3 - 24x^2 + 16x}{4x}$
18. $\dfrac{40x^4y^5 - 32x^3y^6 + 12x^2y}{-8x^3y^4}$
19. $(3x^3 + 7x^2 + 4x - 5) \div (x + 3)$
20. $\dfrac{x^3 + 1}{x - 1}$

1. _____

2. _____

3. _____

4. _____

5. _____

6. _____

7. _____

8. _____

9. _____

10. _____

11. _____

12. _____

13. _____

14. _____

15. _____

16. _____

17. _____

18. _____

19. _____

20. _____

21. $(2x - 5)^2$
22. $(3x - 5y)(3x + 5y)$
23. $\left(\dfrac{1}{4}x - y\right)\left(\dfrac{1}{3}x + y\right)$
24. $(x + 0.5y)^2$
25. $(x - 0.5)(2x - 0.4)$

21. _____

22. _____

23. _____

24. _____

25. _____

6 Linear Equations and Inequalities in One Variable

Objectives The objectives for this chapter are listed below along with sample problems for each objective. By the end of this chapter you should be able to find the solutions to the given problems.

1. Determine if a given value is the solution of an equation *(Section 6.1/Idea 2)*.
   a. $x - 0.37 = 0.2$, $x = 0.57$ b. $20 = 8 - 2(9 - 3x)$, $x = -3$

2. Solve a linear equation by using the addition principle *(Section 6.2/Idea 1)*.
   a. $x - 3 = 1$ b. $x - 1.1 = -3.35$

3. Solve a linear equation by using the multiplication or division principle *(Section 6.2/Ideas 2–3)*.
   a. $\dfrac{x}{-6} = 3$ b. $15x = 25$

4. Solve a linear equation containing more than one operation *(Section 6.3/Ideas 1–2)*.
   a. $\dfrac{5}{6}x + 2 = 17$ b. $6x - 2 = 8 - 14x$

5. Solve a linear equation containing a grouping symbol *(Section 6.4/Idea 1)*.
   a. $0.2(x - 0.1) = 3$ b. $4x - 6 = 2(x + 5)$

6. Solve a linear equation containing fractions *(Section 6.4/Idea 2)*.
   a. $5x - 3 = \dfrac{1}{10} + \dfrac{3}{4}$ b. $\dfrac{2x}{3} + x = 5$

7. Solve a literal equation *(Section 6.5/Idea 1)*.
   a. $ax + bx = c$, for x b. $P = a(x + d)$, for x

8. Solve a linear inequality using the addition principle *(Section 6.6/Idea 2)*.
   a. $x + 11 < -8$ b. $x + 1 \leq -8$

9. Solve a linear inequality by using the multiplication or division principle *(Section 6.6/Ideas 3–4)*.
   a. $-0.5x > 15$ b. $\dfrac{2}{3}x \geq 10$

10. Solve a linear inequality containing more than one operation *(Section 6.7/Idea 1)*.
    a. $x + 10 \leq 7x + 20$ b. $15 \geq 2(x + 3)$ c. $\dfrac{4 - 2x}{3} \leq \dfrac{21}{12} + \dfrac{5x - 3}{4}$

6.1 Linear Equations

IDEA 1

An **equation** is a mathematical statement that two quantities are equal. A **linear** or **first-degree** equation containing one variable is an equation in which the variable has an exponent of one. Some examples of linear equations in one variable are:

$$2x + 5 = 7 \qquad 6x - 2 = 4x + 8 \qquad 3(x - 6) = 4x - 9$$

Each of the above equations contains three parts. The equal sign ($=$), the expression to the left of the equal sign called the left side or left member of the equation, and the expression to the right called the right side or right member of the equation.

IDEA 2

A **solution** or **root** of an equation is a value that can be substituted for the variable so that the equation becomes a true statement. For example, 3 is a solution of the equation $x + 6 = 9$, since when x is replaced by 3 in the equation $x + 6 = 9$ we obtain a true statement.

$$x + 6 = 9$$
$$3 + 6 = 9 \qquad \text{Replace } x \text{ with 3}$$
$$9 = 9 \qquad \text{This is a true statement}$$

The procedure for determining if a value is a solution of an equation is outlined below.

To Determine if a Value is a Solution of an Equation:

1. Replace the variable with its given (or obtained) value.
2. Do the indicated operations on both sides of the equation.
3. If the result is a true statement, the value is a solution. Otherwise, it is not a solution.

Example 1 Determine if the given value is a solution.

a. $2x - 9 = -15, x = -3$

b. $3x + 2.11 = 5x - 3.2, x = 0.5$

c. $3(x - 2) - 1 = 2 - 5(x + 5), x = -2$

d. $5x - 3 = \frac{1}{10} + \frac{3}{4}x, x = \frac{1}{2}$

Solution To determine if a given value is a solution of an equation, replace the variable with its given value and then determine if the equation is a true or false statement.

1a. $\qquad 2x - 9 = -15, x = -3$
$\qquad 2(-3) - 9 \stackrel{?}{=} -15 \qquad$ Replace x with -3
$\qquad -6 - 9 \stackrel{?}{=} -15$
$\qquad -15 = -15 \qquad$ True

Thus, -3 is a solution.

1b. $3x + 2.11 = 5x - 3.2, x = 0.5$

$3(0.5) + 2.11 \stackrel{?}{=} 5(0.5) - 3.2$ Replace x with 0.5
$1.5 + 2.11 \stackrel{?}{=} 2.5 - 3.2$
$3.61 = -0.7$ False

Thus, 0.5 is not a solution.

1c. $3(x - 2) - 1 = 2 - 5(x + 5), x = -2$

$3(-2 - 2) - 1 \stackrel{?}{=} 2 - 5(-2 + 5)$ Replace x with -2
$3(-4) - 1 \stackrel{?}{=} 2 - 5(3)$
$-12 - 1 \stackrel{?}{=} 2 - 15$
$-13 = -13$ True

Thus, -2 is a solution.

1d. $5x - 3 = \frac{1}{10} + \frac{3}{4}x, x = \frac{1}{2}$

$5\left(\frac{1}{2}\right) - 3 \stackrel{?}{=} \frac{1}{10} + \frac{3}{4} \cdot \frac{1}{2}$ Replace x with $\frac{1}{2}$
$\frac{5}{2} - 3 \stackrel{?}{=} \frac{1}{10} + \frac{3}{8}$
$\frac{5}{2} - \frac{6}{2} \stackrel{?}{=} \frac{4}{40} + \frac{15}{40}$
$-\frac{1}{2} = \frac{19}{40}$ False

Thus, $\frac{1}{2}$ is not a solution.

Practice Problem 1 *Determine if the given value is the solution.*

a. $8x - 3 = -2x + 4, x = -2$

b. $3.7x - 5 = -6.87, x = 3.1$

c. $\frac{5n}{6} + \frac{49}{27} = \frac{12n + 7}{27}, n = -4$

IDEA 3

Up to this point we have examined conditional equations. A **conditional equation** is an equation that is a true statement for only certain values of the variable. For example,

$$x + 8 = 11$$

is a conditional equation since 3 is the only solution of the equation.

Not every linear equation in one variable is a conditional equation. For example, the equation $2x = x + x$ is not a conditional equation since if we simplify both sides of the equation we obtain

$$2x = x + x$$
$$2x = 2x \quad \text{Combine like terms}$$

that is a true statement for every value of x. An equation that is a true statement for every value of the variable is called an **identity**. An identity has an infinite number of solutions.

Similarly, the equation

$$x = 0.5(2x + 6)$$

is not a conditional equation since if we simplify both sides of the equation we obtain

$$x = 0.5(2x + 6)$$
$$x = x + 3 \quad \text{Multiply to remove ( )}$$

that is a false statement for every value of x. An equation that is a false statement for every value of the variable is called an **inconsistent equation** or a **contradiction.** Inconsistent equations have no solution.

Example 2   Determine whether the given equation is an identity or an inconsistent equation.

   a. $5(x - 4) - 5x + 6 = x - (x - 6)$ **b.** $2(x + 5) = 2x + 10$

Solution   To determine if an equation is an identity or an inconsistent equation, first simplify both sides. Then, determine if the result is a true or a false statement.

$$\begin{aligned}\text{2a.}\ 5(x - 4) - 5x + 6 &= x - (x - 6) \\ 5x - 20 - 5x + 6 &= x - x + 6 \\ 14 &= 6 \quad \text{False}\end{aligned}$$

Since $14 = 6$ is a false statement, $5(x - 4) - 5x + 6 = x - (x - 6)$ is an inconsistent equation and has no solution.

$$\begin{aligned}\text{2b.}\ 2(x + 5) &= 2x + 10 \\ 2x + 10 &= 2x + 10 \quad \text{True}\end{aligned}$$

Since $2x + 10 = 2x + 10$ is always a true statement, $2(x + 5) = 2x + 10$ is an identity and has an infinite number of solutions.

6.1 Exercises

Identify the linear equations in one variable.

1. $3x + 2y = 6$
2. $x + 6 = 8$
3. $x^2 + 9 = 16$
4. $3n + 2 = 5n + 2$
5. $x^2 + y^2 = 16$

Determine if the given value is a solution.

6. $x + 2 = 6, x = 4$
7. $2x + 5 = 9, x = 4$
8. $10x = 25, x = \dfrac{5}{2}$
9. $x - 8 = 10, x = -2$
10. $4x - 2(3 - x) = 12, x = 3$
11. $y - 8.4 = -6.5, y = 1.9$
12. $\dfrac{1}{3} + n = \dfrac{5}{6}, n = \dfrac{4}{3}$
13. $b + \dfrac{5}{12} = \dfrac{1}{10}, b = -\dfrac{3}{8}$
14. $4 = \dfrac{2x}{5} - 6, x = 25$
15. $\dfrac{3a}{5} + \dfrac{4}{7} = \dfrac{a}{5} + \dfrac{3}{8}, x = 4$
16. $0.023x = -46, x = -20$
17. $-36 + x = 8.3, x = 11.9$
18. $9x + 7 = 5x - 13, x = -5$
19. $5x + 4 = 3x - 2, x = 3$
20. $0.35n - 1.7 = 1.85n + 4, n = -3.8$
21. $6x + 2 = -5x, x = -3$
22. $4x = 2(12 - 2x), x = 3$
23. $5(3x - 2) = 35, x = -2$

24. $10 - 7x = 4(11 - 6x)$, $x = 4$

25. $4x - 5(3 + 2x) = 3$, $x = 5$

26. $\dfrac{x}{3} - \dfrac{x}{2} = 2$, $x = 12$

27. $\dfrac{x}{6} - \dfrac{2x}{9} = -\dfrac{x}{18}$, $x = 5$

Fill in the blanks.

28. A linear or _____ equation in one variable is an equation in which the highest power of the variable is one.

29. If a value for a variable makes an equation a true statement, this value is called a root or _____ of the equation.

30. To determine if a given value is the solution of an equation, replace the _____ with its given value. If this results in a _____ statement, the solution is _____. Otherwise, it is _____.

Determine if the given value is a solution.

31. $x + 5.6 = 2.8$, $x = -2.8$

32. $x + 3.04 = 2.96$, $x = -0.08$

33. $x + \dfrac{1}{2} = \dfrac{5}{2}$, $x = 4$

34. $x + \dfrac{3}{4} = \dfrac{23}{4}$, $x = 20$

35. $2.5x - 3.8 = -7.9$, $x = -1.64$

36. $3.75x + 0.125 = -0.125$, $x = -0.0067$

37. $-73 = 24x + 31$, $x = -4\dfrac{1}{3}$

38. $-48 = 36x + 42$, $x = -2\dfrac{1}{2}$

39. $7 = \dfrac{2x}{5} + 3$, $x = -10$

40. $9 = \dfrac{3x}{4} + 6$, $x = -4$

41. $7 - 9x - 12 = 3x + 5 - 8x$, $x = -\dfrac{5}{2}$

42. $13 - 11x - 17 = 5x + 4 - 10x$, $x = -\dfrac{4}{3}$

43. $10x - 2(3 + 4x) = 7 - (x - 2)$, $x = -5$

44. $7x - 3(2x - 5) = 6(2 + 3x) - 31$, $x = -2$

45. $\dfrac{2(x - 3)}{5} - \dfrac{3(x + 2)}{2} = \dfrac{7}{10}$, $x = -3$

46. $\dfrac{5(x - 4)}{6} - \dfrac{2(x + 4)}{9} = \dfrac{5}{18}$, $x = -5$

Determine whether the given equation is an identity or an inconsistent equation.

47. $2(x + 4) = 2x + 8$

48. $-3(x - 1) = -3x + 1$

49. $2x = 2x + 6$

50. $3a = 3a + 5$

51. $3(2x - 5) - 6x = 9$

52. $4(3x - 7) - 12x = -8$

53. $2(x + 2) - 2x = 3(x + 1) - (3x - 1)$

54. $3(x - 2) - 3x = 4(x - 3) - 4x + 12$

Answers to Practice Problems **1a.** -2 is *not* the solution **b.** 3.1 is *not* the solution **c.** -4 is the solution

6.2 Solving Linear Equations Containing One Operation

IDEA 1 To solve (or find the solution to) a linear equation containing one variable, your goal is to isolate the variable on one side of the equation and a constant on the opposite side. This can be done by using the addition, multiplication, and/or division principles of equality to produce a series of *equivalent equations*.

Equivalent equations are equations that have the same solution. For example,

$$3x + 1 = 7$$
$$3x = 6$$
$$x = 2$$

are all equivalent equations since 2 is the only solution for each equation.

The first type of linear equation we will consider can be solved by using the *addition principle of equality*.

Addition Principle of Equality

> If the same quantity is added to both sides of an equation, the resulting equation is equivalent to the original equation,
>
> or
>
> If $A = B$
> Then $A + C = B + C$
>
> where A, B, and C are algebraic expressions.

Using this principle, we can solve equations of the form

$$x + a = b$$

where a and b are constants by adding the additive inverse of a to both sides of the equation. For example, to solve for x in the equation $x - 7 = 11$, our goal is to isolate x on the left side of the equation and some constant on the opposite side. To do this, we will remove the term -7 from the left side of the equation by adding its additive inverse (7) to both sides of the equation.

$$\begin{aligned} x - 7 &= -11 \\ x - 7 + 7 &= -11 + 7 \quad \text{Add 7 to both sides} \\ & \quad\quad\quad\quad\quad\quad\quad\;\, \text{Do this mentally} \\ x + 0 &= -4 \\ x &= -4 \end{aligned}$$

To check, replace x with -4 in the equation

$$\begin{aligned} x - 7 &= -11 \\ -4 - 7 &\stackrel{?}{=} -11 \\ -11 &\stackrel{?}{=} -11 \quad \text{True} \end{aligned}$$

Thus, -4 is the solution.

To Solve a Linear Equation of the Form $x + a = b$:

1. Add the additive inverse of a to both sides of the equation.
2. Simplify both sides of the resulting equation.

Example 3 Solve.

a. $x + 9 = -11$ **b.** $x - 0.11 = -0.5$ **c.** $x - \dfrac{2}{9} = \dfrac{1}{6}$

Solution To solve an equation of the form $x + a = b$, add the additive inverse of a to both sides of the equation.

3a.
$$x + 9 = -11$$
$$x + 9 + (-9) = -11 + (-9)$$
$$x = -20$$

Check:
$$x + 9 = -11$$
$$-20 + 9 \stackrel{?}{=} -11$$
$$-11 = -11$$

3b.
$$x - 0.11 = -0.5$$
$$x - 0.11 + 0.11 = -0.5 + 0.11$$
$$x = -0.39$$

Check:
$$x - 0.11 = -0.5$$
$$-0.39 - 0.11 \stackrel{?}{=} -0.5$$
$$-0.5 = -0.5$$

3c.
$$x - \frac{2}{9} = \frac{1}{6}$$
$$x - \frac{2}{9} + \frac{2}{9} = \frac{1}{6} + \frac{2}{9}$$
$$x = \frac{3}{18} + \frac{4}{18}$$
$$x = \frac{7}{18}$$

Check:
$$x - \frac{2}{9} = \frac{1}{6}$$
$$\frac{7}{18} - \frac{2}{9} \stackrel{?}{=} \frac{1}{6}$$
$$\frac{7}{18} - \frac{4}{18} \stackrel{?}{=} \frac{1}{6}$$
$$\frac{3}{18} \stackrel{?}{=} \frac{1}{6}$$
$$\frac{1}{6} = \frac{1}{6}$$

Practice Problem 2 **Solve.**

a. $x + 5 = -9$ **b.** $x + 0.35 = 0.1$ **c.** $x - \frac{3}{10} = -\frac{5}{6}$

IDEA 2

The second type of equation we will discuss can be solved by using the *division principle of equality*.

Division Principle of Equality

> If both sides of an equation are divided by the same nonzero quantity, the resulting equation is equivalent to the original equation,
>
> or
>
> If $A = B$
>
> Then $\frac{A}{C} = \frac{B}{C}$
>
> where A, B, and C are algebraic expressions, and C is not equal to zero.

Using this principle, we can solve equations of the form

$$ax = b$$

where a and b are constants by dividing both sides of the equation by a (the coefficient of x). For example, to solve for x in the equation $2x = 6$, our goal is to isolate x on the left side of the equation and some constant on the opposite side. To do this, we will remove the coefficient 2 from the left side of the equation by dividing both sides of the equation by 2.

$$2x = 6$$
$$\frac{2x}{2} = \frac{6}{2} \quad \text{Divide both sides by 2}$$
$$1x = 3 \quad \text{Do this mentally}$$
$$x = 3$$

Check:
$$2x = 6$$
$$2(3) \stackrel{?}{=} 6$$
$$6 = 6$$

> **To Solve a Linear Equation of the Form $ax = b$:**
>
> 1. Divide both sides of the equation by a.
> 2. Simplify both sides of the resulting equation.

Example 4 Solve.

  a. $-9 = 3x$ b. $-0.6x = 3.6$ c. $-27x = -36$

Solution To solve an equation of the form $ax = b$, divide both sides by a.

4a. $-9 = 3x$ Check: $-9 = 3x$
$\dfrac{-9}{3} = \dfrac{3x}{3}$ $-9 \stackrel{?}{=} 3(-3)$
$-3 = x$ $-9 = -9$

We could rewrite $-9 = 3x$ as $3x = -9$ and then solve for x. Also, remember that the equation $x = -3$ and the equation $-3 = x$ both mean that x is equal to -3.

4b. $-0.6x = 3.6$ Check: $-0.6x = 3.6$
$\dfrac{-0.6x}{-0.6} = \dfrac{3.6}{-0.6}$ $-0.6(-6) \stackrel{?}{=} 3.6$
$x = -6$ $3.6 = 3.6$

4c. $-27x = -36$ Check: $-27x = -36$
$\dfrac{-27x}{-27} = \dfrac{-36}{-27}$ $-27\left(\dfrac{4}{3}\right) \stackrel{?}{=} -36$
$x = \dfrac{4}{3}$ $-36 = -36$

Practice Problem 3 *Solve.*

  a. $-6x = 12$ b. $0.2x = 0.84$ c. $9x = 15$

IDEA 3

The third type of equation we will examine can be solved by using the *multiplication principle of equality*.

Multiplication Principle of Equality

> If both sides of an equation are multiplied by the same nonzero quantity, the resulting equation is equivalent to the original equation,
>
> or
>
> If $A = B$
>
> then $A \cdot C = B \cdot C$
>
> where A, B, and C are algebraic expressions, and C is not equal to zero.

204 Linear Equations and Inequalities in One Variable

Using this principle, we can solve equations of the form

$$\frac{x}{a} = b$$

by multiplying both sides of the equation by a (the denominator of the fraction). For example, to solve for x in the equation

$$\frac{x}{4} = 6$$

our goal is to isolate x on the left side of the equation and some constant on the opposite side. To do this, we will remove the denominator 4 from the left side of the equation by multiplying both sides of the equation by 4.

$$\frac{x}{4} = 6 \qquad\qquad\qquad\qquad Check: \quad \frac{x}{4} = 6$$

$$\boxed{4} \cdot \frac{x}{4} = \boxed{4} \cdot 6 \quad \text{Multiply both sides by 4} \qquad \frac{24}{4} \stackrel{?}{=} 6$$

$$\frac{\cancel{4} \cdot x}{\cancel{4}} = 24 \quad \text{Do this mentally} \qquad\qquad\qquad 6 = 6$$

$$x = 24$$

To Solve a Linear Equation of the Form $\frac{x}{a} = b$:

1. Multiply both sides of the equation by a.
2. Simplify both sides of the resulting equation.

Example 5   Solve.

a. $\frac{x}{5} = -6$ **b.** $-8 = \frac{x}{-2}$

c. $\frac{n}{0.2} = 0.3$ **d.** $\frac{x}{14} = \frac{6}{7}$

Solution   To solve an equation of the form $\frac{x}{a} = b$, multiply both sides by a.

5a. $\frac{x}{5} = -6$ \qquad\qquad Check: $\frac{x}{5} = -6$

$\boxed{5}\left(\frac{x}{5}\right) = \boxed{5}(-6)$ \qquad\qquad $\frac{-30}{5} \stackrel{?}{=} -6$

$x = -30$ \qquad\qquad\qquad $-6 = -6$

5b. $-8 = \frac{x}{-2}$ \qquad\qquad Check: $-8 = \frac{x}{-2}$

$\boxed{-2}(-8) = \boxed{-2}\left(\frac{x}{-2}\right)$ \qquad\qquad $-8 \stackrel{?}{=} \frac{16}{-2}$

$16 = x$ \qquad\qquad\qquad $-8 = -8$

5c.
$$\frac{n}{0.2} = 0.3$$
$$\boxed{0.2}\left(\frac{n}{0.2}\right) = \boxed{0.2}(0.3)$$
$$n = 0.06$$

Check:
$$\frac{n}{0.2} = 0.3$$
$$\frac{0.06}{0.2} \stackrel{?}{=} 0.3$$
$$0.3 = 0.3$$

5d.
$$\frac{x}{14} = \frac{6}{7}$$
$$\boxed{14} \cdot \frac{x}{14} = \boxed{14} \cdot \frac{6}{7}$$
$$x = 12$$

Check:
$$\frac{x}{14} = \frac{6}{7}$$
$$\frac{12}{14} \stackrel{?}{=} \frac{6}{7}$$
$$\frac{6}{7} = \frac{6}{7}$$

Practice Problem 4 **Solve.**

a. $\frac{x}{-6} = -2$ **b.** $\frac{x}{4} = -0.5$ **c.** $\frac{n}{12} = \frac{7}{2}$

The multiplication principle can also be used to solve equations of the form

$$ax = b$$

where a and b are constants by multiplying both sides of the equation by the reciprocal of a (the coefficient of x). For example, to solve for x in the equation $\frac{1}{2}x = 5$, our goal is to isolate x on one side of the equation and some constant on the opposite side. To do this, we will remove the coefficient $\frac{1}{2}$ from the left side of the equation by multiplying both sides of the equation by the reciprocal of $\frac{1}{2}$ which is $\frac{2}{1}$ or 2.

$$\frac{1}{2}x = 5$$
$$\boxed{2}\left(\frac{1}{2}x\right) = \boxed{2}(5) \quad \text{Multiply both sides by 2}$$
$$1x = 10 \quad \text{Do this mentally}$$
$$x = 10$$

Check:
$$\frac{1}{2}x = 5$$
$$\frac{1}{2}(10) \stackrel{?}{=} 5$$
$$5 = 5$$

Example 6 Solve.

a. $\frac{3}{4}x = -18$ **b.** $-\frac{6}{25}x = \frac{21}{10}$ **c.** $3x = -15$

Solution To solve an equation of the form $ax = b$, multiply both sides by the reciprocal of a. Remember, to find the reciprocal of a number, just invert that number.

6a.
$$\frac{3}{4}x = -18$$
$$\boxed{\frac{4}{3}}\left(\frac{3}{4}x\right) = \boxed{\frac{4}{3}}(-18)$$
$$1x = -24 \quad \text{Do this mentally}$$
$$x = -24$$

Check:
$$\frac{3}{4}x = -18$$
$$\frac{3}{4}(-24) \stackrel{?}{=} -18$$
$$-18 = -18$$

6b. $-\dfrac{6}{25}x = \dfrac{21}{10}$

$-\dfrac{25}{6}\left(-\dfrac{6}{25}x\right) = -\dfrac{25}{6}\left(\dfrac{21}{10}\right)$

$x = -\dfrac{\overset{5}{\cancel{25}} \cdot \overset{7}{\cancel{21}}}{\underset{2}{\cancel{10}} \cdot \underset{2}{\cancel{6}}}$

$x = -\dfrac{35}{4}$

Check: $-\dfrac{6}{25}x = \dfrac{21}{10}$

$-\dfrac{6}{25}\left(-\dfrac{35}{4}\right) \overset{?}{=} \dfrac{21}{10}$

$+\dfrac{\overset{3}{\cancel{6}} \cdot \overset{7}{\cancel{35}}}{\underset{5}{\cancel{25}} \cdot \underset{2}{\cancel{4}}} \overset{?}{=} \dfrac{21}{10}$

$\dfrac{21}{10} = \dfrac{21}{10}$

6c. $3x = -15$

$\dfrac{1}{3}(3x) = \dfrac{1}{3}(-15)$

$x = -5$

Check: $3x = -15$

$3(-5) \overset{?}{=} -15$

$-15 = -15$

When the coefficient of the variable is -1, we can use the multiplication principle to solve the equation.

Example 7 Solve: $-x = 2$

Solution To solve for x, multiply by -1.

$-x = 2$

$(-1)(-x) = (-1)(2)$

$x = -2$

Check: $-x = 2$

$-(-2) \overset{?}{=} 2$

$2 = 2$

Practice Problem 5 Solve.

a. $\dfrac{1}{2}x = -4$ **b.** $-\dfrac{4}{9}x = \dfrac{10}{12}$ **c.** $-n = -2$

At this point note that we can solve equations of the form $ax = b$ by using the multiplication or the division principle. This is because dividing an expression by the constant a is the same as multiplying that expression by $\dfrac{1}{a}$ (the reciprocal of a). For example, we can solve the equation $4x = 12$ by writing

$4x = 12$ or $4x = 12$

$\dfrac{4x}{4} = \dfrac{12}{4}$ $\dfrac{1}{4}(4x) = \dfrac{1}{4}(12)$

$x = 3$ $x = 3$.

In this book to solve equations of the form $ax = b$, we will use the division principle when the coefficient of the variable is an integer ($3x = 6$) or a signed decimal ($-0.3x = 6$). However, when the coefficient of the variable is a signed fraction $\left(\dfrac{1}{2}x = 6\right)$ or a negative one ($-n = 6$), we will use the multiplication principle.

6.2 Exercises

Solve by using the addition principle and then check your answer.

1. $x + 5 = 10$ **2.** $x - 6 = -10$ **3.** $x - 7 = -13$ **4.** $x - 9 = -2$

5. $x + 8 = 3$ **6.** $8 = x + 13$ **7.** $x - 8 = -3$ **8.** $5 + x = -9$

9. $3 + x = -15$
10. $x + 5.2 = 4.8$
11. $x - 3 = -4.1$
12. $x + 0.7 = -0.9$
13. $x - 2.4 = -5.11$
14. $x + 9 = -11$
15. $x + \frac{1}{2} = -\frac{1}{2}$
16. $x - \frac{5}{6} = \frac{7}{10}$
17. $x + \frac{3}{10} = -\frac{5}{8}$
18. $\frac{5}{18} = x + \frac{7}{12}$
19. $x + 9 = -35$
20. $x + 5.7 = 9$
21. $x - \frac{1}{14} = -\frac{1}{21}$

Solve by using the multiplication or division principle and then check your answer.

22. $5x = 10$
23. $-6x = -12$
24. $4x = -16$
25. $-15x = 105$
26. $10x = 30$
27. $8x = -12$
28. $\frac{x}{3} = 5$
29. $\frac{x}{-4} = -7$
30. $\frac{x}{5} = -8$
31. $\frac{x}{0.3} = -0.3$
32. $\frac{x}{-1.2} = -5$
33. $\frac{3}{4}x = -18$
34. $\frac{4}{9}x = 20$
35. $-\frac{3}{5}x = \frac{6}{25}$
36. $\frac{1}{2}x = -10$
37. $1.2x = -3.6$
38. $0.1x = 5$
39. $-0.5x = -1.25$
40. $25x = -35$
41. $-2.4x = 96$
42. $91x = 35$

Fill in the blanks.

43. To solve an equation of the form $x + a = b$, add the _____ of a to both sides.
44. To solve an equation of the form $ax = b$, _____ both sides by a or multiply both sides by the _____ of a.

Solve.

45. $x + 3.5 = 9$
46. $0.3x = -3$
47. $\frac{x}{-4} = 6$
48. $\frac{9}{16}x = -\frac{6}{10}$
49. $-x = -6$
50. $45x = 135$
51. $\frac{x}{-0.2} = 10$
52. $\frac{9}{25} = x + \frac{1}{35}$
53. $3 = -\frac{1}{5}x$
54. $8 = -\frac{1}{4}x$
55. $\frac{2}{3}x = 10$
56. $\frac{3}{4}x = 9$
57. $-\frac{2}{7}x = -6$
58. $-\frac{3}{5}x = -9$
59. $\frac{3}{8}x = \frac{3}{8}$
60. $\frac{2}{9}x = \frac{2}{7}$
61. $\frac{4}{9}x = \frac{10}{27}$
62. $\frac{6}{49}x = \frac{15}{14}$
63. $-\frac{1}{5}x = \frac{3}{10}$
64. $-\frac{1}{4}x = \frac{7}{10}$
65. $\frac{x}{-7} = -5$
66. $\frac{x}{-3} = -8$
67. $\frac{x}{1.2} = 5$
68. $\frac{x}{1.4} = 5$
69. $\frac{x}{9} = -\frac{7}{6}$
70. $\frac{x}{6} = -\frac{11}{9}$
71. $\frac{x}{35} = \frac{15}{25}$
72. $\frac{x}{49} = \frac{14}{21}$
73. $\frac{x}{-0.3} = 0.05$
74. $\frac{x}{-0.2} = -0.08$
75. $\frac{x}{5} = -8.2$
76. $\frac{x}{3} = -7.4$
77. $0.12x = -13.2$
78. $0.11x = -12.1$
79. $-5x = 18.5$
80. $-8x = 33.6$
81. $x + 0.4 = -\frac{3}{10}$
82. $x + 0.3 = -\frac{1}{10}$
83. $0.4x = -\frac{4}{25}$
84. $0.8x = -\frac{8}{25}$
85. $\frac{8}{35}x = \frac{12}{45}$
86. $\frac{12}{49}x = \frac{18}{35}$
87. $\frac{x}{-20} = -\frac{3}{8}$
88. $\frac{x}{-18} = -\frac{5}{27}$
89. $\frac{n}{23.45} = -14.2$
90. $\frac{m}{-14.8} = 15.5$
91. $5y = -120.7$
92. $9.5x = -174.8$

Answers to Practice Problems 2a. -14 b. -0.25 c. $-\dfrac{8}{15}$ 3a. -2 b. 4.2 c. $\dfrac{5}{3}$ 4a. 12 b. -2 c. 42 5a. -8 b. $-\dfrac{15}{8}$ c. 2

6.3 Solving Linear Equations Containing More Than One Operation

IDEA 1 Sometimes you will need to use more than one principle to solve an equation. When this occurs, you should usually use the addition principle first.

> **To Solve a Linear Equation Containing Variables on Only One Side of the Equation:**
>
> 1. Combine like terms, yielding an equation of the form $ax + b = c$ or $\dfrac{x}{a} + b = c$.
>
> 2. Add the additive inverse of b to both sides of the equation and simplify.
>
> 3. Solve the resulting equation by using the multiplication or division principle.

Example 8 Solve.

a. $-3x + 5 = -10$ b. $\dfrac{x}{2} - 10 = -30$

c. $\dfrac{3}{5}x + 8 = 2$ d. $-2x + 6x - 8 = -18$

Solution To solve an equation of the form $ax + b = c$ or $\dfrac{x}{a} + b = c$, first add the additive inverse of b to both sides. Next, solve the resulting equation.

8a.
$$-3x + 5 = -10$$
$$-3x + 5 + (-5) = -10 + (-5)$$
$$-3x = -15$$
$$\dfrac{-3x}{-3} = \dfrac{-15}{-3}$$
$$x = 5$$

Check:
$$-3x + 5 = -10$$
$$-3(5) + 5 \stackrel{?}{=} -10$$
$$-15 + 5 \stackrel{?}{=} -10$$
$$-10 = -10$$

8b.
$$\dfrac{x}{2} - 10 = -30$$
$$\dfrac{x}{2} - 10 + 10 = -30 + 10$$
$$\dfrac{x}{2} = -20$$
$$2\left(\dfrac{x}{2}\right) = 2(-20)$$
$$x = -40$$

Check:
$$\dfrac{x}{2} - 10 = -30$$
$$\dfrac{-40}{2} - 10 \stackrel{?}{=} -30$$
$$-20 - 10 \stackrel{?}{=} -30$$
$$-30 = -30$$

8c.
$$\frac{3}{5}x + 8 = 2$$
$$\frac{3}{5}x + 8 + (-8) = 2 + (-8)$$
$$\frac{3}{5}x = -6$$
$$\frac{5}{3}\left(\frac{3}{5}x\right) = \frac{5}{3}(-6)$$
$$x = -10$$

Check:
$$\frac{3}{5}x + 8 = 2$$
$$\frac{3}{5}(-10) + 8 \stackrel{?}{=} 2$$
$$-6 + 8 \stackrel{?}{=} 2$$
$$2 = 2$$

8d. When like terms appear on the same side of an equation, combine them before applying any of the principles.

$$-2x + 6x - 8 = -18$$
$$4x - 8 = -18$$
$$4x - 8 + 8 = -18 + 8$$
$$4x = -10$$
$$\frac{4x}{4} = -\frac{10}{4}$$
$$x = -\frac{5}{2}$$

Check:
$$-2x + 6x - 8 = -18$$
$$-2\left(-\frac{5}{2}\right) + 6\left(-\frac{5}{2}\right) - 8 \stackrel{?}{=} -18$$
$$5 + (-15) - 8 \stackrel{?}{=} -18$$
$$-18 = -18$$

Practice Problem 6 *Solve.*

a. $-4x + 9 = -3$ b. $\frac{1}{2}x - 8 = -13$

c. $x - 0.3x + 5 = -2$

IDEA 2

Quite often, variables will appear on both sides of an equation. When this occurs, our goal is still to isolate the variable on one side of the equation by using the addition principle. For example, to solve the equation

$$4x = 18 + 2x$$

we first eliminate the term $2x$ from the right side of the equation by adding $-2x$ to both sides of the equation.

$$4x = 18 + 2x$$
$$4x + (-2x) = 18 + 2x + (-2x)$$
$$2x = 18$$

Next, we solve the resulting equation.

$$2x = 18$$
$$\frac{2x}{2} = \frac{18}{2}$$
$$x = 9$$

The procedure for solving equations containing variables on both sides is summarized below.

To Solve a Linear Equation Containing Variables on Both Sides of the Equation:

1. Combine like terms on each side of the equation.
2. Isolate all terms containing the variable on one side of the equation and all constants on the opposite side by using the addition principle. Combine like terms again.
3. Solve the resulting equation by using the multiplication or division principle.

Example 9 Solve.

a. $6x = 2x + 8$ **b.** $x + 20 = 7x + 10$

c. $1.9x - 7.8 + 5.3x = 3 + 1.8x$

Solution To solve a linear equation containing variables on both sides, combine like terms on each side. Next, use the addition principle to get all variables on the same side and then solve the resulting equation.

9a.
$$6x = 2x + 8$$
$$6x + (-2x) = 2x + 8 + (-2x)$$
$$4x = 8$$
$$\frac{4x}{4} = \frac{8}{4}$$
$$x = 2$$

Check: $6x = 2x + 8$
$6(2) \stackrel{?}{=} 2(2) + 8$
$12 \stackrel{?}{=} 4 + 8$
$12 = 12$

9b.
$$x + 20 = 7x + 10$$
$$x + 20 + (-x) = 7x + 10 + (-x)$$
$$20 = 6x + 10$$
$$20 + (-10) = 6x + 10 + (-10)$$
$$10 = 6x$$
$$\frac{10}{6} = \frac{6x}{6}$$
$$\frac{5}{3} = x$$

Check: $x + 20 = 7x + 10$
$\frac{5}{3} + 20 \stackrel{?}{=} 7\left(\frac{5}{3}\right) + 10$
$1\frac{2}{3} + 20 \stackrel{?}{=} \frac{35}{3} + 10$
$21\frac{2}{3} \stackrel{?}{=} 11\frac{2}{3} + 10$
$21\frac{2}{3} = 21\frac{2}{3}$

9c.
$$1.9x - 7.8 + 5.3x = 3 + 1.8x$$
$$7.2x - 7.8 = 3 + 1.8x \quad \text{Combine like terms}$$
$$7.2x - 7.8 + (-1.8x) = 3 + 1.8x + (-1.8x)$$
$$5.4x - 7.8 = 3$$
$$5.4x - 7.8 + 7.8 = 3 + 7.8$$
$$5.4x = 10.8$$
$$\frac{5.4x}{5.4} = \frac{10.8}{5.4}$$
$$x = 2$$

Check: $1.9x - 7.8 + 5.3x = 3 + 1.8x$
$1.9(2) - 7.8 + 5.3(2) \stackrel{?}{=} 3 + 1.8(2)$
$3.8 - 7.8 + 10.6 \stackrel{?}{=} 3 + 3.6$
$6.6 = 6.6$

Practice Problem 7 **Solve.**

a. $13x + 5 = 8x + 40$ **b.** $0.4x = x + 8$

c. $x + 2x - 7 = 3x + x - 5$

6.3 Exercises

Solve. Check your answers.

1. $2x + 9 = 189$
2. $10x + 8 = -17$
3. $0.2x - 0.9 = 0.27$
4. $\dfrac{x}{2} - 10 = -20$
5. $\dfrac{1}{2}x + 4 = 1$
6. $\dfrac{3}{5}x + 7 = -14$
7. $-\dfrac{6}{7}x + 5 = -5$
8. $\dfrac{x}{0.3} - 0.55 = 0.2$
9. $\dfrac{x}{-2} - 8 = -10$
10. $2x + 3x = -15$
11. $7x - 9x = -4$
12. $x + 0.3x = 39$
13. $4x = 2x - 12$
14. $8x + 40 = 2x$
15. $x = 13.2 + 0.2x$

Solve. Check your answers.

16. $13x + 5 = 8x + 40$
17. $6x - 2 = 8 - 14x$
18. $x + 2x - 7 = 4x - 5$
19. $x + 1 = 16 - x$
20. $10x + 4 = 7x + 12$
21. $10x - 12 = 6x + 12$
22. $0.8x + 0.6 = 0.2 - x$
23. $4.2x + 0.4 = 6.4 - 16.8x$
24. $8x + 9 = -2x - 16$
25. $6x + 5 = -3x - 10$
26. $2.2x - 5 + 4.5x = 1.7x - 20$
27. $0.17x - 1.7 + 0.18x = 1.85x + 4$
28. $10x + 1 = -15$
29. $8x + 2 = -12$
30. $x + 0.1x = 0.33$
31. $x + 0.2x = 0.48$
32. $-0.2x + 0.42 = 0.8$
33. $-0.3x + 0.24 = 0.6$
34. $5x - 8x = -15$
35. $4x - 6x = -12$
36. $15x - 5 = -20$
37. $18x - 6 = -24$
38. $\dfrac{x}{3} - 8 = -3.2$
39. $\dfrac{x}{2} - 6 = -4.4$
40. $\dfrac{x}{0.2} + 3.1 = -6.9$
41. $\dfrac{x}{0.3} + 9.3 = -20.7$
42. $\dfrac{2}{5}x + 3 = 7$
43. $\dfrac{3}{4}x + 6 = 9$
44. $13 = -\dfrac{2}{9}x + 3$
45. $41 = -\dfrac{4}{5}x + 25$
46. $6x + 4 = 4x - 2$
47. $5x + 3 = 2x - 9$
48. $4x - 7 - x = 8 - 2x$
49. $5x - 10 - x = 4 - 3x$
50. $y + 15 = 18y - 19$
51. $y + 18 = 15y - 24$
52. $10x - 8 = 4x$
53. $8x - 10 = 2x$
54. $6.4 - 0.4x = 0.4x$
55. $1.2 - 0.3x = 0.3x$
56. $25x - 91 = -10x$
57. $39x - 77 = -10x$
58. $10x - 6 - 8x = 7 - x + 2$
59. $11x - 9 - 8x = 8 - x + 3$
60. $-9x - 5 = -5x + 5$
61. $-6x - 8 = -2x + 10$
62. $0.7x - 0.8 = 0.8 - 0.9x$
63. $0.5x - 0.7 = 0.7 - 0.9x$
64. $3.4x = -18 + 3x$
65. $4.3x = -15 + 4x$
66. $0.3x - 0.6 = -0.485$
67. $0.2x - 0.4 = -0.382$
68. $\dfrac{1}{2}x + 0.3 = -0.2$
69. $\dfrac{1}{4}x + 0.2 = -0.6$
70. $0.37x + 0.013 = -0.098$
71. $0.22x + 0.011 = -0.143$
72. $-3x + 8x + 17 = -8$
73. $-2x + 6x + 11 = -5$
74. $-36 = 18x - 9$
75. $-24 = 12x - 8$
76. $-1.2x - 8 = 16$
77. $1.5x - 7 = 23$
78. $3.5x - 17 = 18.5x + 40$
79. $2.3x - 15 = 15.3x + 39.6$
80. $x + 0.6 - 0.2x = 2x$
81. $x + 0.24 - 0.4x = 3x$

Fill in the blanks.

82. To solve an equation of the form $ax + b = c$ or $\dfrac{x}{a} + b = c$, first add the _____ of b to both sides of the equation. Next, solve the resulting equation.

212 Linear Equations and Inequalities in One Variable

83. To solve an equation having variables on both sides, _____ like terms on each side of the equation. Next, use the _____ principle to get all variables on the same side and then solve the resulting equation.

Solve. Check your answers.

84. $5.17p - 4.89 = 1.4174$

85. $-0.574x + 1.493 = 4.65$

86. $0.291x + 3.715 = -0.874x + 1.9675$

87. $1.043x = 0.359x + 1.4364$

Answers to Practice Problems **6a.** 3 **b.** -10 **c.** -10 **7a.** 7 **b.** $-\dfrac{40}{3}$ **c.** -2

6.4 Solving Linear Equations Containing Grouping Symbols or Fractions

IDEA 1   Sometimes an equation may contain grouping symbols.

> **To Solve a Linear Equation Containing Grouping Symbols:**
> 1. Eliminate the grouping symbol by performing the indicated operation.
> 2. Solve the resulting equation.

Example 10   Solve.

a. $8x = 4(12 - 3x)$ **b.** $0.5(x + 2) = 0.25$

c. $7(5x - 1) - 18x = 12x - (3 - x)$

Solution   To solve an equation containing a grouping symbol, perform the indicated operation to eliminate the grouping symbol, and then solve the resulting equation.

10a.
$$8x = 4(12 - 3x)$$
$$8x = 48 - 12x \quad \text{Multiply}$$
$$8x + \boxed{12x} = 48 - 12x + \boxed{12x}$$
$$20x = 48$$
$$\frac{20x}{\boxed{20}} = \frac{48}{\boxed{20}}$$
$$x = \frac{12}{5}$$

Check:
$$8x = 4(12 - 3x)$$
$$8\left(\frac{12}{5}\right) \stackrel{?}{=} 4\left(12 - 3 \cdot \frac{12}{5}\right)$$
$$\frac{96}{5} \stackrel{?}{=} 4\left(12 - \frac{36}{5}\right)$$
$$\frac{96}{5} \stackrel{?}{=} 4\left(\frac{24}{5}\right)$$
$$\frac{96}{5} = \frac{96}{5}$$

10b.
$$0.5(x + 2) = 0.25$$
$$0.5x + 1 = 0.25$$
$$0.5x + 1 + \boxed{(-1)} = 0.25 + \boxed{(-1)}$$
$$0.5x = -0.75$$
$$\frac{0.5x}{\boxed{0.5}} = \frac{-0.75}{\boxed{0.5}}$$
$$x = -1.5$$

Check:
$$0.5(x + 2) = 0.25$$
$$0.5(-1.5 + 2) \stackrel{?}{=} 0.25$$
$$0.5(0.5) \stackrel{?}{=} 0.25$$
$$0.25 = 0.25$$

10c.
$$7(5x - 1) - 18x = 12x - (3 - x)$$
$$35x - 7 - 18x = 12x - 3 + x$$
$$17x - 7 = 13x - 3 \quad \text{Combine like terms}$$
$$17x - 7 + \boxed{(-13x)} = 13x - 3 + \boxed{(-13x)}$$
$$4x - 7 = -3$$
$$4x - 7 + \boxed{7} = -3 + \boxed{7}$$
$$4x = 4$$
$$\frac{4x}{\boxed{4}} = \frac{4}{\boxed{4}}$$
$$x = 1$$

Check:
$$7(5x - 1) - 18x = 12x - (3 - x)$$
$$7(5 \cdot 1 - 1) - 18 \cdot 1 \stackrel{?}{=} 12 \cdot 1 - (3 - 1)$$
$$7(5 - 1) - 18 \stackrel{?}{=} 12 - 2$$
$$7(4) - 18 \stackrel{?}{=} 10$$
$$28 - 18 \stackrel{?}{=} 10$$
$$10 = 10$$

Practice Problem 8 *Solve.*

a. $4x - 6 = 2(x + 5)$ **b.** $20 = 8 - 2(9 - 3x)$ **c.** $0.4(x - .1) = 6$

IDEA 2

Frequently you will need to solve an equation containing one or more fractions.

> **To Solve a Linear Equation Containing Fractions:**
>
> 1. Multiply every term on both sides of the equation by the LCD of all terms of the equation.
> 2. Solve the resulting equation.

Example 11 Solve.

a. $\dfrac{x}{4} + \dfrac{x}{3} + \dfrac{x}{2} = 26$ **b.** $\dfrac{2x}{3} = 6 + 2x$ **c.** $\dfrac{3}{4}x + \dfrac{1}{3}x = x - \dfrac{5}{3} - \dfrac{2}{3}x$

Solution To solve an equation containing fractions, remove all fractions by multiplying both sides by the LCD of all denominators. Solve the resulting equation.

11a.
$$\frac{x}{4} + \frac{x}{3} + \frac{x}{2} = 26$$
$$*\boxed{12}\left(\frac{x}{4} + \frac{x}{3} + \frac{x}{2}\right) = \boxed{12}(26)$$
$$12\left(\frac{x}{4}\right) + 12\left(\frac{x}{3}\right) + 12\left(\frac{x}{2}\right) = 12(26)$$
$$3x + 4x + 6x = 312$$
$$13x = 312$$
$$\frac{13x}{\boxed{13}} = \frac{312}{\boxed{13}}$$
$$x = 24$$

Check:
$$\frac{x}{4} + \frac{x}{3} + \frac{x}{2} = 26$$
$$\frac{24}{4} + \frac{24}{3} + \frac{24}{2} \stackrel{?}{=} 26$$
$$6 + 8 + 12 \stackrel{?}{=} 26$$
$$26 = 26$$

Multiplying both sides of an equation by the LCD means multiplying each *term* of the equation by the LCD.

11b.
$$\frac{2x}{3} = 6 + 2x$$
$$\boxed{3}\left(\frac{2x}{3}\right) = \boxed{3}(6 + 2x)$$
$$2x = 18 + 6x$$
$$2x + (\boxed{-6x}) = 18 + 6x + (\boxed{-6x})$$
$$-4x = 18$$
$$\frac{-4x}{\boxed{-4}} = \frac{18}{\boxed{-4}}$$
$$x = -\frac{9}{2}$$

Check:
$$\frac{2x}{3} = 6 + 2x$$
$$\frac{2\left(-\frac{9}{2}\right)}{3} \stackrel{?}{=} 6 + 2\left(-\frac{9}{2}\right)$$
$$\frac{-9}{3} \stackrel{?}{=} 6 + (-9)$$
$$-3 = -3$$

11c.
$$\frac{3}{4}x + \frac{1}{3}x = x - \frac{5}{3} - \frac{2}{3}x$$
$$\boxed{12}\left(\frac{3}{4}x + \frac{1}{3}x\right) = \boxed{12}\left(x - \frac{5}{3} - \frac{2}{3}x\right)$$
$$9x + 4x = 12x - 20 - 8x$$
$$13x = 4x - 20$$
$$13x + (\boxed{-4x}) = 4x - 20 + (\boxed{-4x})$$
$$9x = -20$$
$$\frac{9x}{\boxed{9}} = -\frac{20}{\boxed{9}}$$
$$x = -\frac{20}{9}$$

Check:
$$\frac{3}{4}x + \frac{1}{3}x = x - \frac{5}{3} - \frac{2}{3}x$$
$$\frac{3}{4}\left(-\frac{20}{9}\right) + \frac{1}{3}\left(-\frac{20}{9}\right) \stackrel{?}{=} -\frac{20}{9} - \frac{5}{3} - \frac{2}{3}\left(-\frac{20}{9}\right)$$
$$-\frac{\cancel{3}\cdot \cancel{20}}{\cancel{4}\cdot \cancel{9}} + \left(-\frac{20}{27}\right) \stackrel{?}{=} -\frac{20}{9} - \frac{5}{3} + \frac{40}{27}$$
$$-\frac{5}{3} + \left(-\frac{20}{27}\right) \stackrel{?}{=} -\frac{20}{9} - \frac{5}{3} + \frac{40}{27}$$
$$-\frac{45}{27} + \left(-\frac{20}{27}\right) \stackrel{?}{=} -\frac{60}{27} - \frac{45}{27} + \frac{40}{27}$$
$$-\frac{65}{27} = -\frac{65}{27}$$

Practice Problem 9 *Solve for x.*

a. $\frac{3x}{4} + \frac{3}{4} = \frac{2x}{3}$

b. $3x + \frac{13}{25} = 13$

c. $\frac{7}{8}x + \frac{1}{4} + \frac{3}{4}x = \frac{1}{16} + x$

IDEA 3

You may not always be able to find only one solution for a linear equation in one variable.

Example 12 Solve.

a. $x = x + 2$ b. $2x - 2(x - 6) = 12$

Solution   12a.
$$x = x + 2$$
$$x + (-x) = x + 2 + (-x)$$
$$0 = 2 \quad \text{False}$$

Since the variable was eliminated and we obtained a false statement, the original equation is an inconsistent equation and has no solution.

12c.
$$2x - 2(x - 6) = 12$$
$$2x - 2x + 12 = 12$$
$$12 = 12$$
$$12 + (-12) = 12 + (-12)$$
$$0 = 0 \quad \text{True}$$

Since the variable was eliminated and we obtained a true statement, the original equation is an identity and every number is a solution.

IDEA 4 To summarize solving linear equations, below is a general procedure to follow.

To Solve a Linear Equation:

1. Simplify both sides of the equation by eliminating fractions, eliminating grouping symbols, and/or collecting like terms.

2. Use the addition principle of equality to isolate all terms containing the variable on one side of the equation and all constants on the other. Combine like terms again.

3. Use the multiplication or division principle of equality to solve the resulting equation.

6.4 Exercises

Solve. Check each answer.

1. $3(x - 6) = -15$
2. $5(x - 2) = -10$
3. $2x - 3(2 + 2x) = -6$
4. $9x - 2(5 + 5x) = 8$
5. $10x + 5 = 3(4x + 5)$
6. $6x + 6 = 2(6x + 9)$
7. $9 + x = 5(9 - 7x)$
8. $10 - 3x = 4(11 - 5x)$
9. $3(2x + 3) = 27$
10. $4(2x - 3) = 28$
11. $45 = 5(3x + 2)$
12. $12 = 3(5x - 2)$
13. $2(3 + 4x) - 9 = 53$
14. $3(5 + 3x) - 8 = 38$
15. $5x - (2x + 8) = 24$
16. $7x - (3x + 8) = 16$
17. $6x - 6 = 6(7 - x)$
18. $6x - 20 = 10(x - 4)$
19. $12 - 4(3x - 1) = 4$
20. $20 - 6(2x - 1) = 2$
21. $7(x - 2) = 5(x + 4)$
22. $9(x + 2) = 3(x - 2)$
23. $4(2x + 1) = 2(7x + 7)$
24. $5(2x + 3) = 3(6x + 6)$
25. $2x + 2 = -0.5(4x - 8)$
26. $2x + 12 = -0.25(12x - 72)$
27. $\frac{1}{2}(4x - 6) = -9$
28. $\frac{1}{4}(4x - 8) = -8$
29. $\frac{3}{4}(8x - 4) = \frac{1}{3}(6x - 6)$
30. $\frac{1}{5}(5x - 10) = \frac{3}{2}(4x - 6)$
31. $\frac{4}{5}(10x - 10) = -4x$
32. $\frac{3}{7}(14x - 21) = -9x$
33. $\frac{x}{2} + \frac{x}{3} = -15$

34. $\frac{x}{3} + \frac{x}{4} = -21$
35. $\frac{x}{3} = 7 - \frac{x}{4}$
36. $\frac{x}{5} = 6 - \frac{x}{3}$
37. $\frac{x}{2} - \frac{x}{5} - 6 = 0$
38. $\frac{x}{3} - \frac{x}{7} = 12$
39. $\frac{x}{4} = \frac{x}{6} + 3$
40. $\frac{x}{6} = \frac{x}{8} + 9$
41. $\frac{x}{4} = 3 - \frac{x}{2}$
42. $\frac{x}{5} = 14 - \frac{x}{2}$
43. $\frac{x}{6} - \frac{x}{4} = \frac{x}{3}$
44. $\frac{x}{6} - \frac{x}{9} = \frac{x}{3}$
45. $\frac{x}{10} + \frac{x}{15} = \frac{x}{2} - 10$
46. $\frac{x}{8} + \frac{x}{6} = \frac{x}{4} - 1$
47. $\frac{x+3}{10} = \frac{x+3}{4}$
48. $\frac{x+2}{8} = \frac{x+2}{12}$
49. $\frac{x}{3} - \frac{1}{3} = -6$
50. $\frac{x}{4} - \frac{1}{4} = -3$
51. $\frac{2x-2}{5} = 10 + 3x$
52. $\frac{3x-3}{4} = 8 + 2x$
53. $\frac{3}{8}x - 5 = \frac{1}{4}$
54. $\frac{1}{6}x - 4 = \frac{1}{2}$
55. $\frac{5x}{2} + \frac{49}{9} = \frac{12x+7}{9}$
56. $\frac{3x}{2} + \frac{25}{6} = \frac{10x+8}{6}$
57. $\frac{1}{3}x + \frac{3}{4}x + 5 = \frac{1}{6}x$
58. $\frac{1}{2}x + \frac{3}{5}x + 1 = \frac{1}{10}x$
59. $\frac{1}{4}x + \frac{1}{3}x = \frac{1}{6}x - 5$
60. $\frac{1}{5}x + \frac{1}{3}x = \frac{1}{30}x - 1$
61. $\frac{1}{3}(2x + 5) = \frac{3}{5}$
62. $\frac{1}{4}(2x + 6) = \frac{2}{3}$

Fill in the blanks.

63. To solve an equation containing grouping symbols, _____ to remove the grouping symbols and then solve the resulting equation.

64. To solve an equation containing fractions, multiply both sides by the _____ of all terms of the equation. Next, solve the resulting equation.

Solve.

65. $5(x + 4) - 4(x + 3) = 2$
66. $4x - 6 = 2(x + 5)$
67. $5(x - 3) = 15$
68. $3x = 5(12 + 3x)$
69. $7(3x + 6) = 11 - (x + 2)$
70. $\frac{1}{3}(6y - 9) = \frac{1}{2}(8y - 4)$
71. $2 = 5 - 8(3 + 2x)$
72. $0.4(x - 0.1) = 6$
73. $34x - 6(x - 5) = 4 + 8(3x + 3)$
74. $20 - 6(2x - 1) = -10$
75. $\frac{x}{2} + \frac{x}{4} = -6$
76. $\frac{3x}{4} - \frac{3}{4} = \frac{2x}{3}$
77. $\frac{x-2}{5} + \frac{x}{3} = \frac{1}{5}$
78. $6x - \frac{2}{5} = 8$
79. $\frac{2x-6}{8} = 10$
80. $\frac{11}{4} = 2 - \frac{3}{10}x$
81. $\frac{3}{8} = \frac{x-1}{3}$
82. $\frac{x}{-3} + 15 = 2$
83. $\frac{6x}{7} + \frac{8}{5} = \frac{6x}{5} + \frac{4}{7}$
84. $\frac{4-2}{3} = \frac{21}{12} - \frac{5x-3}{4}$
85. $\frac{2}{3}x + 40 = \frac{x}{2} - \frac{x}{6} - \frac{2x}{9}$
86. $\frac{1}{2}\left(\frac{1}{3}x + 4\right) + \frac{x}{4} = 17$
87. $1 - 3(x + 1) = 2(2 - 3x)$
88. $20 - 6(2x - 1) = 2$
89. $-3.2(1.9x + 0.8) - (2.9x - 0.3) = 1.32(2x - 1.8) - 16.028$
90. $1.53(4x - 3) - 2.8(3x + 1) = 0.74(5x + 6) - 6.448$

Answers to Practice Problems   **8a.** 8   **b.** 5   **c.** 15.1   **9a.** -9 **b.** $\frac{104}{25}$ **c.** $\frac{1}{2}$

6.5 Literal Equations

IDEA 1   A **literal equation** is an equation in which a letter may represent a variable or a constant. For example,

$$2x + 3y = 7 \quad \text{and} \quad ax + b = c$$

are literal equations. We will solve literal equations by using the same methods discussed in Sections 6.2, 6.3, and 6.4.

To Solve a Literal Equation for a Particular Letter:

1. Treat the letter you are solving for as a variable and all other letters as constants.
2. Solve the equation using the same principles and procedures used to solve a linear equation.

Example 13   Solve.

  a. $ax - b = c$, for x **b.** $A = LW$, for W

  c. $P = 2(L + W)$, for L **d.** $A = \dfrac{h}{2}(a + b)$, for b

  e. $ax + bx = c$, for x **f.** $a(x - d) = cx + b$, for x

Solution   To solve a literal equation, apply the same principles and procedures used to solve a linear equation in one variable.

13a.
$$ax - b = c$$
$$ax - b + b = c + b \qquad \text{Add } b$$
$$ax = c + b$$
$$\frac{ax}{a} = \frac{c + b}{a} \qquad \text{Divide by } a$$
$$x = \frac{c + b}{a}$$

13b.
$$A = LW$$
$$\frac{A}{L} = \frac{LW}{L} \qquad \text{Divide by } L$$
$$\frac{A}{L} = W$$

13c.
$$P = 2(L + W)$$
$$P = 2L + 2W \qquad \text{Multiply}$$
$$P + (-2W) = 2L + 2W + (-2W) \qquad \text{Add } -2W$$
$$P - 2W = 2L$$
$$\frac{P - 2W}{2} = \frac{2L}{2} \qquad \text{Divide by } 2$$
$$\frac{P - 2W}{2} = L$$

13d.
$$A = \frac{h}{2}(a + b)$$
$$2 \cdot A = 2 \cdot \frac{h}{2}(a + b) \quad \text{Multiply by 2}$$
$$2A = ha + hb$$
$$2A + (-ha) = ha + hb + (-ha) \quad \text{Add } -ha$$
$$2A - ha = hb$$
$$\frac{2A - ha}{h} = \frac{hb}{h} \quad \text{Divide by } h$$
$$\frac{2A - ha}{h} = b$$

13e.
$$ax + bx = c$$
$$(a + b)x = c \quad \text{Combine like terms}$$
$$\frac{(a + b)x}{(a + b)} = \frac{c}{(a + b)} \quad \text{Divide by } (a + b)$$
$$x = \frac{c}{a + b}$$

13f.
$$a(x - d) = cx + b$$
$$ax - ad = cx + b \quad \text{Multiply}$$
$$ax - ad + (-cx) = cx + b + (-cx) \quad \text{Add } -cx$$
$$(a - c)x - ad = b \quad \text{Combine like terms}$$
$$(a - c)x - ad + ad = b + ad \quad \text{Add } ad$$
$$\frac{(a - c)x}{(a - c)} = \frac{b + ad}{(a - c)} \quad \text{Divide by } a - c$$
$$x = \frac{b + ad}{a - c}$$

Practice Problem 10 *Solve for x.*

a. $3x + 2y = 6$ **b.** $abx = 1$

c. $\frac{x}{a} + \frac{1}{2} = b$ **d.** $ax + b = cx + d$

IDEA 2

Formulas are often expressed as literal equations. For example, $d = rt$ is a distance formula (d = distance, r = rate, and t = time). But sometimes it may be more convenient to express a formula in a different form. For example, suppose we know the distance d and the time t and we wish to find the rate r. In this case, we could solve the equation $d = rt$ for r to obtain a more useful formula.

Example 14   Solve for r; $d = rt$.

Solution
$$d = rt$$
$$\frac{d}{t} = \frac{rt}{t} \quad \text{Divide by } t$$
$$\frac{d}{t} = r$$

Practice Problem 11 *Solve.*

a. $A = \frac{1}{2}bh$, for h (formula for area of a triangle)

b. $P = 2(L + W)$, for W (formula for perimeter of a rectangle)

6.5 Exercises

Solve.

1. $d = rt$, for t
2. $I = Prt$, for t
3. $A = \frac{1}{2}bh$, for b
4. $P = 2b + f$, for b
5. $h = \frac{11}{2}(p + 40)$, for p
6. $S = (a + l)\frac{n}{2}$, for n
7. $S = (a + l)\frac{n}{2}$, for l
8. $F = \frac{9}{5}C + 32$, for C
9. $ax - bx = 2$, for x
10. $ax + x = 9$, for x
11. $l = \frac{E}{R + r}$, for r
12. $\frac{1}{F} = \frac{1}{U} + \frac{1}{V}$, for U
13. $A = P(1 + rt)$, for r
14. $C = \frac{5}{9}(F - 32)$, for F
15. $4(x - a) = d$, for x
16. $A = 1 + abx$, for x
17. $V = lwh$, for l
18. $F = \frac{W}{2} + \frac{2Wx}{L}$, for x
19. $ax - b = c(x + d)$, for b
20. $L = \frac{Mt - g}{t}$, for t
21. $acx - ax - a = 1$, for x
22. $V = \frac{1}{3}bh$, for b

Solve for x.

23. $x + y = 5$
24. $7x - y = 7$
25. $\frac{x}{3} + \frac{y}{4} = 1$
26. $3x + 5y = -7x - 9$
27. $3x - 5a = c$
28. $3bx + ax = 9$
29. $5(x - a) + 2 = 3(x + 4)$
30. $p(x + b) = c$

Solve.

31. $\frac{1}{a} + \frac{1}{b} = \frac{1}{c}$, for b
32. $\frac{1}{x} + \frac{1}{y} = \frac{1}{z}$, for y
33. $I = Prt$, for r
34. $I = Prt$, for P
35. $ax + bx = 6$, for x
36. $ay + by = 8$, for y
37. $a(b + x) = 3$, for x
38. $b(a + x) = 4$, for x
39. $A = \frac{1}{2}h(a + b)$, for a
40. $A = \frac{1}{2}h(a + b)$, for b

Fill in the blanks.

41. A _____ equation is an equation in which a letter may represent a variable or a _____.
42. To solve a literal equation, remember that the letter you are solving for is treated as a _____. All other letters are treated as _____.

Answers to Practice Problems   10a. $\frac{6 - 2y}{3}$ b. $\frac{1}{ab}$ c. $ab - \frac{a}{2}$ d. $\frac{b - d}{a - c}$ 11a. $\frac{2A}{b}$ b. $\frac{P - 2L}{2}$

6.6 Solving Linear Inequalities Containing One Operation

IDEA 1 An **inequality** is a mathematical statement that one quantity is less than or greater than another quantity. A **linear inequality in one variable** is an inequality of the form

$$ax + b < c$$

where a, b, and c are real numbers with $a \neq 0$. Some examples are;

$x + 2 < 5$	This is read as "x plus two is less than five." The $<$ means *less than*.
$2x + 3 \leq 9$	$\leq$ means *less than or equal to*.
$x - 7 > 6$	$>$ means *greater than*.
$2x \geq 10$	$\geq$ means *greater than or equal to*.

As with equations, a **solution** or **root** of an inequality is a value that can be substituted for the variable so that the inequality becomes a true statement. For example, -3 is a solution of the inequality $x - 3 < -4$ since

$$x - 3 < -4$$
$$-3 - 3 < -4 \qquad \text{Replace } x \text{ with } -3$$
$$-6 < -4 \qquad \text{True}$$

Remember that $a < b$ means that a is the left of b on the number line. Thus, since -6 is to the left of -4, -3 is a solution of $x - 3 < -4$.

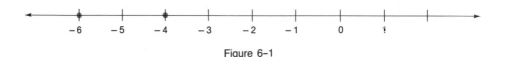

Figure 6-1

At this point we should also note that $b > a$ means that b is to the right of a on the number line. Thus, in Figure 6-1 we can see that $-6 < -4$ and $-4 > -6$ are equivalent statements. In general, $a < b$ and $b > a$ are equivalent statements.

IDEA 2 The methods used to solve linear inequalities are very similar to the procedures we used to solve linear equations in Sections 6.2 and 6.3. That is, to solve a linear inequality our goal is to isolate the variable on one side of the inequality symbol and a constant on the opposite side. This can be done by using the addition, multiplication, and/or division principles of inequality to produce a series of equivalent inequalities. **Equivalent inequalities** are inequalities that have the same solutions.

Before stating the addition principle of inequality, let us examine what happens when the same quantity is added to both sides of an inequality. For example, consider the inequality $3 < 5$. If we add 2 to both sides, we obtain

$$3 < 5$$
$$3 + 2 < 5 + 2$$
$$5 < 7 \qquad \text{True}$$

Similarly, if we add -2 to both sides, we obtain

$$3 < 5$$
$$3 + (-2) < 5 + (-2)$$
$$1 < 3 \quad \text{True}$$

The above results suggest that when we add the same quantity to both sides of a true inequality the result is a true inequality. This property is called the *addition principle of inequality*.

Addition Principle of Inequality

> If we add the same number to both sides of an inequality, the resulting inequality is equivalent to the original one,
>
> or
>
> If $A < B$
> Then $A + C < B + C$
>
> where A, B, and C are algebraic expressions.

NOTE: The principles of inequality will be stated by using the $<$ symbol, but they are valid for the $>$, $\geq$, or $\leq$ symbol.

We will solve linear inequalities in the same manner that we solved linear equations. For example, to solve for x in the inequality $x + 2 < 9$, we isolate x on the left side of the inequality by adding -2 to both sides of the inequality.

$$x + 2 < 9$$
$$x + 2 + (-2) < 9 + (-2) \quad \text{Add } -2$$
$$x < 7$$

The solution $x < 7$ means that all numbers less than 7 (6, 4.5, ½, -2, and so on) are solutions of $x + 2 < 9$. That is, if we replace x with any number less than 7, we will have a true statement. For example, since $1 < 7$,

$$x + 2 < 9$$
$$1 + 2 < 9$$
$$3 < 9 \quad \text{True}$$

Thus, 1 is a solution.

We can represent all the solutions to an inequality by using a number line. This is called **graphing the solution**. For example, the solution $x < 7$ can be represented by Figure 6-2. The hollow circle means that 7 is not a solution. The arrow to the left of the circle indicates that all numbers to the left of 7 are solutions.

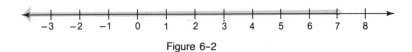

Figure 6-2

Example 15 Solve.

a. $x - 9 \geq -11$ b. $8 < x + 4$

Solution To eliminate a term from one side of an inequality, add the additive inverse of that term to both sides of the inequality.

15a.
$$x - 9 \geq -11$$
$$x - 9 + \boxed{9} \geq -11 + \boxed{9} \quad \text{Add 9}$$
$$x \geq -2$$

The solid dot in Figure 6-3 means that -2 is a solution. The arrow to the right of the dot implies that numbers to the right of -2 are solutions.

Figure 6-3

15b.
$$8 < x + 4$$
$$8 + \boxed{(-4)} < x + 4 + \boxed{(-4)}$$
$$4 < x$$
$$x > 4 \qquad \text{Rewrite the inequality}$$

We will usually write the variable on the left side. Thus, remember that $a < x$ and $x > a$ mean the same thing. The solution to this problem is shown in Figure 6-4.

Figure 6-4

Practice Problem 12 *Solve. Graph each solution.*

a. $x + 5 \leq -9$ **b.** $x - 2 > -8$ **c.** $18 \geq x + 19$

IDEA 3

Before stating the division principle of inequality, consider the true inequality

$$-6 < 8$$

If we divide both sides by 2, we get another true inequality:

$$-6 < 8$$
$$\frac{-6}{2} < \frac{8}{2}$$
$$-3 < 4$$

However, to obtain a true inequality when we divide both sides by -2, we must also reverse the direction of the inequality symbol:

$$-6 < 8$$
$$\frac{-6}{-2} > \frac{8}{-2} \qquad \text{Change} < \text{to} >$$
$$3 > -4$$

Based on the above results, we can now state the *division principle of inequality*.

Division Principle of Inequality

If both sides of an inequality are divided by the same positive quantity the resulting inequality is equivalent to the original inequality. If both sides of an inequality are divided by the same negative quantity and the direction (or sense) of the inequality symbol is reversed, the resulting inequality is equivalent to the original inequality,

or

If $A < B$

Then $\dfrac{A}{C} < \dfrac{B}{C}$ when C is positive

and $\dfrac{A}{C} > \dfrac{B}{C}$ when C is negative

where A, B, and C are algebraic expressions.

The division principle of inequality is used to eliminate the numerical coefficient of a term containing a variable.

Example 16 Solve.

    a. $3x < 18$ **b.** $-10x \leq 25$

    c. $3.2x \leq -13.76$ **d.** $18 > -6x$

Solution To eliminate the numerical coefficients of a variable, divide both sides of the inequality by that numerical coefficient.

16a. $3x < 18$

$\dfrac{3x}{\boxed{3}} < \dfrac{18}{\boxed{3}}$ Divide by 3

$x < 6$ See Figure 6-5

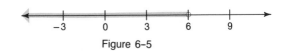

Figure 6-5

16b. $-10x \leq 25$

$\dfrac{-10x}{\boxed{-10}} \geq \dfrac{25}{\boxed{-10}}$ Divide by -10 and reverse the $\leq$

$x \geq -\dfrac{5}{2}$ or $-2\dfrac{1}{2}$ See Figure 6-6

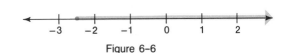

Figure 6-6

16c. $3.2x \leq -13.76$

$\dfrac{3.2x}{\boxed{3.2}} \leq \dfrac{-13.76}{\boxed{3.2}}$ Divide by 3.2

$x \leq -4.3$ See Figure 6-7

Figure 6-7

16d. $18 > -6x$

$\dfrac{18}{-6} < \dfrac{-6x}{-6}$ Divide by -6 and reverse the $>$

$-3 < x$

$x > -3$ Rewrite the inequality. See Figure 6-8

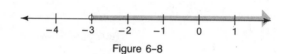
Figure 6-8

Practice Problem 13 **Solve. Graph each solution.**

a. $6x < -12$ **b.** $-0.3x < 9$ **c.** $-8x \leq 12$

IDEA 4

The multiplication principle of inequality is similar to the division principle of inequality. To illustrate this, consider the true inequality

$$-3 < 5$$

If both sides are multiplied by 2, we will obtain another true inequality. However, to obtain a true inequality when both sides are multiplied by -2, we must also reverse the direction of the inequality symbol:

$$-3 < 5$$
$$(-2)(-3) > (-2)(5)$$
$$6 > -10$$

Based on the above statements, we can now state the *multiplication principle of inequality*.

Multiplication Principle of Inequality

If both sides of an inequality are multiplied by the same positive quantity, the resulting inequality is equivalent to the original inequality. If both sides of an inequality are multiplied by the same negative quantity and the direction of the inequality symbol is reversed, the resulting inequality is equivalent to the original inequality,

or

If $A < B$
Then $AC < BC$ when C is positive
and $AC > BC$ when C is negative

where A, B, and C are algebraic expressions.

The multiplication principle of inequality is used to clear an inequality of fractions and to eliminate the numerical coefficient of a term containing a variable.

Example 17 Solve.

a. $\dfrac{x}{3} < 10$ b. $\dfrac{x}{-0.2} \geq 5$

c. $\dfrac{4}{9}x \leq \dfrac{10}{21}$ d. $-3x > 12$

Solution To eliminate the denominator of a fraction or a numerical coefficient, use the multiplication principle.

17a. $\dfrac{x}{3} < 10$

$\boxed{3} \cdot \dfrac{x}{3} < \boxed{3} \cdot 10$ Multiply by 3

$x < 30$ See Figure 6-9

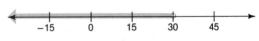

Figure 6-9

17b. $\dfrac{x}{-0.2} \geq 5$

$\boxed{-0.2}\left(\dfrac{x}{-0.2}\right) \leq \boxed{-0.2}(5)$ Multiply by -0.2 and reverse the $\geq$

$x \leq -1$ See Figure 6-10

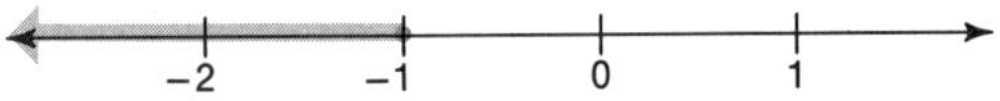

Figure 6-10

17c. $\dfrac{4}{9}x \leq \dfrac{10}{21}$

$\boxed{\dfrac{9}{4}} \cdot \dfrac{4}{9}x \leq \boxed{\dfrac{9}{4}} \cdot \dfrac{10}{21}$ Multiply by $\dfrac{9}{4}$

$x \leq \dfrac{\overset{3}{\cancel{9}} \cdot \overset{5}{\cancel{10}}}{\underset{2}{\cancel{4}} \cdot \underset{7}{\cancel{21}}}$

$x \leq \dfrac{15}{14}$ or $1\dfrac{1}{14}$ See Figure 6-11

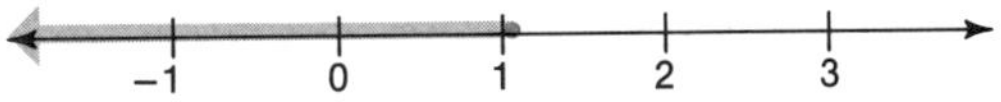

Figure 6-11

17d. $-3x > 12$

$\boxed{-\dfrac{1}{3}}(-3x) < \boxed{-\dfrac{1}{3}}(12)$ Multiply by $-\dfrac{1}{3}$ and reverse the $>$

$x < -4$ See Figure 6-12

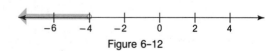

Figure 6-12

Practice Problem 14 *Solve. Graph each solution.*

a. $\dfrac{x}{-5} > -4$ b. $6x \le 15$ c. $\dfrac{35}{18} < \dfrac{25}{42}x$

6.6 Exercises

Solve by using the addition principle. Graph each solution.

1. $x + 7 < -2$
2. $x - 8 > 2$
3. $x + 8 \le 2$
4. $8 < x - 2$
5. $x + 0.7 \le 0.11$
6. $x - 0.2 \ge -0.5$
7. $x - 1.5 < 3.2$
8. $x - 5 \le -9$
9. $7 + x \ge -3$
10. $x + \dfrac{2}{3} < \dfrac{5}{3}$
11. $x - \dfrac{5}{6} \ge \dfrac{1}{10}$
12. $x + \dfrac{1}{18} \le -\dfrac{3}{10}$

Solve by using the multiplication or division principle. Graph each solution.

13. $4x < 8$
14. $5x < -10$
15. $-9x \ge -18$
16. $15x \ge -25$
17. $-6x \le 33$
18. $18x \le -9$
19. $0.3x \le 3$
20. $-2.3x > 0.69$
21. $1.2x < 1.44$
22. $\dfrac{x}{5} \le -3$
23. $\dfrac{x}{-6} \le 5$
24. $\dfrac{x}{-2} \ge -8$
25. $\dfrac{2}{3}x < -10$
26. $\dfrac{1}{2}x < -7$
27. $-\dfrac{3}{4}x > 2$
28. $\dfrac{4}{9}x \le \dfrac{6}{7}$
29. $-\dfrac{1}{5}x > \dfrac{3}{5}$
30. $\dfrac{x}{-0.1} \le 1$
31. $\dfrac{x}{0.2} > 0.1$
32. $0.5 \le \dfrac{x}{-5}$
33. $-0.5x \le 0.85$

Solve. Graph each solution.

34. $x + 0.2 < 0.7$
35. $-0.3x \le 6$
36. $-0.4x \le 8$
37. $\dfrac{3}{4}x > 9$
38. $\dfrac{3}{5}x > 6$
39. $-\dfrac{1}{4}x < 2.5$
40. $-\dfrac{1}{5}x < 3.2$
41. $\dfrac{x}{-3} \le -5$
42. $\dfrac{x}{-4} \le -2$
43. $x - \dfrac{3}{8} < \dfrac{1}{6}$
44. $x - \dfrac{3}{4} < \dfrac{5}{6}$
45. $x + 0.3 \le 2$
46. $x + 0.4 \le 3$
47. $x + 1 < -2$
48. $x + 4 < -6$
49. $x + \dfrac{3}{4} > -0.25$
50. $x + \dfrac{7}{4} > -0.25$
51. $6x \le 1.2$
52. $4x \le 1.6$
53. $-2x \ge 10$
54. $-5x \ge 20$
55. $14x < 20$
56. $12x < 30$
57. $-0.3x \le 0.36$
58. $-0.2x \le 0.24$
59. $\dfrac{x}{2} > -5.4$
60. $\dfrac{x}{4} > -3.5$
61. $\dfrac{x}{-5} \le 2$
62. $\dfrac{x}{-4} \le 6$
63. $\dfrac{x}{1.2} > 0.53$
64. $\dfrac{x}{3.4} > 0.48$
65. $6 < \dfrac{x}{2}$
66. $8 < \dfrac{x}{2}$
67. $\dfrac{x}{-2} \ge 4.3$
68. $\dfrac{x}{-3} \ge 2.4$
69. $\dfrac{x}{0.3} < -0.4$
70. $\dfrac{x}{0.6} < -0.3$
71. $\dfrac{3}{7}x < 6$
72. $\dfrac{5}{11}x < 10$
73. $\dfrac{1}{2}x > 4$
74. $\dfrac{1}{3}x > 2$
75. $-\dfrac{3}{4}x \ge 6$
76. $-\dfrac{4}{5}x \ge 8$
77. $\dfrac{6}{25}x \le \dfrac{9}{35}$

78. $\frac{4}{25}x \le \frac{6}{45}$ 　　 79. $-2x \le \frac{4}{7}$ 　　 80. $-3x < \frac{6}{7}$ 　　 81. $0.2x \le \frac{1}{5}$

82. $0.3x \le \frac{20}{3}$ 　　 83. $-0.1x \le \frac{1}{10}$ 　　 84. $-0.3x \le \frac{3}{10}$

Fill in the blanks.

85. An _____ is a mathematical statement that one quantity is less than or _____ than another.

86. To solve an inequality, proceed as if you were solving a _____ in one variable. The only exception is that if you multiply or _____ both sides of the inequality by a _____ number the _____ symbol must be reversed.

87. The solution $x > 1$ is represented by the following figure.

The hollow circle means that 1 is _____ a solution. The arrow to the right of the circle implies that all numbers to the right of _____ are _____ .

Answers to Practice Problems

12a. $x \le -14$ 　　 b. $x > -6$ 　　 c. $x \le -1$

13a. $x < -2$ 　　 b. $x > -30$ 　　 c. $x \ge -\frac{3}{2}$

14a. $x < 20$ 　　 b. $x \le \frac{5}{2}$ 　　 c. $x > \frac{49}{15}$

6.7 Solving Linear Inequalities Containing More Than One Operation

IDEA 1 Quite often you may need to use more than one principle to solve a linear inequality. Proceed as you would for linear equations.

> **To Solve a Linear Inequality:**
>
> 1. Apply the same procedures and principles used to solve a linear equation, *excepting*
>
> 2. If both sides of the inequality are multiplied or divided by the same negative quantity, reverse the direction of the inequality symbol.

Example 18 Solve.

a. $-12x - 1 < -17$
b. $7x + 4 > 2x - 6$
c. $7(3x + 6) \leq 11 - (x + 2)$
d. $\dfrac{x + 3}{4} - \dfrac{1}{4} > \dfrac{x - 2}{3}$

Solution To solve a linear inequality, proceed as if you were solving a linear equation. The only exception is that if you multiply or divide both sides of the inequality by a negative number, the inequality symbol must be reversed.

18a.
$$-12x - 1 < -17$$
$$-12x - 1 \boxed{+1} < -17 \boxed{+1} \quad \text{Add 1}$$
$$-12x < -16$$
$$\dfrac{-12x}{\boxed{-12}} > \dfrac{-16}{\boxed{-12}} \quad \text{Divide by } -12 \text{ and reverse the sign}$$
$$x > \dfrac{4}{3} \text{ or } 1\dfrac{1}{3} \quad \text{See Figure 6–13}$$

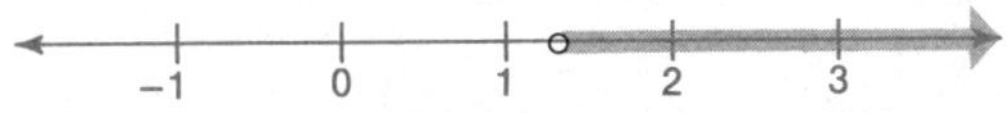

Figure 6–13

18b.
$$7x + 4 > 2x - 6$$
$$7x + 4 + \boxed{(-2x)} > 2x - 6 + \boxed{(-2x)} \quad \text{Add } -2x$$
$$5x + 4 > -6$$
$$5x + 4 + \boxed{(-4)} > -6 + \boxed{(-4)} \quad \text{Add } -4$$
$$5x > -10$$
$$\dfrac{5x}{\boxed{5}} > \dfrac{-10}{\boxed{5}} \quad \text{Divide by 5}$$
$$x > -2 \quad \text{See Figure 6–14}$$

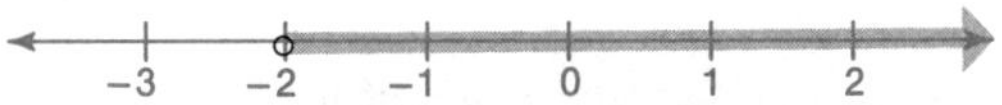

Figure 6–14

18c.
$$7(3x + 6) \leq 11 - (x + 2)$$
$$21x + 42 \leq 11 - x - 2 \quad \text{Multiply}$$
$$21x + 42 \leq 9 - x \quad \text{Combine like terms}$$
$$21x + 42 \boxed{+x} \leq 9 - x \boxed{+x} \quad \text{Add } x$$
$$22x + 42 \leq 9$$
$$22x + 42 + \boxed{(-42)} \leq 9 + \boxed{(-42)} \quad \text{Add } -42$$
$$22x \leq -33$$
$$\dfrac{22x}{\boxed{22}} \leq \dfrac{-33}{\boxed{22}} \quad \text{Divide by 22}$$
$$x \leq -\dfrac{3}{2} \text{ or } -1\dfrac{1}{2} \quad \text{See Figure 6–15}$$

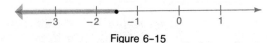

Figure 6–15

18d. $\dfrac{x+3}{4} - \dfrac{1}{4} > \dfrac{x-2}{3}$

$12\left(\dfrac{x+3}{4} - \dfrac{1}{4}\right) > 12\left(\dfrac{x-2}{3}\right)$ Multiply by LCD = 12

$3x + 9 - 3 > 4x - 8$

$3x + 6 > 4x - 8$ Combine like terms

$3x + 6 + (-3x) > 4x - 8 + (-3x)$ Add $-3x$

$6 > x - 8$

$6 + 8 > x - 8 + 8$ Add 8

$14 > x$

$x < 14$ Rewrite the inequality. See Figure 6-16

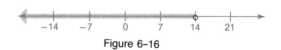

Figure 6-16

Practice Problem 15 *Solve. Graph each solution.*

a. $6x + 5 \geq -13$ b. $x \leq 13.2 + 0.2x$

c. $20 > 8 - 2(9 + 4x)$ d. $\dfrac{2x}{4} + \dfrac{x}{3} \geq 2$

IDEA 2

The inequalities that we have considered up to this point are called **conditional inequalities**. A **conditional inequality** is an inequality that is true for some but not all values of the variable. However there are two other types of inequalities. An **identity** or **absolute inequality** is a true statement for every value of the variable. An **inconsistent inequality** or **contradiction** is a false statement for every value of the variable.

Example 19 Solve.

a. $x < x + 2$ b. $x > x + 2$

Solution 19a. $x < x + 2$

$x + (-x) < x + 2 + (-x)$

$0 < 2$ True

Since the variable was eliminated and we obtained a true statement, every number is a solution of the inequality and $x < x + 2$ is an identity.

19b. $x > x + 2$

$x + (-x) > x + 2 + (-x)$

$0 > 2$ False

Since the variable was eliminated and the resulting statement was false, $x > x + 2$ is an inconsistent inequality and has no solution.

6.7 Exercises

Solve. Graph each solution.

1. $4x + 1 \leq 9$
2. $5x + 2 \leq 12$
3. $6x + 1 > 13$
4. $7x + 1 > 8$

5. $\frac{x}{2} + 1 < 4$

6. $\frac{x}{3} + 2 < 5$

7. $\frac{x}{-4} - 2 \geq -4$

8. $\frac{x}{-3} - 3 \geq -2$

9. $\frac{2}{5}x + 3 < 7$

10. $\frac{3}{4}x + 6 < 9$

11. $\frac{7}{5}x - 4 \leq -11$

12. $\frac{8}{7}x - 14 \leq -54$

13. $7x \leq 5x - 4$

14. $9x \leq 6x - 9$

15. $7x + 7 > 3 + 9x$

16. $5x + 28 > 7 + 2x$

17. $0.3x - 0.7 - 0.1x \geq 0.9 - 0.2x$

18. $0.5x - 0.2 - 0.1x \geq -2.3 - 0.3x$

19. $-0.9x - 0.5 < -0.5x + 0.5$

20. $-1.1x - 0.4 < -0.5x + 0.4$

21. $12 - 5x \leq x$

22. $16 - 3x \leq x$

23. $4(x + 2) < 16$

24. $3(x - 5) < 9$

25. $6(x + 2) \leq -12$

26. $2(x - 3) \leq 6$

27. $3(x + 3) > -3(2x + 3)$

28. $2(2x + 2) > -2(x + 2)$

29. $-4 \geq 3(2x + 1) + 11$

30. $8 \geq 2(3x - 4) + 2x$

31. $9 - 4x < 5(0 - 8x)$

32. $10 - 7x < 4(11 - 6x)$

33. $\frac{x}{6} + \frac{x}{4} \leq \frac{7}{2}$

34. $\frac{x}{15} + \frac{x}{9} \leq \frac{8}{5}$

35. $\frac{x}{2} > \frac{x}{2} + 6$

36. $\frac{x}{3} > \frac{x}{7} + 12$

37. $\frac{x-2}{10} + \frac{x}{6} \geq \frac{1}{10}$

38. $\frac{x+2}{8} + \frac{x}{10} \geq \frac{1}{8}$

39. $\frac{2x-1}{3} + \frac{3x}{4} < \frac{5}{6}$

40. $\frac{3x-2}{4} + \frac{3x}{8} < \frac{3}{4}$

41. $\frac{x}{3} + 20 < \frac{x}{4} - \frac{x}{12} - \frac{x}{9}$

42. $\frac{x}{5} + 10 < \frac{x}{3} - \frac{x}{10} - \frac{x}{5}$

43. $6(5 - 4x) \leq 3(4x - 2) - 7(6 + 8x)$

44. $5(3 - 2x) \geq 8(3x - 4) - 4(1 + 7x)$

45. $5x + 1 < -11$

46. $-7x + 14 \leq -7$

47. $10x + 4 > -2$

48. $-12x + 40 \geq 10$

49. $5x + 5x \leq -20$

50. $-7x + x \leq -12$

51. $0.1x + 0.11x \leq 21$

52. $8x + 56 < 14 + 2x$

53. $42 - 30x \geq -16x - 14$

54. $0.35x - 1.7 \leq 1.85x + 4$

55. $16x + 1 > 4x + 25$

56. $x \leq 15 + 4x$

57. $4x < 2(12 - 2x)$

58. $2(2x + 3) > 14$

59. $40 > 5(-3x + 2)$

60. $7(5x - 2) < 6(6x - 1)$

61. $\frac{x}{5} + 3 \geq 5$

62. $7 \leq \frac{x}{3} + \frac{x}{4}$

63. $\frac{2x+4}{2} - \frac{4}{3} < \frac{2x-4}{3}$

64. $\frac{x}{7} - 10 \leq -\frac{x}{3}$

65. $\frac{x}{3} + 2 \leq 2 + \frac{2x}{6}$

66. $\frac{x+2}{7} > \frac{x-3}{2}$

Solve. Graph each solution.

67. $12.85y - 15.49 \geq 22.06(9.66y - 12.74)$

68. $7.12(3.65y - 8.09) < 5.76 - 3.39y$

69. $0.582x + 7.43 \leq -1.748x + 3.935$

70. $2.086x - 5.632 \leq 0.718 - 8.5048$

Answers to Practice Problems **15a.** $x \geq -3$ **b.** $x \leq 16.5$ **c.** $x > -\dfrac{15}{4}$

d. $x \geq \dfrac{12}{5}$

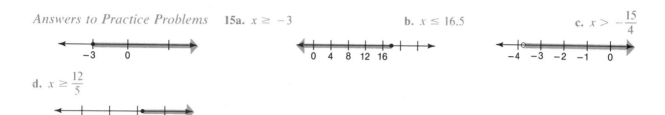

Chapter 6 Summary

Important Terms

An **equation** is a statement that two quantities are equal. $2 + 2 = 4$ is an equation that is a **true statement** and $3 + 1 = 6$ is an equation that is a **false statement**. [Section 6.1/Idea 1]

A **linear** or **first-degree equation in one variable** is an equation in which the variable has an exponent of one. $3x = 6$, $2n + 2 = 8$, and $0.3x - x + 4$ are linear equations in one variable. [Section 6.1/Idea 1]

A **solution or root** of an equation is a value that can be substituted for the variable so that the equation becomes a true statement. 3 is the solution of the equation $x + 4 = 7$. [Section 6.1/Idea 2]

A **conditional equation** is an equation that is true for only certain values of the variable. $x + 6 = -2$ and $2x = 12$ are conditional equations. [Section 6.1/Idea 3]

An **identity** is an equation that is a true statement for every value of the variable. $2x = x + x$ and $2(x + 5) = 2x + 10$ are identities. [Section 6.1/Idea 3]

An **inconsistent equation** or a **contradiction** is an equation that is a false statement for every value of the variable. $x = x + 6$ and $2(x - 3) - 2x = 5$ are inconsistent equations. [Section 6.1/Idea 3]

Equivalent equations are equations that have the same solutions $2x - 5 = 0$ and $2x = 5$ are equivalent equations. [Section 6.2/Idea 1]

The **addition principle of equality** says that if the same quantity is added to both sides of an equation, the resulting equation is equivalent to the original equation. If $A = B$, then $A + C = B + C$. [Section 6.2/Idea 1]

The **division principle of equality** says that if both sides of an equation are divided by the same nonzero quantity, the resulting equation is equivalent to the original equation. If $A = B$, then $\dfrac{A}{C} = \dfrac{B}{C}$. [Section 6.2/Idea 2]

The **multiplication principle of equality** says that if both sides of an equation are multiplied by the same nonzero quantity, the resulting equation is equivalent to the original equation. If $A = B$, then $AC = BC$. [Section 6.2/Idea 3]

A **literal equation** is an equation in which a letter may represent a variable or a constant. $ax + b = c$ is a literal equation. [Section 6.5/Idea 1]

An **inequality** is a statement that one quantity is less than or greater than another quantity. $2 < 3$ and $5 > 3$ are inequalities. [Section 6.6/Idea 1]

A **linear inequality in one variable** is an inequality that can be written in the form $ax + b < c$, where a, b, and c are constants and a is not equal to zero. $2x + 4 < 3$ and $2x < x + 5$ are linear inequalities in one variable. [Section 6.6/Idea 1]

A **solution** or **root** of a linear inequality in one variable is a value that can be substituted for the variable so that the inequality becomes a true statement. 4 is a solution of the inequality $2x + 3 < 15$. [Section 6.6/Idea 1]

Equivalent inequalities are inequalities that have the same solutions. $x < 2$ and $2x < 4$ are equivalent inequalities. [Section 6.6/Idea 2]

The **addition principle of inequality** says that if the same quantity is added to both sides of an inequality, the resulting inequality is equivalent to the original inequality. If $A < B$, then $A + C < B + C$ [Section 6.6/Idea 2]

The **division principle of inequality** says that if both sides of an inequality are divided by the same positive quantity, the resulting inequality is equivalent to the original inequality; if both sides of an inequality are divided by the same negative quantity, and the direction of the inequality symbol is reversed, the resulting inequality is equivalent to the original inequality. If $A < B$, then $\frac{A}{C} < \frac{B}{C}$ when C is positive, and $\frac{A}{C} > \frac{B}{C}$ when C is negative. [Section 6.6/Idea 3]

The **multiplication principle of inequality** says that if both sides of an inequality are multiplied by the same positive quantity, the resulting inequality is equivalent to the original inequality; if both sides of an inequality are multiplied by the same negative quantity and the direction of the inequality symbol is reversed, the resulting inequality is equivalent to the original inequality. If $A < B$, then $AC < BC$ when C is positive, and $AC > BC$ when C is negative. [Section 6.6/Idea 4]

A **conditional inequality** is an inequality that is a true statement for only certain values of the variable. $6x + 9 < 14$ is a conditioned inequality. [Section 6.7/Idea 2]

An **identity** or **absolute inequality** is an inequality that is a true statement for every value of the variable. $x < x + 6$ is an identity. [Section 6.7/Idea 2]

An **inconsistent inequality** or **contradiction** is an inequality that is a false statement for every value of the variable. $x > x + 6$ is an inconsistent inequality. [Section 6.7/Idea 2]

Important Skills

Determining if Values are Solutions of an Equation

To determine if a value is a solution of an equation, replace the variable with its given (or obtained) numerical value and simplify both sides of the equation. If the result is a true statement, the given value is a solution of the equation. [Section 6.1/Idea 2]

Solving Linear Equations

To solve an equation of the form $x + a = b$, add the additive inverse of a to both sides of the equation and simplify. [Section 6.2/Idea 1]

To solve an equation of the form $ax = b$, divide both sides of the equation by a or multiply both sides of the equation by the reciprocal of a and simplify. [Section 6.2/Idea 2]

To solve an equation of the form $\frac{x}{a} = b$, multiply both sides of the equation by a and simplify. [Section 6.2/Idea 3]

To solve a linear equation containing variables on only one side of the equation, combine like terms, yielding an equation of the form $ax + b = c$ or $\frac{x}{a} + b = c$. Add the additive inverse of b to both sides of the equation, simplify, and solve the resulting equation by using the multiplication or division principle of equality. [Section 6.3/Idea 1]

To solve a linear equation containing variables on both sides of the equation, combine like terms on each side of the equation. Isolate all terms containing the variable on one side and constant terms on the other by using the addition principle of equality. Combine like terms again, and use the division or multiplication principle of equality to solve the resulting equation. [Section 6.3/Idea 2]

To solve a linear equation containing grouping symbols, eliminate the grouping symbols by performing the indicated operations. Then solve the resulting equation. [Section 6.4/Idea 1]

To solve a linear equation containing fractions, multiply every term of the equation by the LCD of all terms. Then solve the resulting equation. [Section 6.4/Idea 2]

To solve a linear equation in general, simplify both sides of the equation by eliminating fractions, eliminating grouping symbols, and/or collecting like terms. Use the addition principle of equality to isolate terms containing the variable on one side of the equation and the constants on the other. Combine like terms again. Finally, use the multiplication or division principle of equality to solve the resulting equation. [Section 6.4/Idea 4]

Solving Literal Equations

To solve a literal equation, treat the letter you are solving for as a variable and all other letters as constants. Then apply the same procedures and principles used to solve a linear equation. [Section 6.5/Idea 1]

Solving Linear Inequalities

To solve a linear inequality, apply the same procedures and principles used to solve a linear equation, excepting if both sides of the equation are multiplied or divided by a negative quantity, reverse the direction of the inequality symbol. [Section 6.7/Idea 1]

Chapter 6 Review Exercises

Is the given value the solution?

1. $3x - 9 = -39$, $x = -10$
2. $3x + 2.11 = 5x - 3.2$, $x = 0.2$
3. $7(5x + 1) - 18x = 12x$, $x = -\dfrac{7}{5}$

Solve for x. For all inequalities, graph the solution.

4. $x + 5 = -2$
5. $-5x < 10$
6. $\dfrac{2}{3}x = -18$
7. $-6x = 18$
8. $\dfrac{x}{3} - 9 = 2$
9. $\dfrac{x}{6} - \dfrac{x}{10} = 1$
10. $x - 0.7 = -0.31$
11. $\dfrac{x}{-3} + 4 < 1$
12. $ax + b = c$
13. $x = 0.2x - 16$
14. $8(x + 3) = 4$
15. $12x - \dfrac{3}{4} = \dfrac{4x}{3}$
16. $\dfrac{1}{x} = \dfrac{1}{y} + \dfrac{1}{z}$
17. $x + 7 < 1$
18. $\dfrac{1}{5}x \geq -5$
19. $x - 0.1 \geq 3.52$
20. $4(3x - 5) - (x - 1) = 5x - 5$
21. $2(x - 7) \leq -2x + 8$
22. $\dfrac{x}{4} + \dfrac{x}{3} < 26 - \dfrac{x}{2}$
23. $ax + b = cx + 2b$
24. $\dfrac{3}{4}x + \dfrac{1}{3}x = 6\left(x + \dfrac{5}{36} - \dfrac{1}{72}x\right)$
25. $4(x + 2) - 4 = 4x - 1$
26. $2(x - 3) = 2x - 6$
27. $5x - 3(x - 2) < 2x + 8$

28. $4x - (x - 7) \leq 3x + 4$
29. $-3(x - 5) + 8 < -4(x + 2)$
30. $-\dfrac{14}{25}x \leq \dfrac{21}{35}$
31. $\dfrac{2}{3}(2x + 5) = \dfrac{6}{5}$
32. $0.35x - 1.7 = 1.85x + 4$
33. $0.6x \leq x - 18$
34. $2[3 - 5(x - 4)] = 10 - 5x$
35. $3[5 - 2(7 - x)] = 6(x - 7)$

Chapter 6 Test

Name: _____

Class: _____

Determine if the given value is a solution.
1. $2x - 6 = 15$, $x = 8$
2. $\frac{2x}{3} = 6 + 2x$, $x = -\frac{9}{2}$

Solve.
3. $x - 5 = -9$
4. $x - \frac{1}{10} = \frac{-5}{18}$
5. $0.5x = 5$
6. $\frac{x}{-1.2} = 5$
7. $\frac{2}{3}x = -10$
8. $5x + 25 = 21 - 4x$
9. $0.4x - 1.8 = 0.54$
10. $8x - 3(x - 7) = 4(3x + 2)$
11. $\frac{1}{2}(2x + 6) = \frac{4}{3}$
12. $\frac{1}{3}n + \frac{3}{4}n + 5 = \frac{1}{6}n$
13. $\frac{x}{5} = 14 - \frac{x}{2}$
14. $ax - ad = cx + b$, for x
15. $h = \frac{11}{2}(x + 40)$ for x
16. $x - 0.7 \leq 0.55$
17. $-15x < 40$
18. $\frac{8}{35}y \geq \frac{12}{45}$
19. $x + 5 > 0.3x - 2$
20. $x - \frac{5}{3} - \frac{2}{3}x < \frac{13}{12}x$

1. _____
2. _____
3. _____
4. _____
5. _____
6. _____
7. _____
8. _____
9. _____
10. _____
11. _____
12. _____
13. _____
14. _____
15. _____
16. _____
17. ←_____→
18. ←_____→
19. ←_____→
20. ←_____→

7 Solving Word Problems by Using Linear Equations in One Variable

Objectives The objectives for this chapter are listed below along with sample problems for each objective. By the end of this chapter you should be able to find the solutions to the given problems.

1. Express a written phrase as an algebraic expression *(Section 7.1/Ideas 1-2)*.
   If the sum of two numbers is 12, how would you represent each number?

2. Solve a word problem by using equations *(Section 7.1/Idea 4)*.
   If ten less than twice a number is 50, find the number.

3. Write a written phrase as a ratio *(Section 7.2/Idea 1)*.
   Find the ratio of eight inches to eight feet.

4. Solve a ratio and proportion problem *(Section 7.2/Idea 5)*.
   If eight pens cost $10, how much would 20 pens cost?

5. Solve a percent problem *(Section 7.3/Ideas 1-2)*.
   Jane got 90% of the questions correct on a 50-item test. How many items did she get wrong?

6. Solve a simple interest problem *(Section 7.4/Ideas 1-2)*.
   Dan invested $7000 at 10%. How much money must he invest at 16% so that his total investment will yield an annual income of 12%?

7. Solve a mixture problem *(Section 7.5/Idea 1)*.
   How many liters of pure acid must be added to 200 liters of a 10% acid solution to obtain a solution that is 20% acid?

8. Solve a uniform motion problem *(Section 7.6/Ideas 1-2)*.
   A truck leaves a depot at 35 mph. Two hours later, a car leaves the same depot at 55 mph. How many hours will it take the car to pass the truck?

7.1 Solving Word Problems

IDEA 1 Many applied or word problems can be solved by using equations. However, before we learn how to solve word problems, it is important to review how to change written statements to algebraic expressions. For example, to translate "the total of 30 and a number" into a mathematical expression, you should be thinking:

1. Assign a variable to the unknown.　　$x = $ a number
2. Identify the phrase that indicates a mathematical operation.　　total means add ($+$)
3. Use given numbers to complete the algebraic expression.　　$30 + x$

Thus, "the total of 30 and a number" is written as $30 + x$.

Example 1　Change each phrase to an algebraic expression. Let x be the number.

a. six more than a number.

b. a number decreased by 3

c. five less than a number

d. three-fourths of a number

e. the quotient of a number and six

f. the product of a number and two

g. the sum of a number and four

h. a number increased by seven

i. four times the sum of a number and eight

Solution

1a. $x + 6$　　More than indicates addition

1b. $x - 3$　　Decreased by indicates subtraction

1c. $x - 5$　　Less than indicates subtraction

1d. $\frac{3}{4}x$　　A fractional part of a number indicates multiplication

1e. $\frac{x}{6}$　　Quotient indicates division

1f. $2x$　　Product indicates multiplication

1g. $x + 4$　　Sum indicates addition

1h. $x + 7$　　Increased by indicates addition

1i. $4(x + 8)$　　Times indicates multiplication

When translating written statements into algebraic expressions, you may find it helpful to use a chart. For example, you can translate the phrase "six more than a number" into an algebraic expression by thinking:

Number	Six more than a number	Answer
2	$2 + 6$	8
3	$3 + 6$	9
x	$x + 6$	$x + 6$

Practice Problem 1 *Change each phrase to an algebraic expression. Let x be the number.*

    a. one-half of a number   **b.** a number squared

    c. the sum of a number and nine   **d.** six times a number

    e. a number decreased by two   **f.** four more than twice a number

    g. the reciprocal of a number

IDEA 2

To solve word problems, you must also be able to write two or more unknown quantities in terms of the same variable.

Example 2 Express each unknown in terms of the same variable.

    a. If x = the length of a piece of board, how would you represent the length of a second piece of board that is two feet longer than the first?

    b. If d = the cost of one tire, how would you represent the cost of four tires?

    c. Let x = an integer. How would you represent the next consecutive integer?

    d. Let n = an even integer. How would you represent the next consecutive even integer?

    e. Let n = an integer. How would you represent the next two consecutive integers?

    f. Let x = the number of men in the class. How would you represent the number of women in the class if the number of women is five less than twice the number of men?

Solution

2a. $x + 2$ = length of second piece

2b. $4d$ = cost of four tires

2c. The numbers 2 and 3 are examples of two consecutive integers. This implies that given any integer, the next consecutive integer is found by adding 1 to that number. Thus, $x + 1$ = the next consecutive integer.

2d. The numbers 6 and 8 are examples of two consecutive even integers. This implies that given any even integer, the next consecutive even integer is found by adding 2 to that number. Thus, $n + 2$ = the next consecutive even integer.

2e. n = an integer
$n + 1$ = the next consecutive integer
$n + 2$ = the third consecutive integer

2f. $2x - 5$ = the number of women

Always check to see if you have represented the unknowns correctly by replacing the variable with a number. For example, x and $x + 1$ represent two consecutive integers since if $x = 12$, then $x + 1 = 13$.

Practice Problem 2 *Express each unknown in terms of the given variable.*

    a. If x = an odd integer, how would you represent the next two consecutive odd integers?

    b. If b = Pam's age, how would you represent her age 10 years from now?

c. Let m = the measure of the first angle of a triangle. If the second angle is three times as large as the first angle and the third angle is 30° less than the second angle, how would you represent the measure of the second and third angles?

Example 3   Represent each unknown in terms of the same variable.

a. If a house costs $18,000 more than a lot, how could you represent the cost of each?

b. The sum of two numbers is 35. How would you represent the two numbers?

c. The ABT construction company has 23 employees. If some of the workers are skilled and the rest are unskilled, how would you represent the number of skilled workers and the number of unskilled workers?

d. Train A leaves Detroit at 9:00 A.M. for Chicago, Train B leaves Chicago at 10:00 A.M. for Detroit. When the trains pass each other, how many hours has each train traveled?

Solution   To represent two or more unknowns, let any letter represent one of the unknowns. Next, express all other unknowns in terms of that letter.

3a. x = cost of the lot
      $x + 18,000$ = cost of the house

3b. n = first number
      $35 - n$ = second number

It may be helpful to think of two or more numbers to represent these unknowns.

First number	Second number	Check
2	$35 - 2 = 33$	$2 + 33 = 35$
5	$35 - 5 = 30$	$5 + 30 = 35$
n	$35 - n$	$n + (35 - n) = 35$

As the table helps to show, when we are given the sum of two numbers (or quantities) and we represent one number with a variable, the other is represented by the given sum minus that variable.

3c. x = number of skilled workers
      $23 - x$ = number of unskilled workers

3d. t = hours train A travels before passing train B. Since train B leaves one hour later, it travels one hour less than train A. Thus, $t - 1$ = hours train B travels before passing train A.

!   The concept presented in Example 3b will be helpful in solving many word problems. Thus, we suggest that you review the solution to Example 3b carefully.

Practice Problem 3   *Represent each unknown in terms of the same variable.*

a. Richard is two years older than Reggie. How would you represent each man's age?

b. Dan invested $9000 in two real estate developments. How would you represent the amount of money he invested in each development?

c. Ernie leaves his house at 11:00 A.M. on a bike. Two hours later, Jerry leaves Ernie's house in a car. How would you represent the number of hours each person has traveled when Jerry catches up to Ernie?

IDEA 3

The third skill that will be reviewed before we show the complete solution to a word problem is how to change written statements into equations. For example, to translate "the sum of a number and half that number is 30" into an equation, you should be thinking:

1. Represent the unknowns — x = a number; $\frac{x}{2}$ = half a number
2. Rewrite the problem as a short phrase. — a number plus half a number is 30
3. Change the words to algebraic expressions or symbols.

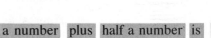

$$x + \frac{x}{2} = 30$$

Example 4 Write an equation for each problem.

a. If 10 plus three times a number is 40, what is the number?

b. If two-thirds of a number is 20, what is the number?

c. If the sum of two consecutive integers is 19, what are the numbers?

d. Four times a number is equal to the product of three and that number decreased by four. What is the number?

e. Jim earned twice as much money as Joe. If their earnings totaled $960, how much money did each man earn?

f. A class contains 35 students. The number of girls is five less than three times the number of boys. How many boys and how many girls are in the class?

Solution To change a word problem to an equation, first represent all unknowns in terms of the same letter. Next, rewrite the original problem as a short statement and then translate the words to algebraic expressions or symbols.

4a. x = the number

ten plus three times a number is 40

$$10 + 3x = 40$$

4b. x = the number

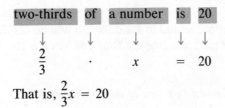

That is, $\frac{2}{3}x = 20$

4c. x = an integer; $x + 1$ = the next consecutive integer

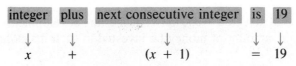

4d. x = the number

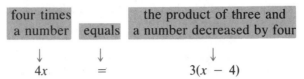

4e. x = Joe's earnings; $2x$ = Jim's earnings

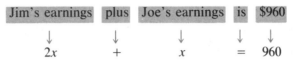

4f. x = number of boys; $3x - 5$ = number of girls

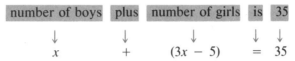

Practice Problem 4 *Write an equation for each of the following.*

a. If 12 more than twice a number is 50, what is the number?

b. The sum of two consecutive even integers is 70. What are the numbers?

c. Jane types 55 words per minute. If Jane only types two-fifths as fast as Jim, how many words per minute can Jim type?

IDEA 4

Now that we have reviewed the basic skills required to solve word problems, we can present several steps that can be used to solve word problems.

> **To Solve a Word Problem:**
>
> 1. **Represent unknowns.** Read the problem until you understand what is given (the knowns) and what must be found (the unknowns). Represent each unknown in terms of the same variable.
>
> 2. **Write an equation.** Reread the problem and then change the written statements into an equation.
>
> 3. **Solve the equation and answer the question asked.** Solve the equation written in step 2 and then use this answer to find the quantity or quantities asked for in the original problem.
>
> 4. **Check your solution.** Determine if the values obtained in step 3 meet the conditions given in the original problem.

Example 5 Solve by using equations.

a. Three times a number decreased by four is equal to twice the sum of that number and six. What is the number?

b. The sum of two consecutive even integers is 70. Find the integers.

c. A basketball coach wants to buy practice jerseys for her team. If each jersey costs $9.50, how many jerseys can she buy with $136?

d. Cazzie makes hats and sells them to earn extra money. The cost for making one hat is $2 and he has fixed costs of $120 a month. How many hats can he make next month for $270?

Solution To solve a word problem, read the problem carefully and then (1) represent the unknowns, (2) write an equation, (3) solve it and find all solutions, and (4) check your answers.

5a. Step 1: $x =$ the number

Step 2: 3 times a number decreased by 4 is 2 times the sum of that number and 6
$$3x \quad - \quad 4 \quad = \quad 2(x + 6)$$

Step 3: $3x - 4 = 2(x + 6)$
$3x - 4 = 2x + 12$
$x - 4 = 12$
$x = 16$
The number is 16.

Step 4: The answer checks since 3 times 16 decreased by 4 is 44 and 2 times the sum of 16 and 6 is 44.

5b. Step 1: $x =$ an even integer
$x + 2 =$ next consecutive even integer

Step 2: first integer plus second integer is 70
$$x \quad + \quad (x + 2) \quad = \quad 70$$

Step 3: $x + (x + 2) = 70$
$2x + 2 = 70$
$2x = 68$
$x = 34$
$x + 2 = 36$
The numbers are 34 and 36.

Step 4: The answers check since 34 and 36 are consecutive even integers and $34 + 36 = 70$.

5c. Step 1: $n =$ the number of jerseys the coach can buy

Step 2: total cost of jerseys equals amount she can spend
$$9.50n \quad = \quad 136$$

Step 3: $9.50n = 136$
$n \doteq 14.3$ Divided both sides by 9.50

The coach can buy 14 jerseys.

Step 4: The answer checks since 14 jerseys will cost 14 ($9.50) = $133 and the remaining $3 is not enough to buy another jersey.

5d. Step 1: $n =$ number of hats that can be made

Step 2: fixed cost plus cost for making n hats is $270
$$120 \quad + \quad 2n \quad = \quad 270$$

Step 3: $120 + 2n = 270$
$2n = 150$
$n = 75$
Cazzie can make 75 hats.

Step 4: The answer checks since the cost of making 75 hats is $120 + 2 ($75) = $120 + $150 = $270.

Practice Problem 5 **Solve by using equations.**

    a. Joan has a piece of material that is 48 inches long. If it must be cut into two pieces so that one piece is 12 inches longer than the other, how long should each piece be?

    b. Two-fifths of a number plus one-half of that same number is equal to -9. What is the number?

    c. Davis Sporting Goods makes basketball uniforms. The manufacturing cost of each uniform is $10. If the company has fixed costs of $100,000 a year, how many uniforms can be made next year for $250,000?

7.1 Exercises

Change each phrase to an algebraic expression. Let x be the number.

1. 10 more than a number
2. a number decreased by eight
3. two-thirds of a number
4. the quotient of a number and six
5. the product of a number and two
6. the sum of a number and twice that number
7. six less than a number
8. eight more than a number
9. the product of a number and six
10. the product of a number and 10
11. eight less than a number
12. nine less than a number
13. 50 less than a number
14. 60 less than a number
15. the difference between a number and seven
16. the difference between a number and nine
17. the difference between a number and 80
18. the difference between a number and 60
19. the quotient of a number and 12
20. the quotient of a number and 20
21. a number divided into 10
22. a number divided into eight
23. subtract two from a number
24. subtract nine from a number
25. a number increased by 90
26. a number increased by 80
27. nine times a number
28. six times the sum of a number and two
29. the sum of three and the quotient of a number and two
30. the difference of a number and 30
31. the sum of four times a number and seven
32. five times the difference of a number and six
33. twice a number subtracted from two divided by four
34. twice a number subtracted from one divided by five
35. eight more than twice a number
36. six more than three times a number
37. 10 less than half a number
38. eight less than one-third of a number
39. nine times the sum of a number and seven
40. eight times the sum of a number and six
41. five more than one-fourth a number
42. six more than one-fifth a number
43. the sum of a number and the quotient of two and six
44. the difference of a number and the quotient of two and three

Express each unknown in terms of the same variable.

45. If $x =$ John's monthly salary and Jim earns $130 more a month than John, how would you represent Jim's earnings?

46. If b = the cost of one chair, how would you represent the cost of six chairs?
47. If l = the length of a piece of board, how would you represent the length of a second piece of board that is 13 feet shorter than the first piece?
48. If t = the hours train A traveled, and train B traveled 11 hours longer than train A, how would you represent the hours train B traveled?
49. If n = an odd integer, how would you represent the next consecutive odd integer?
50. If n = an even integer, how would you represent the next consecutive even integer?
51. If d = the balance remaining in Steve's checking account, how would you represent his new balance if he withdrew $35?
52. If c = the cost of a coat, how would you represent the cost of a second coat that costs $100 more than the first coat?
53. If s = Joe's salary, how would you represent Sue's salary if she earns $75 more than Joe?
54. If b = Jill's bank balance, how would you represent her new balance after a deposit of $525?
55. If the sum of two numbers is 10 and x = the first number, how would you represent the second number?
56. If the sum of two numbers is 30 and x = the first number, how would you represent the second number?
57. If the sum of two numbers is 40 and x = the first number, how would you represent the second number?
58. If the sum of two numbers is 11 and x = the first number, how would you represent the second number?
59. If x = the number of players on Team A, how would you represent the number of players on Team B if Team B has half as many players as Team A?
60. If n = the number of people on Team A, how would you represent the number of people on Team C if Team C has one-third as many players as Team A.
61. If one number is twice as large as another, how would you represent the two numbers?
62. If one number is four times as large as another, how would you represent the two numbers?
63. If Joe is eight inches taller than John, how would you represent each man's height?
64. If Sue is two inches taller than Jane, how would you represent each woman's height?
65. Tom's savings account contains $1025 less than Dan's savings account. How would you represent the amount of money each man has in his account?
66. Frank's car is worth $587 less than Ralph's car. How would you represent how much each man's car is worth?
67. The width of a rectangle is three inches longer than its length. How would you represent the length and the width of the rectangle?
68. The length of a rectangle is eight inches longer than its width. How would you represent the length and the width of the rectangle?
69. The sum of two numbers is 60. How would you represent each number?
70. If the sum of two numbers is eight, how would you represent the two numbers?
71. Joe has $1000. If he gives part of the money to his brother and the rest to his sister, how much money did each person receive?
72. Rich has $80,000. If Guerin inherits part of the money and Cazzie inherits the rest, how much money should each man receive?
73. If b = the length of the base of a triangle, how would you represent its altitude if the altitude is five less than twice the length of the base?
74. If the sum of two numbers is 75, how would you represent the two numbers?
75. If one number is four times as large as another, how would you represent each number?
76. If the sum of two numbers is 10, how would you represent each number?
77. The ABC nursery school employs 50 people. If some of the employees are teachers and the rest are teacher-aides, how would you represent the number of teachers and the number of teacher-aides?

78. John is ten years older than Sue. How would you represent each person's age?
79. Vanessa has a balance of $300 in her savings account. If she deposits x dollars into her account, how would you represent her new balance?
80. One number exceeds another by 12. How would you represent the two numbers?
81. Ernie invested $5000. If he invested some of his money in stocks and the rest in real estate, how would you represent the amount he invested in stocks and the amount he invested in real estate?
82. Two cars travel in opposite directions from the state university. If in one hour they are 50 miles apart, how would you represent the distance each car traveled?
83. If x gallons of water is drained from a tank containing 10 gallons of water, how would you represent the amount of water remaining in the tank?
84. The second angle of a triangle is five times as large as the measure of the first angle. The third angle is two degrees smaller than the first angle. How would you represent the measure of the three angles of the triangle?
85. How would you represent three consecutive integers if x = the first integer?

Solve by using equations.

86. If seven is added to three times a number and the result is 22, what is the number?
87. The sum of two numbers is 60. If one number exceeds the other by 10, what are the numbers?
88. A number plus twice that number is 30. What is the number?
89. The sum of two consecutive integers is 35. Find the numbers.
90. The sum of two consecutive even numbers is 70. What are the numbers?
91. Four times the sum of nine and a number is 16. Find the number.
92. Four times a number is equal to the product of three and that number decreased by four. What is the number?
93. When eight is subtracted from six times a number, the result is 76. Find the number.
94. When 37 is subtracted from three times a number, the result is -22. Find the number.
95. If seven is added to three-fifths of a number, the result is 13. Find the number.
96. If eight is added to two-fifths of a number, the result is two. Find the number.
97. The current price of a chair is $75 less than three times its cost in 1970. If the current price is $1200, what was the cost in 1970?
98. The current price of a table is $18 less than three times its cost in 1965. If the current price is $207, what was the cost in 1965?
99. The sum of two consecutive integers is 79. Find the numbers.
100. The sum of two consecutive integers is 91. Find the numbers.
101. The sum of two consecutive odd numbers is 188. Find the numbers.
102. The sum of two consecutive odd numbers is 168. Find the numbers.
103. Find three consecutive even numbers whose sum is 36.
104. Find three consecutive even numbers whose sum is 60.
105. The first of two numbers is 10 more than the other. If their sum is 56, find the numbers.
106. The first of two numbers is eight more than the other. If their sum is 46, find the numbers.
107. In a school election, Howard received 140 more votes than Gene. If the total number of votes cast was 384, how many votes did each candidate get?
108. Laura earns twice as much money as Dick. If their combined salaries total $615, how much does each person earn?
109. A class contains 40 students. If three-fourths of the students are women, how many women are in the class?
110. A house and lot cost $55,000. If the lot cost $15,000 less than the house, how much does each cost?

111. The second angle of a triangle is five times as large as the measure of the first angle. The third angle is 40° greater than the first angle. Find the measure of the three angles. (*Hint:* the sum of the measures of the three angles must be 180°.)

112. Guerin makes model airplanes and sells them to make extra money. The cost for making one plane is $5, and he has fixed costs of $50 a month. How many planes can he make for $155?

113. The ABC car rental company charges $20.95 per day plus $0.17 (17 cents) per mile to rent a Ford. If Jim rented a car for one day and his total bill was $50.02, how many miles did he travel?

114. The Carlson car rental company rents compact cars for $19.95 per day and $0.18 per mile. If a car was rented for two days, how many miles was it driven if the total bill was $75.90?

115. Rich has 125 shares of the PDQ paint company. If Rich has only one-fifth as many shares as Ron, how many shares of PDQ stock does Ron have?

116. A class contains 39 students. If the number of girls is five less than three times the number of boys, how many boys and how many girls are in the class?

117. The sum of three consecutive odd numbers is 117. What are the numbers?

118. A record-breaking total of 1100 people attended a local basketball game. The admission was $2 for students and $3 for nonstudents. If total gate receipts were $2800, how many student tickets and how many nonstudent tickets were sold?

119. The daily payroll of the Pace Education Center is $775. The teachers earn $50 per day and the tutors earn $25 per day. If the center employs 21 people, find the number of teachers and the number of tutors employed.

Answers to Practice Problems 1a. $\frac{1}{2}x$ or $\frac{x}{2}$ b. x^2 c. $x + 9$ d. $6x$ e. $x - 2$ f. $2x + 4$ g. $\frac{1}{x}$
2a. $x + 2, x + 4$ b. $b + 10$ c. $3m, 3m - 30$ 3a. Reggie's age $= x$ b. $x, 9000 - x$
Richard's age $= x + 2$
c. Number of hours Jerry traveled $= x$ 4a. $2x + 12 = 50$ b. $x + (x + 2) = 50$ c. $55 = \frac{2}{5}x$ 5a. 18 inches,
Number of hours Ernie traveled $= x + 2$
30 inches   b. -10 c. 15,000

7.2 Ratio and Proportion Problems

IDEA 1 We can compare numbers in different ways. One way is by using a ratio. A **ratio** is the quotient of two numbers or quantities, or a fraction that expresses a relationship between two quantities. For example, the ratio of 4 feet to 6 feet is

$$\frac{4 \text{ feet}}{6 \text{ feet}} = \frac{4}{6} = \frac{2}{3} \text{ or } 4 \text{ feet}: 6 \text{ feet} = 4:6 = 2:3$$

In general, the ratio of a to b is written as $\frac{a}{b}$ or $a:b$.

When a ratio of two quantities contains the same unit of measurement, write the ratio without the unit of measure. Also, always reduce ratios to their lowest terms.

Example 6   Class A contains 25 boys and class B contains 35 boys.

  a. What is the ratio of 25 boys to 35 boys?

  b. What is the ratio of 35 boys to 25 boys?

Solution   The ratio of a to b is written as $\frac{a}{b}$ or $a:b$.

6a. The ratio of 25 boys to 35 boys is $\frac{25 \text{ boys}}{35 \text{ boys}} = \frac{5}{7}$. The ratio can also be written as 5:7. This means that for every five boys in class A there are seven boys in class B. It also means that class A has $\frac{5}{7}$ as many boys as class B.

6b. The ratio of 35 boys to 25 boys is $\frac{35 \text{ boys}}{25 \text{ boys}} = \frac{7}{5}$.

When finding a ratio expressed in different units of measure, first express the quantities in the same unit.

Example 7
a. What is the ratio of six hours to three days?

b. What is the ratio of 3 dimes to 8 nickels?

Solution   First express the quantities in the same unit.

7a. Since one day = 24 hours, three days = 3(24 hours) = 72 hours. Thus the ratio of six hours to three days is $\frac{6 \text{ hours}}{3 \text{ days}} = \frac{6 \text{ hours}}{72 \text{ hours}} = \frac{1}{12}$.

7b. Since 1 dime = 2 nickels, 3 dimes = 6 nickels. Thus, the ratio of 3 dimes to 8 nickels is $\frac{3 \text{ dimes}}{8 \text{ nickels}} = \frac{6 \text{ nickels}}{8 \text{ nickels}} = \frac{3}{4}$

IDEA 2

One very useful example of a ratio is a rate. A **rate** is the ratio of two quantities having different units of measure. For example, if you traveled 30 miles in 2 hours, then your average rate of speed is the ratio of miles to hours. The ratio is

$$\frac{30 \text{ miles}}{2 \text{ hours}} = \frac{15 \text{ miles}}{1 \text{ hour}} \qquad \text{Divide both 30 and 15 by 2}$$

This is usually expressed as $15 \frac{\text{miles}}{\text{hour}}$ or 15 miles per hour.

Example 8
a. A pipe can fill a 125 gallon tank with water in 2 hours. What is the rate in gallons per hour?

b. A car travels 45.6 miles on 2.4 gallons of gas. What is the rate in miles per gallon.

Solution
8a. The ratio of gallons to hours is

$$\frac{125 \text{ gallons}}{2 \text{ hours}} = 62.5 \frac{\text{gallons}}{\text{hour}} = 62.5 \text{ gallons per hour}$$

The tank is filled at a rate of 62.5 gallons per hour.

8b. The ratio of miles to gallons is

$$\frac{45.6 \text{ miles}}{2.4 \text{ gallons}} = 19 \frac{\text{miles}}{\text{gallon}} = 19 \text{ miles per gallon}$$

The gas mileage of the car is 19 miles per gallon.

Practice Problem 6   *Write each statement as a ratio.*

a. The ratio of 11 to 13

b. The ratio of 16 boys to 24 girls

c. The ratio of $30 to three hours

The concept of a rate is very useful for determining unit prices. A **unit price** is the ratio of price to quantity and has a denominator of 1. For example, if a 100 ounce (oz) box of a dry bleach costs $4, the unit price of the bleach (cost per ounce) is the ratio of $4 to 100 ounces.

$$\frac{\$4}{100 \text{ ounces}} = \frac{4 \text{ dollars}}{100 \text{ oz}} = 0.25 \frac{\text{dollars}}{\text{oz}} = \$0.25 \text{ per oz}$$

Thus, the unit cost of the bleach is $0.25 per oz or 25 cents per ounce.

Example 9 Solve.

a. A 3-oz (ounce) jar of lemon pepper costs 98¢. Find the cost per ounce.

b. A 32-oz box of rice costs $1.80, but a 48-oz box of the same brand of rice costs $2.40. Which quantity is the better buy?

Solution To find the unit price of an item, express the ratio of cost to weight as a fraction, divide, and write the units with the ratio.

9a. The ratio of cost to weight is 98¢ to 3 oz.

$$\frac{98¢}{3 \text{ oz}} \doteq 32.7 \frac{¢}{\text{oz}} \quad \text{Round to the nearest tenth}$$

This implies that the ratio is approximately 32.7¢ to 1 oz. That is, the cost of lemon pepper is 32.7¢ per oz.

9b. *Ratio of $1.80 to 32 oz*   *Ratio of $2.40 to 48 oz*

$$\frac{\$1.80}{32 \text{ oz}} \doteq 0.06 \frac{\$}{\text{oz}} \qquad \frac{\$2.40}{48 \text{ oz}} = 0.05 \frac{\$}{\text{oz}}$$

The cost is approximately $0.06 (6¢) per oz.   The cost is $0.05 per oz.

The 48-oz box is a better buy since it costs less per unit.

IDEA 3

Ratios are helpful when solving certain types of word problems.

Example 10 Solve.

a. Two numbers have a ratio of 3 to 5. If this sum is 64, what are the numbers?

b. If Paula and Joan divide $5400 in a ratio of 5:4, how much money should each woman receive?

Solution When a ratio problem involves an equation, use the same methods for solving word problems presented in Section 7.1. However, remember that two numbers in the ratio a to b can be expressed as ax and bx, where x is any nonzero number. This can be done since

$$a{:}b = \frac{a}{b} = \frac{ax}{bx} = ax{:}bx$$

where x is not zero.

10a. Step 1: $3x$ = the first number $5x$ = the second number

Step 2: first number + second number = 64
$$3x + 5x = 64$$

Step 3:
$$3x + 5x = 64$$
$$8x = 64$$
$$x = 8$$
$$3x = 24$$
$$5x = 40$$

The numbers are 24 and 40.

Step 4: The answer checks since the ratio of 24 to 40 is 3:5 and $24 + 40 = 64$.

10b. Step 1: $5x$ = dollars Paula should receive
$4x$ = dollars Joan should receive

Step 2: Paula's share + Joan's share = $5400
$$5x + 4x = 5400$$

Step 3:
$$5x + 4x = 5400$$
$$9x = 5400$$
$$x = 600$$
$$5x = 3000$$
$$4x = 2400$$

Paula and Joan should receive $3000 and $2400, respectively.

Step 4: The answer checks since the ratio of $3000 to $2400 is 5 to 4, and $3000 + $2400 = $5400.

Practice Problem 7 **Solve.**

a. An 8-oz jar of mayonnaise costs 99¢. Find the cost per ounce.

b. A 16-oz jar of honey costs $1.50 and a 24-oz jar of the same brand of honey costs $2.40. Which is the better buy?

c. Reggie and Matt divided the profits from the sale of their business in a ratio of 5:8. If the total profits were $29,250, how much money did each man receive?

IDEA 4

Now that we understand the concept of ratio, we can discuss proportion. A **proportion** is a statement that two ratios are equal. For example,

$$\frac{3}{4} = \frac{6}{8} \quad \text{and} \quad \frac{100}{500} = \frac{1}{5}$$

are proportions.

Proportions are usually written in the form $\frac{a}{b} = \frac{c}{d}$. However, they can also be written as $a:b = c:d$ (read as "a is to b as c is to d").

In the proportion $\frac{a}{b} = \frac{c}{d}$, a and d are the first and fourth **terms** of the proportion. These terms are called the **extremes**. The second and third terms of the proportion are b and c. These are called the **means**.

Example 11 Identify the means and the extremes.

a. $\dfrac{1}{3} = \dfrac{4}{12}$ **b.** $5:6 = 10:12$

Solution   The first and fourth terms are the extremes. The second and third terms are the means.

11a. Extremes	Means	**11b.** Extremes	Means
1, 12	3, 4	5, 12	6, 10

Practice Problem 8   *Identify the means and the extremes.*

a. $\dfrac{7}{15} = \dfrac{21}{45}$ **b.** $5:10 = 1:2$

Every proportion obeys the property that the product of the extremes is equal to the product of the means. For example, in the proportion

$$\dfrac{2}{3} = \dfrac{4}{6}$$

the product of the extremes ($2 \cdot 6 = 12$) is equal to the product of the means ($3 \cdot 4 = 12$). We used this property in Chapter 2 to determine if two fractions are equal. This property is called the *fundamental property of proportions*.

Fundamental Property of Proportions

> In a proportion, the product of the extremes is equal to the product of the means,
>
> or
>
> if $\dfrac{a}{b} = \dfrac{c}{d}$
>
> then $a \cdot d = b \cdot c$

Example 12   Determine if the following ratios are equal.

a. $\dfrac{3}{4} = \dfrac{6}{9}$ **b.** $4:11 = 8:22$

Solution   Two ratios are equal if the product of the extremes is equal to the product of the means.

12a. The product of the extremes ($3 \cdot 9 = 27$) is not equal to the product of the means ($4 \cdot 6 = 24$). Thus, $\dfrac{3}{4}$ is not equal to $\dfrac{6}{9}$.

12b. The product of the extremes ($4 \cdot 22 = 88$) is equal to the product of the means ($11 \cdot 8 = 88$). Thus, $4:11 = 8:22$.

Practice Problem 9   *Determine if the ratios are equal.*

a. $\dfrac{35}{91} = \dfrac{5}{13}$ **b.** $2:8 = 3:6$

We can also use the property of proportions to solve a proportion. To *solve a proportion* means to find the value of the letter that makes the two ratios equal.

7.2 Ratio and Proportion Problems

> **To Solve a Proportion:**
>
> 1. Set the product of the extremes equal to the product of the means.
> 2. Solve the resulting equation.

Example 13 Solve each proportion.

a. $\dfrac{6}{13} = \dfrac{x}{26}$ b. $\dfrac{3}{5} : \dfrac{6}{25} = 7 : x$ c. $\dfrac{3x}{x+8} = \dfrac{3}{5}$

Solution To solve a proportion, use the property of the proportions and then solve the resulting equations.

13a.
$$\dfrac{6}{13} = \dfrac{x}{26}$$
$$6 \cdot 26 = 13 \cdot x \quad \text{Cross-multiply}$$
$$156 = 13x$$
$$12 = x$$

Check:
$$\dfrac{6}{13} = \dfrac{x}{26}$$
$$\dfrac{6}{13} \stackrel{?}{=} \dfrac{12}{26}$$
$$6 \cdot 26 \stackrel{?}{=} 13 \cdot 12$$
$$156 = 156$$

13b.
$$\dfrac{3}{5} : \dfrac{6}{25} = 7 : x$$
$$\dfrac{\frac{3}{5}}{\frac{6}{25}} = \dfrac{7}{x}$$
$$\dfrac{6}{25} \cdot \dfrac{7}{1} = \dfrac{3}{5} \cdot x$$
$$\dfrac{42}{25} = \dfrac{3}{5}x$$
$$\dfrac{5}{3} \cdot \dfrac{42}{25} = \dfrac{5}{3} \cdot \dfrac{3}{5}x$$
$$\dfrac{\overset{1}{\cancel{5}} \cdot \overset{14}{\cancel{42}}}{\underset{1}{\cancel{3}} \cdot \underset{5}{\cancel{25}}} = x$$
$$\dfrac{14}{5} = x$$

Check:
$$\dfrac{3}{5} : \dfrac{6}{25} = 7 : x$$
$$\dfrac{\frac{3}{5}}{\frac{6}{25}} \stackrel{?}{=} \dfrac{7}{\frac{14}{5}}$$
$$\dfrac{6}{25} \cdot \dfrac{7}{1} \stackrel{?}{=} \dfrac{3}{5} \cdot \dfrac{14}{5}$$
$$\dfrac{42}{25} = \dfrac{42}{25}$$

13c.
$$\dfrac{3x}{x+8} = \dfrac{3}{5}$$
$$3(x+8) = 5(3x)$$
$$3x + 24 = 15x$$
$$24 = 12x$$
$$2 = x$$

Check:
$$\dfrac{3x}{x+8} = \dfrac{3}{5}$$
$$\dfrac{3(2)}{2+8} \stackrel{?}{=} \dfrac{3}{5}$$
$$\dfrac{6}{10} \stackrel{?}{=} \dfrac{3}{5}$$
$$10 \cdot 3 \stackrel{?}{=} 6 \cdot 5$$
$$30 = 30$$

Practice Problem 10 *Solve each proportion.*

a. $\dfrac{6}{10} = \dfrac{x}{20}$ b. $0.3 : 3 = x : 30$

Solving Word Problems by Using Linear Equations in One Variable

IDEA 5 Word problems involving proportions can be solved in much the same way that we solved word problems in Section 7.1.

> **To Solve a Word Problem Involving Proportions:**
>
> 1. Represent the unknown by a letter.
> 2. Write a proportion. Make sure that the units of measure occupy corresponding positions in the two ratios. For example,
>
> $$\frac{\text{hours}}{\text{dollars}} = \frac{\text{hours}}{\text{dollars}}$$
>
> 3. Eliminate the units of measure, solve the proportion, and answer the question asked.
> 4. Check your answer.

Example 14 Solve.

    a. Rich drove 120 miles on eight gallons of gas. At this rate, how many gallons of gas will he need to travel 250 miles?

    b. One mile is equal to 1.6 kilometers. If Jim traveled 88 kilometers, how many miles did he travel?

    c. Curtis won $513 in a lottery that pays 171 to 2 odds (for every $2, you bet you can win $171). How much money did he bet?

    d. Cazzie and Guerin formed a real estate corporation. Cazzie invested $11,000 and Guerin invested $9000. If the corporation is now worth $90,000, what is Guerin's share in dollars?

Solution To solve a word problem involving proportions, (1) represent the unknown, (2) write a proportion, and (3) solve the proportion, and determine the solution.

14a. Step 1: x = gallons of gas needed to travel 250 miles

    Step 2: $\dfrac{8 \text{ gallons}}{120 \text{ miles}} = \dfrac{x \text{ gallons}}{250 \text{ miles}}$ Ratio of units must be the same on both sides of the equation.

    Step 3: $\dfrac{8}{120} = \dfrac{x}{250}$

    $8 \cdot 250 = 120 \cdot x$
    $2000 = 120x$
    $16.6 \doteq x$ Round to the nearest tenth

Rich needs approximately 16.6 gallons of gas to travel 250 miles.

    Step 4: The answer checks since $8 \cdot 250 \doteq 120 \cdot 16.5$.

14b. Step 1: x = distance traveled in miles

    Step 2: $\dfrac{1 \text{ mile}}{1.6 \text{ kilometers}} = \dfrac{x \text{ miles}}{88 \text{ kilometers}}$

Step 3: $\dfrac{1}{1.6} = \dfrac{x}{88}$

$(1)(88) = (1.6)(x)$
$88 = 1.6x$
$55 = x$

Jim traveled 55 miles.

Step 4: The answer checks since $1 \cdot 88 = 1.6 \cdot 55$.

14c. Step 1: $x =$ the number of dollars Curtis bet

Step 2: $\dfrac{\$171 \text{ pay-off}}{\$2 \text{ bet}} = \dfrac{\$513 \text{ pay-off}}{x \text{ bet}}$

Step 3: $\dfrac{171}{2} = \dfrac{513}{x}$

$171 \cdot x = 2 \cdot 513$
$171x = 1026$
$x = 6$

Curtis bet $6.

Step 4: The answer checks since $171 \cdot 6 = 2 \cdot 513$.

14d. Step 1: $x =$ Guerin's share of the business in dollars

Step 2: $\dfrac{\$9000 \text{ (Guerin's share)}}{\$20{,}000 \text{ (total)}} = \dfrac{x \text{ (Guerin's share)}}{\$90{,}000 \text{ (total)}}$

Step 3: $\dfrac{9000}{20{,}000} = \dfrac{x}{90{,}000}$

$\dfrac{9}{20} = \dfrac{x}{90{,}000}$ Reduce

$9 \cdot 90{,}000 = 20 \cdot x$
$810{,}000 = 20x$
$40{,}500 = x$

Guerin's share of the business is $40,500.

Step 4: The answer checks since $9000 \cdot 90{,}000 = 20{,}000 \cdot 40{,}500$.

Practice Problem 11 **Solve.**

a. A basketball team won eight of its first 11 games. At this rate, how many games would you expect them to win out of their first 22 games?

b. A recipe for 50 ice cream sodas requires four quarts of ice cream. How many quarts of ice cream would be needed to make 275 ice cream sodas?

c. If one inch represents 78 miles on a map, how many miles would 4.5 inches represent?

7.2 Exercises

Write each phrase as a ratio in fractional form.

1. 20 miles to 30 miles
2. 60 feet to 90 feet
3. 73 cents to 110 cents
4. 80 men to 120 men
5. 30 inches to 14 inches
6. five days to five hours
7. 12 minutes to three hours
8. five dollars to eight quarters
9. 14 feet to two minutes
10. 70 miles to five hours
11. 30 dollars to two hours
12. 90 dogs to 45 boys

Write each ratio using colon notation (a:b).

13. ratio of 5 to 9   **14.** ratio of 6 to 15   **15.** $\dfrac{3}{7}$ **16.** $\dfrac{35}{91}$

Find the unit price.

17. A 16-oz can of beans costs 40¢. What is the cost per ounce?

18. A 10-oz can of soup costs 30¢. What is the cost per ounce?

19. A three-pound box of rice costs $2.40. What is the cost per pound?

20. A five-pound bag of flour costs $1.32. What is the cost per pound?

Identify the extremes and the means.

21. $\dfrac{5}{10} = \dfrac{15}{30}$ **22.** $\dfrac{7}{9} = \dfrac{21}{27}$ **23.** $12:13 = 60:65$ **24.** $3:5 = 9:15$

Determine if the following ratios are equal.

25. $\dfrac{13}{9} = \dfrac{26}{18}$ **26.** $\dfrac{18}{32} = \dfrac{36}{96}$ **27.** $2:5 = 4:25$ **28.** $7:13 = 49:91$

Find x in the following proportions.

29. $\dfrac{2}{3} = \dfrac{6}{x}$ **30.** $\dfrac{5}{7} = \dfrac{x}{21}$ **31.** $\dfrac{9}{18} = \dfrac{x}{54}$ **32.** $\dfrac{8}{x} = \dfrac{12}{8}$

33. $\dfrac{15}{25} = \dfrac{x}{100}$ **34.** $\dfrac{x}{9} = \dfrac{30}{24}$ **35.** $2:3 = x:6$ **36.** $7:14 = 21:x$

37. $\dfrac{4}{9} \cdot \dfrac{6}{35} = 14:x$ **38.** $\dfrac{5}{6} : 14 = x : \dfrac{4}{15}$ **39.** $\dfrac{0.3}{3} = \dfrac{x}{5}$ **40.** $\dfrac{4}{x} = \dfrac{0.48}{5.4}$

41. $\dfrac{7.7}{11} = \dfrac{x}{5}$ **42.** $\dfrac{x}{1} = \dfrac{19.05}{7.5}$ **43.** $6:9 = 3\dfrac{1}{3}:x$ **44.** $\dfrac{24}{18} \cdot \dfrac{30}{28} = x:42$

45. $\dfrac{x+1}{x+2} = \dfrac{14}{8}$ **46.** $\dfrac{9}{2x+7} = \dfrac{10}{4x+6}$ **47.** $\dfrac{x-2}{2} = \dfrac{3x-5}{7}$ **48.** $\dfrac{3x+4}{8x-2} = \dfrac{4}{6}$

49. $13:26 = 39:x$

Fill in the blanks.

50. A _____ is the quotient of two numbers or quantities.

51. When comparing quantities expressed in different units of measure, first express the quantities in the _____ unit. If this is not possible, write the _____ with the ratio.

52. A _____ is a statement that two ratios are equal.

53. The property of _____ states that the _____ of the extremes is equal to the _____ of the means.

Solve.

54. A 64-oz bottle of a cola beverage costs $1.50, while a 48-oz bottle of the same brand of cola costs $1.20. Which is the better buy?

55. Two numbers have a ratio of 4:5. If their sum is 45, what are the numbers?

56. A farmer wants to plant 80 acres of corn and soybeans in a ratio of 7 to 9. How many acres of each must he plant?

57. Jim earned $25.50 in six hours. At this rate, how much could he earn in 15 hours?

58. John used 12 gallons of gas on a 420-mile trip. At this rate, how many gallons of gas can he expect to use on a 1050-mile trip?

59. In keeping with equal opportunity employment guidelines, a company tries to hire people on a nondiscriminatory basis. If the ratio of men to women must be 7:4 and the company must hire 165 people, how many men and how many women must be hired?

60. Dr. Cowsen believes that the ideal student-teacher ratio for a day care center is 17:2. If she opens a day care center with an enrollment of 187 students, how many teachers should she employ if she maintains her ideal student-teacher ratio?

61. Joyce wants to invest $15,000 in stocks, bonds, and real estate in a ratio of 3:4:5. How much money should she invest in each?

62. In a certain city of Illinois, the ratio of Republicans to Democrats is 2:5. If there are 1662 Republicans, how many Democrats are there?

63. A profit of $22,000 on the sale of a painting was split between Phyllis and Pam in a ratio of 4 to 7. How much did each woman receive?

64. The real estate tax rate in a certain city is $98.19 per $1000 of assessed valuation. Find the real estate tax on a piece of property assessed at $80,000.

65. A father divided $120,000 among his three children in the ratio 3:5:7. How much did each child receive?

66. A recipe for 60 people uses four cups of milk. How many cups of milk are needed for a recipe for 150 people?

67. A recipe for 40 people uses eight tablespoons of honey. How many tablespoons of honey are needed for a recipe for 100 people?

68. John drove 246 miles on 12 gallons of gas. At this rate, how many gallons of gas will he need to travel 861 miles?

69. Bill drove 170 miles on 10 gallons of gas. At this rate, how far could he drive on 35 gallons of gas?

70. One mile is equal to 1.6 kilometers. If Linda traveled 24 kilometers, how many miles did she travel?

71. One mile is equal to 1.6 kilometers. If Dave traveled 144 miles, how many kilometers did he travel?

72. In four days, Betty ran 50 miles. At this rate, how long would it take her to run 125 miles?

73. In six days, Art drove 2100 miles. At this rate, how many days will it take him to drive 5250 miles?

74. Vanessa used 44 quarts of ice cream to make 550 milk shakes. How many quarts of ice cream are needed to make 100 milk shakes?

75. Ed used 12 quarts of sherbet to make a dessert for 100 people. How many quarts of sherbet are needed to make dessert for 350 people?

76. A basketball team won eight of its first 12 games. At this rate, how many will it win out of its first 30 games?

77. A basketball player made eight out of the first 10 free throws he attempted. At this rate, how many free throws should he make out of the first 65 he attempts?

78. There are 20 women in a class containing 32 students. At this rate, how many of the college's 2576 students would be women?

79. There are 12 men in a class containing 34 students. At this rate, how many of the college's 2737 students would be men?

80. A store sells four candy bars for 86 cents. At this rate, how much will 10 candy bars cost?

Solve.

81. $\dfrac{576}{192} = \dfrac{288}{x}$

82. $\dfrac{x}{2128} = \dfrac{1344}{4256}$

83. $\dfrac{x}{88.5} = \dfrac{23.59}{18.6}$

84. $\dfrac{7.093}{1.025} = \dfrac{x}{8.23}$

85. Four out of 25 eligible voters will not vote in the next election in Theta, Tennessee. If there are 38,414 eligible voters in Theta, how many eligible voters will not vote in the next election.

Answers to Practice Problems 6a. $\dfrac{11}{13}$ b. $\dfrac{2 \text{ boys}}{3 \text{ girls}}$ c. $10 \dfrac{\$}{\text{hour}}$ 7a. 12.4¢ per oz   b. 16-oz jar   c. Reggie's share = $11,250   8a. Extremes: 7, 45   b. Extremes: 5, 2   9a. equal   b. not equal   10a. 12   Matt's share = $18,000   Means: 15, 21   Means: 10, 1   b. 3   11a. 16   b. 22   c. 351

7.3 Percent Problems

IDEA 1 There are three basic types of percent problems. In each of these problems, there are three quantities involved:

1. The base (*B*) is the original quantity and usually follows the word "of" in a percent problem.
2. The rate (*R*) is a percent that is expressed as a fraction or decimal before solving a percent problem.
3. The amount (*A*) is a percent of the base.

In a percent problem, two quantities will be given and we will find the third quantity by using the *percent formula* or percent equation.

Percent Formula

> The amount is equal to the product of the rate times the base,
>
> or
>
> $$A = R \cdot B$$
>
> where A = amount, R = rate, and B = base.

When solving percent problems, it will be very important for you to be able to identify *A*, *R*, and *B*. To do this, it will be helpful to rewrite all percent problems in the form "the amount is some percent of the base."

Example 15 Identify the amount, the base, and the rate.

a. What number is 80% of 1000?

b. 18 is what percent of 20?

c. 40 is 37% of what number?

d. The Johnsons have $8000 that can be used as the down payment on a house. If the minimum down payment must be 20% of the total cost of the house, what is the highest price that they can pay for a house?

e. Mark attempted 20 shots in a game and made 70% of them. How many shots did he make?

f. Jimmy earns $1000 a month. If he pays $400 a month for rent, what percent of his earnings is spent on rent?

Solution To identify *A*, *B*, and *R* in a percent problem, rewrite the problem as the amount (*A*) is some percent (*R*) of the base (*B*).

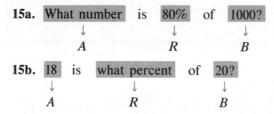

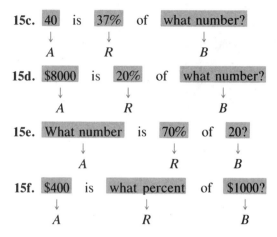

15c. 40 is 37% of what number?
     A   R   B

15d. $8000 is 20% of what number?
       A   R   B

15e. What number is 70% of 20?
            A   R   B

15f. $400 is what percent of $1000?
       A   R   B

Practice Problem 12 **Identify A, B, and R.**

   a. 14 is what percent of 20?

   b. What number is 30% of 50?

   c. Hondo made 80% of his free throws. If he made eight free throws, how many shots did he attempt?

   d. Adrienne earns a 10% commission on everything she sells. If her total sales for one week were $1200, what was her commission?

IDEA 2 One procedure that can be used to solve percent problems involves translating a written statement that contains a percent into an equation that can be solved for the amount, rate, or base. For example, if a salesman's commission is 8% of a $60,000 sale, to find the amount that the salesman receives requires that we answer the question "What number is 8% of $60,000?"

$$\text{What number is } 8\% \text{ of } \$60,000$$
$$n = (0.08) \cdot (60,000)$$
$$n = 4800$$

The salesman's commission is $4800.

The procedure for solving percent problems by using the percent formula is summarized below.

To Solve Percent Problems by Using Equations:

1. Represent the unknown.

2. Write an equation using the percent formula $A = R \cdot B$.

3. Solve the equation and answer the question asked.

4. Check your answer.

Example 16 Solve.

a. 40 is 50% of what number?

b. 20 is what percent of 80?

c. What number is $33\frac{1}{3}\%$ of $\frac{6}{10}$?

d. A $80,000 house is rented for $640 per month. What percent of the value of the house is the monthly rental?

Solution To solve a percent problem by using equations, solve the percent formula $A = R \cdot B$ for the unknown quantity and answer the question asked.

16a. Step 1: x = the number (base)

Step 2: 40 is 50% of what number
$\downarrow\ \ \downarrow\ \ \downarrow\ \ \downarrow\ \ \ \ \ \ \ \ \downarrow$
40 = (0.5) · x

Step 3: $40 = 0.5x$
$\frac{40}{0.5} = \frac{0.5x}{0.5}$
$80 = x$
40 is 50% of 80.

Step 4: The answer checks since $40 = 0.5 \cdot 80$.

16b. Step 1: x = the rate

Step 2: 20 is what percent of 80
$\downarrow\ \ \downarrow\ \ \ \ \downarrow\ \ \ \ \ \ \ \ \downarrow\ \ \downarrow$
20 = r · 80

Step 3: $20 = 80r$
$\frac{20}{80} = \frac{80r}{80}$
$\frac{1}{4} = r$
$25\% = r$ Think: $\frac{1}{4} = 0.25 = 25\%$
20 is 25% of 80.

Step 4: The answer checks since $20 = 80 \cdot 0.25$.

16c. Step 1: x = the number (amount)

Step 2: What number is $33\frac{1}{3}\%$ of $\frac{6}{5}$
$\downarrow\ \ \ \ \ \ \ \ \ \ \ \ \ \ \downarrow\ \ \downarrow\ \ \ \ \ \downarrow\ \ \downarrow$
x = $\frac{1}{3}$ · $\frac{6}{5}$

Step 3: $x = \frac{1}{3} \cdot \frac{6}{5}$

$x = \frac{1 \cdot \overset{2}{\cancel{6}}}{\underset{1}{\cancel{3}} \cdot 5}$

$x = \frac{2}{5}$

$\frac{2}{5}$ is $33\frac{1}{3}\%$ of $\frac{6}{5}$

Step 4: The answer checks since $\frac{2}{5} = \frac{1}{3} \cdot \frac{6}{5}$.

 In most percent problems, the percent should be written as a decimal before solving the percent equation. However, percents such as

$$33\tfrac{1}{3}\% = \tfrac{1}{3},\ 66\tfrac{2}{3}\% = \tfrac{2}{3},\text{ and } 83\tfrac{1}{3}\% = \tfrac{5}{6}$$

should be expressed as fractions before solving the percent equation.

16d. Step 1: r = percent of value of the house that is monthly rental.

Step 2: Rewording:

$640	is	what percent	of	$80,000
↓	↓	↓	↓	↓
640	=	r	·	80,000

Step 3:
$$640 = 80{,}000r$$
$$\frac{640}{80{,}000} = \frac{80{,}000r}{80{,}000}$$
$$0.008 = r$$
$$0.8\% = r$$

Step 4: The answer checks since $640 = (8\%)(80{,}000)$.
The monthly rental is 0.8% of the value of the house.

Practice Problem 13 **Solve.**

a. 5 is what percent of 20.

b. What number is 12% of 85.

IDEA 3

The percent problems in Example 16 were solved by using the percent formula. Those same problems could also be solved by setting up a proportion. To set up this proportion, we use the fact that

$$\text{Amount} = \text{Rate} \cdot \text{Base}$$
$$\frac{\text{Amount}}{\text{Base}} = \frac{\text{Rate} \cdot \cancel{\text{Base}}}{\cancel{\text{Base}}}$$
$$\frac{\text{Amount}}{\text{Base}} = \text{Rate}$$

Now, since the rate is a percent ($P\%$) which can be expressed as a fraction, we can replace $P\%$ with $\frac{P}{100}$ since $P\% = \frac{P}{100}$, where P is any number.

$$\frac{\text{Amount}}{\text{Base}} = \text{Rate}$$
$$\frac{\text{Amount}}{\text{Base}} = P\%$$
$$\frac{\text{Amount}}{\text{Base}} = \frac{P}{100}$$

The above proportion is called the *percent proportion*.

Percent Proportion

> The ratio of the amount to the base is equal to the rate, expressed as a fraction,
>
> or
>
> $$\frac{A}{B} = \frac{P}{100}$$
>
> where A = Amount, B = Base, and $P\%$ = Rate.

To solve percent problems by using the percent proportion we identify A, P, and B, solve the proportion for the unknown quantity, and answer the question asked. For example, to answer the question "What number is 8% of $60,000?" we observe that $A = ?$, $B = 60,000$, and $P = 8$.

$$\frac{A}{B} = \frac{P}{100}$$
$$\frac{A}{60,000} = \frac{8}{100}$$
$$100A = 480,000$$
$$A = 4800$$

Thus, $4800 is 8% of $60,000.

To Solve a Percent Problem by Using the Percent Proportion:

1. Identify A, B, and P.
2. Replace the known quantities in the formula with their numerical values.
3. Solve for the unknown quantity and answer the question asked.

Example 17 Solve by using the percent proportion.

a. 40 is 50% of what number?

b. 20 is what percent of 80?

c. What number is 80% of 10?

d. An examination contains 20 problems. If Barbara answers 75% of the problems correctly, how many problems has she answered correctly?

e. A baseball team won 25 out of 30 games. What percent of the games did they win?

Solution To solve a percent problem, solve the percent proportion for the unknown quantity and answer the question asked.

17a. $A = 40$, $B = ?$, $P = 50$

$$\frac{A}{B} = \frac{P}{100}$$
$$\frac{40}{B} = \frac{50}{100}$$
$$50B = 4000$$
$$B = 80$$

40 is 50% of 80.

17b. $A = 20, B = 80, P = ?$

$$\frac{A}{B} = \frac{P}{100}$$
$$\frac{20}{80} = \frac{P}{100}$$
$$80P = 2000$$
$$P = 25$$

20 is 25% of 80.

17c. $A = ?, B = 10, P = 80$

$$\frac{A}{B} = \frac{P}{100}$$
$$\frac{A}{10} = \frac{80}{100}$$
$$100A = 800$$
$$A = 8$$

8 is 80% of 10.

17d. Rewrite: What number is 75% of 20?
$A = ?, B = 20, P = 75$

$$\frac{A}{B} = \frac{P}{100}$$
$$\frac{A}{20} = \frac{75}{100}$$
$$100A = 1500$$
$$A = 15$$

Barbara answered 15 questions correctly.

17e. Rewrite: 25 is what percent of 30?

$A = 25, B = 30, P = ?$
$$\frac{A}{B} = \frac{P}{100}$$
$$\frac{25}{30} = \frac{P}{100}$$
$$30P = 2500$$
$$P \doteq 83 \quad \text{Round to nearest one}$$

The team won 83% of its games.

Practice Problem 14 **Solve.**

a. 210 is 15% of what number?

b. What number is 25% of 40?

c. Jim earns $1400 per month. If his rent is $210 per month, what percent of his income does he spend on rent?

d. There are 200 employees in the ABT company. If 80% of the employees are students, how many employees are students?

Sometimes the numbers given in a percent problem are not the values that you must substitute for *A, P,* or *B*.

Example 18 Solve.

a. Rich wants to buy a pair of jeans that cost $50. If the owner gives him a 20% discount, how much should Rich pay for the jeans?

b. Susan earned $20,000 last year. This year her salary is $24,000. What was the percent of increase?

Solution

18a. Since Rich gets a 20% discount, he must pay only 80% (100% − 20%) of the original cost. Rewrite: What number is 80% of $50?

$$A = ?, B = 50, P = 80\%$$
$$\frac{A}{B} = \frac{P}{100}$$
$$\frac{A}{50} = \frac{80}{100}$$
$$100A = 4000$$
$$A = 40$$

The jeans will cost Rich $40.

 When working percent problems, remember that (1) the whole or original amount is 100% and (2) $100\% - N\% = (100 - N)\%$. For example, $100\% - 70\% = (100 - 70)\% = 30\%$.

Alternate Solution

Use the general method of solving word problems.

Step 1: c = cost of jeans

Step 2: cost of jeans   is $50   minus   20% of 50
$$c = 50 - (0.20)(50)$$

Step 3: $c = 50 - (0.20)(50)$
$c = 50 - 10$
$c = 40$

The jeans will cost Rich $40.

Step 4: The answer checks since $40 = $50 − (0.20)($50).

18b. To determine the percent of increase, we must find the amount of her salary increase. This amount is $4000 ($24,000 − $20,000). Rewrite: $4000 is what percent of $20,000?

$$A = 4000, B = 20{,}000, P = ?$$
$$\frac{4000}{20{,}000} = \frac{P}{100}$$
$$20{,}000P = 400{,}000$$
$$P = 20$$

Susan received a 20% salary increase.

Alternate Solution

Use the general method of solving word problems.

Step 1: r = rate (percent) of increase

Step 2: $4000   is   what percent   of $20,000
$$4000 = r \cdot 20{,}000$$

Step 3:
$$4000 = 20{,}000r$$
$$\frac{4000}{20{,}000} = \frac{20{,}000r}{20{,}000}$$
$$0.2 = r$$
$$20\% = r$$

Susan received a 20% salary increase.

Step 4: The answer checks since $4000 = (\$20{,}000)(0.20)$.

Practice Problem 15 **Solve.**

a. Phil earns $12,000 a year and his wife $13,000 a year. If they can afford to buy a car that is 40% of their total income, what is the highest price that they can pay for a car?

b. Carver must pay a salesman a 5% commission to sell his car. If he wants to have $7600 after paying the commission, at what price must the car be sold?

7.3 Exercises

Solve. If necessary, round to the nearest hundredth.

1. 15 is 25% of what number?
2. What number is 25% of 60?
3. 15 is what percent of 60?
4. What number is 63% of 48?
5. 12.5 is 50% of what number?
6. What number is 3.2% of 300?
7. What number is 113% of 200?
8. 18 is what percent of 24?
9. 35 is what percent of 50?
10. 62 is 31% of what number?
11. 78 is 6% of what number?
12. What number is 9% of 90?
13. 16 is what percent of 24?
14. 92 is 50% of what number?
15. 210 is 15% of what number?
16. 52 is what percent of 217?
17. What number is 0.4% of 360?
18. 252 is 70% of what number?
19. 900 is what percent of 1220?
20. What number is 38% of 145?

Solve.

21. John won up 70% of his games. If he won 14 games, how many games has he played?
22. In a class of 50 students, 30 people passed a test. What percent of the students passed?
23. Joe earned $200 last week. If he always saves 15% of his weekly salary, how much money did he save last week?
24. Wayne won 12 out of 15 games. What percent of the games did he win?
25. In an algebra class, 20% of the students earned A's. If 5 students received an A, how many students are in the class?
26. Pam bought a shirt that costs $40. If she must pay a 6% sales tax, how much sales tax will she pay?
27. Bill earns $14,000 per year. If he wants to save 20% of his annual salary, how much should he save?
28. In a poll, 15 out of 60 people wanted to see a play. What percent of the people wanted to see the play?
29. Barbara earns $1000 per month. If she spends $80 per month to repay a loan, what percent of her salary does she spend to repay the loan?
30. Jane received a 10% increase in pay. If she is now earning an extra $140 per month, what was her monthly salary before the raise?

31. Vanessa changed her driving habits and increased the mpg rating on her car by four miles. If this increase represents a 20% boost in the car's mpg rating, what was the original rating?

32. Fifty out of 200 applicants for a job are women. What percent of the applicants are women?

33. In business, the selling price equals the cost plus the markup. If Gatsby's marks up all its merchandise by 60% of its original cost, what is the markup on a coat that cost Gatsby's $150? What is the selling price of the coat?

34. John earns $1600 per month. If $320 is deducted from his check, what percent of his salary is deducted from his monthly earnings?

35. John wants to buy a house that costs $80,000. If the owner wants at least a 20% down payment, how much money does John need as a down payment?

36. Buford paid $300 for a $400 stereo. What percent decrease in cost is this?

37. Ron weighed 200 pounds in college. If he now weighs 220 pounds, what percent increase in weight is this?

38. Kathy paid $6000 for a new car. Two years later the car is worth $4000. What percent decrease in value is this? (Round to the nearest percent.)

39. The value of a house increased 40% in 4 years. If the amount of the increase was $20,000, find the original cost and the new cost of the house.

40. Wayne won 12 out of 15 games. What percent of the games did he lose?

41. Bill earns $14,000 per year. If he saves 20% of his income, how much money does that leave him for living expenses?

42. Reggie paid $8616 for a new car. If this price is 7% more than the list price, find the list price. (*Hint*: The price he paid is 107% of the list price.)

43. A union contract called for a 11% increase in the hourly rate of pay. If the new hourly wage is $11.10, what was the original hourly wage?

44. The owner of a clothing store always gives Rich a 20% discount on any item he buys. If he paid $40 for a sweater, what was its original cost?

45. Phil wants to have $73,600 after paying Ron an 8% commission to sell his house. What must his asking price be?

46. Find 10% of $8\frac{3}{4}$.

47. Find 15% of $\frac{80}{9}$.

48. Find 20% of $\frac{15}{4}$.

49. Find $8\frac{1}{3}$% of 40.

50. Find $66\frac{2}{3}$% of 300.

51. Find $33\frac{1}{3}$% of $2194.35.

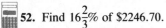
52. Find $16\frac{2}{3}$% of $2246.70.

53. A car priced at $5299 is marked down 10%. If the sales tax is $7\frac{1}{2}$%, find the total cost of the car. (Round answer to nearest cent.)

54. If Lorie is paid $12\frac{3}{4}$% commission on every item she sells, find her commission of the sale of a painting that costs $375,000.

55. John marks up every antique car he sells by $40\frac{3}{4}$%. If John purchased a car for $875,000, what is the selling price of the car?

Answers to Practice Problems 12a. $A = 14, B = 20, R = ?$ b. $A = ?, B = 50, R = 30\%$ c. $A = 8, B = ?, R = 80\%$ d. $A = ?, B = 1200, R = 10\%$ 13a. 25% b. 10.2 14a. 1400 b. 10 c. 15% d. 160 15a. $10,000 b. $8000

7.4 Simple Interest

IDEA 1

Interest is the amount of money paid for the use of money. For example, if you borrowed money from a bank, the interest is the amount of money you pay the bank for the privilege of using their money. Similarly, if you deposited money into a savings account, the interest is the amount of money you earn for letting the bank use your money.

Interest computed on the original amount of money borrowed or invested is called **simple interest**. To compute simple interest, we use the *interest formula*.

Interest Formula

> The interest equals the product of the principal times the rate times the time,
>
> or
>
> $$I = P \cdot R \cdot T$$
>
> where P = principal borrowed or invested, R = rate of annual interest, T = time in years the principal is borrowed or invested, I = interest or amount of money charged or earned.

For example, to compute the simple interest on $5000 deposited for 2 years at an annual interest rate of 8.5%, we write

$$\begin{aligned} I &= PRT \\ &= (5{,}000)(0.085)(2) \\ &= 850 \end{aligned}$$

Thus, the interest earned is $850.

To solve a simple interest problem by using the formula $I = PRT$ use the following procedure.

> **To Solve a Simple Interest Problem by Using the Interest Formula:**
>
> 1. Identify I, P, R, and T.
> 2. Replace the known quantities in the formula with their numerical values.
> 3. Solve for the unknown quantity and answer the question asked.
> 4. Check your answer.

Example 19   Solve by using the formula $I = PRT$.

   a. John invested $900 in a savings plan that pays 8% interest. At the end of six months, how much interest will his account earn?

   b. How much money must Ron invest at a 6.5% interest rate in order to earn $390 in interest in one year?

   c. How many months will it take to earn $600 in interest when $8000 is invested at 10%?

d. Jan wants to deposit $1000 in a savings plan that will earn her $120 in interest in one year. At what interest rate must she invest her money?

Solution To solve a simple interest problem, solve the formula $I = PRT$ for the unknown quantity and answer the question asked in the original problem. However, remember that T should be given in years and $P\% = P(.01)$.

19a. Step 1: $I = ?, P = \$900, R = 8\% = 8(.01) = .08, T = 6 \text{ mo} = \frac{1}{2} \text{ yr}$

Step 2: $I = PRT$
$I = (900)(.08)\left(\frac{1}{2}\right)$

Step 3: $I = 36$
John will earn $36 in interest.

Step 4: The answer checks since $36 = \$900(8\%)\left(\frac{1}{2}\right)$.

19b. Step 1: $I = \$390, P = ?, R = 6.5\% = 6.5(.01) = .065, T = 1 \text{ yr.}$

Step 2: $I = PRT$
$390 = (P)(.065)(1)$

Step 3: $390 = .065P$
$6000 = P$ Divide by .065
Ron must invest $6000.

Step 4: The answer checks since $390 = \$6000 (6.5\%)(1)$.

19c. Step 1: $I = \$600, P = \$8000, R = 10\% = 10(.01) = .10, T = ?$

Step 2: $I = PRT$
$600 = (8000)(.10)(T)$

Step 3: $600 = 800T$
$\frac{3}{4} = T$ Divide by 800

It will take $\frac{3}{4}$ of a year, which is $\frac{3}{4}(12 \text{ months}) = 9$ months.

Step 4: The answer checks since $\$600 = \$8000 (10\%)\left(\frac{3}{4}\right)$.

19d. Step 1: $I = \$120, P = \$1000, R = ?, T = 1 \text{ yr.}$

Step 2: $I = PRT$
$120 = (1000)(R)(1)$

Step 3: $120 = 1000R$
$.12 = R$ Divide by 1000
The rate should be .12 or 12%.

Step 4: The answer checks since $\$120 = \$1000 (12\%)(1)$.

Practice Problem 16 **Solve by using the formula $I = PRT$.**

a. Find the amount of interest earned on $500 at 6.25% for nine months (round the answer to the nearest hundredth).

b. How much money must Rich invest at 12% to earn $600 in interest in one year?

c. If Pam wants to earn $50 in interest in six months on a $500 investment, at what rate should she invest her money?

IDEA 2

A simple interest problem may involve more than one interest rate. To solve this type of problem, we will use the same basic steps presented in Section 7.1.

To Solve a Simple Interest Problem Involving More Than One Interest Rate:

1. Use the four-step procedure to solve word problems.
2. In step 2, writing an equation, it may be helpful to construct a table of the following information:

Investments	Principal (P)	Rate (R)	Time (T)	Interest (I) = $P \cdot R \cdot T$

Example 20 Solve.

a. Of a total investment of $10,000, Vanessa invested part at 9% and the rest at 10.5%. If her total interest for the year was $990, how much was invested at each rate?

b. Rochelle has $20,000 to invest. She invested $10,000 in a certificate of deposit that pays 15%, and $3000 in a savings plan that pays 8.5%. At what rate should she invest the remaining money if she wants to earn an annual income of $3155 from these three investments?

Solution To solve a simple interest problem involving two or more interest rates, (1) represent the unknowns, (2) write an equation, (3) solve the equation and determine all solutions, and (4) check.

20a. Step 1:

x = dollars invested at 9%
$10,000 - x$ = dollars invested at 10.5%

Step 2:

	Principal (P) (in dollars)	Rate (R) (in percent)	Time (T) (in years)	Interest (I) = $P \cdot R \cdot T$ (in dollars)
Investment 1	x	9% = .09	1	.09x
Investment 2	$10,000 - x$	10.5% = .105	1	.105($10,000 - x$)
Total	10,000			990

$$\text{Interest at 9\%} + \text{Interest at 10\%} = \text{Total interest}$$
$$.09x + .105(10,000 - x) = 990$$

Step 3:
$$.09x + .105(10{,}000 - x) = 990$$
$$.09x + 1050 - .105x = 990$$
$$-.015x + 1050 = 990$$
$$-.015x = -60 \quad \text{Added } -1050 \text{ to both sides}$$
$$x = 4000 \quad \text{Divided both sides by } -0.15x$$
$$10{,}000 - x = 6000$$

Vanessa invested $4000 at 9% and $6000 at 10.5%.

Step 4: The answer checks since the total invested = $4000 + $6000 = $10,000, and .09($4000) + .105($6000) = $360 + $630 = $990.

20b. Step 1: x = rate at which $7000 must be invested

Step 2:

	Principal (P) (in dollars)	Rate (R) (in percent)	Time (T) (in years)	Interest (I) = $P \cdot R \cdot T$ (in dollars)
Savings Plan	3,000	8.5% = .085	1	255
Certificate	10,000	15% = .15	1	1500
3rd Investment	7,000	$x\%$ = .01x	1	70x
Total	20,000			3155

$$\boxed{\text{Interest on Savings Plan}} + \boxed{\text{Interest on Certificate}} + \boxed{\text{Interest on 3rd Investment}} = \boxed{\$3155}$$
$$255 + 1500 + 70x = 3155$$

Step 3:
$$255 + 1500 + 70x = 3155$$
$$1755 + 70x = 3155$$
$$70x = 1400$$
$$x = 20$$

The $7000 must be invested at a rate of 20%.

Step 4: The answer checks since the total interest is $255 + $1500 + .20($7000) = $255 + $1500 + $1400 = $3155.

Practice Problem 17 **Solve.**

a. Barbara has $1500. She deposited some of the money into a savings account that pays 6.25%, and the rest into an account that pays 13%. If the total annual interest earned from the two accounts was $168, how much did she deposit into each account?

b. Carver invested $4000 in a Ford and a BMW and then he leased both cars. His earned annual interest income from leasing each car was 20% and 60%, respectively. If his income from the total investment was 50%, how much did Carver pay for each car?

7.4 Exercises

Solve by using the formula I = PRT.

1. Find the interest on a $1000 loan at 8% for one year.
2. Find the interest on a $1000 loan at 10% for one year.
3. Find the interest on a $2000 loan at 12% for one year.
4. Find the interest on a $500 loan at 18% for six months.
5. If $500 is deposited into a savings account that pays 8% interest, how much interest will be earned in nine months?
6. If $50 in interest is paid on a loan of $500 for one year, what is the interest rate?
7. If $500 in interest is paid on a loan of $8000 for one year, what is the interest rate?
8. If the interest on an $800 loan for six months is $40, what is the interest rate?
9. If the interest on a $2400 charge account is $30 for one month, what is the interest rate?
10. If $50 in interest is paid for a 10% loan for one year, what is the principal?
11. If $600 in interest is paid for an 8% loan for six months, what is the principal?
12. If $30 in interest is paid for an 18% loan for one month, what is the principal?
13. If the total interest paid on a $400 loan at 10% was $40, how many months was the principal used?
14. If the total interest paid on a $2000 loan at 15% was $100, how many months was the principal used?
15. If the total interest paid on a $3000 loan at 14.5% was $145, how many months was the principal used?
16. Ron wants to invest some money at 15%. If he wants to earn $4500 in interest in one year, how much money should he invest?
17. Penny's monthly interest charge on a charge card is computed on a closing balance of $1080. If the annual interest rate is 18%, how much interest is due at the end of the month?
18. Reggie deposited $3000 into a savings account that pays 9.75%. How much interest will he earn in six months?
19. The Second National Bank has a savings plan that pays interest by mail. If the interest rate is 12%, how much would you have to deposit in order to receive a monthly check of $200?
20. The local furniture store charged Mary $60 in interest for a $200 sofa. If she paid the bill over a nine-month period, what was the annual interest rate?
21. John deposited $2400 in an account paying 10.5% in interest. How many months will it take him to earn $189 in interest?
22. Don inherited $1800 from his sister. Part of the money was placed in a savings plan that earns 8% interest, and the rest was deposited in an account earning 9%. If his total annual interest was $152, how much money was deposited in each account?
23. Barbara received a check last week. She invested part of it at 15% and $1000 more than this amount at 16%. If she will earn $780 in interest in one year, how much was invested at each rate?
24. Cazzie earned $25,000 from the sale of his book. He invested $10,000 at 12.5% and $7000 at 15%. At what rate must he invest the remaining $8000 so that he can earn $3740 in interest in one year?
25. Betty invested $1000 at 11.5%. How much additional money must she deposit in a savings plan at 10% so that her annual interest income from both investments will be $200?
26. Guerin earned $25,000 from the sale of a house. He invested part at 15% and the rest at 25%. If the annual interest income from the 25% investment is $250 more than the 15% investment, how much was invested at each rate?

27. Van wants to earn an annual interest income of 12.5% from his total investments. If he has $5000 invested at 11%, how much additional money must he invest at 15%?

28. Pam invested a total of $40,000 in two business ventures. The first investment earned an annual interest income of 18%, and the second investment earned 5%. If the interest from the 18% investment is six times as much as the 5% investment, how much did she invest at each rate?

29. Dr. Johnson has $6000. He invested part of the money in a savings plan at 10% and the rest in a certificate of deposit at 12%. If he has a total of $340 in interest after six months, how much did he invest at each rate?

30. Matthew invested $4000 in two savings plans. If the first plan has an interest rate of 15% and the second plan has a rate of 9%, how much must be invested at each rate so that Matthew will earn a total of $420 in 10 months?

Answers to Practice Problems   **16a.** $23.44   **b.** $5000   **c.** 20%   **17a.** $400 at 6.25%   **b.** $1000 for the Ford
$1100 at 13% $3000 for the BMW

7.5 Mixture Problems

IDEA 1 A mixture problem involves the mixing of two or more quantities. We will solve this type of problem in much the same way we solved simple interest problems. The difference is that the equation is based on the fact that the amount (or value) of a substance present before mixing must equal the amount (or value) of that substance after mixing. To solve mixture problems, use the *mixture formulas*.

Mixture Formula 1

The amount of a pure substance is equal to its concentration times the total volume of the mixture,

or

$$A = C \cdot T$$

where A = amount of pure substance, C = concentration of pure substance, and T = total volume of mixture.

Mixture Formula 2

The total cost is equal to the unit price times the number of units,

or

$$T = P \cdot N$$

where T = total cost, P = unit price, and N = number of units.

7.5 Mixture Problems

To Solve Mixture Problems:

1. Use the four-step procedure to solve word problems.
2. In step 2, writing an equation, it may be helpful to construct one of the following tables:

Item	Unit Price (P)	Number of Units (N)	Total Cost (T) = P · N

Mixture	Concentration of Substance (C)	Total Volume of Mixture (T)	Amount of Pure Substance (A) = C · T

Example 21 Solve.

a. John needs 12 quarts of a 10% salt solution. If he has one container that is a 5% salt solution and another that is a 25% salt solution, how many quarts of each must be mixed to obtain the desired mixture?

b. How many ounces of pure acid must be added to 200 ounces of a 10% acid solution to obtain a solution that is 20% acid?

c. Lou has 10 pounds of walnuts that cost $5 per pound and 15 pounds of cashews that cost $6 per pound. How many pounds of pecans costing $8 per pound should be added to these nuts to yield a mixture worth $6 per pound?

Solution To solve a mixture problem, (1) represent the unknown(s), (2) write an equation, (3) solve the equation and determine all solutions, and (4) check.

21a. Step 1: x = volume of 5% salt mixture in quarts
  $12 - x$ = volume of 25% salt mixture in quarts

Step 2:

	Concentration of Salt (C) (in percent)	Total Volume of Mixture (T) (in quarts)	Amount of Salt (A) = C · T (in quarts)
Mixture 1	5% = .05	x	$(.05)(x) = .05x$
Mixture 2	25% = .25	$12 - x$	$.25(12 - x)$
Final Mixture	10% = .10	12	$.10(12) = 1.2$

	Amount of salt in 5% mixture	+	Amount of salt in 25% mixture	=	Amount of salt in final mixture
	.05x	+	.25(12 − x)	=	1.2

Step 3: .05x + .25(12 − x) = 1.2
.05x + 3 − .25x = 1.2
−.20x + 3 = 1.2
−.20x = −1.8   Added −3 to both sides
x = 9   Divided both sides by −.20
12 − x = 3

John must mix nine quarts of the 5% salt solution with three quarts of the 25% salt solution.

Step 4: The answer checks since:
(a) 9 qt + 3 qt = 12 qt
(b) Salt before mixing = Salt after mixing
.05(9 qt) + .25(3 qt) = .10(12 qt)
.45 qt + .75 qt = 1.20 qt
1.20 qt = 1.20 qt

21b. Step 1: x = ounces (oz) of pure acid

Step 2:

	Concentration of Acid (in percent)	Total Volume of Mixture (in oz)	Amount of Acid (in oz)
Solution #1	10% = .10	200	.10(200) = 20
Solution #2	100% = 1.00	x	1(x) = x
Mixture	20% = .20	x + 200	.20(x + 200)

	Amount of acid in 10% mixture	+	Amount of acid in 100% mixture	=	Amount of acid in final mixture
	20	+	x	=	.20(x + 200)

Step 3: 20 + x = .20(x + 200)
20 + x = .20x + 40
.80x + 20 = 40   Added −.20x to both sides
.80x = 20   Added −20 to both sides
x = 25   Divided both sides by .80

25 oz of pure acid must be added.

Step 4: The answer checks since:
Acid before mixing = Acid after mixing
.10(200 oz) + 25 oz = .20(25 + 200) oz
20 oz + 25 oz = .20(225 oz)
45 oz = 45 oz

21c. Step 1: x = pounds (lb) of pecans

Step 2:

	Unit price (P) (in dollars per lb)	Number of units (N) (in lb)	Total cost $(T) = P \cdot N$ (in dollars)
Walnuts	5	10	$5 \cdot 10 = 50$
Cashews	6	15	$6 \cdot 15 = 90$
Pecans	8	x	$8 \cdot x = 8x$
Mixture	6	$25 + x$	$6(25 + x)$

$$\boxed{\text{Total cost of walnuts}} + \boxed{\text{Total cost of cashews}} + \boxed{\text{Total cost of pecans}} = \boxed{\text{Total cost of mixture}}$$

$$50 + 90 + 8x = 6(25 + x)$$

Step 3:
$$50 + 90 + 8x = 6(25 + x)$$
$$140 + 8x = 150 + 6x$$
$$140 + 2x = 150 \quad \text{Added } -6x \text{ to both sides}$$
$$2x = 10 \quad \text{Added } -140 \text{ to both sides}$$
$$x = 5 \quad \text{Divided both sides by 2}$$

Lou must add five pounds of pecans.

Step 4: The answer checks since:
$$\text{Value before mixing} = \text{Value after mixing}$$
$$\$5(10) + \$6(15) + \$8(5) = \$6(25 + 5)$$
$$\$50 + \$90 + \$40 = \$6(30)$$
$$\$180 = \$180$$

Practice Problem 18 *Solve.*

a. A bottle contains 100 cubic centimeters (cc) of a 20% acid solution. How much of this solution must be removed and replaced with pure acid in order for the new solution to be 100 cc of a 30% acid solution?

b. Susan mixes some dried fruit that costs $0.75 per pound with roasted soybeans costing $1.50 per pound. If she wants a mixture of 90 pounds worth $1.24 per pound, how many pounds of each item must she use?

7.5 Exercises

Solve.

1. How many gallons of solution that is 15% salt must be mixed with a solution that is 75% salt to obtain 12 gallons of a 30% salt solution?

2. How many gallons of a 20% acid solution should be mixed with a 40% acid solution to obtain 10 gallons of a mixture that is 30% acid?

3. Betty needs six ounces of a medicine that is 10% alcohol. If she has one bottle of this medicine that is a 12% alcohol solution and another bottle that is a 7% alcohol solution, how many ounces of each must be mixed to obtain six ounces of a 10% alcohol solution?

4. How many pounds of a dried fruit costing $0.75 per pound should be mixed with another dried fruit costing $1.50 per pound to obtain a 50-pound mixture worth $1.00 per pound?

5. Ed needs a coolant that is 40% antifreeze. How many gallons of a solution that is 60% antifreeze should he mix with 10 gallons of a solution that is 6% antifreeze to obtain the desired coolant?

6. How much pure glycerin must be added to 30 quarts of a solution that is 40% glycerin to obtain a mixture that is 60% glycerin?

7. A bottle contains 25 ounces of a solution that is 30% acid. How much of this solution must be removed and replaced with distilled water in order to obtain 25 ounces of a solution that is 20% acid?

8. A farmer has 81 pounds of grass seed that costs $1.20 per pound. How many pounds of clover seed costing $1.60 per pound must be added to the grass seed to obtain a mixture costing $1.24 per pound?

9. How much water must be evaporated from 1000 gallons of 12% dye solution to obtain a solution that is 20% dye?

10. How much pure acid must be added to 200 cc of an 8% acid solution to obtain a solution that is 20% acid?

11. How many quarts of pure nitric acid must be added to three quarts of a 12% solution to obtain a solution that is 28% nitric acid?

12. How many liters of a cocktail that is 37% alcohol must be combined with five liters of a cocktail that is 45% alcohol in order to obtain a mixture that is 42% alcohol?

13. A janitor wishes to make a cleaning solution that is 40% disinfectant. How many gallons of water must be added to 20 gallons of a solution that is 50% disinfectant to obtain the desired solution?

14. Marcia mixed 20 quarts of a chocolate ice cream containing 20% butterfat with 10 quarts of a coffee ice cream containing 15% butterfat to make mocha ice cream. What percent of the mocha ice cream is butterfat?

15. How much cream that is 15% butterfat must be combined with 10 pints of milk that is 1.5% butterfat to obtain a cream that is 10% butterfat?

16. A chemistry student has 50 ounces of a solution that is 60% acid. How much of this solution must be removed and replaced with pure acid in order to bring the acid strength up to 70%?

17. How many ounces of a perfume costing $84 per ounce must be combined with 24 ounces of a perfume costing $60 per ounce to make a perfume worth $68 per ounce?

18. How many pounds of peanuts costing $4 per pound should be mixed with cashews costing $7 per pound to obtain a 30-pound mixture worth $6 per pound?

19. How many barrels of a type of glue costing $80 per barrel should be mixed with glue costing $100 per barrel to obtain 50 barrels of a glue worth $92 per barrel?

20. How many cases of a candy costing $40 per case must be mixed with a candy costing $60 a case to obtain 80 cases of a candy worth $48 per case?

Answers to Practice Problems **18a.** 12.5 cc **b.** 31.2 pounds of fruit, 58.8 pounds of soybeans

7.6 Uniform Motion Problems

IDEA 1 A uniform motion problem deals with distance, rate, and time. To solve uniform motion problems, use the *distance formula*.

Distance Formula

> Distance traveled is equal to the rate of travel times the time traveled,
>
> or
>
> $$d = r \cdot t$$
>
> where d = distance traveled, r = rate of travel, and t = time traveled.

7.6 Uniform Motion Problems

> **To Solve a Uniform Motion Problem by Using the Distance Formula:**
>
> 1. Identify d, r, and t.
> 2. Replace the known quantities in the formula with their numerical values.
> 3. Solve for the unknown quantity and answer the questions asked.
> 4. Check your answer.

 When using the distance formula, make the units of measure consistent. That is, if the rate is expressed as miles per hour, then be sure to express the distance in miles and the time in hours.

Example 22 Solve.

a. If you drove four hours at a rate of 55 mph, how many miles would you travel?

b. Peter traveled 400 miles in six hours and 15 minutes. How fast was he traveling?

Solution To solve a uniform motion problem, solve the formula $d = rt$ for the unknown quantity and answer the question asked in the original problem. Remember that the units of measure must be consistent.

22a. Step 1: $d = ?, r = 55$ mph, $t = 4$ hr

Step 2: $d = rt$
$d = (55)(4)$

Step 3: $d = 220$
You would travel 220 miles.

Step 4: The answer checks since 220 miles = $\left(55\dfrac{\text{miles}}{\text{hour}}\right)(4 \text{ hour})$.

22b. Step 1: $d = 400$ miles, $r = ?$, $t = 6$ hrs 15 min = 6 hrs + $15\left(\dfrac{1}{60} \text{ hr}\right)$ = $6\dfrac{1}{4}$ hr

Step 2: $d = rt$
$400 = r\left(6\dfrac{1}{4}\right)$

Step 3: $400 = \dfrac{25}{4}r$
$64 = r$ Multiply both sides by $\dfrac{4}{25}$
Peter was traveling at 64 mph.

Step 4: The answer checks since 400 miles = $\left(64\dfrac{\text{miles}}{\text{hour}}\right)\left(6\dfrac{1}{4} \text{ hour}\right)$

Practice Problem 19 *Solve.*

a. If a bus travels 45 mph for four hours, how many miles will it travel?

b. A boat traveled 300 miles at a rate of 40 mph. How many hours did the boat travel?

IDEA 2

Many uniform motion problems involve a relationship between two objects (or one object under two different conditions). We can solve such problems in much the same manner that we solved simple interest and mixture problems. The difference is that the equation is usually based on how the distances are related.

To Solve a Uniform Motion Problem Involving Two Objects:

1. Use the four-step procedure to solve word problems.

2. In step 2, writing an equation, it may be helpful to construct a table of the following information:

Object	Rate (r)	Time (t)	Distance (d) = $r \cdot t$

Example 23 Solve.

a. Betty leaves school on a bike and travels at a rate of 15 mph. One hour later, Rochelle follows Betty in a car and travels 30 mph. How long will it take Rochelle to catch Betty?

b. Dan and Frank leave the gym and travel in opposite directions on a highway. If they are 30 miles apart in 20 minutes and Frank's car is traveling 5 mph faster than Dan's, how fast was each man driving?

c. It takes a boat two hours to travel downstream with a 6 mph current. The return trip upstream against the same current takes three hours. Find the rate of the boat in still water.

Solution To solve a uniform motion problem involving two objects, (1) represent the unknowns, (2) write an equation, (3) solve the equation and determine all solutions, and (4) check. For step 2, it may be helpful to make a sketch.

23a. Step 1: x = hours it takes Rochelle to catch Betty

Step 2:

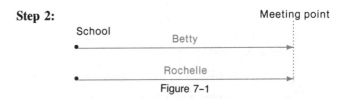

Figure 7-1

	Rate (r) (in mph)	Time (t) (in hours)	Distance (d) = $r \cdot t$ (in miles)
Betty	15	$x + 1$	$15(x + 1)$
Rochelle	30	x	$30x$

Figure 7-1 shows you that

	Distance Rochelle travels	=	Distance Betty travels
	$30x$	=	$15(x + 1)$

Step 3: $30x = 15(x + 1)$
$30x = 15x + 15$
$15x = 15$ Added $-15x$ to both sides
$x = 1$ Divided both sides by 15

It will take Rochelle one hour to catch Betty.

Step 4: The answer checks since distances traveled must be the same.
Rochelle's distance = (30 mph)(1 hr) = 30 miles
Betty's distance = (15 mph)(2 hr) = 30 miles

23b. Step 1: x = rate (in mph) of Dan's car
$x + 5$ = rate (in mph) of Frank's car

Step 2:

```
         Gym
  Dan   ■   Frank
←———————————————→
    30 miles
   Figure 7-2
```

	Rate (in mph)	Time (in hours)	Distance (in miles)
Dan	x	20 min = $\frac{1}{3}$ hr	$\frac{1}{3}x$
Frank	$x + 5$	20 min = $\frac{1}{3}$ hr	$\frac{1}{3}(x + 5)$

Figure 7-2 shows you that

Frank's distance	+	Dan's distance	=	30 miles
$\frac{1}{3}(x + 5)$	+	$\frac{1}{3}x$	=	30

Step 3: $\frac{1}{3}(x + 5) + \frac{1}{3}x = 30$
$x + 5 + x = 90$ Multiplied both sides by 3
$2x + 5 = 90$
$2x = 85$
$x = 42\frac{1}{2}$
$x + 5 = 47\frac{1}{2}$

Dan's rate is $42\frac{1}{2}$ mph and Frank's is $47\frac{1}{2}$ mph.

Step 4: The answer checks since the sum of distance traveled must be 30 miles.

Frank's distance = $\left(47\frac{1}{2} \text{ mph}\right)\left(\frac{1}{3} \text{ hr}\right) = \frac{95}{6}$ miles

Dan's distance = $\left(42\frac{1}{2} \text{ mph}\right)\left(\frac{1}{3} \text{ hr}\right) = \frac{85}{6}$ miles

Total distance = 30 miles

23c. **Step 1:** x = rate (mph) of the boat in still water

Step 2:

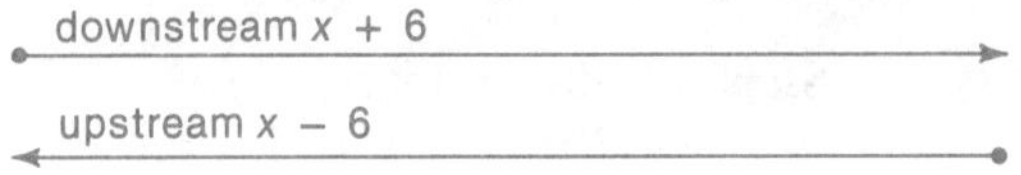

downstream $x + 6$

upstream $x - 6$

Figure 7-3

	Rate (in mph)	Time (in hours)	Distance (in miles)
Downstream	$x + 6$	2	$2(x + 6)$
Upstream	$x - 6$	3	$3(x - 6)$

Figure 7-3 shows you that

Distance upstream = Distance downstream
$$3(x - 6) = 2(x + 6)$$

Step 3: $3(x - 6) = 2(x + 6)$
$3x - 18 = 2x + 12$
$x - 18 = 12$ Added $-2x$ to both sides
$x = 30$ Added 18 to both sides

The boat traveled 30 mph in still water.

Step 4: The answer checks since the distances traveled must be the same.
Distance upstream = (36 mph)(2 hr) = 72 miles
Distance downstream = (24 mph)(3 hr) = 72 miles

Practice Problem 20 **Solve.**

a. Train A leaves Dallas at 10:00 A.M. and travels south at a rate of 45 mph. Train B leaves Dallas at 1:00 P.M. on a parallel track and travels south at a rate of 75 mph. How long will it take train B to catch up to train A?

b. Reggie made a trip of 360 miles to his summer home in eight hours. He traveled for six hours on an interstate highway and the rest of the time on an unpaved road. If his rate on the highway was 20 mph more than on the unpaved road, find his rate on the highway and the road.

In some uniform motion problems, the equation may not always be based on distance.

Example 24   Train A leaves Chicago for Boston (a distance of 1050 miles) at the same time that train B leaves Boston for Chicago. If train A travels at a rate of 60 mph and train B travels at 40 mph, how far has each train traveled when the two trains pass?

Solution   **Step 1:** x = miles train A travels before passing train B
$1050 - x$ = miles train B travels before passing train A

Step 2:

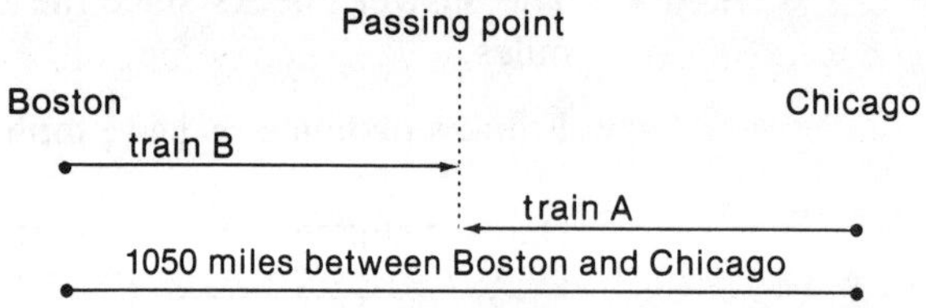

Passing point

Boston ← train B → | ← train A → Chicago

1050 miles between Boston and Chicago

Figure 7-4

	Rate (in mph)	Time (in hours)	Distance (in miles)
Train A	60	$\dfrac{x}{60}$	x
Train B	40	$\dfrac{1050 - x}{40}$	$1060 - x$

Use the fact that, since $d = r \cdot t$, $t = \dfrac{d}{r}$.

The equation is based on the fact that

Time train A travels = Time train B travels

$$\dfrac{x}{60} = \dfrac{1050 - x}{40}$$

Step 3:
$\dfrac{x}{60} = \dfrac{1050 - x}{40}$

$40x = 60(1050 - x)$ Cross-multiplied

$40x = 63{,}000 - 60x$

$100x = 63{,}000$ Added $60x$ to both sides

$x = 630$ Divided both sides by 100

$1050 - x = 420$

Train A travels 630 miles and train B travels 420 miles.

Step 4: The answer checks since the time traveled by each train must be the same.

Time train A travels $= \dfrac{630 \text{ miles}}{60 \text{ mph}} = 10.5$ hr

Time train B travels $= \dfrac{420 \text{ miles}}{40 \text{ mph}} = 10.5$ hr

Practice Problem 21 *Solve.*

A plane traveling 480 mph and a train traveling 64 mph leave the same point at the same time and travel north. In less than an hour, the plane has traveled 390 miles more than the train. How far has each vehicle traveled?

7.6 Exercises

Solve.

1. If you drove at a rate of 55 mph for two hours, how many miles would you travel?
2. John rode his bike at a rate of 20 mph for 45 minutes. How far did he travel?
3. Jim drove 150 miles in three hours. How fast was he driving?
4. How many hours will it take to travel 500 miles if you drive at a constant rate of 40 mph?
5. How many hours will it take a jet traveling at a rate of 550 mph to travel 1650 miles?
6. A jet traveled 3000 miles in five hours. How fast was the jet traveling?
7. Ricky completed a car race in three hours. If the race was 330 miles, how fast was Ricky traveling?
8. Denny rode his motorcycle at a constant rate of 45 mph for 40 minutes. How far did he travel?

9. June ran 28 miles in two hours and 14 minutes. What was her rate?
10. A speedboat travels at a constant rate of 90 mph for one hour and 12 minutes. How far did the boat travel?
11. Reggie leaves his house and runs at a rate of 5 mph. One hour later, his brother leaves the house on a bike and travels at a rate of 15 mph. How long will it take Reggie's brother to catch him?
12. Ed took 4 minutes to finish a race and Mike took $4\frac{1}{2}$ minutes to finish the same race. If the rate of the faster runner is three feet per second more than that of the slower runner, find each man's rate.
13. At 7:00 A.M. a boat traveling 30 mph leaves station A for station B, a distance of 150 miles. At 8:00 A.M. a boat traveling 20 mph leaves station B for station A. At what time will they pass each other?
14. A plane leaves an airport and travels 450 mph. Three hours later a second plane leaves the same airport and travels 475 mph. How long will it take the second plane to catch the first?
15. Two trains start from the same point at the same time and travel in the same direction on parallel tracks. If one train travels at 90 mph and the other at 68 mph, how many hours will it take before they will be 66 miles apart?
16. The pilot of a plane sets out to overtake the pilot of a helicopter that is 12 miles ahead. If the rate of the plane is 210 mph and the rate of the helicopter is 165 mph, how long will it take the plane to overtake the helicopter?
17. From a point on a highway, Sam and Tim ride their motorcycles in opposite directions. If they are 38 miles apart in 30 minutes and Tim's rate is 8 mph less than Sam's rate, how fast is each man traveling?
18. Ernie drove 760 miles in 14 hours. He traveled eight hours on a highway and six hours on a dirt road. If his rate on the highway was 25 mph more than on the dirt road, find the rate he traveled on both parts of his trip.
19. At 9:00 A.M., Rich leaves the train station on a bike and travels south at 8 mph. One hour later Joe leaves the same train station in a car and travels south at 20 mph. At what time will Joe overtake Rich?
20. At 11:00 A.M., Sylvia left the office to deliver a package to a restaurant. She stayed at the restaurant for one hour to eat lunch, and then she returned directly to the office, arriving at 3:00 P.M. She traveled to the restaurant at a rate of 45 mph, but her speed returning to the office was 35 mph. If she traveled the same route both ways, how far is it from her office to the restaurant?
21. Two steamboats are 50 miles apart. Both start at the same time and travel toward each other in a straight canal. If one averages 18 mph and the other averages 12 mph, how far does each travel before passing the other?

Answers to Practice Problems   **19a.** 180 miles   **b.** $7\frac{1}{2}$ hours   **20a.** $4\frac{1}{2}$ hours   **b.** 30 mph on the road, 50 mph on the highway   **21.** Plane traveled 450 miles; train traveled 60 miles.

Chapter 7 Summary

Important Terms

A **ratio** is a fraction that expresses a relationship between two quantities. The ratio of 3 to 4 is $\frac{3}{4}$. [Section 7.2/Idea 1]

A **rate** is a ratio of two quantities having different units of measure. The ratio of 30 miles to 2 hours is the rate $15 \frac{\text{miles}}{\text{hour}}$. [Section 7.2/Idea 2]

A **unit price** is a rate of cost. It is the ratio of price to quantity and has a denominator of 1. The ratio of 99 cents to 3 lbs is the unit price $33 \frac{\text{cents}}{\text{lb}}$. [Section 7.2/Idea 2]

A **proportion** is a statement that two ratios are equal, usually written in the form $\frac{a}{b} = \frac{c}{d}$. The numbers a, b, c, and d are called **terms**. The first and fourth terms (a and d) are called the **extremes**. The second and third terms (b and c) are called the **means**. The statement $\frac{3}{4} = \frac{6}{8}$ is a proportion. The numbers 3, 4, 6, and 8 are called terms. The numbers 3 and 8 are the extremes, and the numbers 4 and 6 are the means. [Section 7.2/Idea 4]

The **fundamental property of proportions** says that, in a proportion, the product of the extremes is equal to the product of the means. If $\frac{a}{b} = \frac{c}{d}$ then $ad = bc$.

[Section 7.2/Idea 4]

The **percent formula** says that the amount equals the product of the rate times the base. $A = R \cdot B$. [Section 7.3/Idea 1]

The **percent proportion** says that the ratio of the amount to the base is equal to the rate, expressed as a fraction. $\frac{A}{B} = \frac{P}{100}$. [Section 7.3/Idea 3]

Interest is the amount of money paid for the use of money. **Simple interest** is the interest computed on the original amount of money borrowed or invested. [Section 7.4/Idea 1]

The simple **interest formula** says that interest is equal to the product of the principal times the rate times the time. $I = P \cdot R \cdot T$. [Section 7.4/Idea 1]

The **mixture formulas** say that (1) the amount of a pure substance is equal to its concentration times the total volume of the mixture $A = C \cdot T$; (2) the total cost is equal to the unit price times the number of units $T = P \cdot N$. [Section 7.5/Idea 1]

The **distance formula** says that the distance traveled is equal to the rate of travel times the time traveled $d = r \cdot t$. [Section 7.6/Idea 1]

Important Skills

Solving Word Problems

To solve a word problem, represent the unknowns; write an equation, solve the equation, and answer the question asked; then check your solution. [Section 7.1/Idea 4]

Solving Proportions

To solve a proportion for an unknown term, set the product of the extremes equal to the product of the means, and solve the resulting equation. [Section 7.2/Idea 4]

To solve a word problem by using proportions, represent the unknowns. Write a proportion. Make sure the units occupy corresponding positions in the ratios. Finally, solve the proportion, and answer the question asked. [Section 7.2/Idea 5]

Solving Percent Problems

To solve a percent problem by using equations, represent the unknown, write an equation by using the formula $A = R \cdot B$, and solve the equation and answer the question asked. Then check your answer. [Section 7.3/Idea 2]

To solve percent problems by using the percent proportion formula $\frac{A}{B} = \frac{P}{100}$, identify A, P, and B. Replace the known quantities in the formula with their numerical values. Finally, solve for the unknown quantity and answer the question asked. [Section 7.3/Idea 3]

Solving Simple Interest Problems

To solve a simple interest problem (involving only one interest rate) by using the interest formula $I = PRT$, identify I, P, R, and T. Replace the known quantities in the formula with their numerical values. Solve for the unknown quantity, and answer the question asked. [Section 7.4/Idea 1]

To solve a simple interest problem involving two or more interest rates, use the four-step procedure used to solve word problems. In step 2, it may be helpful to construct the following table:

Investment	Principal (P)	Rate (R)	Time (T)	Interest (I) = $P \cdot R \cdot T$

[Section 7.4/Idea 2]

Solving Mixture Problems

To solve a mixture problem, use the four-step procedure used to solve word problems. In step 2, it may be helpful to construct one of the following tables:

Mixture	Concentration of Substance (C)	Total Volume of Mixture (T)	Amount of Pure Substance (A) = C · T

or

Item	Unit Price (P)	Number of Units (N)	Total Cost (T) = P · N

[Section 7.5/Idea 1]

Solving Uniform Motion Problems

To solve a uniform motion problem (involving only one object) by using the distance formula $d = rt$, identify d, r, and t. Replace the known quantities in the formula with their numerical values. Solve for the unknown quantity and answer the question asked. [Section 7.6/Idea 1]

To solve a uniform motion problem involving two objects, use the four-step procedure to solve word problems. In step 2, it may be helpful to construct the following table:

Object	Rate (R)	Time (t)	Distance (d) = r · t

[Section 7.6/Idea 2]

Chapter 7 Review Exercises

Solve.

1. Express "five less than a number" as an algebraic expression.
2. The sum of two numbers is 50. Represent each number in terms of the same variable.
3. The sum of two consecutive integers is 263. Find the integers.
4. John won 18 out of 30 games. What percent of the games did he win?
5. What is the ratio of eight inches to three feet?
6. If $54 in interest is paid on an 18% loan for six months, what is the principal of the loan?
7. How many liters of a 40% salt solution must be added to 10 liters of a 70% salt solution to obtain a solution that is 50% salt?
8. After paying an 8% sales commission. Rich received $73,600 from the sale of his summer home. What was the actual selling price?
9. Jim earned $5200 in five months. At this rate, how much can he expect to earn in one year?
10. Two cars moving in opposite directions from the same point are 19 miles apart in 15 minutes. If one car's rate is 8 mph less than the other, find the rate of each car.
11. Jane invested $8000. Part was invested at 11% and the rest at 10%. If her total annual interest was $850, how much was invested at each rate?
12. Solve for x. $6:4\frac{1}{2} = x:30$

13. If a car travels 500 miles in 6 hours 15 minutes, how fast was the car traveling?
14. How much distilled water must be added to 50 ounces of a 12% acid solution to obtain a solution that is 6% acid?
15. A car rental company charges $19.95 per day plus 18¢ per mile to rent a compact car. If Joe rented a car for one day and his total bill was $34.44, how many miles did he travel?
16. The profit from the sale of a boat must be divided among three men in a ratio of 5:7:8. If the total profit from the sale of the boat was $22,400, how much money should each man receive?
17. A 3 cup serving of green beans contains 26 calories. A 1⅓ cup serving of broccoli contain 58 calories. For the same size serving, which contains more calories the broccoli or the beans?
18. The price of a dessert rose from 75 cents to 99 cents. What was the percent increase in price?
19. One liter equals 1.06 quarts. How many liters are there in 2 gallons?
20. Sixty people attended a play. A general admission ticket cost $6.25 and a senior citizen ticket costs $4.75. If the box office collected $319.50, how many of each type of ticket did they sell?
21. How many liters of a 30% alcohol solution must be mixed with 50 liters of a 70% alcohol solution to produce a 55% alcohol solution?
22. Steve invested $6000. Part was invested in a money market account paying 7% interest per year and the remainder was invested in stocks paying 11% interest per year. If the total yearly interest from the two investments is $492, how much was invested in the money market account.
23. Ferry can grade 12 tests per hour while Dan can grade 16 tests per hour. If Ferry starts grading tests at 1:00 P.M. and Dan starts at 2:20 P.M., at what time will Dan have graded as many tests as Ferry? (*Hint*: number of tests graded = rate · time.)
24. A jacket that sells for $228 costs the dealer $120. What percent profit does the dealer make when he sells this jacket?
25. It takes 1½ gallon of paint to cover 500 square feet of wall space. How much paint do you need to cover 1020 square feet of wall space?

Chapter 7 Test

1. If the sum of two numbers is 15, how would you represent each number.

Solve.

2. Find three consecutive even numbers whose sum is 60.

3. Find the unit price of a 32 oz bottle of juice that cost $1.28.

4. Cazzie and Don divided $9000 in the ratio of 5:3. How much money did each person receive?

5. If you paid $16.56 for 18 gallons of gas, find the cost of 8.5 gallons of gas.

6. On an examination having 120 questions, Shirley got 96 correct answers. What percent of the problems did she answer correctly.

7. Ruth earned a $636 commission on the sale of a camper. If her commission rate is 12% find the original cost of the camper.

8. A clothing store owner had $6000 invested in inventory. The profit on women's clothes was 35%, while the profit on men's clothing was 25%. If the profit on the entire stock was 28%, how much was invested in each type of clothing?

9. How many gallons of a 25% alcohol solution must be added to 50 gallons of a 55% alcohol solution to obtain a solution that is 45% alcohol.

10. Jill finished a race in 3 minutes, while Joe took 4 minutes to finish the same race. If Jill's rate is 5 feet per second more than Joe's find the rates of each person.

1. _____
2. _____
3. _____
4. _____
5. _____
6. _____
7. _____
8. women's _____
    men's _____
9. _____
10. Jill _____
    Joe _____

8 Graphing

Objectives The objectives for this chapter are listed below along with sample problems for each objective. By the end of the chapter you should be able to find the solution to the given problems.

1. Graph an ordered pair in the rectangular coordinate system *(Section 8.1/Idea 2)*.
   a. (0, 3)   b. (−2, 3)
   c. (−1, −3)   d. (0, 0)
   e. (−3, 0)   f. (2, −3)

2. Find the coordinates of a point *(Section 8.1/Idea 3)*.

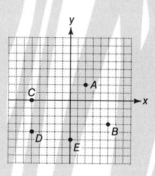

3. Graph a linear equation in the rectangular coordinate system by using the intercept method *(Section 8.2/Idea 2)*.
   a. $x - y = 3$ b. $2x + y = 4$
   c. $6x - 4y = 8$ d. $x - 5 = -1$

4. Graph a linear equation in the rectangular coordinate system by using the slope-intercept method *(Section 8.3/Idea 2)*.
   a. $y = 2x - 2$ b. $3x - 2y = -6$

5. Graph a linear inequality in the rectangular coordinate system *(Section 8.4/Idea 2)*.
   a. $x + y < 4$ b. $2x - y \geq 4$
   c. $x - 5y \leq 0$ d. $2y - 8 > -2$

8.1 The Rectangular Coordinate System

IDEA 1 The **rectangular coordinate system** is a two-dimensional plane formed by two intersecting perpendicular number lines. The point of intersection is called the **origin** (see Figure 8–1).

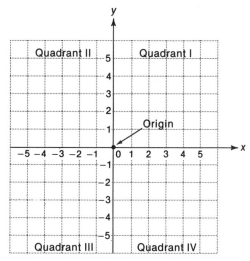

Figure 8–1

The rectangular coordinate system is divided into four regions called **quadrants.** The quadrants are numbered (with Roman numerals) counterclockwise beginning with quadrant I in the upper right region. The horizontal line is called the **x-axis** and the vertical line is called the **y-axis.** An arrow on each axis is used to show which direction is positive, and the origin is labeled "O."

Points in the rectangular coordinate system correspond to **ordered pairs** of numbers (x, y). For example, $(0, 2)$, $(-4, 2)$, $(-0.5, -3)$, and $(2\frac{1}{2}, 5)$ are ordered pairs that correspond to points in the coordinate system. The numbers in the ordered pair are called **coordinates** of the point. The first number is the **x-coordinate** or **abscissa** of the point. The second number is the **y-coordinate** or **ordinate** of the point. In an ordered pair the x-coordinate always appears before the y-coordinate.

IDEA 2 To locate a point in the rectangular coordinate system, we will need to know the x-coordinate and the y-coordinate. For example, given the ordered pair $(3, 2)$, 3 is the x-coordinate and 2 is the y-coordinate. To locate this point we find 3 on the x-axis and draw a vertical line, and then we find 2 on the y-axis and draw a horizontal line. As shown in Figure 8–2, the intersection of these two lines is the point $(3, 2)$.

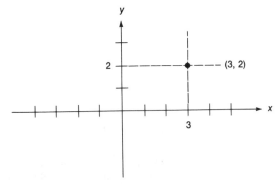

Figure 8–2

Finding points in the coordinate system is called **graphing** or **plotting points.** When you are given the coordinates of a point and are asked to find the point (x, y) in the coordinate system, use the following procedure.

> **To Graph a Point (x, y):**
>
> 1. Locate the x-coordinate on the x-axis and draw a vertical line through this point. This line will be parallel to the y-axis.
>
> 2. Locate the y-coordinate on the y-axis and draw a horizontal line through this point. This line will be parallel to the x-axis.
>
> 3. The intersection of these two lines is the graph of the ordered pair (x, y).

Example 1 Graph the following ordered pairs.

a. (2, 4) b. (−2, 3) c. (−4, −2) d. (3, −4)

Solution Points in the coordinate system can be seen as the intersection of two lines (see Figure 8-3).

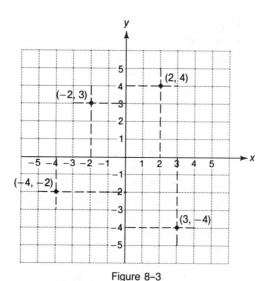

Figure 8-3

With practice you should be able to plot points without actually drawing the intersecting lines.

Example 2 Graph the following ordered pairs.

a. (0, 3) b. (−2, 0) c. (3, 0) d. (0, −3)

Solution All points having an x-coordinate of zero are on the y-axis. All points having a y-coordinate of zero are on the x-axis (see Figure 8-4).

 The order in which the coordinates of a point are written is important. For example, the ordered pairs (0, 3) and (3, 0) are not the same; they represent two different points.

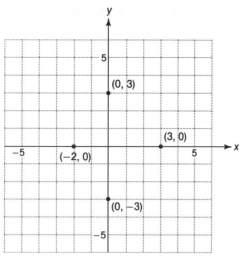

Figure 8-4

Practice Problem 1   *Graph the following ordered pairs.*

    a. (3, 4)   **b.** (−2, 0)   **c.** (3, −4)   **d.** (0, 4)   **e.** (−2, −3)

IDEA 3

So far we have plotted points, given their coordinates. Let us now find the coordinates of a given point. To find the coordinates of a given point, use the following procedure.

To Find the Coordinates of a Given Point:

1. Draw a vertical line through the point. This line will intersect the x-axis at the x-coordinate.

2. Draw a horizontal line through the point. This line will intersect the y-axis at the y-coordinate.

3. Write the coordinates as the ordered pair (x, y).

Example 3   Find the coordinates of each point in Figure 8-5.

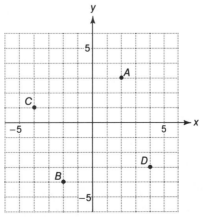

Figure 8-5

290   Graphing

Solution   The Solution is shown in Figure 8-6.

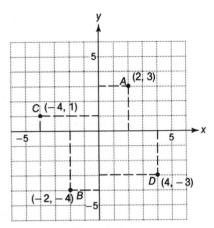

Figure 8-6

With practice you should be able to find the coordinates of a point without drawing the vertical and horizontal lines.

Example 4   Find the coordinates of each point in Figure 8-7.

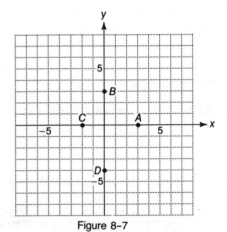

Figure 8-7

Solution   As shown in Figure 8-8 every point on the x-axis is of the form $(x, 0)$. Every point on the y-axis is of the form $(0, y)$.

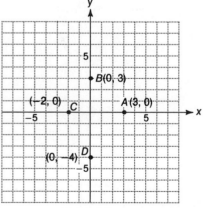

Figure 8-8

Practice Problem 2 *Find the coordinates of each point in the figure below.*

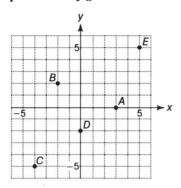

8.1 Exercises

1. Graph (2, 3), (−3, 2), (−3, −3), and (−2, 4).
2. Graph (3, −2), (−3, −1), (2, 2), and (−5, 1).
3. Graph (0, 3), (3, 0), (−4, 0), and $\left(0, -2\frac{1}{2}\right)$.
4. Graph (−6, 0), (0, −6), (2.5, 0), and (0, 6).
5. Plot the points having the following coordinates: (0, 4), $\left(-2, 3\frac{1}{2}\right)$, (−5, 0), and (−3, −4).
6. Plot the points having the following coordinates: (3, −5), (0, −1), (0, 0), and (5, −2).

Find the coordinates of points A, B, C, D, E, and F.

7.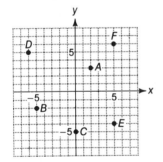

8.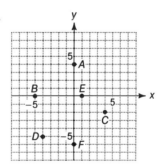

9.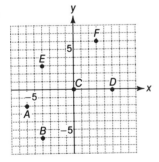

10.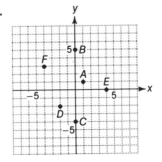

Fill in the blanks.

11. The rectangular coordinate system is divided into four regions called _____.
12. In the rectangular coordinate system, the horizontal line is called the _____ and the vertical line is called the _____.
13. Points in the rectangular coordinate system correspond to _____ of numbers.

292 Graphing

14. The numbers in an ordered pair are called the _____ of the point.

15. The first number in an ordered pair is the _____ or _____ and the second number is the _____ or _____ of the point.

16. To graph an ordered pair (x, y), first locate x on the x-axis and draw a _____ line through this point. Next, locate y on the y-axis and draw a _____ line through this point. Finally, the _____ of these two lines is the graph of the ordered pair (x, y).

17. To find the coordinates of a given point, first draw a _____ line through the point. This line will intersect the x-axis at _____, which is the first _____ of the point. Next, draw a _____ line through the point. This line will intersect the y-axis at _____, which is the _____ coordinate of the point. Finally, write the _____ as the _____ (x, y).

18. The _____ is assigned the coordinates (0, 0).

Answers to Practice Problems   1.   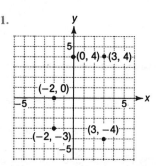   2. $A = (3, 0)$, $B = (-2, 2)$, $C = (-4, -5)$, $D = (0, -2)$, $E = (5, 5)$

8.2 Linear Equations in Two Variables

IDEA 1   An equation that can be written in the form $ax + by = c$, where x and y are variables and a, b, and c are constants (a and b are not both zero), is called a **linear** or **first-degree equation in two variables** or, simply, a **linear equation.** Some examples of linear equations in two variables are:

$$x + y = 4 \qquad 3x - 5y = 10 \qquad x - 5y = 0$$

A **solution of a linear equation is two variables** is an ordered pair of numbers that makes the equation a true statement. For example, (1, 3) is a solution of the equation $x + y = 4$ because $x + y = 4$ is true when x is 1 and y is 3. We can also say that (1, 3) *satisfies* the equation $x + y = 4$.

$$x + y = 4$$
$$1 + 3 = 4 \qquad \text{Replace } x \text{ with 1 and } y \text{ with 3}$$
$$4 = 4 \qquad \text{True}$$

The equation $x + y = 4$ has an infinite number of solutions. Some of these solutions are (0, 4), (4, 0), (2, 2), and (6, −2).

The **graph of a linear equation in two variables** is the collection of points whose coordinates satisfy the equation.

Example 5   Find and graph three ordered pairs that are solutions of $x + y = 4$.

Solution   By inspection, you have found that (0, 4), (4, 0), and (2, 2) are solutions of $x + y = 4$ since these ordered pairs make the equation $x + y = 4$ a true statement. Figure 8.9 shows that the ordered pairs that are solutions of a linear equation lie on a straight

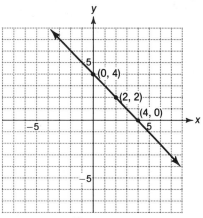

Figure 8-9

line called the graph of the equation. All ordered pairs associated with the points on this line will satisfy the equation.

Practice Problem 3 *Find and graph four ordered pairs that are solutions of $x - y = 5$.*

IDEA 2

Quite often you will not be able to find the solution of a linear equation by inspection. When this occurs, either (1) choose a convenient value of x and find the corresponding value for y, or (2) choose a value for y and find the corresponding value for x. For example, to graph the equation $2x - 3y = 6$, let $x = 0$ and solve for y, let $y = 0$ and solve for x, and let $x = -3$ and solve for y. Next, graph the three solutions and draw a straight line through these three points.

$$\text{Let } x = 0$$
$$2x - 3y = 6$$
$$2(0) - 3y = 6$$
$$-3y = 6$$
$$y = -2 \to (0, -2)$$

Thus, $(0, -2)$ is a solution of $2x - 3y = 6$. The y-coordinate of the point $(0, -2)$ is called the **y-intercept** since this is the point where the graph of the equation crosses the y-axis (see Figure 8-10).

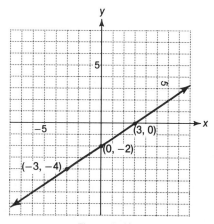

Figure 8-10

$$\text{Let } y = 0$$
$$2x - 3y = 6$$
$$2x - 3(0) = 6$$
$$2x = 6$$
$$x = 3 \rightarrow (3, 0)$$

Thus, (3, 0) is a solution of $2x - 3y = 6$. The *x*-coordinate of the point (3, 0) is called the **x-intercept** since this is the point where the graph of the equation crosses the *x*-axis (see Figure 8–10).

$$\text{Let } x = -3$$
$$2x - 3y = 6$$
$$2(-3) - 3y = 6$$
$$-6 - 3y = 6$$
$$-3y = 12$$
$$y = -4 \rightarrow (-3, -4)$$

Thus, $(-3, -4)$ is a solution of $2x - 3y = 6$ (see Figure 8–10).

 Although you need only two points to draw a straight line, it is advisable to find at least one other point as a check.

The procedure we have just reviewed is called the *intercept method* for graphing a straight line. This procedure is summarized below.

To Graph a Linear Equation in Two Variables by Using the Intercept Method:

1. Find the *x*-intercept by letting $y = 0$ and solving for *x*.
2. Find the *y*-intercept by letting $x = 0$ and solving for *y*.
3. Find a third ordered pair that satisfies the equation by letting *x* or *y* be any convenient number and then solving for the other variable.
4. Graph the three ordered pairs that satisfy the equation and connect them with a straight line. This line is the graph of the equation.

Example 6   Graph $4x - 3y = 9$.

Solution   To graph a linear equation, first find at least three ordered pairs that satisfy the equation. Next, graph these ordered pairs. Finally, connect the points with a straight line.

x-intercept:
$$4x - 3y = 9 \qquad \text{Let } y = 0 \text{ and solve for } x$$
$$4x - 3(0) = 9$$
$$4x = 9$$
$$x = \frac{9}{4}$$
$$x = 2\frac{1}{4}$$

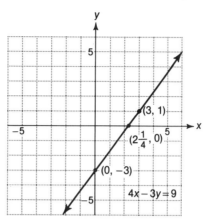

Figure 8-11

The x-intercept is $2\frac{1}{4}$ (see Figure 8-11). Thus we graph $(2\frac{1}{4}, 0)$.

y-intercept:
$$4x - 3y = 9 \quad \text{Let } x = 0 \text{ and solve for } y$$
$$4(0) - 3y = 9$$
$$-3y = 9$$
$$y = -3$$

The y-intercept is -3 (see Figure 8-11). Thus we graph $(0, -3)$.

Check:
$$4x - 3y = 9 \quad \text{Let } x = 3 \text{ and solve for } y$$
$$4(3) - 3y = 9$$
$$12 - 3y = 9$$
$$-3y = -3$$
$$y = 1$$

The checkpoint to graph is $(3, 1)$ (see Figure 8-11).

Example 7 Graph $x = 2y$.

Solution

x-intercept:
$$x = 2y \quad \text{Let } y = 0 \text{ and solve for } x$$
$$x = 2(0)$$
$$x = 0$$

The x-intercept is 0 (see Figure 8-12). Thus we graph $(0, 0)$.

y-intercept:
$$x = 2y \quad \text{Let } x = 0 \text{ and solve for } y$$
$$0 = 2y$$
$$0 = y$$

The y-intercept is 0 (see Figure 8-12). Thus we graph $(0, 0)$. Since the x and y-intercept are the same point, we must find two other ordered pairs that satisfy the equation.

Third ordered pair:
$$x = 2y \quad \text{Let } y = 1 \text{ and solve for } x$$
$$x = 2(1)$$
$$x = 2$$

296 Graphing

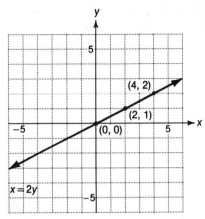

Figure 8-12

The third point is (2, 1) (see Figure 8-12).

Check: $x = 2y$ Let $x = 4$ and solve for y
            $4 = 2y$
            $2 = y$

The checkpoint is (4, 2) (see Figure 8-12).

Practice Problem 4 *Graph each equation.*

   a. $2x + 3y = 9$ **b.** $2x + y = 0$

IDEA 3

Some linear equations contain only one variable. The graph of such equations is either a horizontal line or a vertical line.

Example 8 Graph each equation.

   a. $x = -3$ **b.** $y = 2$

Solution **8a.** Since there is no y, the equation $x = -3$ is equivalent to $x + 0y = -3$ and every ordered pair $(-3, y)$ satisfies the equation. That is, $(-3, 0)$, $(-3, 3)$, and $(-3, -2)$ satisfy the equation and correspond to points on the graph of the line $x = -3$ (see Figure 8-13).

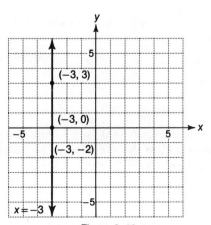

Figure 8-13

8b. Since there is no x, the equation $y = 2$ is equivalent to $0x + y = 2$, and every ordered pair $(x, 2)$ is a solution. That is, $(0, 2)$, $(-2, 2)$, and $(3, 2)$ satisfy the equation and correspond to points on the graph of the line $y = 2$ (see Figure 8-14).

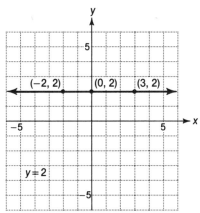

Figure 8-14

In general, the graph of $x = a$, where a is a constant, is a vertical line that passes through the point $(a, 0)$. Similarly, the graph of $y = b$, where b is a constant, is a horizontal line that passes through the point $(0, b)$.

Practice Problem 5 **Graph each equation.**

   a. $x = 3$ **b.** $y = -\dfrac{7}{2}$ **c.** $x - 5 = -9$

8.2 Exercises

Find the missing coordinate by using the given equation, and then graph the ordered pairs.

1. $y = 3x + 1$ $(0, \ )$, $(-2, \ )$, $(\ , 4)$ **2.** $2x + y = 6$ $(\ , 0)$, $(1, \ )$, $(5, \ )$

3. $3x - 7y = 14$ $(0, \ )$, $(\ , 1)$, $(5, \ )$ **4.** $x = 5$ $(\ , 2)$, $(\ , -1)$, $(\ , 0)$

5. $y + 3 = 1$ $(0, \ )$, $(1, \ )$, $(-3, \ )$

Find the x-intercept and the y-intercept.

6. $x + y = 6$ **7.** $3x - y = 6$ **8.** $3x + 4y = 12$ **9.** $y = 3x + 6$

10. $5x - 10y = 20$ **11.** $\dfrac{x}{3} + \dfrac{y}{2} = 1$ **12.** $2x = 3y$ **13.** $y = \dfrac{x}{2} - 6$

Graph each equation using the intercept method.

14. $x + y = 3$ **15.** $2x + y = 4$ **16.** $2x - 3y = 6$ **17.** $x - 3y = 6$

18. $x - y = 0$ **19.** $2x + 5y = 10$ **20.** $x = -2$ **21.** $y = -\dfrac{3}{2}$

22. $x - 5 = -3$ **23.** $y + 4 = 6$ **24.** $x = -2y$ **25.** $x = \dfrac{y}{2} - 4$

26. $x = \dfrac{3}{4}y - 3$ **27.** $x = \dfrac{2}{3}y - 4$ **28.** $2x = 5(y + 2)$ **29.** $3x = 5(y + 3)$

30. $4y = 16 - 8x$ **31.** $3y = 6 - 2x$ **32.** $y = \dfrac{3}{5}x$ **33.** $y = \dfrac{3}{4}x$

34. $0.2x + 0.3y = 0.6$ **35.** $0.1x - 0.1y = 0.3$ **36.** $x - 2.5 = -5$ **37.** $x - 3.5 = -8$
38. $y = -3.5$ **39.** $y = -2.5$

Fill in the blanks.

40. A _____ for a linear equation in two variables is an _____ that makes the equation a true statement.

41. The graph of a linear equation consists of ordered pairs that lie on a _____.

42. The ordered pairs associated with the points on a line that is the graph of an equation will _____ the equation.

43. To graph a linear equation using the intercept method, find at least three _____ that _____ the equation. Next, plot these _____ and connect them with a _____.

44. The graph of $x = a$, where a is a constant, is a _____ line that passes through the point _____.

Solve.

45. Jim charges $10 per hour plus $10 for materials to repair a lawn mower.

   a. Write an equation that determines Jim's total fee for a service call. Let y stand for the number of dollars he earns for working x hours.

   b. Find Jim's fee for working two hours, four hours, and six hours. That is, find the value of y in the ordered pairs $(2, y)$, $(4, y)$, and $(6, y)$.

   c. Graph the equation that you wrote for 45a by plotting the ordered pairs that you found for 45b on the graph at the right.

   d. Determine from the graph in 45c the number of hours it will take Jim to earn $85.

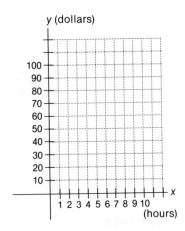

46. Using a straight-line depreciation method, a machine costing $2000 will have a book value of $b = 2000 - 200n$ dollars n years after purchase.

   a. Find the book value of the machine after one year, three years, and five years. That is, find b when $n = 1$, $n = 3$, and $n = 5$.

   b. Using the fact that a solution of the equation $b = 2000 - 200n$ is the ordered pair (n, b), graph $b = 2000 - 200n$ at the right.

   c. Determine from the graph in 46b the number of years it will take for the book value of the machine to be $1200.

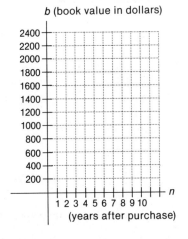

Answers to Practice Problems

3. 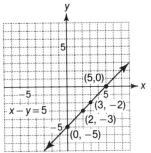   4a. 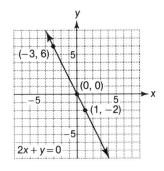   4b.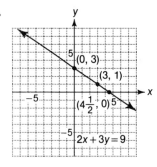

5a. 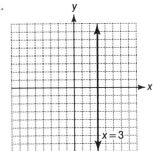   5b. 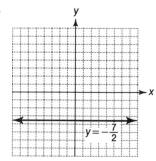   5c.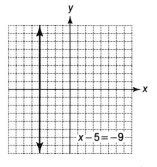

8.3 Slope-Intercept Form of Linear Equations

IDEA 1

Until now we have discussed linear equations written in the form $ax + by = c$. However, linear equations can be written in many different forms. One useful form is the **slope-intercept form** $y = mx + b$, where m is the slope of the line and b is the y-intercept. The **slope** m of a line is the ratio of the rise (vertical rise) to the run (horizontal run).

Example 9   A slope of $\frac{4}{3}$ means that for every three units of run in a horizontal direction, there is a corresponding rise of four units. If a line slants *up* to the right (see Figure 8-15) we say that it has a positive slope.

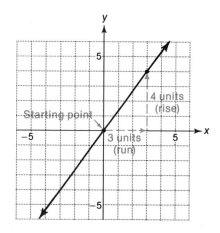

The slope $m = \dfrac{\text{rise}}{\text{run}}$

$= \dfrac{4}{3}$

If a line slants *down* to the right (see Figure 8–16) we say that it has a negative slope.

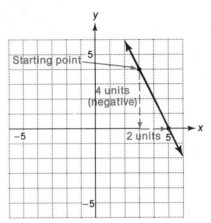

Figure 8–16

The slope $m = \dfrac{\text{rise}}{\text{run}}$

$= \dfrac{-4}{2}$

$= -2$

A slope of $-2 \left(\dfrac{-2}{1}\right)$ means that for every run of 1 unit in a horizontal direction there is a corresponding fall of 2 units.

IDEA 2 The slope-intercept form is useful because it lets you identify the slope *m* and the *y*-intercept *b* directly from the equation.

Example 10 Write each equation in the slope-intercept form and identify the slope and the *y*-intercept.

a. $3x + y = 4$ **b.** $5x - 4y = 8$

Solution To write a linear equation in the slope-intercept form $y = mx + b$, first solve the equation for *y* and write the *x* term to the left of the constant. Then, the coefficient of *x* is the slope *m*, and the constant *b* is the *y*-intercept.

10a. $3x + y = 4$
$y = -3x + 4$ Add $-3x$ to both sides

The slope is -3 and the *y*-intercept is 4.

10b. $5x - 4y = 8$
$-4y = -5x + 8$ Add $-5x$ to both sides
$y = \dfrac{5}{4}x - 2$ Divide both sides by -4

The slope is $\dfrac{5}{4}$ and the *y*-intercept is -2.

Practice Problem 6   *Write each equation in slope-intercept form and identify the slope and the y-intercept.*

   a. $4x + y = 4$ **b.** $5x - 3y = 8$

When an equation is in the slope-intercept form, the slope and y-intercept can be used to graph the equation.

To Graph a Linear Equation in Two Variables by Using the Slope-Intercept Method:

1. Rewrite the original equation in the slope-intercept form $y = mx + b$, and identify the slope m and the y-intercept b.
2. Graph the point $(0, b)$.
3. Starting at $(0, b)$ and remembering that the slope is the $\frac{\text{rise}}{\text{run}}$, find a second point on the graph.
4. Draw a line that passes through $(0, b)$ and the point found in step 3.
5. Find at least one other point as a check.

Example 11   Use the slope-intercept method to graph each equation.

   a. $2x + y = 3$ **b.** $3x - 4y = 10$

Solution   **11a.** $2x + y = 3$
          $y = -2x + 3$

The y-intercept is 3. Thus we graph $(0, 3)$. The slope is $-2 \left(\frac{\text{rise}}{\text{run}} = \frac{-2}{1} \right)$.

Next, starting at $(0, 3)$, measure two units down and then one unit to the right to obtain another point $(1, 1)$. Finally, draw a line passing through $(0, 3)$ and $(1, 1)$. It is advisable to find a third point as a check. Thus, let $x = 3$ to find that

$$y = -2x + 3$$
$$= -2(3) + 3$$
$$= -3 \qquad (3, -3)$$

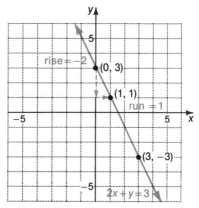

Figure 8-17

You can see in Figure 8-17 that $(3, -3)$ is on the line. Therefore, the graph is correct.

11b. $3x - 4y = 10$
$-4y = -3x + 10$
$y = \frac{3}{4}x - \frac{5}{2}$

The y-intercept is $-2\frac{1}{2}$. Thus we graph $\left(0, -2\frac{1}{2}\right)$. The slope is $\frac{3}{4}\left(\frac{\text{rise}}{\text{run}} = \frac{3}{4}\right)$.

Next, starting at $\left(0, -2\frac{1}{2}\right)$, measure three units up and then four units to the right to obtain another point $\left(4, \frac{1}{2}\right)$. Finally, draw a line passing through $\left(0, -2\frac{1}{2}\right)$ and $\left(4, \frac{1}{2}\right)$.

Check:
$y = \frac{3}{4}x - \frac{5}{2}$ Let $x = 2$ and solve for y
$y = \frac{3}{4}(2) - \frac{5}{2}$
$y = \frac{3}{2} - \frac{5}{2}$
$y = -1$ $(2, -1)$

You can see in Figure 8-18 $(2, -1)$ is on the line. Therefore, the graph is correct.

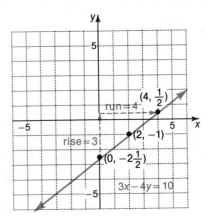

Figure 8-18

Practice Problem 7   *Use the slope-intercept method to graph each equation.*

    a. $y = 3x - 1$ **b.** $3x + 2y = 8$ **c.** $4x - 6y = 8$

8.3 Exercises

Write each equation in the slope-intercept form and identify the slope and the y-intercept.

1. $2x + y = 6$
2. $2x + 3y = 10$
3. $2x - 5y = 30$
4. $4x + 6y = 12$
5. $x - 4y = 8$
6. $3x + 5y = 15$
7. $3x + 2y = 6$
8. $3x + 4y = 12$
9. $3x + y = 6$
10. $2x + y = 4$
11. $5x - y = 8$
12. $4x - y = 9$

13. $6x + 4y = 10$ 14. $4x + 6y = 14$ 15. $3x - 5y = 11$ 16. $2x - 3y = 5$

Use the slope-intercept method to graph each equation.

17. $6x + 2y = 4$ 18. $10x - 8y = 16$ 19. $y = 2x - 1$ 20. $3x - 2y = 6$

21. $4x + 6y = 15$ 22. $3x - y = 0$ 23. $3x - 4y = -12$ 24. $y = \frac{3}{2}x - 4$

25. $\frac{x}{3} + \frac{y}{2} = 1$ 26. $4x - 5y = 15$ 27. $5x + 2y = 10$ 28. $2x + 3y = 6$

29. $4x + 3y = 12$ 30. $3x - 4y = -12$ 31. $2x - y = 4$ 32. $3x - y = 6$

33. $\frac{x}{2} + \frac{y}{2} = 1$ 34. $\frac{x}{3} + \frac{y}{3} = 1$ 35. $2x = 5(y + 2)$ 36. $3x = 5(y + 3)$

37. $4y = 16 - 8x$ 38. $3y = 6 - 2x$ 39. $y = 2x$ 40. $y = 3x$

41. $x = -2y$ 42. $x = -3y$

Fill in the blanks.

43. The slope-intercept form of an equation is _____.

44. When an equation is in the slope-intercept form $y = mx + b$, the slope is _____ and the _____ is $(0, b)$.

45. If a line has a positive slope, it slants up to the _____.

46. If a line has a negative slope, it slants down to the _____.

47. The slope m of a line is the ratio of the _____ to the _____.

Answers to Practice Problems **6a.** $y = -4x + 4$, slope $= -4$, y-intercept $= 4$ **b.** $y = \frac{5}{3}x - \frac{8}{3}$, slope $= \frac{5}{3}$, y-intercept $= -\frac{8}{3}$

7a. 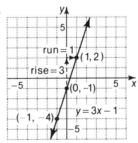   7b. 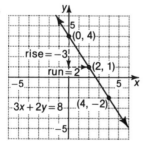   7c.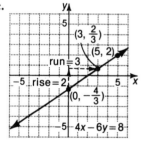

8.4 Linear Inequalities in Two Variables

IDEA 1 A **linear inequality in two variables** is an inequality that can be written in the form $ax + by < c$. You can graph linear inequalities in two variables in much the same manner that you graph linear equations in two variables. The only distinction is that the **graph of a linear inequality in two variables** is a half-plane that may or may not include the *boundary line*. The **boundary line** is the graph of the equation obtained when the inequality symbol is replaced with an equal sign.

Example 12 Graph the following.

　　a. $x + y \leq 3$ b. $x + y > 3$

Solution

12a. Since the original inequality includes an equality symbol, the boundary line $x + y = 3$ is part of the solution. This is indicated by a solid line (see Figure 8-19).

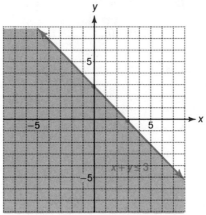

Figure 8-19

12b. Since the original inequality does *not* include an equality symbol, the boundary line $x + y = 3$ is *not* part of the solution. This is indicated by a dotted line (see Figure 8-20).

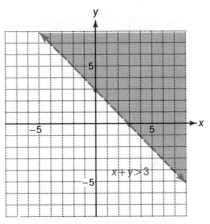

Figure 8-20

In each case, the shaded region comprises the solutions of the inequality. That is, every point in the shaded region is a solution to the original inequality.

IDEA 2

Example 12 illustrates that the boundary line divides the plane into two regions and that the points in only one of these regions are solutions of the inequality. To determine which is the correct region, choose any point that is not on the boundary line and substitute its coordinates into the inequality. If the resulting inequality is true, the solutions are the half-plane containing the point selected. If the resulting inequality is false, the solutions are the half-plane *not* containing the point selected. For example, in the inequality $x + y \leq 3$ (see Example 12), if you select the origin $(0, 0)$ as a test point, you can see that

$$x + y \leq 3$$
$$0 + 0 \leq 3$$
$$0 \leq 3 \quad \text{True}$$

Therefore, the half-plane that contains the point (0, 0) is the solution of the inequality.

Based on the above statements, graph linear inequalities in two variables using the following procedure.

To Graph a Linear Inequality in Two Variables:

1. Graph the boundary line. The boundary line is a *solid* line when the inequality includes an equality symbol ($\leq$ or $\geq$) but it is a *dotted* line when the inequality does not include an equality symbol ($<$ or $>$).

2. Select a point that is not on the line to determine which half-plane comprises the solutions.

3. Shade the half-plane identified in step 2.

Example 13 Graph each inequality.

a. $2x + 3y < 6$ **b.** $5x - 4y \geq 10$

Solution **13a.** $2x + 3y < 6$.

Step 1: Graph the boundary line $2x + 3y = 6$ (see Figure 8–21).

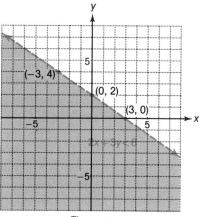

Figure 8–21

x-intercept:
$2x + 3y = 6$
$2x + 3(0) = 6$
$x = 3; (3, 0)$

y-intercept:
$2x + 3y = 6$
$2(0) + 3y = 6$
$y = 2; (0, 2)$

Check: $2x + 3y = 6$
$2(-3) + 3y = 6$
$y = 4; (-3, 4)$

Step 2: Select the correct half-plane by using the test point (0, 0).

$2x + 3y < 6$
$2(0) + 3(0) < 6$
$0 < 6$ True

Since this statement is true, the half-plane containing (0, 0) comprises the solutions.

Step 3: Shade the half-plane containing the point (0, 0).

13b. $5x - 4y \geq 10$

Step 1: Graph the boundary line $5x - 4y = 10$ (see Figure 8-22).

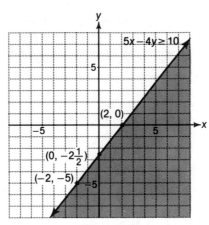

Figure 8-22

x-intercept:	y-intercept:
$5x - 4y = 10$	$5x - 4y = 10$
$5x - 4(0) = 10$	$5(0) - 4y = 10$
$x = 2; (2, 0)$	$y = -\frac{5}{2}; \left(0, -2\frac{1}{2}\right)$

Check: $5x - 4y = 10$
$5(-2) - 4y = 10$
$y = -5; (-2, -5)$

Step 2: Select the correct half-plane by using the test point (0, 0).

$$5x - 4y \geq 10$$
$$5(0) - 4(0) \geq 10$$
$$0 \geq 10 \quad \text{False}$$

Since the statement is false, the half-plane not containing (0, 0) comprises the solutions.

Step 3: Shade the half-plane not containing (0, 0).

Example 14 Graph $x \geq 2y$.

Solution **Step 1:** Graph the boundary line $x = 2y$ (see Figure 8-23).

x-intercept:	y-intercept:	Third point:
$x = 2y$	$x = 2y$	$x = 2y$
$x = 2(0)$	$0 = 2y$	$x = 2(-1)$
$x = 0; (0, 0)$	$0 = y; (0, 0)$	$x = -2; (-2, -1)$

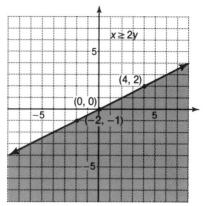

Figure 8-23

$$\begin{aligned} \text{Check:}\\ x &= 2y \\ x &= 2(2) \\ x &= 4; (4, 2) \end{aligned}$$

Step 2: You cannot use $(0, 0)$ as the test point since this point is on the line. Thus, use $(3, 0)$.

$$\begin{aligned} x &\geq 2y \\ 3 &\geq 2(0) \\ 3 &\geq 0 \quad \text{True} \end{aligned}$$

Since the statement is true, the half-plane containing $(3, 0)$ comprises the solutions.

Step 3: Shade the half-plane containing $(3, 0)$.

Practice Problem 8 *Graph each inequality.*

a. $3x + y \leq 6$ **b.** $4x - 3y > 12$ **c.** $2x - 3y < 0$

IDEA 3

Some inequalities contain only one variable. The boundary line of such inequalities are either a vertical or a horizontal line.

Example 15 Graph $x - 3 > -5$

Solution **Step 1:** Graph the boundary line $x - 3 = -5$
$$\begin{aligned} x - 3 &= -5 \\ x &= -2 \quad \text{(see Figure 8-24)} \end{aligned}$$

Step 2: Select the correct half-plane by using the test point $(0, 0)$.

$$\begin{aligned} x - 3 &> -5 \\ 0 - 3 &> -5 \\ -3 &> -5 \quad \text{True (see Figure 8-25)} \end{aligned}$$

Since the statement is true, the half-plane containing $(0, 0)$ comprises the solutions.

Step 3: Shade the region containing the point $(0, 0)$.

308   Graphing

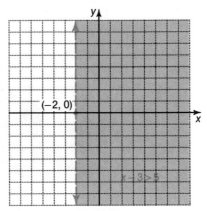

Figure 8-24

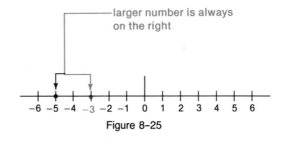

Figure 8-25

Practice Problem 9   *Graph each inequality.*

   a. $x \leq 4$ **b.** $3y - 2 > 10$

8.4 Exercises

Determine if the given ordered pair is a solution of the inequality.

1. $5x + y \leq 10$, $(2, 0)$ **2.** $x < 5y$, $(-3, -4)$ **3.** $3x + 2y > 9$, $\left(-2, \frac{3}{4}\right)$ **4.** $7x - 3y \leq 8$, $(1, 4)$

In problems 5–10, complete the graph of the inequality by shading the correct half-plane.

5. $x - y < 3$

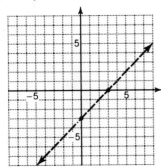

6. $2x + 5y \geq 10$

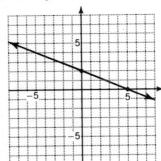

7. $3x - 4y \leq 12$

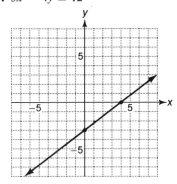

8. $x > 3y$

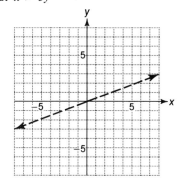

9. $x \geq -3$

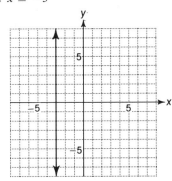

10. $y - 3 < -5$

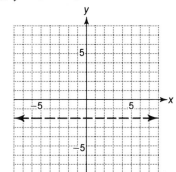

Graph each inequality.

11. $x - y < 4$ **12.** $x - 4y \geq 2$ **13.** $2x + 3y > 9$ **14.** $2x - 3y \leq -9$

15. $5x - 3y < 15$ **16.** $2x - y \geq 6$ **17.** $x > -4$ **18.** $y \leq -4$

19. $x \leq 2y + 4$ **20.** $x - y \leq 0$ **21.** $y \geq 4x + 3$ **22.** $2x - y < 0$

23. $5x + 2 \leq 12$ **24.** $4x + 5y > -10$ **25.** $4x + 5y > 20$ **26.** $3x + 5y \leq 15$

27. $\dfrac{x}{5} + \dfrac{y}{3} > 1$ **28.** $5x - 2y \geq 10$ **29.** $\dfrac{1}{2}x - \dfrac{1}{3}y > 1$ **30.** $2(x + 5) \leq 2$

Fill in the blanks.

31. When graphing an inequality, the boundary line is the graph of the _____ obtained when the inequality symbol is replaced with an _____ sign.

32. When the original inequality contains a $\leq$ or a $\geq$ symbol, the boundary line is a _____ line. In this case, the boundary line is _____ of the solution.

33. When the original inequality contains a $<$ or $>$ symbol, the boundary line is a _____ line. In this case, the boundary line is _____ of the solution.

34. To determine which half-plane is the graph of an inequality, choose any point not on the boundary line and substitute its _____ into the inequality. If the resulting inequality is true, the solutions are the half-plane _____ the point selected. If the resulting inequality is false, the solutions are the half-plane _____ the point selected.

35. To graph a linear inequality, first graph the _____. Next, select a point not on the _____ to determine which half-plane _____. Finally, _____ the half-plane identified in step 2.

Answers to Practice Problems

8a.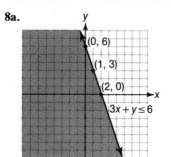

8b.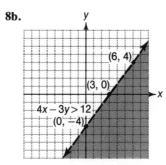

8c.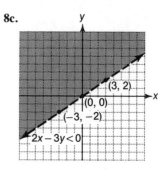

9a.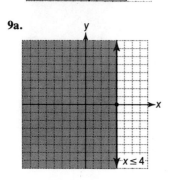

9b.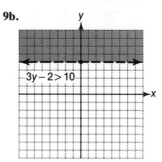

Chapter 8 Summary

Important Terms

The **rectangular coordinate system** is a two-dimensional plane formed by two intersecting perpendicular lines. The horizontal line is called the **x-axis,** and the vertical line is called the **y-axis.** The point of intersection is called the **origin.** The four regions formed are called **quadrants.** [Section 8.1/Idea 1]

in the rectangular coordinate system. The numbers in the ordered pair are called the **coordinates** of the point. (2, 3) and (−3, 2) are points in the rectangular coordinate system. The **abscissa** or **x-coordinate** is the first number in an ordered pair. The **ordinate** or **y-coordinate** is the second number in an ordered pair.

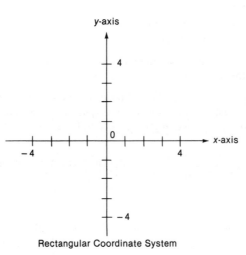

Rectangular Coordinate System

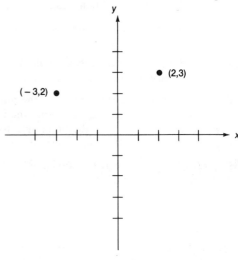

Given the ordered pair (1, 5), the *x*-coordinate is 1 and the *y*-coordinate is 5. [Section 8.1/Idea 1]

A **linear** or **first-degree equation in two variables** is an equation of the form $ax + by = c$, where a, b, and c are constants. $3x + y = 6$ and $2x - 3y = 9$ are linear equations in two variables. [Section 8.2/Idea 1]

A **solution of a linear equation in two variables** is an ordered pair of numbers that makes the equation a true statement. (3, 2) is a solution of the equation $x + y = 5$. [Section 8.2/Idea 1]

The **graph of a linear equation in two variables** is the line of points whose coordinates satisfy the equation. [Section 8.2/Idea 1]

The ***x*-intercept** of a graph is the *x*-coordinate of the point where the graph crosses the *x*-axis. The ***y*-intercept** of a graph is the *y*-coordinate of the point where the graph crosses the *y*-axis. [Section 8.2/Idea 2]

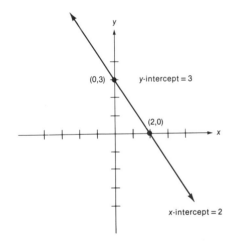

The **slope** of a line is the ratio of the rise to the run. [Section 8.3/Idea 1]

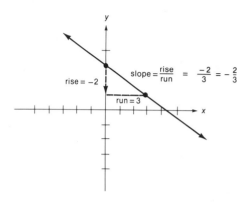

A linear equation in two variables written in the form $y = mx + b$ is said to be in **slope-intercept form,** where m is the slope and b is the *y*-intercept. $y = 3x - 7$ is in slope-intercept form, where 3 is the slope and -7 is the *y*-intercept. [Section 8.3/Idea 1]

A **linear inequality in two variables** is an inequality that can be written in the form $ax + by < c$, where a, b, and c are constants. $3x + y < 6$ and $x - 7 > 10$ are linear inequalities. [Section 8.4/Idea 1]

The **graph of a linear inequality in two variables** is the half-plane of points whose coordinates satisfy the inequality. The **boundary line** of the graph of $ax + by < c$ is the graph of $ax + by = c$. [Section 8.4/Idea 1]

Important Skills

Graphing Points

To graph a point (x, y), locate the *x*-coordinate on the *x*-axis and draw a vertical line. Locate the *y*-coordinate on the *y*-axis and draw a horizontal line. The intersection of those two lines is the graph of the ordered pair (x, y). [Section 8.1/Idea 2]

Finding Coordinates of Points

To find the coordinates of a point, draw a vertical line through the point. This line intersects the *x*-axis at the *x*-coordinate. Draw a horizontal line through the point. This line intersects the *y*-axis at the *y*-coordinate. Write the coordinates as the ordered pair (x, y). [Section 8.1/Idea 3]

Graphing Linear Equations

To graph a linear equation in two variables by using the intercept method, (a) find the *x*-intercept by letting $y = 0$ and solving for x; (b) find the *y*-intercept by letting $x = 0$

and solving for y; (c) find a third ordered pair that satisfies the equation by letting x or y be any convenient number and then solving for the other variable; (d) graph the three ordered pairs that satisfy the equation and connect them with a straight line. [Section 8.2/Idea 2]

To graph a linear equation in two variables by using the slope-intercept method, (a) write the original equation in the slope-intercept form $y = mx + b$ and identify the slope m and the y-intercept b; (b) graph the point $(0, b)$; (c) starting at $(0, b)$ and remembering that the slope $= \frac{\text{rise}}{\text{run}}$, find a second point on the graph; (d) draw a line that passes through $(0, b)$ and the point found in step C; (e) find at least one other point as a check. [Section 8.3/Idea 2]

Graphing Linear Inequalities To graph a linear inequality in two variables, graph the boundary line. The boundary line is a solid line when the inequality includes equality symbol ($\leq$ or $\geq$), but it is a dotted line when the inequality does not include an equality symbol ($<$ or $>$). Select a point that is not on the boundary line to determine which half-plane comprises the solutions. Shade the half-plane of the solutions. [Section 8.4/Idea 2]

Chapter 8 Review Exercises

1. Graph $(1, 3)$, $(0, 1)$, $(-2, -3)$, $(-2, 4)$.

2. Plot $(0, 4)$, $(0, 0)$, $(3, -4)$, $(-3, 0)$.

Find the coordinates of points A, B, C, D, and E.

3.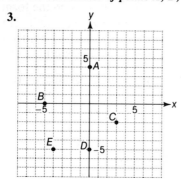

4.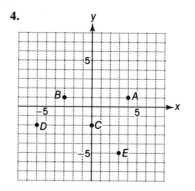

Graph.

5. $x + y = 2$
6. $3x - 2y = 6$
7. $4x - 5y < 10$
8. $3x - 4y \leq 6$
9. $x - 4y = 0$
10. $y - 3 = 1$
11. $-3x < 6$
12. $8x + 6y = -24$
13. $x < -y$
14. $6x + 4y = -12$

Use the slope-intercept method to graph each equation.

15. $3x - y = 4$
16. $6x + 4y = -8$
17. $\frac{x}{2} - \frac{y}{3} = 1$
18. $y - 3x = 0$

Solve.

19. A machine costing \$4000 will have a book value of $b = 4000 - 400n$ dollars n years after purchase.

    a. Find the book value of the machine after two years, three years, and five years. That is, find b when $n = 2$, $n = 3$, and $n = 5$.

    b. Graph $b = 4000 - 400n$. Let n represent the horizontal axis and let b represent the vertical axis.

    c. Interpret the meaning of the ordered pair $(1, 3600)$.

d. Using the graph in problem 19b, determine when the book value of the machine will be $800.

Graph.

20. $y = 25x + 250$
21. $2y = 4x + 1$
22. $y = 250$
23. $x < -200$
24. $x - y = 0$
25. $y = -2x - 1$
26. $x = 4y$
27. $8x + 3y < 24$
28. $y - 5 \leq 0$
29. $3x - y \geq 5$
30. $9x - 4y \leq 18$
31. $5x + 2 \leq 3x - 6$

Name: _____

Class: _____

Chapter 8 Test

1. Graph (3, 2), (−3, 0), (2, −3), and (0, 3)

1.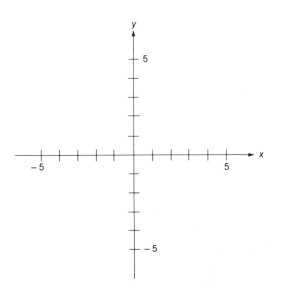

2. Find the coordinates of each point.

2. A _____ B _____

C _____ D _____

E _____ F _____

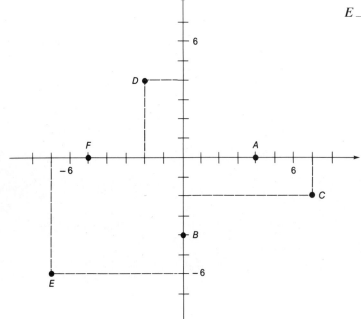

3. Find the *x*-intercept and the *y*-intercept of $2x - y = 4$.

3. *x*-intercept _____

   y-intercept _____

Graph each equation by using the intercept method.

4. $3x - 2y = 6$

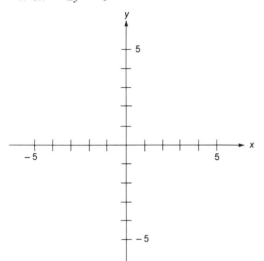

5. $x - 3 = 5$

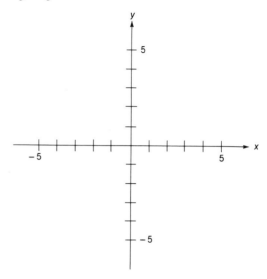

6. $x = 3y$

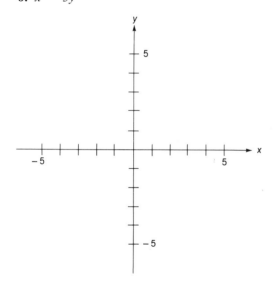

316 Graphing

Use the slope-intercept method to graph each equation.

7. $y = 2x - 3$

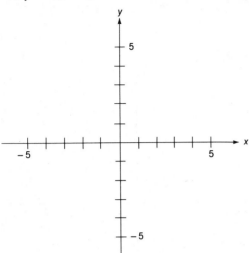

8. $x - 2y = 0$

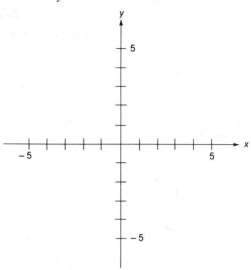

Graph each inequality.

9. $x - y > 3$

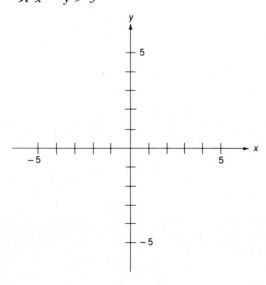

10. $3x + 2y \leq 8$

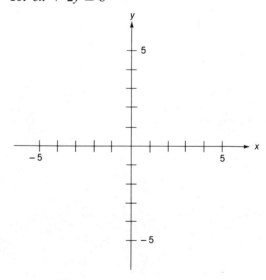

11. $3x - 2 \leq 7$

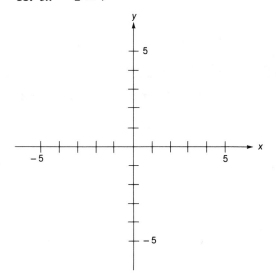

12. $x \leq 4y$

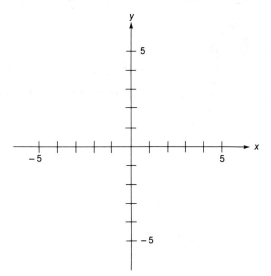

9 Linear Systems of Equations

Objectives The objectives for this chapter are listed below along with sample problems for each objective. By the end of this chapter you should be able to find solutions to the given problems.

1. Solve a system of two linear equations in two variables graphically *(Section 9.1/Idea 2)*.

   a. $x + y = 5$
      $x - y = 1$
   b. $x + y = 6$
      $2x + 2y = 12$
   c. $x + 2y = 4$
      $2x + y = -1$
   d. $2x + y = 6$
      $4x + 2y = 8$

2. Solve a system of two linear equations in two variables by using the addition method *(Section 9.2/Idea 1)*.

   a. $x + 2y = 1$
      $-x - 2y = 2$
   b. $2x + y = 5$
      $2x - y = -1$
   c. $4x - 3y = 12$
      $6x - 5y = 19$
   d. $\dfrac{x}{5} - \dfrac{y}{2} = 1$
      $x + y = 6$

3. Solve a system of two linear equations in two variables by using the substitution method *(Section 9.3/Idea 1)*.

   a. $2x + y = 4$
      $y = -1 - 2x$
   b. $3x + y = 1$
      $9x + 3y = 3$
   c. $3x + y = 6$
      $2x + 3y = 3$
   d. $2x + 3y = -8$
      $5x + 4y = -20$
   e. Cazzie earns twice as much money as Guerin. If their combined salaries total $615, how much does each man earn?

9.1 Intersecting, Parallel, and Equal Lines

IDEA 1 Two equations such as

$$x + y = 4$$
$$x - y = 2$$

are called a **system of two linear equations in two variables**. A **solution** of this kind of system is an ordered pair that is a solution of both equations in the system. For example, the ordered pair (3, 1) is the solution of the above system since

$$\begin{array}{lll} x + y = 4 & \text{and} & x - y = 2 \\ 3 + 1 = 4 & & 3 - 1 = 2 \\ 4 = 4 & & 2 = 2 \end{array}$$

We can also say that the ordered pair (3, 1) *satisfies* the system $x + y = 4$ and $x - y = 2$.

In Section 8.2 you learned that the graph of a linear equation in one variable is a straight line. What if you have a system of *two* linear equations in *two* variables? If you graph the equations in the same plane, one of three situations may occur:

1. The graphs of the equations may intersect at only one point (see Figure 9–1). In this case, the lines have one and only one point in common, and the ordered pair representing this point is the solution of the system.

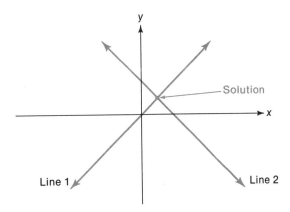

Figure 9–1

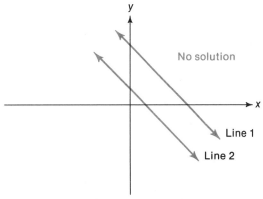

Figure 9–2

2. The graphs of the equations may be two parallel lines (see Figure 9-2). In this case, the lines have no point in common since they do not intersect. Thus, the system has no solution since no ordered pair satisfies both equations.

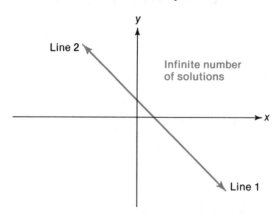

Figure 9-3

3. The graphs of the equations may be the same line (see Figure 9-3). In this case, every point on one line is also on the other line. The system has an infinite number of solutions since every ordered pair that satisfies one equation will also satisfy the other equation.

IDEA 2

Based on the information provided in Idea 1 to find the solution of a system of two linear equations in two variables graphically, we will graph both equations on the same set of axes. For example, to find the solution of the system

$$x + y = 4$$
$$x - y = 2$$

we first graph both equations by using the intercept method (see Figure 9-4).

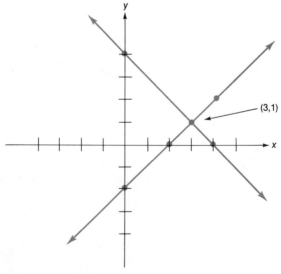

Figure 9-4

Line 1: $x + y = 4$

The x-intercept point = $(4, 0)$; let $y = 0$ and solve for x.
The y-intercept point = $(0, 4)$; let $x = 0$ and solve for y.
The checkpoint = $(2, 2)$; let $x = 2$ and solve for y.

Line 2: $x - y = 2$

The x-intercept point $= (2, 0)$; let $y = 0$ and solve for x.
The y-intercept point $= (0, -2)$; let $x = 0$ and solve for y.
The checkpoint $= (4, 2)$; let $y = 2$ and solve for x.

Since the two lines intersect at (3, 1), the solution for the system is (3, 1). To check this solution, substitute $x = 3$ and $y = 1$ into both equations.

$$\begin{array}{ll} \textbf{Check:} \quad x + y = 4 & \quad x - y = 2 \\ \qquad\qquad\; 3 + 1 = 4 & \quad 3 - 1 = 2 \\ \qquad\qquad\qquad\; 4 = 4 \;\;\text{True} & \quad\;\; 2 = 2 \;\;\text{True} \end{array}$$

The procedure for solving a system of two linear equations in two variables is summarized below.

To Solve a System of Linear Equations Graphically:

1. Graph each equation on the same coordinate system.

2a. If the lines *intersect*, the ordered pair representing the point of intersection is the solution to the system.

2b. If the lines are *parallel*, the system has no solution.

2c. If the graph of each equation is the *same line*, every ordered pair corresponding to a point on that line is a solution of the system.

Example 1 Solve each system graphically.

a. $x + y = 3$ 　 b. $2x + 3y = 6$ 　 c. $x + y = 3$
　 $x - y = 1$ 　　　 $4x + 6y = 12$ 　　 $2x + 2y = -4$

Solution To solve a system of two linear equations in two variables graphically, first graph each equation on the same coordinate system. Then, interpret the results.

1a.

	x-intercept point:	y-intercept point:	Checkpoint:
Line 1: $x + y = 3$	(3, 0)	(0, 3)	(1, 2)
Line 2: $x - y = 1$	(1, 0)	(0, -1)	(3, 2)

Since the lines intersect at (2, 1), this ordered pair is the system's solution (see Figure 9-5).

Check: 　　　　　　　　$x + y = 3$ 　　$x - y = 1$
　　　　　　　　　　　　$2 + 1 = 3$ 　　$2 - 1 = 1$
　　　　　　　　　　　　　　$3 = 3$ 　　　　$1 = 1$

Since the graphs of the equations in Solution 1a intersect at only one point, we say that the equations are **consistent**.

1b.

	x-intercept point:	y-intercept point:	Checkpoint:
Line 1: $2x + 3y = 6$	(3, 0)	(0, 2)	(-3, 4)
Line 2: $4x + 6y = 12$	(3, 0)	(0, 2)	(-3, 4)

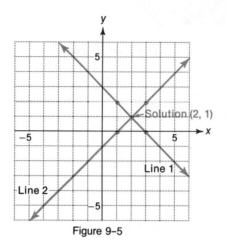

Figure 9-5

Since every point on one line also lies on the other line (see Figure 9-6) every ordered pair corresponding to a point on the line is a solution. Thus, the system has an infinite number of solutions.

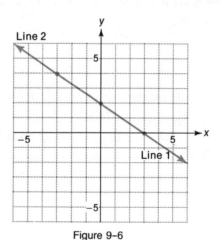

Figure 9-6

Since the graphs of the equations in Solution 1b are the same line, we say that the equations are **dependent** or **equivalent**.

1c.

	x-intercept point:	y-intercept point:	Checkpoint:
Line 1: $x + y = 3$	(3, 0)	(0, 3)	(2, 1)
Line 2: $2x + 2y = -4$	(−2, 0)	(0, −2)	(1, −3)

Since the lines are parallel and they will never intersect (see Figure 9-7), the system has no solution.

Note that in Example 1c, if you rewrite each equation in the slope-intercept form, the equations will have the same slope and different y-intercepts. That is,

$$x + y = 3 \quad \text{becomes} \quad y = -x + 3$$
$$2x + 2y = 4 \quad \text{becomes} \quad y = -x + 2$$

In general, this means that the graphs of the equations are distinct parallel lines and that the equations have no common solution. When the graphs of the equations are distinct parallel lines, we say that the equations are **inconsistent**.

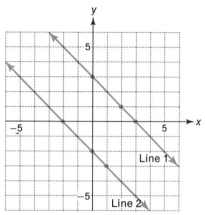

Figure 9-7

Practice Problem 1 **Solve each system graphically.**

a. $x + y = 5$
   $x - y = 1$

b. $x + 2y = 4$
   $2x + 4y = -4$

c. $2x + 2y = 4$
   $x + y = 2$

IDEA 3

When solving a system of two linear equations in two variables, you can save time if you know whether the equations are consistent, inconsistent, or dependent. That is, given the system

$$a_1 x + b_1 y = c_1$$
$$a_2 x + b_2 y = c_2$$

where a_1, a_2, b_1, c_1, c_2 are constants, a_1 and b_1 are not both zero, and a_2 and b_2 are not both zero, the following will be true.

To Determine If a System Is Consistent, Inconsistent, or Dependent:

Case 1: If $\dfrac{a_1}{a_2} \neq \dfrac{b_1}{b_2}$ ("$\neq$" means *not equal to*), then the original equations are *consistent* and their graphs are a pair of lines that intersect at one point.

Case 2: If $\dfrac{a_1}{a_2} = \dfrac{b_1}{b_2} = \dfrac{c_1}{c_2}$, then the original equations are *dependent* and their graphs are the same line.

Case 3: If $\dfrac{a_1}{a_2} = \dfrac{b_1}{b_2} \neq \dfrac{c_1}{c_2}$, then the original equations are *inconsistent* and their graphs are two distinct parallel lines.

where $a_1 x + b_1 y = c_1$
      $a_2 x + b_2 y = c_2$

is a system of equations, a_1 and b_1 are not both equal to zero, and a_2 and b_2 are not both equal to zero.

Example 2 Determine if the equations are consistent, inconsistent, or dependent.

a. $2x + y = 4$
   $4x + 2y = 8$

b. $6x = 12y - 8$
   $3x - 6y = 4$

c. $2x - 3y = 12$
   $x + y = 1$

Solution To determine if two equations are consistent, inconsistent, or dependent, first write the equations in the form $ax + by = c$. If $\frac{a_1}{a_2} \neq \frac{b_1}{b_2}$, the equations are *consistent*. If $\frac{a_1}{a_2} = \frac{b_1}{b_2} \neq \frac{c_1}{c_2}$, the equations are *inconsistent*. If $\frac{a_1}{a_2} = \frac{b_1}{b_2} = \frac{c_1}{c_2}$, the equations are *dependent*.

2a. Since $\frac{2}{4} = \frac{1}{2} = \frac{4}{8} \left(\frac{a_1}{a_2} = \frac{b_1}{b_2} = \frac{c_1}{c_2} \right)$, the equations are dependent and their graphs are the same line. Thus, there are an infinite number of solutions.

2b. $6x = 12y - 8$ becomes $6x - 12y = -8$
$3x - 6y = 4$ $3x - 6y = 4$

Since $\frac{6}{3} = \frac{-12}{-6} \neq \frac{-8}{4} \left(\frac{a_1}{a_2} = \frac{b_1}{b_2} \neq \frac{c_1}{c_2} \right)$, the equations are inconsistent and their graphs are two distinct parallel lines. Thus, the system has no solution.

2c. Since $\frac{2}{1} \neq \frac{-3}{1} \left(\frac{a_1}{a_2} \neq \frac{b_1}{b_2} \right)$, the equations are consistent and their graphs are a pair of lines that intersect at one point. Thus, the system has a unique solution (x, y).

Practice Problem 2 Determine if the following equations are consistent, inconsistent, or dependent.

a. $3x - 4 = 2y$
   $6x - 4y = -8$

b. $2x + 5y = 10$
   $3x + 4y = 1$

c. $\frac{x}{3} + \frac{y}{4} = 1$
   $8x + 6y = 24$

9.1 Exercises

Determine whether or not the given ordered pair is the solution of the given system.

1. $(3, 2)$; $x + y = 5$
   $x - y = 1$

2. $(4, 2)$; $x + 2y = 8$
   $2x - 3y = 2$

3. $(2, -3)$; $2x + y = 4$
   $4x + 2y = 8$

4. $(3, -1)$; $2x + 3y = -1$
   $3x + 5y = -2$

5. $(12, 10)$; $x + y = 22$
   $0.05x + 0.10y = 1.7$

6. $(1, -4)$; $3x + \frac{1}{4}y = 2$
   $\frac{x}{2} + \frac{3}{4}y = -\frac{5}{2}$

7. $(-4, -8)$; $3x = y - 4$
   $2x - 5y = 15$

8. $\left(\frac{5}{2}, -\frac{7}{4}\right)$; $x - 2y = 6$
   $3x + 2y = 4$

Solve each system graphically. If the lines are parallel, write "no solution." Write "infinite number of solutions" if the graph is a single straight line.

9. $3x + y = 3$
   $6x + 2y = 6$

10. $x + y = 6$
    $x - y = 2$

11. $2x + 2y = 24$
    $2x - 2y = 8$

12. $4x = y + 10$
    $2x + 3y = 12$

13. $x - 3y = 6$
    $2x - 6y = -18$

14. $3x + 4y = 12$
    $-6x - 8y = -24$

15. $2x + 3y = -6$
    $2x - 4y = 8$

16. $4x - y = 8$
    $x = y + 5$

17. $x + y = 8$
    $x - 2y = 2$

18. $x = 3y - 6$
    $2x - 6y = 12$

19. $x = 4$
    $3x - y = 6$

20. $2x - 3y = 6$
    $y = -4$

21. $x - 3y = 6$
    $x + 3y = -12$

22. $x + 2y = 8$
    $6x + 2y = 18$

23. $\dfrac{x}{2} + \dfrac{y}{3} = 1$
    $x - y = 2$

Determine if the following equations are consistent, inconsistent, or dependent (see Example 2). Also, tell whether the system has one solution, no solution, or an infinite number of solutions.

24. $x + y = 6$
    $x - y = 5$

25. $8x + y = 30$
    $x - 3y = 30$

26. $x + 3y = 6$
    $2x + 6y = 12$

27. $3x - 4y = 8$
    $6x - 8y = 8$

28. $2x + 2y = 9$
    $2x - 2y = 9$

29. $-5x + y = 7$
    $10x - 2y = -14$

30. $3x = y + 4$
    $6x - 2y = 12$

31. $2x = 23 - 5y$
    $2x + 3y = 29$

32. $x = 2y$
    $3x + 2y = 29$

33. $4x - y = 3$
    $-2x + \dfrac{y}{2} = \dfrac{-3}{2}$

34. $5x + 2y = -2$
    $y = -2x$

35. $3x + \dfrac{1}{4}y = 2$
    $\dfrac{1}{2}x + \dfrac{3}{4}y = \dfrac{5}{2}$

Fill in the blanks.

36. A solution of two linear equations in two variables is an _____ that is a _____ of _____ equations in the system.

37. If the graphs of two linear equations in two variables intersect at one and only one point, the equations are said to be _____ and the system has _____ solution.

38. If the graphs of two linear equations in two variables produce two distinct parallel lines, the equations are said to be _____ and the system has _____ solution(s).

39. If two linear equations in two variables have the same line for their graph, the equations are said to be _____ and the system has _____ solution(s).

40. Given the system $a_1x + b_1y = c_1$ and $a_2x + b_2y = c_2$, we know that if:

	Condition	Type of Equation	Number of Solutions
a.	$\dfrac{a_1}{a_2} \neq \dfrac{b_1}{b_2}$	_____	_____
b.	_____	dependent	_____
c.	_____	_____	no solution

41. To solve a system of two equations in two variables graphically, first graph each _____ on the same coordinate system. Then, _____ the results.

Determine if the following equations are consistent, inconsistent, or dependent. If they are consistent, find the solution graphically.

42. $x + y = 6$
    $x - y = 2$

43. $x + y = 7$
    $x - y = 3$

44. $2x + 3y = 8$
    $4x + 6y = 5$

45. $3x + 4y = 7$
    $6x + 8y = 3$

46. $x + y = 1$
    $3x + 3y = 3$

47. $x + y = -2$
    $2x + 2y = -4$

48. $x = 2y - 1$
    $3x - 6y = 3$

49. $x = 3y + 2$
    $2x - 6y = -4$

50. $x = 3 - 3y$
    $2x - 3y = -12$

51. $x = 2 - 2y$
    $2x - 2y = 10$

52. $x = 2y - 5$
    $2x - 2y = -10$

53. $x = 3y - 2$
    $2x - 6y = -4$

54. $\dfrac{x}{2} + \dfrac{y}{3} = 1$
    $y = 6$

55. $\dfrac{x}{2} - \dfrac{y}{3} = 1$
    $y = -3$

56. $\dfrac{1}{2}x + \dfrac{1}{2}y = 3$
    $\dfrac{1}{3}x + \dfrac{1}{3}y = -3$

57. $\dfrac{5}{3}x + \dfrac{10}{3}y = \dfrac{10}{3}$
    $\dfrac{1}{4}x + \dfrac{1}{2}y = \dfrac{1}{2}$

58. $\dfrac{2}{3}x + \dfrac{1}{3}y = 2$
    $\dfrac{1}{3}x + \dfrac{1}{2}y = 1$

59. $\dfrac{3}{4}x + \dfrac{1}{2}y = 3$
    $\dfrac{1}{2}x - \dfrac{2}{3}y = 2$

Answers to Practice Problems

1a.
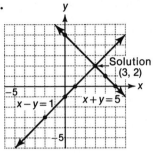
Solution is (3, 2)

1b.
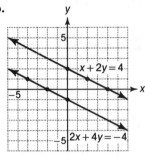
No solution

1c.
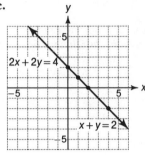
Infinite number of solutions

2a. inconsistent   b. consistent   c. dependent

9.2 Solving Linear Systems of Equations by Addition

IDEA 1 Solving systems of linear equations graphically can be time-consuming and inaccurate especially when the coordinates of the point of intersection are not integers. Therefore, to obtain an exact solution, we should use an algebraic method.

One such method is based on the addition property of equality. Recall from Chapter 6 that this property states that if equals are added to equals, the results are equal. Using this property, we can eliminate one variable so that only one equation containing one variable will remain.

Example 3 Solve each system.

a. $x + y = 3$
   $x - y = 1$

b. $3x - 4y = -5$
   $x + 2y = 5$

Solution 3a. Since each equation is a statement of equality, add the two equations to obtain one equation in one variable.

$$\begin{array}{r} x + y = 3 \\ x - y = 1 \\ \hline 2x = 4 \end{array}$$

Now, solve this equation.

$$2x = 4$$
$$x = 2$$

Next, find y by substituting 2 for x in one of the original equations.

$$x + y = 3$$
$$2 + y = 3$$
$$y = 1$$

Thus, the solution of the system is (2, 1).

Check: $x + y = 3$ $x - y = 1$
          $2 + 1 = 3$ $2 - 1 = 1$
               $3 = 3$ True $1 = 1$ True

3b. Adding these equations will not result in one equation containing one variable. Thus, before adding, multiply both sides of the second equation by 2 in order to make the coefficients of the y-terms additive inverses.

$$\begin{array}{ll} 3x - 4y = -5 & \text{remains} \quad 3x - 4y = -5 \\ 2(x + 2y) = 2(5) & \text{becomes} \quad 2x + 4y = 10 \end{array}$$

Now, add the resulting equation to the first equation to obtain one equation in one variable.

$$\begin{array}{r} 3x - 4y = -5 \\ \underline{2x + 4y = 10} \\ 5x = 5 \end{array}$$

Solve this equation.

$$5x = 5$$
$$x = 1$$

Next, find y by substituting 1 for x in one of the original equations.

$$\begin{aligned} x + 2y &= 5 \\ 1 + 2y &= 5 \\ 2y &= 4 \\ y &= 2 \end{aligned}$$

Thus, the solution of the system is the ordered pair $(1, 2)$.

Check:
$$\begin{array}{ll} 3x - 4y = -5 & \quad x + 2y = 5 \\ 3(1) - 4(2) = -5 & \quad 1 + 2(2) = 5 \\ 3 - 8 = -5 & \quad 1 + 4 = 5 \\ -5 = -5 & \quad 5 = 5 \end{array}$$

The method of solving a system of linear equations in two variables by using the addition property of equality is called the **addition method** or the **elimination method** (see the box on page 328).

Example 4 Solve each system by the addition method.

a. $5x - y = 10$
$2x + y = 4$

b. $3x + y = 4$
$-3x - y = -4$

c. $6x + 9y = 18$
$-6x - 12y = -16$

Solution To solve a system of two linear equations in two variables by the addition method, first determine if the equations are consistent, inconsistent, or dependent.

4a. Since $\dfrac{5}{2} \neq \dfrac{-1}{1} \left(\dfrac{a_1}{a_2} \neq \dfrac{b_1}{b_2} \right)$, the equations are consistent (the graph is two intersecting lines) and there is only one solution. Now, since the coefficients of y are additive inverses, add and solve the resulting equation.

$$\begin{array}{r} 5x - y = 10 \\ \underline{2x + y = 4} \\ 7x = 14 \\ x = 2 \end{array}$$

328 Linear Systems of Equations

> **To Solve a System of Linear Equations by Using the Addition Method:**
>
> 1. Write both equations in the form $ax + by = c$.
> 2. If the equations in step 1 are inconsistent, write "no solution." If they are dependent, write "infinite number of solutions."
> 3. If the equations in step 1 are consistent:
>   a. Multiply (if necessary) one or both of the equations by a number that will make the coefficients of x (or y) additive inverses of each other.
>   b. Add the resulting two equations to obtain one equation in one variable.
>   c. Solve the equation obtained in step 3b.
>   d. Substitute the solution obtained in step 3c into one of the original equations and solve the resulting equation.
>   e. Write the values obtained in steps 3c and 3d as an ordered pair. This is the solution of the system.
>   f. Check the solution in both of the original equations.

Next, find y by substituting 2 for x in one of the original equations.

$$2x + y = 4$$
$$2(2) + y = 4$$
$$4 + y = 4$$
$$y = 0$$

The solution is (2, 0).

Check: $5x - y = 10$ $2x + y = 4$
            $5(2) - 0 = 10$ $2(2) + 0 = 4$
            $10 = 10$ $4 = 4$

4b. Since $\dfrac{3}{-3} = \dfrac{1}{-1} = \dfrac{4}{-4} \left(\dfrac{a_1}{a_2} = \dfrac{b_1}{b_2} = \dfrac{c_1}{c_2} \right)$, the equations are dependent (graph is a single straight line). Thus, the system has an infinite number of solutions.

Alternate method
Since the coefficients of x are additive inverses, add.

$$3x + y = 4$$
$$-3x - y = -4$$
$$\overline{\ 0 = 0}$$

By doing this, you have eliminated both variables and the resulting equation is a true statement. In general, this means that the original equations are dependent and their graph is a straight line. Therefore, the system has an infinite number of solutions.

4c. Since $\dfrac{6}{-6} \neq \dfrac{9}{-12} \left(\dfrac{a_1}{a_2} \neq \dfrac{b_1}{b_2} \right)$, the equations are consistent and there exists one and only one solution (x, y). Next, since the coefficients of x are additive inverses, add.

$$6x + 9y = 18$$
$$\underline{-6x - 12y = -16}$$
$$-3y = 2$$
$$y = -\frac{2}{3}$$

Now, find x by substituting $-\frac{2}{3}$ for y in one of the original equations.

$$6x + 9y = 18$$
$$6x + 9\left(-\frac{2}{3}\right) = 18$$
$$6x - 6 = 18$$
$$6x = 24$$
$$x = 4$$

The solution is $\left(4, -\frac{2}{3}\right)$.

Check:
$$6x + 9y = 18 \qquad -6x - 12y = -16$$
$$6(4) + 9\left(-\frac{2}{3}\right) = 18 \qquad -6(4) - 12\left(-\frac{2}{3}\right) = -16$$
$$24 - 6 = 18 \qquad -24 + 8 = -16$$
$$18 = 18 \qquad -16 = -16$$

Practice Problem 3 **Solve each problem by the addition method.**

a. $x + y = 8$
   $x - y = 4$

b. $-8x = -10y + 6$
   $8x - 10y = 6$

c. $2x + 5y = 12$
   $-2x - 6y = -14$

Sometimes you must multiply one of the equations by a number to make the coefficients of x (or y) additive inverses.

Example 5 Solve each system by the addition method.

a. $x + y = 2$
   $3x + 2y = -9$

b. $10x - 9y = 13$
   $4x + 3y = 14$

c. $x + 3y = 6$
   $x = -2y + 5$

Solution To solve a system of two linear equations in two variables, first determine if the equations are consistent, inconsistent, or dependent.

5a. Since $\frac{1}{3} \neq \frac{1}{2}\left(\frac{a_1}{a_2} \neq \frac{b_1}{b_2}\right)$, the equations are consistent and one and only one solution exists. Now, multiply both sides of the first equation by -2. This will make the coefficients of y in the equations additive inverses. Then, add.

$$-2(x + y) = -2(2) \quad \text{becomes} \quad -2x - 2y = -4$$
$$3x + 2y = -9 \quad \text{remains} \quad \underline{3x + 2y = -9}$$
$$x = -13$$

Next, find y by substituting -13 for x in one of the original equations.

$$x + y = 2$$
$$-13 + y = 2$$
$$y = 15$$

The solution is $(-13, 15)$.

Check:
$$x + y = 2 \qquad 3x + 2y = -9$$
$$-13 + 15 = 2 \qquad 3(-13) + 2(15) = -9$$
$$2 = 2 \qquad -9 = -9$$

Alternate method
By multiplying both sides of the first equation by -3, we could have obtained the same solution.

5b. Since $\dfrac{10}{4} \neq \dfrac{-9}{3} \left(\dfrac{a_1}{a_2} \neq \dfrac{b_1}{b_2}\right)$, the equations are consistent and there exists only one solution (x, y). Now, multiply both sides of the second equation by 3. This will make the coefficients of y in the equations additive inverses. Then, add.

$$\begin{array}{rl} 10x - 9y = 13 & \text{remains} \\ 3(4x + 3y) = 3(14) & \text{becomes} \end{array} \quad \begin{array}{r} 10x - 9y = 13 \\ 12x + 9y = 42 \\ \hline 22x = 55 \end{array}$$

$$x = \dfrac{55}{22}$$
$$x = \dfrac{5}{2}$$

Next, find y by substituting $\dfrac{5}{2}$ for x in one of the original equations.

$$10x - 9y = 13$$
$$10\left(\dfrac{5}{2}\right) - 9y = 13$$
$$25 - 9y = 13$$
$$-9y = -12$$
$$y = \dfrac{12}{9}$$
$$y = \dfrac{4}{3}$$

The solution is $\left(\dfrac{5}{2}, \dfrac{4}{3}\right)$.

Check:
$$10x - 9y = 13 \qquad 4x + 3y = 14$$
$$10\left(\dfrac{5}{2}\right) - 9\left(\dfrac{4}{3}\right) = 13 \qquad 4\left(\dfrac{5}{2}\right) + 3\left(\dfrac{4}{3}\right) = 14$$
$$25 - 12 = 13 \qquad 10 + 4 = 14$$
$$13 = 13 \qquad 14 = 14$$

5c. Write both equations in the form $ax + by = c$. Thus, in the second equation, add $2y$ to both sides.

$$\begin{array}{rl} x + 3y = 6 & \text{remains} \\ x = -2y + 5 & \text{becomes} \end{array} \quad \begin{array}{r} x + 3y = 6 \\ x + 2y = 5 \end{array}$$

Next, since $\dfrac{1}{1} \neq \dfrac{3}{2} \left(\dfrac{a_1}{a_2} \neq \dfrac{b_1}{b_2}\right)$, the equations are consistent and there exists one and only one solution (x, y). Now, multiply both sides of the first equation by -1. This will make the coefficients of x in the equations additive inverses. Then, add.

$$\begin{array}{ll} -1(x + 3y) = -1(6) & \text{becomes} \\ x + 2y = 5 & \text{remains} \end{array} \qquad \begin{array}{r} -x - 3y = -6 \\ \underline{x + 2y = 5} \\ -y = -1 \\ y = 1 \end{array}$$

Next, find y by substituting 1 for y in one of the original equations.

$$\begin{array}{r} x + 3y = 6 \\ x + 3(1) = 6 \\ x + 3 = 6 \\ x = 3 \end{array}$$

The solution is $(3, 1)$.

Check:
$$\begin{array}{ll} x + 3y = 6 & x = -2y + 5 \\ 3 + 3(1) = 6 & 3 = -2(1) + 5 \\ 6 = 6 & 3 = 3 \end{array}$$

Practice Problem 4 *Solve each system by the addition method.*

a. $3x - 4y = -5$ $\quad$ **b.** $2x + 5y = 1$ $\quad$ **c.** $x = 16 - 3y$
$\phantom{\textbf{a.} } x + 2y = 5$ $\qquad\phantom{\textbf{b.} } 4x + 10y = 2$ $\qquad\phantom{\textbf{c.} } 4x - 2y = 8$

Frequently it is necessary to multiply both equations by the appropriate numbers to make the coefficients of x (or y) additive inverses.

Example 6 Solve each system by the addition method.

a. $3x - 5y = 2$ $\quad$ **b.** $4x + 2y = 1$ $\quad$ **c.** $\dfrac{1}{2}x + \dfrac{5}{6}y = 1$
$\phantom{\textbf{a.} } 2x + 3y = -5$ $\qquad\phantom{\textbf{b.} } 10x + 5y = -3$ $\qquad\phantom{\textbf{c.} } 5x + 3y = 4$

Solution **6a.** The equations are consistent and there is only one solution. Thus, multiply both sides of the first equation by 3 and multiply both sides of the second equation by 5 to make the coefficients of y additive inverses.

$$\begin{array}{ll} 3(3x - 5y) = 3(2) & \text{becomes} \\ 5(2x + 3y) = 5(-5) & \text{becomes} \end{array} \qquad \begin{array}{r} 9x - 15y = 6 \\ \underline{10x + 15y = -25} \\ 19x = -19 \\ x = -1 \end{array}$$

Now, find y by substituting -1 for x in one of the original equations.

$$\begin{array}{r} 2x + 3y = -5 \\ 2(-1) + 3y = -5 \\ -2 + 3y = -5 \\ 3y = -3 \\ y = -1 \end{array}$$

The solution is $(-1, -1)$.

Check:
$$3x - 5y = 2 \qquad 2x + 3y = -5$$
$$3(-1) - 5(-1) = 2 \qquad 2(-1) + 3(-1) = -5$$
$$2 = 2 \qquad -5 = -5$$

Alternate method
By multiplying both sides of the first equation by 2 and multiplying both sides of the second equation by -3, we could have obtained the same solution.

6b. Since $\dfrac{4}{10} = \dfrac{2}{5} \neq \dfrac{1}{3} \left(\dfrac{a_1}{a_2} = \dfrac{b_1}{b_2} \neq \dfrac{c_1}{c_2} \right)$, the equations are inconsistent and their graph is two distinct parallel lines. Thus, the system has no solution.

Alternate method
Multiply both sides of the first equation by -5 and multiply both sides of the second equation by 2 to make the coefficients of y additive inverses. Then, add.

$$-5(4x + 2y) = -5(1) \quad \text{becomes} \quad -20x - 10y = -5$$
$$2(10x + 5y) = 2(-3) \quad \text{becomes} \quad \underline{20x + 10y = -6}$$
$$0 = -11 \quad \text{False}$$

We have eliminated both variables and the resulting equation is a false statement. In general, this means that the equations are inconsistent and their graph is two distinct parallel lines. Thus, the system has no solution.

6c. First, multiply both sides of the first equation by 6 (LCD) to eliminate the fractions.

$$6\left(\tfrac{1}{2}x + \tfrac{5}{6}y\right) = 6(1) \quad \text{becomes} \quad 3x + 5y = 6$$
$$5x + 3y = 4 \quad \text{remains} \quad 5x + 3y = 4$$

Since the equations are consistent, the system has a unique solution. Multiply both sides of the first equation by -5 and multiply both sides of the second equation by 3 to eliminate x.

$$-15x - 25y = -30 \quad \text{Result after multiplying by } -5$$
$$\underline{15x + 9y = 12} \quad \text{Result after multiplying by } 3$$
$$-16y = -18$$
$$y = \dfrac{-18}{-16}$$
$$y = \dfrac{9}{8}$$

To find x, substitute $\dfrac{9}{8}$ for y in one of the original equations.

$$5x + 3y = 4$$
$$5x + 3\left(\dfrac{9}{8}\right) = 4$$
$$5x + \dfrac{27}{8} = 4$$
$$40x + 27 = 32 \quad \text{Multiply by LCD} = 8$$
$$40x = 5$$

$$x = \frac{5}{40}$$
$$x = \frac{1}{8}$$

Check:
$$5x + 3y = 4 \qquad \frac{1}{2}x + \frac{5}{6}y = 1$$
$$5\left(\frac{1}{8}\right) + 3\left(\frac{9}{8}\right) = 4 \qquad \frac{1}{2} \cdot \frac{1}{8} + \frac{5}{6} \cdot \frac{9}{8} = 1$$
$$\frac{5}{8} + \frac{27}{8} = 4 \qquad \frac{1}{16} + \frac{15}{16} = 1$$
$$4 = 4 \qquad 1 = 1$$

Practice Problem 5 *Solve each system by the addition method.*

a. $3x + 5y = -2$
   $2x + 3y = -5$

b. $3x = 6y + 6$
   $5x - 4y = 1$

c. $x + 2y = 4$
   $-2x - 4y = -8$

IDEA 2

When solving word problems involving two unknowns, it is sometimes easier to represent each unknown in terms of a different letter. However, since two variables are used, you must write two equations to find the desired solution.

Example 7 Solve.

a. The sum of two numbers is 21. Their difference is 9. Find the numbers.

b. Princess invested $10,000. She invested part at 7% and the rest at 12%. If she earned $1000 in interest in one year, how much did she invest at each rate?

Solution To solve a word problem involving two unknowns, (1) represent the unknowns, (2) write two equations, (3) solve the system obtained in step 2 and determine the solution, and (4) check your final answers.

7a. Step 1: $x =$ the first number; $y =$ the second number

Step 2: first number plus second number is 21
$\qquad\qquad\quad x \qquad\quad + \qquad\quad y \qquad\qquad\;\; = \;\; 21$
$\qquad$ first number minus second number is 9
$\qquad\qquad\quad x \qquad\quad - \qquad\quad y \qquad\qquad\;\; = \;\; 9$

Step 3: Now, find x.

$$\begin{aligned} x + y &= 21 \\ x - y &= 9 \\ \hline 2x &= 30 \\ x &= 15 \end{aligned}$$

Next, find y.

$$\begin{aligned} x + y &= 21 \\ 15 + y &= 21 \\ y &= 6 \end{aligned}$$

The numbers are 15 and 6.

Step 4: The answers check since $15 + 6 = 21$ and $15 - 6 = 9$.

7b. Step 1: $x = \$$ invested at 7%; $y = \$$ invested at 12%

Step 2:

	Principal (P) (in dollars)	Rate (R) (in percent)	Time (T)	Interest (I) = PRT
Investment 1	x	7% = 0.07	1	$0.07x$
Investment 2	y	12% = 0.12	1	$0.12y$

$\boxed{\$ \text{ invested at } 7\%}$ plus $\boxed{\$ \text{ invested at } 12\%}$ is $\boxed{\$10{,}000}$

$\quad\quad x \quad\quad\quad + \quad\quad y \quad\quad = \quad 10{,}000$

$\boxed{\text{Interest at } 7\%}$ plus $\boxed{\text{Interest at } 12\%}$ is $\boxed{\$1000}$

$\quad\quad 0.07x \quad\quad + \quad\quad 0.12y \quad\quad = \quad 1000$

Step 3: Now, find y. Multiply both sides of the first equation by -0.07.

$$x + y = 10{,}000 \quad \text{becomes} \quad -0.07x - 0.07y = -700$$
$$0.07x + 0.12y = 1{,}000 \quad \text{remains} \quad \underline{0.07x + 0.12y = 1000}$$
$$0.05y = 300$$
$$y = 6000$$

Next, find x.

$$x + y = 10{,}000$$
$$x + 6000 = 10{,}000$$
$$x = 4000$$

Princess invested $4000 at 7% and $6000 at 12%.

Step 4: The answer checks since:
a) Total invested = $4000 + $6000 = $10,000

b) Total interest = 7% of $4000 + 12% of $6000
   = $280 + $720
   = $1000

Practice Problem 6 *Solve.*
a. The sum of two numbers is 80. Their difference is 6. Find the numbers.

b. There are 1380 people at a basketball game. A student ticket costs $1.50 and nonstudent tickets cost $3.00. If the total gate receipts were $2910, how many students and how many nonstudent tickets were sold?

9.2 Exercises

Solve each system by the addition method.

1. $x - y = -5$
   $x + y = 7$

2. $x - y = 3$
   $x + y = -1$

3. $2x + y = 6$
   $-2x - y = 8$

4. $3x + y = 7$
   $6x + 2y = 14$

5. $6x - y = 2$
   $8x + 3y = 7$

6. $5x - y = 10$
   $x - 2y = -7$

7. $3x + 2y = 9$
   $4x - 3y = 12$

8. $8x + 3y = 9$
   $3x + 5y = 16$

9. $4x + 3y = -12$
   $6x - 4y = 1$

10. $x - 3y = 9$
    $x - 5y = 13$

11. $3x - 5y = 2$
    $3x + 5y = 22$

12. $2x + y = 8$
    $4x + 2y = 9$

13. $x - 7y = 1$
    $2x = 2 + 14y$

14. $2x - y = -6$
    $-2x + 3y = 20$

15. $5x + 7y = 3$
    $2x - 3y = 7$

16. $3x + 4y = 1$
    $4x + 3y = 8$

17. $2x + 2y = 4$
    $2x - 3y = -8$

18. $x + 4y = 4$
    $x - 2y = 10$

19. $\dfrac{x}{2} + \dfrac{y}{5} = 1$
    $7x - 5y = 36$

20. $3x = 5y + 2$
    $5y - 3x = 7$

21. $x + y = 8$
    $2x - y = 10$

22. $x + 5y = 8$
    $x - 10y = -29$

23. $3x + y = 7$
    $4x - 10y = -1$

24. $8x - y = 29$
    $2x + y = 11$

25. $\dfrac{x}{4} - \dfrac{y}{3} = \dfrac{5}{12}$
    $\dfrac{x}{10} + \dfrac{y}{5} = \dfrac{1}{2}$

26. $3x + 5y = 2$
    $2x + 3y = -5$

27. $9x - 6y = -18$
    $6x - 9y = -32$

28. $0.3x + 0.2y = 0.1$
    $0.4x - y = 2.6$

29. $x + y = 4$
    $\dfrac{x}{2} + \dfrac{y}{2} = 2$

30. $5x + 3y = 6$
    $3x + 5y = 4$

Fill in the blanks.

31. When solving a system of two linear equations in two variables by the addition method, first write both equations in the form _____. If the resulting equations are inconsistent, write _____. If they are dependent, write _____.

32. If while solving a system of two linear equations in two variables by the addition method you eliminated both variables and the resulting equation is a true statement, the original equations are _____ and the system has an _____ of solutions.

33. If while solving a system of two linear equations in two variables you eliminated both variables and the resulting equation is a false statement, the original equations are _____ and the system has _____ solution.

34. If the equations in a system of two linear equations in two variables are consistent, you can find the solution as follows:
    a. **Step 1:** Multiply (if necessary) one or both of the equations by a number that will make the coefficients of x (or y) _____ of each other.
    b. **Step 2:** Add the resulting equations to obtain _____ equation in _____ variable.
    c. **Step 3:** _____ the equation obtained in step 2.
    d. **Step 4:** Substitute the solution obtained in step 3 into _____ of the _____ equations and then _____ the resulting equation.
    e. **Step 5:** Write the values obtained in steps 3 and 4 as an _____. This is our solution.
    f. **Step 6:** _____ the solution in _____ equations.

Solve.

35. Determine how many gallons of a 7% solution and a 12% solution of acid should be mixed to obtain six gallons of a solution that is 10% acid.

36. Vanessa invested $10,000. Part was invested at 9% and the rest was invested at 10.5%. If her total interest for the year was $990, how much money was invested at each rate?

37. Carver invested $4000 in a Ford and a BMW, and then he leased both cars. His earned annual interest income from leasing each car was 20% and 60%, respectively. If his income from the total investment was 50%, how much did Carver pay for each car?

38. The sum of two numbers is 13. Their difference is -6. What are the numbers?

39. Find two numbers such that six times the first number plus five times the second number is 34, and five times the first number plus three times the second number is 4.

40. There were 1100 people at a concert. The admission was $2 for students and $3 for nonstudents. If the total gate receipts were $2800, how many student tickets and how many nonstudent tickets were sold?

41. The cost of printing 400 brochures was $210 and the cost of printing 700 brochures was $360. If the printer charged a fixed fee for typesetting and an additional fee for each brochure, find the typesetting fee and the fee per brochure.

42. Susan mixed a dried fruit costing $0.75 per pound with roasted soy beans costing $1.50 per pound. If she wants a mixture of 90 pounds worth $1.24 per pound, how many pounds of each item must she use?

43. How many gallons of a 15% acid solution should be mixed with a 40% acid solution to obtain 16 gallons of a mixture that is 30% acid?

44. Train A leaves Chicago for Boston (a distance of 1050 miles) at the same time that train B leaves Boston for Chicago. If train A travels at a rate of 60 mph and train B travels at 40 mph, how far has each train traveled when the two trains pass?

45. The perimeter of a rectangular-shaped garden is 142 feet. If the width of the garden is 6.4 feet greater than the length, find the garden's length and width. (*Hint:* $P = 2L + 2W$)

Solve.

46. $74x - 49y = 197$
$56x + 35y = 77$

47. $33x + 22y = 209$
$18x + 25y = -190$

48. $7.12x + 14.32y = 68$
$2.08x - 5.15y = 56$

49. $2.28x - 6.32y = 19.22$
$8.61x + 5.32y = 0.385$

Answers to Practice Problems **3a.** (6, 2) **b.** no solution **c.** (1, 2) **4a.** (1, 2) **b.** infinite number of solutions **c.** (4, 4) **5a.** (−19, 11) **b.** $\left(-1, -\frac{3}{2}\right)$ **c.** infinite number of solutions **6a.** 43, 37 **b.** 820 student tickets, 560 nonstudent tickets

---

9.3 Solving Linear Systems of Equations by Substitution

IDEA 1 A system of two linear equations in two variables can also be solved by using the **substitution method.** With this method, you still eliminate one of the variables. However, you eliminate it by substitution rather than addition. To illustrate this method, let us solve the system

$$2x + 3y = 40$$
$$y = 2x$$

The equations are consistent. Thus, there is only one solution. Next, since $y = 2x$ in the second equation, we can substitute $2x$ for y in the first equation.

$$2x + 3y = 40$$
$$2x + 3(2x) = 40 \qquad \text{Replace } y \text{ with } 2x$$

We now have one equation in one variable. Solve it.

$$2x + 3(2x) = 40$$
$$2x + 6x = 40$$
$$8x = 40$$
$$x = 5$$

To find y, substitute 5 for x in the second equation.

$$y = 2x$$
$$y = 2(5)$$
$$y = 10$$

The solution is (5, 10).

$$\begin{array}{ll} \text{Check:} & 2x + 3y = 40 \qquad y = 2x \\ & 2(5) + 3(10) = 40 \qquad 10 = 2(5) \\ & \qquad\quad 40 = 40 \qquad\quad 10 = 10 \end{array}$$

Based on the above results to solve a system of two linear equations in two variables by the substitution method, use the following procedure.

To Solve a System of Linear Equations by Using the Substitution Method:

1. Determine (mentally) if the equations are consistent, inconsistent, or dependent. If the equations are inconsistent, write "no solution." If they are dependent, write "infinite number of solutions."

2. If the equations are consistent:
   a. Solve (if necessary) one of the equations for one of the variables.
   b. Substitute the expression obtained in step 3a into the other equation.
   c. You now have one equation in one variable. Solve it.
   d. Find the value of the other variable by substituting the value obtained in step 3c into the equation obtained in step 3a.
   e. Write the values obtained in steps 3c and 3d as an ordered pair. This is the solution.
   f. Check the solution in each equation.

The substitution method is very useful when one of the equations is already solved for a variable or when one of the variables has a coefficient of 1.

Example 8 Solve each system by the substitution method.

a. $8x - y = 29$
   $y = 11 - 2x$

b. $5x - 4y = 1$
   $x - 2y = 2$

c. $x = 3y + 9$
   $2x - 6y = -10$

Solution To solve a system of two linear equations in two variables by the substitution method, first determine (mentally) if the equations are consistent, inconsistent, or dependent.

8a. The equations are consistent $\left(\frac{8}{2} \neq \frac{-1}{1}\right)$. Therefore, a unique solution exists.

Next, since the second equation is solved for y, substitute $11 - 2x$ for y in the first equation and then solve the resulting equation.

$$\begin{array}{ll} 8x - y = 29 & \\ 8x - \boxed{(11 - 2x)} = 29 & \text{Replace } y \text{ with } 11 - 2x \\ 8x - 11 + 2x = 29 & \\ 10x - 11 = 29 & \\ 10x = 40 & \\ x = 4 & \end{array}$$

To find y, substitute 4 for x in the equation

$$\begin{array}{l} y = 11 - 2x \\ y = 11 - 2(4) \\ y = 3. \end{array}$$

The solution is (4, 3).

Check: $8x - y = 29$ $y = 11 - 2x$
         $8(4) - 3 = 29$ $3 = 11 - 2(4)$
         $29 = 29$ $3 = 3$

8b. The equations are consistent. Thus, solve for x in the second equation since its coefficient is 1.

$$x - 2y = 2$$
$$x = 2 + 2y$$

Now substitute $2 + 2y$ for x in the first equation and then solve the resulting equation.

$$5x - 4y = 1$$
$$5(2 + 2y) - 4y = 1 \quad \text{Replace } x \text{ with } 2 + 2y$$
$$10 + 10y - 4y = 1$$
$$10 + 6y = 1$$
$$6y = -9$$
$$y = -\frac{9}{6}$$
$$y = -\frac{3}{2}$$

To find x, substitute $-\frac{3}{2}$ for y in the equation

$$x = 2 + 2y$$
$$x = 2 + 2\left(-\frac{3}{2}\right)$$
$$x = -1$$

The solution is $\left(-1, -\frac{3}{2}\right)$.

Check: $5x - 4y = 1$ $x - 2y = 2$
         $5(-1) - 4\left(-\frac{3}{2}\right) = 1$ $-1 - 2\left(-\frac{3}{2}\right) = 2$
         $-5 + 6 = 1$ $-1 + 3 = 2$
         $1 = 1$ $2 = 2$

8c. If we rewrite our equations in the form $ax + by = c$, then,

$x = 3y + 9$ becomes $x - 3y = 9$
$2x - 6y = -10$ remains $2x - 6y = -10$

which implies that the equations are inconsistent $\left(\frac{1}{2} = \frac{-3}{-6} \neq \frac{9}{-10}\right)$. Thus, the system has no solution.

Alternate method

Since the first equation is solved for x, substitute $3y + 9$ for x in the second equation and then solve the resulting equation.

$$2x - 6y = -10$$
$$2(3y + 9) - 6y = -10 \quad \text{Replace } x \text{ with } 3y + 9$$
$$6y + 18 - 6y = -10$$
$$18 = -10 \quad \text{False}$$

Since you have eliminated both variables and the resulting equation is a false statement, the two equations are inconsistent. Thus, the system has no solution.

Practice Problem 7 **Solve each system by the substitution method.**

    a. $2x + 3y = 12$ **b.** $3x - 6y = 18$ **c.** $2x + y = 5$
         $y = -10 + 4x$ $x = 6 + 2y$ $4x - 2y = 10$

The substitution method, like the addition method, can be used to solve any system of two linear equations in two variables.

Example 9 Solve each system by the substitution method.

    a. $3x + 5y = -2$ **b.** $2x + 3y = 6$ **c.** $\frac{1}{3}x + \frac{1}{4}y = 10$
         $2x + 3y = -5$ $4x + 6y = 12$ $\frac{1}{3}x - \frac{1}{2}y = 4$

Solution **9a.** The equations are consistent $\left(\frac{3}{2} \neq \frac{5}{3}\right)$. Thus, solve for x in the first equation.

$$3x + 5y = -2$$
$$3x = -2 - 5y$$
$$x = \frac{-2 - 5y}{3}$$

Now, substitute $\frac{-2 - 5y}{3}$ for x in the second equation.

$$2x + 3y = -5$$
$$2\left(\frac{-2 - 5y}{3}\right) + 3y = -5 \quad \text{Replace } x \text{ with } \frac{-2 - 5y}{3}$$
$$\frac{-4 - 10y}{3} + 3y = -5 \quad \text{Multiply}$$
$$-4 - 10y + 9y = -15 \quad \text{Multiply by LCD} = 3$$
$$-4 - y = -15$$
$$-y = -11$$
$$y = 11$$

To find x, substitute 11 for y in the equation.

$$x = \frac{-2 - 5y}{3}$$
$$x = \frac{-2 - 5(11)}{3}$$
$$x = \frac{-2 - 55}{3}$$
$$x = -19$$

The solution is $(-19, 11)$.

    Check: $3x + 5y = -2$ $2x + 3y = -5$
                 $3(-19) + 5(11) = -2$ $2(-19) + 3(11) = -5$
                          $-2 = -2$ $-5 = -5$

9b. Since $\frac{2}{4} = \frac{3}{6} = \frac{6}{12}$, the equations are dependent. Thus, the system has an infinite number of solutions. That is, every ordered pair that satisfies one equation will satisfy the other.

Alternate method

First, solve for y in the first equation.

$$2x + 3y = 6$$
$$3y = 6 - 2x$$
$$y = \frac{6 - 2x}{3}$$

Next, substitute $\frac{6 - 2x}{3}$ for y in the second equation.

$$4x + 6y = 12$$
$$4x + \overset{2}{\cancel{6}}\left(\frac{6 - 2x}{\underset{1}{\cancel{3}}}\right) = 12 \qquad \text{Replace } y \text{ with } \frac{6 - 2x}{3}$$
$$4x + 2(6 - 2x) = 12$$
$$4x + 12 - 4x = 12$$
$$12 = 12 \qquad \text{True}$$

Since you eliminated both variables and the resulting equation is a true statement, the equations are dependent. Thus, the system has an infinite number of solutions.

9c. First, eliminate the fractions from both equations by multiplying both equations by the LCD.

$$12\left(\frac{1}{3}x + \frac{1}{4}y\right) = 12(10) \qquad \text{becomes} \qquad 4x + 3y = 120$$
$$6\left(\frac{1}{3}x - \frac{1}{2}y\right) = 6(4) \qquad \text{becomes} \qquad 2x - 3y = 24$$

The equations are consistent. Thus, solve for y in the first equation.

$$4x + 3y = 120$$
$$3y = 120 - 4x$$
$$y = \frac{120 - 4x}{3}$$

Now, substitute $\frac{120 - 4x}{3}$ for y in the second equation.

$$2x - 3y = 24$$
$$2x - \cancel{3}\left(\frac{120 - 4x}{\cancel{3}}\right) = 24 \qquad \text{Replace } y \text{ with } \frac{120 - 4x}{3}$$
$$2x - (120 - 4x) = 24 \qquad \text{Multiply}$$
$$2x - 120 + 4x = 24$$
$$6x - 120 = 24$$
$$6x = 144$$
$$x = 24$$

To find y, substitute 24 for x in the operation.

$$y = \frac{120 - 4x}{3}$$
$$y = \frac{120 - 4(24)}{3}$$
$$y = \frac{120 - 96}{3}$$
$$y = 8$$

The solution is (24, 8).

9.3 Solving Linear Systems of Equations by Substitution 341

Check:
$$\frac{1}{3}x + \frac{1}{4}y = 10 \qquad \frac{1}{3}x - \frac{1}{2}y = 4$$
$$\frac{1}{3}(24) + \frac{1}{4}(8) = 10 \qquad \frac{1}{3}(24) - \frac{1}{2}(8) = 4$$
$$8 + 2 = 10 \qquad 8 - 4 = 4$$
$$10 = 10 \qquad 4 = 4$$

Practice Problem 8 Solve each system by the substitution method.

a. $x - y = 4$
   $2x - 2y = 8$

b. $2x + 3y = 3$
   $3x + 4y = 3$

c. $\frac{3}{10}x + \frac{1}{2}y = -\frac{1}{5}$
   $\frac{x}{6} + \frac{y}{4} = -\frac{5}{12}$

IDEA 2

When you solved word problems involving systems of linear equations in Section 9.2 you used the addition method. You can also solve this type of problem by using the substitution method.

Example 10 Solve.

a. Vanessa needs 12 quarts of a 10% salt solution. If she has one container that holds a 5% salt solution and another that holds a 25% salt solution, how many quarts of each solution must be mixed to obtain the desired mixture?

b. Cazzie and Guerin leave their house and travel in opposite directions on a highway. If they are 30 miles apart in 20 minutes and Cazzie's car is traveling 5 mph faster than Guerin's, how fast was each man driving?

Solution To solve a word problem involving two unknowns, (1) represent the unknowns, (2) write two equations, (3) solve the system derived in step 2 and determine the solutions, and (4) check your final answer.

10a. Step 1: x = quarts (qt) of 5% salt solution
              y = quarts of 25% salt solution

Step 2:

	Concentration of salt (C) (in percent)	Total volume of mixture (T) (in quarts)	Amount of salt (A) = C · T (in quarts)
	5% = 0.05	x	0.05x
	25% = 0.25	y	0.25y
Mixture	10% = 0.10	12	0.10(12) = 1.2

Qt of 5% solution plus Qt of 25% solution = 12 qt
x + y = 12

Amount of salt in 5% mixture plus Amount of salt in 25% mixture = Amount of salt in final mixture
0.05x + 0.25y = 1.2

Step 3: $x + y = 12$
         $0.05x + 0.25y = 1.2$

Solve for x in the first equation.
$$x + y = 12$$
$$x = 12 - y$$

Next, substitute $12 - y$ for x in the second equation.

$$0.05x + 0.25y = 1.2$$
$$0.05(12 - y) + 0.25y = 1.2 \quad \text{Replace } x \text{ with } 12 - y$$
$$0.6 - 0.05y + 0.25y = 1.2$$
$$0.20y = 0.6$$
$$y = 3$$

To find x, substitute 3 for y in the equation.
$$x = 12 - y$$
$$x = 12 - 3$$
$$x = 9$$

Vanessa must mix nine quarts of the 5% salt solution with three quarts of the 25% salt solution.

Step 4: The answer checks since:
a) $9 \text{ qt} + 3 \text{ qt} = 12 \text{ qt}$

b) $\textit{Salt before mixing} = \textit{Salt after mixing}$
$$0.05(9 \text{ qt}) + 0.25(3 \text{ qt}) = 0.10(12 \text{ qt})$$
$$0.45 \text{ qt} + 0.75 \text{ qt} = 1.20 \text{ qt}$$
$$1.20 \text{ qt} = 1.20 \text{ qt}$$

10b. Step 1: x = rate (in mph) of Cazzie's car
y = rate (in mph) of Guerin's car

Step 2:

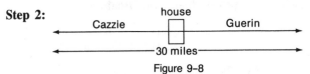

Figure 9-8

	Rate (r) (in mph)	Time (t) (in hours)	Distance (d) = $r \cdot t$ (in miles)
Cazzie	x	20 min = $\frac{1}{3}$ hr	$\frac{1}{3}x$
Guerin	y	20 min = $\frac{1}{3}$ hr	$\frac{1}{3}y$

Cazzie's distance plus Guerin's distance is 30 miles
$$\frac{1}{3}x \quad + \quad \frac{1}{3}y \quad = \quad 30$$

Cazzie's rate equals Guerin's rate plus 5 mph
$$x \quad = \quad y + 5$$

Step 3:
$$\frac{1}{3}x + \frac{1}{3}y = 30$$
$$x = y + 5$$

Substitute $y + 5$ for x in the first equation.

$$\frac{1}{3}x + \frac{1}{3}y = 30$$

$$\frac{1}{3}(y + 5) + \frac{1}{3}y = 30 \quad \text{Replace } x \text{ with } y + 5$$

$$y + 5 + y = 90 \quad \text{Multiply by LCD} = 3$$

$$2y + 5 = 90$$

$$2y = 85$$

$$y = 42\frac{1}{2}$$

To find x, substitute $42\frac{1}{2}$ for y in the equation.

$$x = y + 5$$

$$x = 42\frac{1}{2} + 5$$

$$x = 47\frac{1}{2}$$

Cazzie's rate is $47\frac{1}{2}$ mph and Guerin's is $42\frac{1}{2}$ mph.

Step 4: The answer checks since:

a. Cazzie's rate $\left(47\frac{1}{2} \text{ mph}\right)$ is 5 mph more than Guerin's rate $\left(42\frac{1}{2} \text{ mph}\right)$.

b. Cazzie's distance $= \left(47\frac{1}{2} \text{ mph}\right)\left(\frac{1}{3} \text{ hr}\right) = 15\frac{5}{6}$ miles

Guerin's distance $= \left(42\frac{1}{2} \text{ mph}\right)\left(\frac{1}{3} \text{ hr}\right) = 14\frac{1}{6}$ miles

Total distance $=$ 30 miles

Practice Problem 9 *Solve by using the substitution method.*

a. Kay needs six ounces of a medicine that is 10% alcohol. If she has one bottle of this medicine that is a 12% alcohol solution and another bottle that is a 7% alcohol solution, how many ounces of each must be mixed to obtain the desired mixture?

b. Truck A leaves Chicago for Boston (a distance of 1050 miles) at the same time that truck B leaves Boston for Chicago. If truck A travels at a rate of 60 mph and truck B travels 40 mph, how far has each truck traveled when they pass each other?

9.3 Exercises

Solve each system by substitution.

1. $x + y = 6$
   $x = 6 + y$

2. $x = 6 - 3y$
   $x + 2y = 2$

3. $y = 6 - 2x$
   $3x + y = 6$

4. $y = 9 - 3x$
   $x + 2y = 8$

5. $y = 11 - 2x$
   $x = 13 - 2y$

6. $x + 2y = 4$
   $x + 2y = 10$

7. $2x + y = 10$
   $6x - 2y = 10$

8. $3x + y = 7$
   $2x + 4y = 8$

9. $5x - y = 10$
   $x + y = 7$

10. $2x + 4y = 14$
    $x + y = 2$

11. $3x + 2y = 8$
    $-6x + 4y = -16$

12. $5x + 4y = 12$
    $2x - 5y = 8$

13. $4x + 6y = -8$
    $5x + 4y = -20$

14. $2x + 6y = 14$
    $2x + 5y = 12$

15. $2x + 5y = 10$
    $4x + 10y = 10$

16. $3x - 6y = 8$
    $6x - 12y = 16$

17. $3x + y = -2$
    $-6x + 3y = -14$

18. $5x - 2y = 10$
    $2x + 5y = 8$

19. $5x - 6y = 9$
    $10x - 3y = 6$

20. $y = 2x + 9$
    $y = 3x - 5$

21. $x = 8y + 7$
    $x = 2y - 1$

22. $9x + 3y = 8$
    $y = 3x - 4$

23. $x = -18 - 4y$
    $3x + 5y = -19$

24. $4x = 3y$
    $x + y = 7$

25. $-2x + y = 8$
    $5x + 2y = -20$

26. $5x + 3y = -9$
    $-7x - y = -3$

27. $2x - y = -4$
    $-x - y = 2$

28. $\dfrac{x}{2} + \dfrac{y}{5} = \dfrac{18}{5}$
    $y = 36 - 4x$

29. $\dfrac{3x}{4} + \dfrac{y}{2} = \dfrac{-11}{4}$
    $\dfrac{3x}{2} = 7 + \dfrac{y}{4}$

30. $\dfrac{x}{3} + \dfrac{y}{3} = \dfrac{4}{3}$
    $\dfrac{x}{2} + \dfrac{y}{2} = \dfrac{3}{2}$

31. $0.3x + 0.2y = 0.1$
    $-0.4 + y = -2.6$

32. $0.1x - y = -13.95$
    $0.12x - y = -12.95$

Fill in the blanks.

33. When using the substitution method to solve a system of linear equations, you still _____ one of the variables. However, you _____ the variable by _____ rather than _____.

34. When solving a system of two linear equations in two variables by substitution, first determine if the _____ are consistent, inconsistent, or dependent. If the _____ are inconsistent, write _____. If they are dependent, write _____.

35. If the equations in a system of two linear equations in two variables are consistent, you can find the solution by the substitution method as follows:
    a. Step 1: Solve (if necessary) _____ of the equations for _____ of the variables.
    b. Step 2: _____ the expression obtained in step 1 into the _____.
    c. Step 3: You now have _____ equation in _____ variable. Solve it.
    d. Step 4: Find the value of the other variable by _____ the value obtained in step 3 into the _____ obtained in step 1.
    e. Step 5: Write the values obtained in steps 3 and 4 as an _____. This is the solution.
    f. Step 6: _____ the solution in both equations.

Solve by substitution.

36. The sum of two consecutive even integers is 70. What are the integers?

37. In a school election, Howard received 140 more votes than Gene. If the total number of votes cast was 384, how many votes did each candidate receive?

38. Charles earned twice as much money as Lee. If their combined salaries total $615, how much did each man earn?

39. A house and lot cost $55,000. If the lot costs $15,000 less than the house, how much does each cost?

40. The daily payroll of the Pace Educational Center is $775. The teachers earn $50 per day, and the tutors earn $25 per day. If the center employs 21 people, find the number of teachers and the number of tutors employed.

41. Barbara has $1500. She deposited some of the money into a savings account that pays 6.25% interest and the rest into an account that pays 13% interest. If the total annual interest earned from the two accounts is $168, how much money did she deposit into each account?

42. Matthew invested a total of $4000 in two savings plans. If the first plan has an interest rate of 15% and the second plan has a rate of 9%, how much must be invested at each rate in order for Matthew to earn a total of $420 in 10 months?

43. How many gallons of a solution that is 15% salt must be mixed with a solution that is 75% salt to obtain 12 gallons of a 30% salt solution?

44. How many pounds of caramels costing $1.90 per pound and how many pounds of Hershey bars costing $2.40 per pound must be mixed to obtain 20 pounds of a candy worth $46?

45. Jerry traveled 360 miles to his summer home in eight hours. He drove for six hours on an interstate highway and the rest of the time on an unpaved road. If his rate on the highway was 20 mph more than his rate on the unpaved road, find his rate on the highway and on the road.

46. Tom and Jerry leave school at the same time and travel in opposite directions on the same highway. Tom drives at a rate of 40 mph and Jerry drives at 50 mph. How far has each man traveled when they are exactly 270 miles apart?

47. Two boats are 50 miles apart. Both start at 10:00 A.M. and travel toward each other in a straight canal. If one boat averages 18 mph and the other averages 12 mph, how far has each boat traveled when they pass each other?

Solve.

48. $38x + 55y = -224$
    $45x - 15y = -105$

49. $25x - 30y = -80$
    $40x + 20y = -400$

50. $5.21x - 8.7y = -20.369$
    $9.8x + 4.3y = 23.68$

51. $5.72x - 8.1y = -22.15$
    $5.95x + 6.75y = 45.25$

Answers to Practice Problems **7a.** $(3, 2)$ **b.** infinite numbers of solutions **c.** $\left(\frac{5}{2}, 0\right)$ **8a.** infinite number of solutions **b.** $(-3, 3)$ **c.** $(-19, 11)$ **9a.** 3.6 oz. of the 12% solution 2.4 oz. of the 7% solution **b.** Truck A traveled 630 miles Truck B traveled 420 miles

Chapter 9 Summary

Important Terms

A **system of two linear equations in two variables** consists of two linear equations having the same variables.

$2x + 3y = 6$ and $x + 3y = 6$
$4x + 6y = 8$ $y = 2x + 1$

are systems of linear equations in two variables. [Section 9.1/Idea 1]

A **solution** of a system of two linear equations in two variables is an ordered pair that is a solution of both equations in the system. The solution of the system

$x + y = 6$
$x - y = 2$

is (4, 2). [Section 9.1/Idea 1]

If the graphs of two linear equations in two variables are two intersecting lines, the equations are **consistent**.

$2x + y = 6$
$x + 3y = 4$

are consistent equations. If the graphs of two linear equations in two variables are two distinct parallel lines, the equations are **inconsistent**.

$x + y = 2$
$2x + 2y = 3$

are inconsistent equations. If the graphs of two linear equations in two variables are the same line, the equations are **dependent** or **equivalent**.

$2x + y = 3$
$4x + 2y = 6$

are dependent equations. [Section 9.1/Idea 2]

Important Skills

Solving Systems Graphically   To solve a system of two linear equations in two variables graphically, graph each equation on the same coordinate system. If the lines intersect, the point of intersection is the solution of the system. If the lines are parallel, the system has no solution. If the graphs of the equations are the same line, the system has an infinite number of solutions. [Section 9.1/Idea 2]

Determining Consistency, Inconsistency, or Dependency of Systems	If $a_1x + b_1y = c_1$ $a_2x + b_2y = c_2$ is a system of equations, then (a) if $\frac{a_1}{a_2} \neq \frac{b_1}{b_2}$, the equations are consistent; (b) if $\frac{a_1}{a_2} = \frac{b_1}{b_2} = \frac{c_1}{c_2}$, the equations are dependent; (c) if $\frac{a_1}{a_2} = \frac{b_1}{b_2} \neq \frac{c_1}{c_2}$, the equations are inconsistent. [Section 9.1/Idea 3]
Solving Systems by Addition	To solve a system of two linear equations in two variables by addition, write both equations in the form $ax + by = c$. If the equations are inconsistent, write "no solution." If they are dependent, write "infinite number of solutions." If the equations in step 1 are consistent, (a) multiply one or both of the equations by a number that will make the coefficients of x (or y) additive inverses of each other; (b) add the resulting equations; (c) solve the equation obtained in step (b); (d) substitute the solution obtained in step (c) into one of the original equations; and solve the resulting equation; (e) write the values obtained in steps (c) and (d) as an ordered pair. This is the solution of the system; finally, (f) check your solution. [Section 9.2/Idea 1]
Solving Systems by Substitution	To solve a system of two linear equations in two variables by substitution, determine (mentally) if the equations are consistent, inconsistent, or dependent. If the equations are inconsistent write "no solution." If they are dependent, write "infinite number of solutions." If the equations are consistent, (a) solve (if necessary) one of the equations for one of the variables; (b) substitute the expression in step (a) into the other equation; (c) solve the equation obtained in step (b); (d) find the value of the other variable by substituting the value obtained in step (c) into the equation obtained in step (a); (e) write the values obtained in steps (c) and (d) as an ordered pair. This is the solution of the system; finally, (f) check the solution. [Section 9.3/Idea 1]

Chapter 9 Review Exercises

Solve each system graphically.

1. $x + y = 5$
   $x - y = 3$

2. $x + 2y = 4$
   $2x + 4y = 8$

3. $4x + 3y = 12$
   $y - 3 = -7$

Solve each system using the addition method.

4. $2x + 2y = 9$
   $2x - 2y = -7$

5. $3x + y = 2$
   $6x + 2y = 1$

6. $5x + 7y = 14$
   $3x - 4y = -8$

Solve each system by substitution.

7. $3x - 2y = -5$
   $x = -2y + 9$

8. $6x + 8y = 14$
   $3x + 5y = 8$

9. $4x + 6y = -9$
   $x - 3y = 6$

Solve each system by any method.

10. $3x + y = 6$
    $6x + 2y = -7$

11. $5x + 2y = 10$
    $x - 2y = -4$

12. $x + 4y = 2$
    $x - 2y = -4$

13. $3x + 5y = 10$
    $x = 3y + 8$

14. $0.4x - 0.3y = 1.2$
    $0.6x - 0.5y = 1.9$

15. $\frac{x}{3} + \frac{y}{2} = 1$
    $\frac{x}{5} + \frac{y}{2} = -\frac{7}{10}$

Solve.

16. Find two numbers whose sum is 70 and whose difference is 6.

17. How many pounds of pumpkin seeds costing $0.60 per pound must be mixed with sunflower seeds costing $0.90 per pound to obtain 30 pounds of a mixture worth $24?

18. Jody earned $25,000 from the sale of a house. He invested part at 15% and the rest at 25%. If the annual interest income from the 25% investment is $250 more than the 15% investment, how much did he invest at each rate?

19. Ed took four minutes to finish a race and Mike took $4\frac{1}{2}$ minutes to finish the same race. If the rate of the faster runner is three feet per second more than the slow runner, find each man's rate.

20. Richard received an annual income of $2100 from $10,000 invested in real estate and $6000 invested in stocks. His son received $930 from an investment of $4000 in the same real estate and $3000 in the same stocks. Find the interest rate they received from each investment.

Solve each system.

21. $0.8x + 0.4y = 16$
    $0.7x - 0.3y = 1$

22. $5.2x + 3y = 14$
    $0.6x - 4y = 19$

23. $\frac{x}{2} - 2y = 3$
    $-2x + 8y = 16$

24. $\frac{x - y}{3} = 1$
    $\frac{2x}{5} = 2y - 1$

Name: _____

Class: _____

Chapter 9 Test

Solve each system graphically.

1. $x + y = 2$
   $2x + 2y = 6$

1. _____
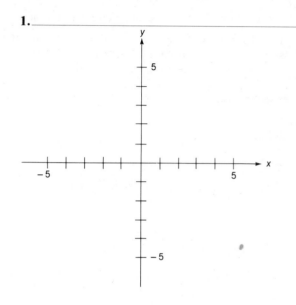

2. $2x + 3y = 6$
   $4x + 6y = 12$

2. _____
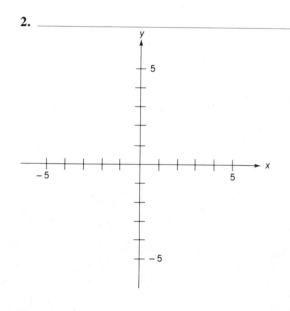

3. $x + y = 6$
   $4x + 6y = 12$

3. _____

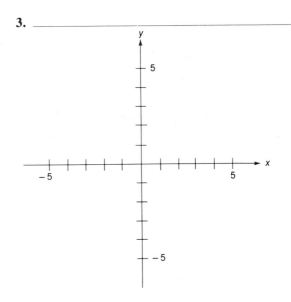

Solve each system by addition.

4. $6x + 2y = 8$
   $4x - 2y = 22$

5. $3x - y = 1$
   $6x - 2y = 5$

6. $4x - 6y = 6$
   $3x + 9y = 9$

7. $x + \frac{5}{4}y = \frac{1}{4}$
   $-\frac{2}{3}x + y = -2$

Solve each system by substitution.

8. $x = 4y + 5$
   $-2x + 8y = 9$

9. $x - 2y = -4$
   $x + 3y = 21$

10. $4x - 6y = -2$
    $10x + 9y = 11$

11. $\frac{x}{2} + \frac{y}{3} = 5$
    $2x - y = -1$

Solve.

12. Find two numbers whose difference is 4 such that the sum of twice the first number and the second number is 5.

13. Tommy inherited $3000. Part of the money was invested at 10% and the rest at 14%. If his total annual interest income was $348, how much was invested at each rate?

4. _____
5. _____
6. _____
7. _____
8. _____
9. _____
10. _____
11. _____
12. _____
13. _____

10 Factoring Polynomials

Objectives The objectives for this chapter are listed below along with sample problems for each objective. By the end of this chapter you should be able to find the solutions to the given problems.

1. Factor a polynomial in which the terms have a common factor *(Section 10.1/Idea 1)*.
   a. $16x^{13} - 18x^5$ b. $15x^6 - 18x^5 + 10x^2$

2. Factor a trinomial of the form $x^2 + px + q$ *(Section 10.2/Idea 1)*.
   a. $x^2 - 6x + 8$ b. $x^2 + 5x - 24$

3. Factor a trinomial of the form $ax^2 + bx + c$ *(Section 10.2/Idea 2)*.
   a. $6x^2 - x - 7$ b. $6x^3 + 26x^2 - 20x$

4. Factor the difference of two squares *(Section 10.3/Idea 1)*.
   a. $x^2 - 16$ b. $16x^2 - 25$

5. Factor a perfect square trinomial *(Section 10.3/Idea 2)*.
   a. $x^2 + 6x + 9$ b. $9x^2 - 30x + 25$

6. Factor a polynomial by grouping *(Section 10.3/Idea 3)*.
   a. $ax - ay + bx - by$ b. $x^2 - 2x + 1 - 81y^2$

10.1 Common Monomial Factors

IDEA 1

In Chapter 6 the distributive law, $a(b + c) = a \cdot b + a \cdot c$ was used to multiply a polynomial by a monomial. For example,

$$2(x + 7) = 2 \cdot x + 2 \cdot 7 = 2x + 14.$$

The distributive law, $ab + ac = a(b + c)$, can also be used to write a polynomial as a product of its factors. This process is called **factoring a polynomial**. For example,

$$2x + 14 = \boxed{2} \cdot x + \boxed{2} \cdot 7 = \boxed{2}(x + 7)$$

In the above example, note that 2 is the largest factor that divides each term of the polynomial $2x + 14$ evenly. Therefore, 2 is called the **greatest common factor (GCF)** of the polynomial $2x + 14$.

If we carefully examine the fact that

$$2x + 14 = 2\,(x + 7)$$
$$\text{GCF}\left(\frac{2x}{2} + \frac{14}{2}\right)$$

we can see that the polynomial $2x + 14$ is factored as the product of the GCF of $2x + 14$ and the polynomial obtained by dividing each term of $2x + 14$ by its GCF. In other words, every polynomial in which the terms have a common monomial factor can be factored as

$$\text{polynomial} = \text{GCF}\left(\frac{\text{polynomial}}{\text{GCF}}\right)$$

For example, to factor $6x^2 + 9x$, we first find the GCF of $6x^2 + 9x$. The GCF is $3 \cdot x = 3x$ since 3 is the largest common factor of the coefficients 6 and 9 and x is the largest common factor of x^2 and x. Now divide each term of the original polynomial $6x^2 + 9x$ by the GCF ($3x$) to obtain the second factor of the polynomial ($2x + 3$).

$$6x^2 + 9x = 3x\,(2x + 3)$$
$$\text{GCF}\left(\frac{6x^2}{3x} + \frac{9x}{3x}\right)$$

Similarly, to factor $9x^4y^4 - 12x^3y^5 + 15x^2y^6$, we first find the GCF. The GCF of $9x^4y^4 - 12x^3y^5 + 15x^2y^6$ is $3 \cdot x^2 \cdot y^4 = 3x^2y^4$. Since 3 is the largest common factor of 9, -12, and 15, x^2 is the largest common factor of x^4, x^3, and x^2, and y^4 is the largest common factor of y^4, y^5, and y^6. Next, we divide each term of the original polynomial $9x^4y^4 - 12x^3y^5 + 15x^2y^6$ by the GCF $3x^2y^4$ to obtain the second factor of the polynomial, which is $3x^2 - 4xy + 5y^2$.

$$9x^4y^4 - 12x^3y^5 + 15x^2y^6 = 3x^2y^4(3x^2 - 4xy + 5y^2)$$
$$\text{GCF}\left(\frac{9x^4y^4 - 12x^3y^5 + 15x^2y^6}{3x^2y^4}\right)$$

352 Factoring Polynomials

The preceding procedure is summarized below.

To Factor a Polynomial Containing a Common Monomial Factor:

1. Find the greatest common monomial factor (GCF) of the polynomial. The greatest common monomial factor of a polynomial is the product of (a) the largest common factor of the numerical coefficients of the polynomial, and (b) the smallest power of every variable that appears in each term of the polynomial.

2. Divide each term of the original polynomial by its greatest common factor (do this mentally).

3. Write the expressions obtained in steps 1 and 2 as a product.

Example 1 Factor.

    a. $10x^3 + 15x^2$ **b.** $8x^5y + 10x^4y^2 - 14x^3y^3$

    c. $12x^3y^2z - 18xy^2$ **d.** $-6x^5 - 12x^4 + 42x^3$

Solution   **1a.** $10x^3 + 15x^2 =$ (first factor)(second factor)

First factor (GCF of $10x^3 + 15x^2$): The GCF of 10 and 15 is 5 while x^2 is the smallest power of the variable which appears in each term. Therefore, the GCF of $10x^3 + 15x^2$ is $5 \cdot x^2 = 5x^2$.

Second factor $(10x^3 + 15x^2 \div GCF)$: $\dfrac{10x^3}{5x^2} + \dfrac{15x^2}{5x^2} = 2x + 3$

Therefore, $10x^3 + 15x^2 = 5x^2(2x + 3)$

1b. $8x^5y + 10x^4y^2 - 14x^3y^3 =$ (first factor)(second factor)

First factor: GCF of polynomial is $2x^3y$

Second factor: $\dfrac{8x^5y}{2x^3y} + \dfrac{10x^4y^2}{2x^3y} - \dfrac{14x^3y^3}{2x^3y} = 4x^2 + 5xy - 7y^2$

Therefore, $8x^5y + 10x^4y^2 - 14x^3y^3 = 2x^3y(4x^2 + 5xy - 7y^2)$

1c. $12x^3y^2z - 18xy^2 =$ (first factor)(second factor)

First factor: GCF of the polynomial is $6xy^2$

Second factor: $\dfrac{12x^3y^2z}{6xy^2} - \dfrac{18xy^2}{6xy^2} = 2x^2z - 3$

Therefore, $12x^3y^2z - 18xy^2 = 6xy^2(2x^2z - 3)$

1d. $-6x^5 - 12x^4 + 42x^3 =$ (first factor)(second factor)

First factor: Since the first term of the polynomial is negative, it will be helpful to include a minus sign in the GCF. Thus, the GCF is $-6x^3$.

Second factor: $\dfrac{-6x^5}{-6x^3} - \dfrac{12x^4}{-6x^3} + \dfrac{42x^3}{-6x^3} = x^2 + 2x - 7$

Therefore, $-6x^5 - 12x^4 + 42x^3 = -6x^3(x^2 + 2x - 7)$

Practice Problem 1 **Factor.**

a. $8x^2 - 28x$ **b.** $36x^4y + 24x^3y^2 - 18x^2y$
c. $24x^5y - 35x^4y^2$ **d.** $-8x^6 - 10x^5 - 12x^3$

IDEA 2

Sometimes you may not be able to recognize the GCF of the numerical coefficients of a polynomial by inspection. When this occurs, it will be helpful to factor each coefficient. For example, to find the GCF of the coefficients of the polynomial $48x^5y + 72x^3y$, we first factor the coefficients 48 and 72.

$$48 = 2 \cdot 2 \cdot 2 \cdot 2 \cdot 3 \qquad 72 = 2 \cdot 2 \cdot 2 \cdot 3 \cdot 3$$
$$= 2^4 \cdot 3 \qquad\qquad\quad = 2^3 \cdot 3^2$$

Next, since 2^3 and 3 are the largest common prime factors of 48 and 72, the GCF of the coefficients 48 and 72 is $2^3 \cdot 3 = 8 \cdot 3 = 24$.

The procedure for finding the GCF of the coefficients of a polynomial is summarized below.

To Find the GCF of the Coefficients of a Polynomial:

1. Write the prime factorization of each coefficient.
2. The GCF is the product of the common prime factors of the coefficients. Each common factor is used the minimum number of times it occurs in any factorization.

Example 2 Find the GCF of the numerical coefficients.

a. $64x^2y + 96xy^2$ **b.** $30x^2 + 75x - 90$

Solution **2a.** $64 = 2 \cdot 2 \cdot 2 \cdot 2 \cdot 2 \cdot 2 \qquad 96 = 2 \cdot 2 \cdot 2 \cdot 2 \cdot 2 \cdot 3$
$\ = 2^6 \qquad\qquad\qquad\qquad\ = 2^5 \cdot 3$

Since the only common prime factor is 2 and the minimum number of times it occurs in any factorization is 5, the GCF of 64 and 96 is $2 \cdot 2 \cdot 2 \cdot 2 \cdot 2 = 32$.

2b. $30 = 2 \cdot 3 \cdot 5 \qquad 75 = 3 \cdot 5 \cdot 5 \qquad -90 = -2 \cdot 3 \cdot 3 \cdot 5$
$= 3 \cdot 5^2 \qquad\quad\ \ = -2 \cdot 3^2 \cdot 5$

Since the only common prime factors are 3 and 5 and the minimum number of times each occurs in any factorization is 1, the GCF of 30, 75, and -90 is $3 \cdot 5 = 15$.

Example 3 Factor: $96x^5 - 144x^4 + 72x^3 - 48x^2$

Solution $96x^5 - 144x^4 + 72x^3 - 48x^2 = $ (first factor)(second factor)

First factor:
The GCF of the coefficients is 24* and x^2 is the smallest power of the variable appearing in each term. Thus, the GCF of the polynomial is $24 \cdot x^2 = 24x^2$

**Note:*
$96 = 2 \cdot 2 \cdot 2 \cdot 2 \cdot 2 \cdot 3 = 2^5 \cdot 3$
$144 = 2 \cdot 2 \cdot 2 \cdot 2 \cdot 3 \cdot 3 = 2^4 \cdot 3^2$
$72 = 2 \cdot 2 \cdot 2 \cdot 3 \cdot 3 = 2^3 \cdot 3^2$
$48 = 2 \cdot 2 \cdot 2 \cdot 2 \cdot 3 = 2^4 \cdot 3$
$\text{GCF} = 2 \cdot 2 \cdot 2 \cdot 3 = 24 = 2^3 \cdot 3$

Second factor:

$$\frac{96x^5}{24x^2} - \frac{144x^4}{24x^2} + \frac{72x^3}{24x^2} - \frac{48x^2}{24x^2} = 4x^3 - 6x^2 + 3x - 2$$

Therefore, $96x^5 - 144x^4 + 72x^3 - 48x^2 = 24x^2(4x^3 - 6x^2 + 3x - 2)$

Practice Problem 2   **Factor.**

$$24a^4b - 60a^3b^2 + 84a^2b^3 + 108ab^4$$

10.1 Exercises

Factor.

1. $2x + 8$
2. $6x^2 - 8x$
3. $9x^2y + 15xy$
4. $2x^3 - 10$
5. $25xy - 35y^2$
6. $12x^4y^2 - 16x^2y^4$
7. $10x^3 - 25x^2 + 20$
8. $16y^4 - 24y^3 + 32y^2$
9. $50x^5 + 100x^3 - 150x^2$
10. $5x^6 + 25x^4 - 20x^2$
11. $16x^3y + 8x^2y + 24xy$
12. $45a^4b^5 - 36a^2b^6 + 81ab^7$
13. $19x^3 + 38x^2y^2$
14. $12x^5 + 14x^4 + 16x^3 + 18$
15. $12a^3d^2 - 18a^2d^3$
16. $45x^3 - 15b^2 + 30$
17. $-27b^4 - 18b^3 + 36b$
18. $-10a^3b^6 - 15a^6b^3$
19. $14x^5y^6 - 42x^3y^7 + 28xy^8$
20. $15x^2yz - 16xy^2$
21. $12a^3b^2 - 18a^2b^3$
22. $15xy^3 - 45x^2y^4$
23. $44a^{14}b^7 - 33a^{10}b^5$
24. $26x^8y^6 - 39x^{12}y^5$
25. $12x^8y^9 + 18x^5y^4 - 20xy^3$
26. $9x^7y^8 - 12xy^5 + 15x^2y^5$
27. $15a^{12}b - 8ab^{12} + 9ab$
28. $10xy^6 + 5x^6y - 4xy$
29. $-30a^3b^4 - 45a^8b^7 - 15a^2b^2$
30. $-18a^3b^4 + 14x^{10}y^6 + 24x^2y^2$
31. $18x^3 - 12y^2 - 48x^4$
32. $32x^5 - 24x^8 + 40y^5$
33. $16x^3y^2 + 24xy^3 - 40x^4y^2$
34. $20x^2y - 30xy^3 - 40xy$
35. $-20x^5 - 10x^3 + 5x^2$
36. $-30x^8 - 10x^5 + 5x^4$
37. $44x^3y^4z - 55x^2y^2z$
38. $66a^6b^4c^2 - 77a^4b^3c^2$
39. $40a^2b^3c^4 - 80a^3b^2c^3$
40. $30x^6y^5z^4 - 60x^3y^2z^2$

Find the GCF (see Example 2).

41. 24, 36, 96
42. 16, 24, 32, 64
43. 26, 39, 65, 91
44. 140, 210, 350
45. 63, 42, 105
46. 48, 60, 84
47. 111, 93, 123
48. 240, 300, 420
49. 75, 90, 120
50. 90, 126, 198
51. 78, 130, 182
52. 66, 110, 132
53. 1001, 1309

Factor.

54. $88x^2 + 121$
55. $65x^3 - 91x^2$
56. $150x^3 - 200x^2 + 250x$
57. $70x^3y - 105x^2y^2 + 140xy^3$
58. $-42a^3b^2c - 66a^2b^3c$
59. $121x^4 - 143x^3 + 187$
60. $24xy^2z + 36xyz + 96xy$
61. $-48x^3y + 60x^2 - 84xy$

Fill in the blanks.

62. The process of writing a polynomial as a product of two or more factors is called _____.

63. The _____ of a polynomial is the largest factor that will divide each term of the polynomial.

64. The greatest common monomial factor of a polynomial is the _____ of the greatest common divisor of

the _____ of the polynomial and the _____ power of every variable that appears in each term of the polynomial.

65. To factor a polynomial in which the terms have a common monomial factor, write the _____ of the polynomial and the expression obtained when each _____ of the polynomial is divided by the _____ of the polynomial as a product.

66. To find the GCF of the numerical coefficients, write the _____ of each coefficient. The GCF is the _____ of the common prime factors of the coefficients. Each common factor is used the _____ number of times it occurs in any factorization.

Answers to Practice Problems   1a. $4x(2x - 7)$ b. $6x^2y(6x^2 + 4xy - 3)$ c. $x^4y(24x - 35y)$
d. $-2x^3(4x^3 + 5x^2 + 6)$ 2. $12ab(2a^3 - 5a^2b + 7ab^2 + 9b^3)$

10.2 Factoring Trinomials

IDEA 1 In Chapter 6 you multiplied two binomials and obtained a trinomial.

$$(x + a)(x + b) = x^2 + (a + b)x + ab$$

In this section, you will do the reverse by factoring trinomials as the product of two binomials.

$$x^2 + (a + b)x + ab = (x + a)(x + b)$$

Note that trinomials that can be factored as $(x + a)(x + b)$ have the following characteristics:

1. The first term is x^2.
2. The product of the last terms of the binomial factors (a and b) equals the last term of the trinomial (ab) and the sum of these same terms ($a + b$) equals the coefficient of the middle term of the trinomial.

For example, to factor $x^2 - 5x + 6$ as the product of two binomial factors we first note that the product of the first terms of the binomials must be equal to the first term of the trinomial. Thus, in order to obtain x^2 as the first term of the trinomial we write

$$x^2 - 5x + 6 = (x \quad)(x \quad)$$

Next, we need to determine the last terms of the binomial factors. To do this, we should remember that the product of the last terms of the binomial factors must equal 6 (last term of trinomial) and the sum of these same factors must be -5 (coefficient of the middle term of the trinomial). Thus, both numbers must be negative since the product of two negatives is a positive and the sum of two negatives is a negative. Therefore, our factors are -3 and -2 since $(-3)(-2) = 6$ and $-3 + (-2) = -5$. Thus,

$$x^2 - 5x + 6 = (x - 3)(x - 2)$$

Always check your factorization by multiplying the factors by using the FOIL method.

Check (mentally): $(x - 3)(x - 2) = x^2 - 3x - 2x + 6 = x^2 - 5x + 6$

Also keep in mind that the order of the factors does not matter. Therefore, $x^2 - 5x + 6$ can be factored as $(x - 3)(x - 2)$ or $(x - 2)(x - 3)$.

Before stating a procedure for factoring trinomials of the form $x^2 + px + q$, let us try some other examples.

Example 4 Factor.

a. $x^2 + 7x + 10$ **b.** $x^2 - 7x + 10$ **c.** $x^2 - 2x - 15$

Solution

4a. To factor $x^2 + 7x + 10$ as $(x + a)(x + b)$, you must find two integers a and b whose product is 10 and whose sum is 7. The numbers are 2 and 5. Therefore, $x^2 + 7x + 10 = (x + 2)(x + 5)$.

4b. To factor $x^2 - 7x + 10$ as $(x + a)(x + b)$, you must find two integers a and b whose product is 10 and whose sum is -7. If the product of the two integers must be positive and their sum must be negative, both integers must be negative. Therefore, the numbers are -2 and -5, and $x^2 - 7x + 10 = (x - 2)(x - 5)$.

4c. To factor $x^2 - 2x - 15$ as $(x + a)(x + b)$, you must find two integers whose product is -15 and whose sum is -2. Since the product of the two integers must be negative, one integer is positive and the other is negative. In this case, it may be helpful to list all pairs of integers whose product is -15.

Factors of -15	Sum of factors	
$-3, 5$	2	
$3, -5$	-2	← Correct sum
$-1, 15$	14	
$1, -15$	-14	

The numbers are 3 and -5. Therefore, $x^2 - 2x - 15 = (x + 3)(x - 5)$.

When factoring a trinomial $x^2 + px + q$ as a product of two binomials, the following information may be helpful.

Sign of the Constant Term q	Coefficient P of the Middle Term	Signs of the Last Terms of the Binomials
Positive	Positive	Both positive
Positive	Negative	Both negative
Negative	Positive or negative	One positive, one negative

Based on the results in Example 4, you can factor a trinomial $x^2 + px + q$ as a product of two binomials using the procedure shown in the box on page 357. In step 2 of the procedure, if the integers do not exist, the trinomial is not factorable. In this case, we say that the trinomial is *prime with respect to the integers*.

Example 5 Factor.

a. $x^2 - 9x + 14$ **b.** $x^2 + 4x - 21$ **c.** $x^2 + 7x + 6$
d. $y^2 - y - 12$ **e.** $a^2 + 7a + 8$ **f.** $x^4 - 7x^2 - 44$

> **To Factor a Trinomial of the Form $x^2 + px + q$:**
>
> 1. Make the first terms of each binomial the same so that their product equals the first term of the trinomial.
> 2. Find two integers (if they exist) whose product is equal to the constant term q of the trinomial and whose sum is equal to the coefficient of the middle term p of the trinomial.
> 3. The last terms of the binomial factors are the integers found in step 2.
> 4. Check the answer mentally. The product of the binomial factors must be the same as the original trinomial.

Solution To factor a trinomial $x^2 + px + q$ as $(x + a)(x + b)$, you must find two integers a and b such that $ab = q$ and $a + b = p$.

5a. Step 1: First term of each binomial must be the same and product must be x^2

$$x^2 - 9x + 14 = (x\ \ )(x\ \ )$$

Step 2: Find two integers whose product is 14 and whose sum is -9.

14	Sum of Factors
$-1, -14$	-15
$-2, -7$	$-9 \leftarrow$ Correct sum

Step 3: Since -9 is the correct sum, the last term of the binomials are -2 and -7.

$$x^2 - 9x + 14 = (x - 2)(x - 7)$$

Check: $(x - 2)(x - 7) = x^2 - 9x + 14$ This is done mentally

5b. $x^2 + 4x - 21 = (x\ \ )(x\ \ )$

-21	Sum of Factors
$3, -7$	-4
$-3, 7$	$4 \leftarrow$ Correct sum
$1, -21$	
$-1, 21$	

Therefore, $x^2 + 4x - 21 = (x - 3)(x + 7)$.

5c. $x^2 + 7x + 6 = (x\ \ )(x\ \ )$

6	Sum of Factors
$2 \cdot 3$	5
$6 \cdot 1$	$7 \leftarrow$ Correct sum

Therefore, $x^2 + 7x + 6 = (x + 6)(x + 1)$

5d. $y^2 - y - 12 = (y\ \ )(y\ \ )$
↓

-12	Sum of Factors
2, −6	−4
−2, 6	4
3, −4	−1 ← Correct sum
−3, 4	
1, −12	
−1, 12	

Therefore, $y^2 - y - 12 = (y + 3)(y - 4)$.

5e. $a^2 + 7a + 8 = (a\ \ )(a\ \ )$
↓

8	Sum of Factors
2, 4	6
1, 8	9
−2, −4	−6
−1, −8	−9

Since there are no two integers whose product is 8 and whose sum is 7, $a^2 + 7a + 8$ is not factorable (with respect to the integers).

5f. $x^4 - 7x^2 - 44 = (x^2\ \ )(x^2\ \ )$
↓

-44	Sum of Factors
2, −22	−20
−2, 22	20
4, −11	−7 ← Correct sum
−4, 11	
1, −44	
−1, 44	

Therefore, $x^4 - 7x^2 - 44 = (x^2 + 4)(x^2 - 11)$.

Practice Problem 3 *Factor.*

   a. $x^2 - 8x + 15$ **b.** $y^2 - 3y - 10$ **c.** $x^2 + 10x - 75$

Practice Problem 4 *Factor mentally. Just write the answer.*

   a. $x^2 + 8x - 20$ **b.** $y^2 - 9y + 20$ **c.** $x^2 - 7x - 30$

IDEA 2   Sometimes it will be necessary to factor trinomials in which the coefficient of the squared term is not equal to 1. When factoring this type of trinomial, we will use the trial-and-error method.

Example 6   Factor: $3x^2 - 7x - 6$

Solution   The product of the first terms of the binomial factors must be $3x^2$ and the product of the last terms must be -6. To obtain $3x^2$, the factors must be $3x$ and x. Therefore, you have $3x^2 - 7x - 6 = (x\ \ )(3x\ \ )$.

Now, identify the possible factors of -6 and list the possible binomial factors of $3x^2 - 7x - 6$.

Possible binomial factors	Resulting middle term	
$(x + 2)(3x - 3)$	$3x$	
$(x + 3)(3x - 2)$	$7x$	
$(x - 2)(3x + 3)$	$-3x$	
$(x - 3)(3x + 2)$	$-7x$	← Correct middle term
$(x + 6)(3x - 1)$	$17x$	
$(x + 1)(3x - 6)$	$-3x$	
$(x - 6)(3x + 1)$	$-17x$	
$(x - 1)(3x + 6)$	$3x$	

Finally, since $-7x$ is the same as the middle term of the trinomial, the correct factors are $(x - 3)(3x + 2)$.

It may be helpful to remember that *if the original trinomial has no common factor, neither of its factors will, either.* For example, $(x + 2)(3x - 3)$ could not be the factors of $3x^2 - 7x - 6$ since $3x - 3$ has a common factor and the original trinomial $3x^2 - 7x - 6$ has no common factor.

Based on the results in Example 6 you can factor a trinomial of the form $ax^2 + bx + c$ as a product of two binomials in the following way.

To Factor a Trinomial of the form $ax^2 + bx + c$:

1. Write two parentheses and fill in any obvious information.
2. Identify all possible factors for the coefficients of the first term and the last term of the trinomial. Also, list all possible binomial factors of the trinomial and their resulting middle term.
3a. Select the correct factors. These are the factors whose resulting middle term is the same as the middle term of the trinomial.
3b. If the factors do not exist, write "not factorable."

In step 2, you need not always list *all* the possible binomial products. Instead, start listing the possible factors along with their resulting middle terms and then stop when you obtain the correct pair.

Example 7 Factor.

   a. $5x^2 - 23x - 10$ b. $7x^2 - 31x + 12$ c. $2x^2 + 3x + 3$
   d. $8y^2 + 14y - 9$ e. $6x^2 - 35x + 50$ f. $4z^2 + 12yz - 7y^2$

Solution   **7a. Step 1:** The only possible factors of $5x^2$ are $5x$ and x.

$$5x^2 - 23x - 10 = (5x\ \ )(x\ \ )$$

Step 2: The factors of -10 are -2 and 5, 2 and -5, 1 and -10, and -1 and 10.

Possible binomial factors	Resulting middle term	
$(5x - 2)(x + 5)$	$23x$	Sign is wrong

Since you want a middle term of $-23x$ (not $23x$), interchange the signs of the last terms of the binomials to obtain the correct factors:

Possible binomial factors	Resulting middle term	
$(5x + 2)(x - 5)$	$-23x$	← Correct middle term

Step 3: $5x^2 - 23x - 10 = (5x + 2)(x - 5)$

7b. Step 1: The only possible factors of $7x^2$ are $7x$ and x.

$$7x^2 - 31x + 12 = (7x\ )(x\ )$$

Step 2: Since the last term is positive and the middle term is negative, both factors of the last terms of the binomials must be negative. Therefore, the possible factors of 12 are -2 and -6, -3 and -4, and -1 and -12.

Possible binomial factors	Resulting middle term	
$(7x - 2)(x - 6)$	$-44x$	
$(7x - 6)(x - 2)$	$-20x$	
$(7x - 3)(x - 4)$	$-31x$	← Correct middle term

Step 3: $7x^2 - 31x + 12 = (7x - 3)(x - 4)$

7c. Step 1: The only possible factors of $2x^2$ are $2x$ and x.

$$2x^2 + 3x + 3 = (2x\ )(x\ )$$

Step 2: Since the last term is positive and the middle term is positive, the signs of both factors of the last term must be positive. Therefore, the possible factors of 3 are 1 and 3.

Possible binomial factors	Resulting middle term
$(2x + 3)(x + 1)$	$5x$
$(2x + 1)(x + 3)$	$7x$

Step 3: Since none of the possible factors yields a middle term of $3x$, the trinomial is prime with respect to the integers and you should write "not factorable."

7d. Step 1: Since both $8x^2$ and -9 have several possible factors, start by writing

$$8y^2 + 14y - 9 = (\quad)(\quad)$$

Step 2: The factors of $8x^2$ are $2x$ and $4x$, and $8x$ and x. The factors of -9 are 3 and -3, 1 and -9, and -1 and 9.

Possible binomial factors	Resulting middle term	
$(2y + 3)(4y - 3)$	$6y$	
$(2y + 9)(4y - 1)$	$34y$	
$(2y - 1)(4y + 9)$	$14y$	← Correct middle term

Step 3: $8y^2 + 14y - 9 = (2y - 1)(4y + 9)$

7e. Step 1: Since $6x^2$ and 50 have several factors, begin by writing

$$6x^2 - 35x + 50 = (\quad)(\quad).$$

Step 2: Since the last term of the trinomial is positive and the coefficient of the middle term is negative, both factors of the last term of the trinomial must be negative. Therefore, the possible factors of 50 are -2 and -25; -5 and -10; and -1 and -50. The factors of $6x^2$ are $3x$ and $2x$; and x and $6x$.

Possible binomial factors	Resulting middle term	
$(3x - 5)(2x - 10)$	$-40x$	This could not be correct since the second binomial has a common factor, while the trinomial does not.
$(3x - 10)(2x - 5)$	$-35x$	← Correct middle term

Step 3: $6x^2 - 35x + 50 = (3x - 10)(2x - 5)$

7f. Step 1: Although this trinomial is not of the form $ax^2 + bx + c$, you can still factor it by using the techniques presented in this section.

$$4z^2 + 12yz - 7y^2 = (\quad)(\quad)$$

Step 2: The possible factors of $4z^2$ are $2z$ and $2z$, or $4z$ and z. The possible factors of $-7y^2$ are y and $-7y$, or $-y$ and $7y$.

Possible binomial factors	Resulting middle term	
$(2z - y)(2z + 7y)$	$12yz$	← Correct middle term

Step 3: $4z^2 + 12yz - 7y^2 = (2z - y)(2z + 7y)$

When factoring trinomials of the form $ax^2 + bx + c$, we will use the trial-and-error method. However, with practice you can make intelligent guesses and eliminate the possibilities in a systematic manner. Also, remember to do as many of the steps as possible mentally.

Practice Problem 5 **Factor.**

a. $3x^2 + 16x + 5$ b. $6x^2 - 13x - 5$ c. $6x^2 + 5x - 6$
d. $8x^2 - 6x - 9$ e. $2x^2 - 17x + 5$ f. $12x^2 + xy - 20y^2$

IDEA 3 Sometimes we will need to factor out a common monomial factor before factoring a trinomial as a product of two binomials.
For example, to factor $3x^3 + 3x^2 - 90x$, first factor out the GCF of $3x$.

$$3x^3 + 3x^2 - 90x = 3x(x^2 + x - 30)$$

Next, factor the trinomial $x^2 + x - 30$.

$$3x^3 + 3x^2 - 90x = 3x(x^2 + x - 30)$$
$$= 3x(x + 6)(x - 5)$$

Example 8 **Factor.**

a. $2x^2 - 20x + 48$ b. $24x^3 - 54x^2 + 30x$

Solution When factoring trinomials, first factor out the GCF (if it exists) and then factor the trinomial obtained when you divide each term of the original trinomial by the GCF. The complete factorization will be the product of the GCF and the binomial factors.

8a. The GCF of $2x^2 - 20x + 48$ is 2.
$$2x^2 - 20x + 48 = 2(x^2 - 10x + 24)$$
Now, factor $x^2 - 10x + 24$.
$$x^2 - 10x + 24 = (x - 6)(x - 4)$$
Finally, $2x^2 - 20x + 48 = 2(x - 6)(x - 4)$

8b. The GCF of $24x^3 - 54x^2 + 30x$ is $6x$.
$$24x^3 - 54x^2 + 30x = 6x(4x^2 - 9x + 5)$$
Now, factor $4x^2 - 9x + 5$ by the trial-and-error method.
$$4x^2 - 9x + 5 = (4x - 5)(x - 1)$$
Finally, $24x^3 - 54x^2 + 30x = 6x(4x - 5)(x - 1)$

Practice Problem 6 *Factor.*

a. $8x^5 - 2x^4 - 6x^3$ **b.** $2x^2 + 30x + 112$

10.2 Exercises

Factor.

1. $x^2 + 12x + 35$
2. $x^2 + 6x + 8$
3. $x^2 - 6x + 5$
4. $x^2 + 4x - 21$
5. $x^2 - x - 30$
6. $x^2 - 9x + 14$
7. $x^2 - 11x + 24$
8. $x^2 + 5x + 8$
9. $x^2 - 9x - 36$
10. $x^2 + 7x + 12$
11. $x^2 + 2x - 35$
12. $x^2 - 3x + 2$
13. $x^2 - 11x + 28$
14. $x^2 + 17x - 60$
15. $x^2 + 17x - 16$
16. $3x^2 + 10x + 3$
17. $5x^2 - 13x - 6$
18. $10x^2 + 37x + 7$
19. $6x^2 - x - 5$
20. $6x^2 + 7x - 3$
21. $2x^2 - 13x + 20$
22. $12x^2 - 7x - 10$
23. $3x^2 - 2x - 8$
24. $10x^2 + 11x - 6$
25. $20x^2 - 19x + 3$
26. $6x^2 + 23x + 20$
27. $12x^2 - 16x - 35$
28. $15x^2 + 2x - 8$
29. $36x^2 - 13x + 1$
30. $4x^2 - 24x + 35$
31. $y^2 + 2y - 15$
32. $2x^2 + 12x + 16$
33. $x^2 + x - 6$
34. $x^2 - 7x + 12$
35. $a^2 - 3a - 18$
36. $6x^2 - 13x - 5$
37. $21x^2 - 29x + 10$
38. $x^2 - 9x + 12$
39. $3x^2 - 48x - 108$
40. $2x^2 - 22x + 20$
41. $27x^2 + 54x + 24$
42. $6x^2 - 13x + 6$
43. $7x^4 + 69x^2 - 10$
44. $12x^2 - 49x + 4$
45. $x^2 - 7xy - 18y^2$
46. $5x^2 - 3x - 14$
47. $3x^2 + 12x + 10$
48. $3x^2 + 10x + 8$
49. $12x^2 - 22x - 20$
50. $3x^3 - 30x^2 + 72x$
51. $x^7 - 5x^6 - 14x^5$
52. $6x^2 - 48x - 120$
53. $x^2 + x - 42$
54. $6y^5 - 36y^4 + 30y^3$
55. $8x^4 - 6x^2 - 9$
56. $28x^4 - 58x^3 - 30x^2$
57. $8x^2 + 13x + 5$
58. $12x^2 - 7x - 12$
59. $6x^2 - 7x - 20$
60. $8x^3 + 20x^2 - 12x$

61. $15x^2 - x - 6$
62. $40x^2 + x - 6$
63. $10x^2 - x - 24$
64. $4y^2 - 5y - 6$
65. $15x^4 - 22x^2 + 8$
66. $15a^2 - 7a - 4$

Fill in the blanks.

67. To factor a trinomial of the form $x^2 + px + q$ as a product of two binomials, do the following:
    a. **Step 1:** Make sure that the first terms of each binomial are the _____ and that their product equals the _____ of the trinomial.
    b. **Step 2:** Find two integers (if they exist) whose _____ is the same as the constant term (q) and whose _____ is the same as the _____ of the middle term (p) of the trinomial.
    c. **Step 3:** The _____ of the binomial factors are the integers found in step 2.
    d. Check the answer mentally. The _____ of the binomial factors must be the _____ as the original trinomial.

68. To factor a trinomial of the form $ax^2 + bx + c$ as a product of two binomials, do the following:
    a. **Step 1:** Write two parentheses and _____ in any obvious information.
    b. **Step 2:** Identify all possible factors for the _____ of the first term and the _____ of the trinomial. Also, list all possible binomial factors of the trinomial and their resulting _____.
    c. **Step 3a:** Select the correct factors. These are the factors whose resulting _____ is the same as the _____ of the _____.
    d. **Step 3b:** If the factors do not exist, write _____.

69. When factoring any trinomial, you should always look first for a _____ factor.

Factor.

70. $6x^2 + 5xy - 6y^2$
71. $25x^2 - 5xy - 2y^2$
72. $9x^2 + 18xy + 8y^2$
73. $4s^2 - 5s - 6$
74. $x^6 + 3x^3 - 7$
75. $16x^2 - 30x + 9$
76. $a^2 + 3ab - 10b^2$
77. $2a^2 + 4ab - 30b^2$
78. $6a^2 - 19ab + 15b^2$
79. $12a^4 - 7a^2b^2 - 10b^4$
80. $16a^4b^2 - 80a^3b^2 + 384a^2b^2$
81. $63c^2 - 31c - 10$

Answers to Practice Problems **3a.** $(x - 5)(x - 3)$ **b.** $(y - 5)(y + 2)$ **c.** $(x + 15)(x - 5)$ **4a.** $(x + 10)(x - 2)$ **b.** $(y - 5)(y - 4)$ **c.** $(x - 10)(x + 3)$ **5a.** $(3x + 1)(x + 5)$ **b.** $(3x + 1)(2x - 5)$ **c.** $(3x - 2)(2x + 3)$ **d.** $(4x + 3)(2x - 3)$ **e.** not factorable **f.** $(3x + 4y)(4x - 5y)$ **6a.** $2x^3(4x + 3)(x - 1)$ **b.** $2(x + 8)(x + 7)$

10.3 Special Factorization

IDEA 1 In Chapter 6, we discussed the following special products:

$$(a + b)(a - b) = a^2 - b^2$$
$$(a + b)^2 = a^2 + 2ab + b^2$$

Since factoring is the reverse of multiplication, we will now consider the following special factorizations:

$$a^2 - b^2 = (a + b)(a - b)$$
$$a^2 + 2ab + b^2 = (a + b)^2$$

The first special factorization involves factoring the difference of two squares, $a^2 - b^2$, as the sum times the difference of the same terms. For example, $x^2 - 25$ is the difference of two squares. Note that in the binomial $x^2 - 25$, x^2 and 25 are called **perfect squares** since they can be expressed as a quantity squared. Also, x and 5 are called the **square roots** of x^2 and 25 since they are the quantities that were squared to obtain x^2 and 25. Similarly, $16x^2 = (4x)^2$ is a perfect square and its square root is $4x$.

To factor $x^2 - 25$, we first observe that it is in the form $a^2 - b^2$, which is the difference of two squares.

$$x^2 - 25 = (x)^2 - (5)^2$$

Therefore, it can be factored as the product of the sum and the difference of the same terms. The first terms of each factor is the square root of the first term (x^2) of the binomial. Thus, the first term of each factor is x. The second term of each factor is the square root of the last term (25) of the binomial. Thus, the second term of each binomial is 5.

$$x - 25 = (x)^2 - 5^2$$
$$= (x + 5)(x - 5)$$

The above example suggests that you can factor the difference of two squares by using the following formula.

To Factor the Difference of Two Squares:

1. Find the square root of each square.
2. Write the first factor as the sum of the square roots.
3. Write the second factor as the difference of the square roots,

or

$$a^2 - b^2 = (a + b)(a - b)$$

Example 9 Factor.

   a. $x^2 - 36$ b. $4x^2 - 49$ c. $50x^2 - 2$ d. $x^4 - 16$

Solution To factor the difference of two squares, use the formula $a^2 - b^2 = (a + b)(a - b)$.

9a. $x^2 - 36 = x^2 - 6^2$ 9b. $4x^2 - 49 = (2x)^2 - 7^2$
$= (x + 6)(x - 6)$ $= (2x + 7)(2x - 7)$

9c. $50x^2 - 2 = 2(25x^2 - 1)$ 2 is a common factor
$= 2(5x + 1)(5x - 1)$

9d. $x^4 - 16 = (x^2)^2 - 4^2$
$= (x^2 + 4)(x^2 - 4)$ $x^2 - 4$ can be factored
$= (x^2 + 4)(x + 2)(x - 2)$

With practice you should be able to factor the difference of two squares mentally. Also, when one of the factors can still be factored, you should factor it. This process of factoring until none of the factors can be factored further is sometimes called **factoring completely**.

Practice Problem 7 *Factor.*

   a. $x^2 - 100$ b. $64x^2 - 1$ c. $2x^2 - 162$

IDEA 2

Another special factorization involves a perfect square trinomial, which is the square of a binomial. To recognize a perfect square trinomial, remember that:

1. The first and last terms must be perfect squares.
2. The middle term is twice the product of the square roots of the perfect squares ($2ab$) or its additive inverse ($-2ab$).

For example, $4x^2 + 20x + 25$ is a perfect square trinomial since $4x^2$ and 25 are perfect squares and the middle term $20x$ is twice the product of the square roots of $4x^2$ and 25.

To factor $4x^2 + 20x + 25$, we first observe that it is of the form $a^2 + 2ab + b^2$, which is a perfect square trinomial.

$$4x^2 + 20x + 25 = (2x)^2 + 2(2x \cdot 5) + (5)^2$$

Therefore, it can be factored as a binomial squared. The first term of the binomial is the square root of the first term of the trinomial ($4x^2$). The second term of the binomial is the square root of the last term of the trinomial (25). Since the middle term is positive, $4x^2 + 20x + 25$ is factored as the square of the sum of the square roots of the first and last terms of the trinomial.

$$4x^2 + 20x + 25 = (2x)^2 + 2(2x \cdot 5) + (5)^2$$
$$= (2x + 5)^2$$

Based on the above results, to factor a perfect square trinomial $a^2 + 2ab + b^2$ as a binomial squared $(a + b)^2$ use the following procedure.

To Factor a Perfect Square Trinomial:

1. Write the square root of the first term.
2. Write the sign of the middle term of the trinomial.
3. Write the square root of the last term and then square the resulting binomial,

or

$$a^2 + 2ab + b^2 = (a + b)^2$$

Example 10 Factor each perfect square trinomial.

    a. $x^2 + 18x + 81$ **b.** $16x^2 - 8x + 1$ **c.** $4x^2 + 10x + 25$

    d. $25x^2 - 30x + 9$ **e.** $x^2 - 14x - 49$

Solution To factor a perfect square trinomial, use the formula $a^2 + 2ab + b^2 = (a + b)^2$.

10a. $x^2 + 18x + 18 = x^2 + 2(x \cdot 9) + 9^2$ is a perfect square trinomial. Therefore,

$$x^2 + 18x + 81 = (x + 9)^2$$

10b. $16x^2 - 8x + 1 = (4x)^2 - 2(4x \cdot 1) + 1^2$ is a perfect square trinomial. Therefore,

$$16x^2 - 8x + 1 = (4x - 1)^2$$

10c. $4x^2 + 10x + 25$ is not a perfect square trinomial since the middle term $10x$ is not twice the product of the square roots of $4x^2$ and 25. Therefore, it cannot be factored as a binomial squared.

10d. $25x^2 - 30x + 9 = (5x)^2 - 2(5x \cdot 3) + 3^2$ is a perfect square trinomial. Therefore,

$$25x^2 - 30x + 9 = (5x - 3)^2$$

10e. $x^2 + 14x - 49$ is not a perfect square trinomial since the last term is negative. Remember that in a perfect square trinomial, the coefficients of the first and last terms must be positive.

Practice Problem 8 *Factor each perfect square trinomial.*

a. $x^2 - 6x + 9$ **b.** $9x^2 + 15x + 25$ **c.** $4x^2 + 12x + 9$

IDEA 3

Until now we have discussed special factorizations of binomials and trinomials. Let us now consider a special factorization technique called *factoring by grouping*, which involves polynomials containing four terms. In this technique, the terms of a polynomial are rearranged into smaller groups that are factorable. For example, to factor

$$x^2 + 4x + xy + 4y$$

we should observe that we can factor out an x from the first two terms and that we can factor out a y from the last two terms.

$$x^2 + 4x + xy + 4y = x(x + 4) + y(x + 4)$$

Now, we can factor out the common binomial factor $x + 4$.

$$x^2 + 4x + xy + 4y = x(x + 4) + y(x + 4)$$
$$= (x + 4)(x + y)$$
$$\uparrow \frac{x(x+4)}{(x+4)} + \frac{y(x+4)}{(x+4)}$$

To Factor a Polynomial by Grouping:

1. Group together terms that have a common factor or groups that are special forms.
2. Factor the resulting groups.
3. Factor the resulting expression by factoring out the GCF.

Example 11 Factor.

a. $ax - ay + bx - by$ **b.** $6x^2 - 2xy - 9x + 3y$

c. $ab + b + a + 1$ **d.** $x^2 + 10x + 25 - 4y^2$

Solution To factor by grouping, first try to group together terms that have a common factor or groups that are special forms. Then, factor the resulting expressions by using the techniques discussed in this chapter.

11a. Although the four terms have no common factor, notice that the first two terms and the last two terms have a common factor.

$ax - ay + bx - by$
$= (ax - ay) + (bx - by)$ Rewrite as two groups
$= a(x - y) + b(x - y)$ Factor each group
$= (x - y)(a + b)$ Factor out GCF of $x - y$

11b. $6x^2 - 2xy - 9x + 3y$
$= (6x^2 - 2xy) - (9x - 3y)$ Rewrite as two groups
$= 2x(3x - y) - 3(3x - y)$ Factor each group
$= (3x - y)(2x - 3)$ Factor out GCF of $3x - y$

11c. $ab + b + a + 1$
$= (ab + b) + (a + 1)$ Rewrite as two groups
$= b(a + 1) + 1(a + 1)$ Factor each group
$= (a + 1)(b + 1)$ Factor out GCF of $a + 1$

11d. The first three terms are a perfect square trinomial.

$x^2 + 10x + 25 - 4y^2$
$= (x^2 + 10x + 25) - 4y^2$ Rewrite as two groups
$= (x + 5)^2 - (2y)^2$ Rewrite as difference of squares
$= [(x + 5) + 2y][(x + 5) - 2y]$ Factor difference of squares
$= (x + 2y + 5)(x - 2y + 5)$ Rearrange terms of factors

Practice Problem 9   **Factor.**

a. $ax + bx + ay + by$ **b.** $ay + a - y - 1$ **c.** $z^2 - x^2 + 4xy - 4y^2$

IDEA 4

In this chapter we have introduced several basic factoring techniques. At this point, it may be helpful to state a general strategy for factoring polynomials.

To Factor a Polynomial—A General Strategy:

1. Factor out all common factors.

2. Factor the resulting polynomial based on the number of terms it contains.
   Two Terms: Try factoring the polynomial as the difference of two squares.
   Three Terms: If it is a perfect square trinomial, use the formula $a^2 + 2ab + b^2 = (a + b)^2$. If it is not, factor the polynomial by using the methods discussed in Section 10.2.
   Four or More Terms: Factor by grouping.

3. Factor completely. If a factor containing two or more terms can be factored, factor it.

10.3 Exercises

Factor each binomial that is the difference of squares.

1. $x^2 - 9$
2. $x^2 - 16$
3. $9x^2 - 4$
4. $x^2 - 36$
5. $x^2 - 64$
6. $4y^2 - 100$
7. $9s^2 - 1$
8. $4x^4 - 9$
9. $x^4 - 16$
10. $x^2 - 12$
11. $x^2 + 9$
12. $4x^2 - 81$
13. $81x^2 - 1$
14. $s^4 - 25$
15. $25y^2 - 4$
16. $36x^2 - 25$

Factor each perfect square trinomial.

17. $x^2 + 4x + 4$
18. $x^2 + 2x + 1$
19. $a^2 - 10a + 25$
20. $x^2 - 14x + 49$
21. $y^2 - 8y + 16$
22. $x^2 - 20x + 100$
23. $x^2 + 8x - 16$
24. $9x^2 - 14x + 25$
25. $4x^2 + 20x - 25$
26. $x^2 - 6x + 9$
27. $9x^2 - 15x + 25$
28. $9x^2 + 30x + 25$
29. $x^2 + 16x + 64$
30. $16x^2 - 8x + 1$
31. $64x^2 - 24x + 9$

Factor.

32. $ap + bp + aq + bq$
33. $cx + cy + dx + dy$
34. $ax - cx - ay + cy$
35. $bx - b + ax - a$
36. $x^2 - y^2 + 2x + 2y$
37. $y^3 + 3y^2 + 4y + 12$
38. $b^2 + ab + b + a$
39. $y^3 - 3y^2 - 9y + 27$
40. $x^2 - 14x + 49 - 9y^2$
41. $z^2 - x^2 + 4xy - 4y^2$
42. $2xy + 3xz - 10y - 15z$
43. $6b^2 - 3bc - 14b + 7c$
44. $x^3 + 5x^2 + 3x + 15$
45. $x^2 + 8x + 16 - 25y^2$
46. $ax + ay + bx + by$
47. $bx + by + cx + cy$
48. $aw - bw - az + bz$
49. $cw - dw - cz + dz$
50. $ay + y - a - 1$
51. $by + y - b - 1$
52. $x^2 - y^2 + 4x - 4y$
53. $x^2 - y^2 + 5x - 5y$
54. $2x^2 - 6xy + 5x - 15y$
55. $4s^2 - 12st + 7s - 21t$
56. $x^2 - 4x + 4 - 25y^2$
57. $x^2 - 6x + 9 - 36y^2$
58. $x^2 - y^2 + 10y - 25$
59. $y^2 - x^2 - 12x - 36$
60. $a^2 - b^2 + a - b$
61. $x^2 - y^2 + x + y$

Fill in the blanks.

62. The difference of two squares $a^2 - b^2$ is factored as $(a + b)(a - b)$, where a and b are the _____ of _____ and _____.

63. To recognize a perfect square trinomial, remember that the first and last terms are _____. Also, the middle term is _____ the product of the square roots of the _____ or its _____.

64. To factor a perfect square trinomial, first write the _____ of the first term. Then, write the sign of the _____ of the trinomial. Finally, write the _____ of the last term and _____ the resulting binomial.

65. To factor by grouping, first try to group together terms that have a _____ factor or groups that are special forms. Then, _____ the resulting expression.

Factor.

66. $2x^3 - 32x$
67. $9x^2 + 48x + 64$
68. $x^4 - 81$
69. $xy - ay + xp - ap$
70. $6x^2 + 3xy + 2x + y$
71. $8x^2 - 24x + 18$
72. $18x^2 - 2y^2$
73. $32x^2 + 16x + 2$
74. $x^2 - z^2 + 10z - 25$
75. $3x^2 - 75$

Answers to Practice Problems **7a.** $(x + 10)(x - 10)$ **b.** $(8x + 1)(8x - 1)$ **c.** $2(x + 9)(x - 9)$ **8a.** $(x - 3)^2$
b. not a perfect square trinomial **c.** $(2x + 3)^2$ **9a.** $(x + y)(a + b)$ **b.** $(y + 1)(a - 1)$
c. $(x - 2y + z)(-x + 2y + z)$

Chapter 10 Summary

Important Terms

Factoring a polynomial is the process of rewriting an expression as the product of two or more factors. The process of rewriting

$$x^2 + 6x \quad \text{as} \quad x(x + 6)$$

and

$$x^2 + 5x + 6 \quad \text{as} \quad (x + 3)(x + 2)$$

is called factoring. [Section 10.1/Idea 1]

The **greatest common factor (GCF)** of a polynomial is the largest factor that divides each term of the polynomial. The GCF of $3x^3 + 6x^2$ is $3x^2$. Similarly, the GCF of $x(x + 2) + y(x + 2)$ is $x + 2$.

[Section 10.1/Idea 1]

An expression is **factored completely** when none of its factors can be factored further. The expression $6a^2 - 18a + 12$ is factored completely as $6(a - 2)(a - 1)$. Similarly, the expression $50x^2 - 2$ is factored completely as $2(5x + 1)(5x - 1)$. [Section 10.3/Idea 1]

A **perfect square** is a number that can be expressed as a quantity squared. A **square root** of a given number is a number whose square is equal to the given number. 36 is a perfect square, and 6 and -6 are square roots of 36. [Section 10.3/Idea 1]

Important Skills

Factoring Polynomials Containing a Common Monomial Factor

To factor a polynomial, (1) find the greatest common monomial factor (GCF) by writing the product of (a) the largest common factor of the numerical coefficients of the polynomial, and (b) the smallest power of every variable that appears in each term of the polynomial; (2) divide each term of the polynomial by the GCF. Write the expressions in (1) and (2) as a product. [Section 10.1/Idea 1]

Finding the GCF of the Coefficients of Polynomials

To find the GCF of the coefficients of a polynomial, write the prime factorization of each coefficient. The GCF is the product of the common factors. Each common factor is used the minimum number of times it occurs in any factorization. [Section 10.1/Idea 2]

Factoring Trinomials

To factor a trinomial of the form $x^2 + px + q$, make the first terms of each binomial factor the same so that their product equals the first term of the trinomial. Next, find two integers whose product is equal to the constant term q of the trinomial and whose sum is equal to the coefficient p of the middle term of the trinomial. The last terms of the binomial factors are these two integers. [Section 10.2/Idea 1]

To factor a trinomial of the form $ax^2 + bx + c$, write ( )( ) and fill in any obvious information. Next, identify all possible factors of the coefficients of the first term and the last term of the trinomial. Also, list all possible binomial factors of the trinomial and their resulting middle term. Select the correct factors (if they exist). These are the factors whose resulting middle term is the same as the middle term of the trinomial. [Section 10.2/Idea 2]

Factoring the Difference of Two Squares $a^2 - b^2 = (a + b)(a - b)$	To factor the difference of two squares $a^2 - b^2$, find the square root of each square. Write the first factor as the sum of the square roots; write the second factor as the difference of the square roots. [Section 10.3/Idea 1]
Factoring Perfect Square Trinomials $a^2 + 2ab + b = (a + b)^2$	To factor a perfect square trinomial $a^2 + 2ab + b$, write the square root of the first term of the trinomial. Next, write the sign of the middle term. Finally, write the square root of the last term of the trinomial and then square the resulting binomial. [Section 10.3/Idea 2]
Factoring Polynomials by Grouping	To factor by grouping, group together terms that have a common factor or groups that are special forms. Factor the resulting groups. Finally, factor the resulting expressions by factoring out the GCF. [Section 10.3/Idea 3]
Factoring Polynomials: A General Strategy	To factor a polynomial, first factor out all common factors. If the resulting polynomial is a binomial, try to factor it as the difference of two squares. If it is a trinomial, try to factor it as a perfect square trinomial or by using the procedure for the factoring $ax^2 + bx + c$. If it contains four or more terms, factor by grouping. Finally, make sure you have factored completely.

Chapter 10 Review Exercises

Factor completely.

1. $4x^2 + 12x + 9$
2. $x^2 - 3x - 10$
3. $x^2 - 11x + 30$
4. $x^4 - 81$
5. $2x^2 + 12x - 14$
6. $25x^2 - 10x + 1$
7. $xy - y - x - 1$
8. $b^3 - 2b^2 - 4b + 8$
9. $x^2 - 10$
10. $32x^2 - 2$
11. $4x^2 + 4x - 3$
12. $18x^6 - 24x^4 + 36x^3$
13. $60x^9 + 84x^8 - 96x^7 + 24x^4$
14. $4x^2 - 12x + 9 - 25y^2$
15. $2x^2 + 5x - 3$
16. $12x^2 - 7x - 12$
17. $6x^2 - 7x - 10$
18. $2x^2 + 22x + 28$
19. $64x^4 - 1$
20. $24x^2 - 10xy - 21y^2$
21. $9x^2 - 42x + 21$
22. $15x^4 - 27x^3 - 9$
23. $x^2 - 12xy + y^2$
24. $x^6 - 4$
25. $18x^2 - 3x - 10$
26. $2a^2 + 17a + 30$
27. $10x^2 + 9x + 2$
28. $3x^2 - 7xy - 20y^2$
29. $3x(a + b) + y(a + b)$
30. $9(x + y) - 4a(x + y)$
31. $x^2 + 9x + xy + 9y$
32. $x^2 - xy - 4x + 4y$
33. $8x^2 - 4xy - 6x + 3y$
34. $8x^3 - 25z^4$
35. $32x^3 - 16x^2 - 30x$
36. $36x^2 + 78x + 40$
37. $4x^3 + 32x^2 + 64x$
38. $a^3b - 25ab^3$
39. $128z^2 + 32z^2 + 2z$
40. $e(5x - 3) - f(5x - 3) + e(3y - 4) - f(3y - 4)$

Chapter 10 Test

Name: _____

Class: _____

Factor completely.
1. $15x^3y^4 - 10x^2y^3$
2. $6x^3y^2z^4 - 15x^2y^3z^2 + 12xy^2z^3$
3. $78x^3 - 130x^2 + 182x$
4. $x^2 - 6x + 12$
5. $x^2 - x - 12$
6. $x^2 - 8x - 20$
7. $2x^2 - 7x + 3$
8. $10x^2 + 7x - 12$
9. $12x^3 + 14x^2 - 40x$
10. $12x^2 - 47xy + 40y^2$
11. $2x^2 - x + 3$
12. $25y^2 - 4$
13. $2x^5 - 2xy^4$
14. $4x^2 - 12x + 9$
15. $x^2 + 6xy + 9y^2$
16. $3x^2 - 6xy + 2x - 4y$
17. $xy + x - y - 1$
18. $x^2 - 10x + 25 - 9y^2$
19. $12x^2 + 6xy + 4x + 2y$
20. $12x^4 + 5x^2 - 2$

1. _____
2. _____
3. _____
4. _____
5. _____
6. _____
7. _____
8. _____
9. _____
10. _____
11. _____
12. _____
13. _____
14. _____
15. _____
16. _____
17. _____
18. _____
19. _____
20. _____

11 Rational Expressions

Objectives The objectives for this chapter are listed below along with sample problems for each objective. By the end of this chapter you should be able to find the solutions to the given problems.

1. Reduce a rational expression *(Section 11.1/Idea 1)*.

    a. $\dfrac{x-2}{5x-10}$ b. $\dfrac{x^2+x-20}{x^2+2x-15}$

2. Multiply and divide rational expressions *(Section 11.2/Ideas 1–2)*.

    a. $\dfrac{x}{2x-6} \cdot \dfrac{4x-12}{x^2}$ b. $\dfrac{x^2-y^2}{x^2+2xy+y^2} \div \dfrac{x+y}{x-y}$

3. Add and subtract rational expressions *(Section 11.3/Ideas 1–2)*.

    a. $\dfrac{8x}{2x-y} - \dfrac{4y}{2x-y}$ b. $\dfrac{2}{x+2} + \dfrac{x}{x^2-x-6}$

4. Simplify a complex fraction *(Section 11.4/Ideas 1–2)*.

    a. $\dfrac{1-\dfrac{1}{x}}{\dfrac{1}{x^2}+\dfrac{1}{x}}$ b. $\dfrac{\dfrac{3}{x+2}-\dfrac{2}{x-1}}{x+2}$

5. Solve an equation containing rational expressions *(Section 11.5/Idea 1)*.

    a. $\dfrac{2}{x-1}+\dfrac{1}{3}=\dfrac{1}{x-1}$ b. $\dfrac{3}{x-1}+\dfrac{2}{x+1}=\dfrac{5}{x^2-1}$

6. Solve a work and a uniform motion problem *(Section 11.6/Ideas 1–2)*.

    a. Jill can paint a room in 3 hours, Ed in 4 hours, and Betty in 2 hours. How long will it take to paint the room if they work together?
    b. Car *A* travels 20 mph faster than Car *B*. If car *A* drives 240 miles in the same time that car *B* drives 200 miles, how fast is each car traveling?

11.1 Simplifying Rational Expressions

IDEA 1

In arithmetic, any number that can be written in the form $\frac{a}{b}$, where a and b are integers and $b \neq 0$, is called a **rational number** or **fraction.** In algebra, expressions that are written in the form $\frac{P}{Q}$, where P and Q are polynomials and $Q \neq 0$ are called **rational expressions** or **algebraic fractions.** For example,

$$\frac{2x^2}{9}, \frac{x-6}{x+6}, \frac{x^2+5x-1}{x-9}, \text{ and } x^2 \left(\text{or } \frac{x^2}{1}\right)$$

are rational expressions. Since division by zero is undefined the value of the polynomial in *the denominator of a rational expression must never be zero.* For example, x cannot be 6 in

$$\frac{x+3}{x-6}$$

since this would result in a denominator of zero. In general, we can determine the value(s) of the variable that will make an expression undefined by setting the denominator equal to zero and solving the resulting equation.

Rather than listing restrictions for every expression, we will assume that all denominators represent nonzero real numbers.

A rational expression represents a number for each value of the variable that does not make the denominator zero. For example, when $x = 2$

$$\frac{2x+6}{x} = \frac{2(2)+6}{2} = \frac{10}{2} = 5$$

For this reason, the properties of fractions also apply to rational expressions. For example, the *fundamental property of fractions* allows us to simplify fractions (rational numbers) and algebraic fractions (rational expressions) to their lowest terms.

Fundamental Property of Fractions

> Multiplying or dividing the numerator and denominator of a fraction by the same nonzero number does not change the value of the fraction,
>
> or
>
> $$\frac{a \cdot k}{b \cdot k} = \frac{a}{b}$$
>
> where b and k are not equal to zero.

The above property has two meanings:

1. Reading from left to right, $\frac{a \cdot k}{b \cdot k} = \frac{a}{b}$ implies that the value of a fraction is not changed if we divide both the numerator and denominator of a fraction by the same nonzero number or rational expression. We use this concept to simplify fractions.

Rational Expressions

Fraction	Algebraic Fraction
$\dfrac{6}{8} = \dfrac{\cancel{2}\cdot 3}{\cancel{2}\cdot 4} = \dfrac{3}{4}$	$\dfrac{3x+6}{3x} = \dfrac{\cancel{3}(x+2)}{\cancel{3}\cdot x} = \dfrac{x+2}{x}$

2. Reading from right to left, $\dfrac{a}{b} = \dfrac{a \cdot k}{b \cdot k}$ implies that the value of a fraction is not changed if we multiply both the numerator and denominator of a fraction by the same nonzero number or rational expression. This concept is used to write equivalent fractions.

Fraction	Algebraic Fraction
$\dfrac{3}{4} = \dfrac{3\cdot 2}{4\cdot 2} = \dfrac{6}{8}$	$\dfrac{x+2}{x} = \dfrac{3(x+2)}{3\cdot x} = \dfrac{3x+6}{3x}$

As stated earlier, we can use the first meaning of the fundamental property of fractions to reduce or simplify algebraic fractions. For example, to reduce $\dfrac{x^2 + 5x + 6}{x^2 - x - 6}$, we first factor the numerator and denominator.

$$\frac{x^2 + 5x + 6}{x^2 - x - 6} = \frac{(x+2)(x+3)}{(x+2)(x-3)}$$

Then, we divide both the numerator and denominator by their common factor $x + 2$.

$$\frac{x^2 + 5x + 6}{x^2 - x - 6} = \frac{\cancel{(x+2)}(x+3)}{\cancel{(x+2)}(x-3)} = \frac{x+3}{x-3}$$

The above example suggests the following procedure.

To Simplify a Rational Expression:

1. Factor the numerator and denominator.
2. Divide the numerator and denominator by any common factors.

Example 1 Reduce each expression.

a. $\dfrac{2x+3}{4x^2 + 12x + 9}$ b. $\dfrac{6x^2 + x - 15}{6x^2 - x - 12}$

c. $\dfrac{9x^3y^2 - 4xy^2}{24x^2y^2 + 13xy^2 - 2y^2}$

Solution To reduce a rational expression, factor the numerator and denominator. Then divide the numerator and denominator by any common factors.

1a. $\dfrac{2x+3}{4x^2 + 12x + 9} = \dfrac{\cancel{(2x+3)}}{\cancel{(2x+3)}(2x+3)} = \dfrac{1}{2x+3}$

1b. $\dfrac{6x^2 + x - 15}{6x^2 - x - 12} = \dfrac{\cancel{(2x-3)}(3x+5)}{\cancel{(2x-3)}(3x+4)} = \dfrac{3x+5}{3x+4}$

11.1 Simplifying Rational Expressions

1c. $\dfrac{9x^3y^2 - 4xy^2}{24x^2y^2 + 13xy^2 - 2y^2} = \dfrac{xy^2(9x^2 - 4)}{y^2(24x^2 + 13x - 2)}$

$= \dfrac{x \cdot \cancel{y^2} \cdot \cancel{(3x+2)}(3x - 2)}{\cancel{y^2} \cdot \cancel{(3x+2)}(8x - 1)}$

$= \dfrac{x(3x - 2)}{8x - 1}$ or $\dfrac{3x^2 - 2x}{8x - 1}$

 You can only divide out factors that are common to the *entire* numerator and denominator. Therefore, you *cannot* write

$$\dfrac{x + 4}{x} = 4 \quad \text{or} \quad 1 + 4 = 5$$

since x is not a factor of x and $x + 4$.

Practice Problem 1 *Reduce each expression.*

a. $\dfrac{2x^3 - 2xy^2}{4x^2 - 4xy}$ b. $\dfrac{5x^3y - 5xy}{3x^3 - 2x^2 - 5x}$

IDEA 2

When reducing rational expressions, it will be helpful to know that *the quotient of any polynomial and its additive inverse is -1*. For example, $x - 6$ and $6 - x$ are additive inverses and

$$\dfrac{x - 6}{6 - x} = \dfrac{\cancel{(x-6)}}{-1\cancel{(x-6)}} = \dfrac{1}{-1} = -1$$

In general,

$$\dfrac{x - y}{y - x} = -1$$

Example 2 Reduce.

$$\dfrac{6x^2 - x^3}{2x^3 - 15x^2 + 18x}$$

Solution $\dfrac{6x^2 - x^3}{2x^3 - 15x^2 + 18x} = \dfrac{x^2(6 - x)}{x(2x^2 - 15x + 18)}$

$= \dfrac{\overset{x}{\cancel{x^2}}\overset{-1}{\cancel{(6-x)}}}{x(2x - 3)\cancel{(x-6)}}$ Think: $\dfrac{6 - x}{x - 6} = -1$

$= -\dfrac{x}{2x - 3}$

$= -\dfrac{x}{2x - 3}$

Practice Problem 2 *Reduce.*

$$\dfrac{10x - 15x^2}{6x^2 - 7x + 2}$$

11.1 Exercises

Reduce each rational expression.

1. $\dfrac{2}{8x + 4}$

2. $\dfrac{3}{6x - 3}$

3. $\dfrac{x^2 - 25}{x - 5}$

4. $\dfrac{y^2 - 36}{y - 6}$

5. $\dfrac{2x^2 + 3x - 2}{6x^3 - x^2 - x}$

6. $\dfrac{3x^2 + x + 2}{2x^3 + x^2 - 3x}$

7. $\dfrac{3x + 2}{6x^2 - 5x - 6}$

8. $\dfrac{2x - 5}{8x^2 - 14x - 15}$

9. $\dfrac{x^2 - 12x + 36}{x^2 - 36}$

10. $\dfrac{x^2 + 10x + 25}{x^2 - 25}$

11. $\dfrac{5a^3b - 5ab}{a^3 + 5a^2 + 4a}$

12. $\dfrac{3x^4y - 12x^2y}{x^4 - x^3 - 6x^2}$

13. $\dfrac{9x^2 - 1}{3x^2 + 11x - 4}$

14. $\dfrac{25y^2 - 4}{5y^2 + 22y + 8}$

15. $\dfrac{4x^2 - 11x + 6}{4x^2 + 5x - 6}$

16. $\dfrac{10x^2 + 19x + 6}{10x^2 - 16x - 8}$

17. $\dfrac{3y^3 - 6y^2 - 24y}{9y^2 + 27y + 18}$

18. $\dfrac{2y^3 + 3y^2 - 14y}{y^2z + 7yz - 18z}$

19. $\dfrac{x + 7}{x^2 + 15x + 56}$

20. $\dfrac{x + 8}{x^2 - x - 72}$

21. $\dfrac{x^3 - 64x}{x^3 + 2x^2 - 48x}$

22. $\dfrac{x^3 - 49x}{x^3 - x^2 - 56x}$

23. $\dfrac{x^2 - 9}{x^2 + 11x + 24}$

24. $\dfrac{x^2 - 49}{x^2 + 2x - 35}$

25. $\dfrac{2x^2 + 4x - 48}{2x^2 - 22x + 56}$

26. $\dfrac{3x^2 + 9x + 6}{3x^2 - 6x - 24}$

27. $\dfrac{n^2 + 3n - 40}{2n + 16}$

28. $\dfrac{n^2 - 2n - 63}{3n + 21}$

29. $\dfrac{9a + 27}{a^4 - 81}$

30. $\dfrac{7a - 14}{a^4 - 16}$

31. $\dfrac{x - 9}{9 - x}$

32. $\dfrac{5 - x}{x - 5}$

33. $\dfrac{x - 1}{5 - 5x}$

34. $\dfrac{8 - 2x}{x - 4}$

35. $\dfrac{3n - 2n^2}{4n^2 - 9}$

36. $\dfrac{2x - 3x^2}{9x^2 - 4}$

37. $\dfrac{x^3 + x^2 - 12x}{3x^2 - x^3}$

38. $\dfrac{x^4 + 2x^3 - 15x^2}{3x^3 - x^4}$

39. $\dfrac{9m^2 + 24mn + 16n^2}{9m^2 - 16n^2}$

40. $\dfrac{16x^2 - 40xy + 25y^2}{4x^2 - 25y^2}$

41. $\dfrac{16a^3b + 24a^2b^2 - 16ab^3}{24a^2b + 12ab^2 - 12b^3}$

42. $\dfrac{4a^2b + 8ab^2 - 12b^3}{18a^3b - 12a^2b^2 - 6ab^3}$

43. $\dfrac{xy - 3x - 2y + 6}{2y + 8 - xy - 4x}$

44. $\dfrac{3xy + 9x - 2y - 6}{2y + 10 - 3xy - 15x}$

45. $\dfrac{x^2 + 10x + 25 - 4y^2}{x^2 + 5x - 4y^2 - 10y}$

46. $\dfrac{x^2 - y^2 + 5x - 5y}{y^2 - x^2 - 10x - 25}$

Fill in the blanks.

47. Expressions that are written in the form $\dfrac{P}{Q}$, where P and Q are polynomials and $Q \ne 0$ are called _____ expressions or _____.

48. To reduce a rational expression to its lowest terms, _____ the numerator and denominator by any _____.

Answers to Practice Problems 1a. $\dfrac{x + y}{2}$ b. $\dfrac{5y(x - 1)}{3x - 5}$ 2. $-\dfrac{5x}{2x - 1}$

11.2 Multiplying and Dividing Rational Expressions

IDEA 1 Since the variables in a rational expression represent real numbers, the rules and procedures used to perform operations with rational expressions are the same as those for performing operations with rational numbers (fractions). For example, we multiply rational expressions in the same manner that we multiply fractions. That is, to multiply $\frac{2x-16}{6x}$ and $\frac{x^2}{x-8}$ we first indicate the product of the numerators over the product of the denominators (do this mentally).

$$\frac{2x-16}{6x} \cdot \frac{x^2}{x-8} = \frac{(2x-16)(x^2)}{(6x)(x-8)}$$

Next, we reduce the resulting expression to its lowest terms by factoring the numerator and denominator and then dividing both the numerator and denominator by their common factors.

$$\frac{2x-16}{6x} \cdot \frac{x^2}{x-8} = \frac{(2x-16)(x^2)}{(6x)(x-8)}$$
$$= \frac{\cancel{2}\cancel{(x-8)}(x \cdot x)}{(\cancel{2} \cdot 3 \cdot \cancel{x})\cancel{(x-8)}}$$
$$= \frac{x}{3}$$

The above example suggests the following rule for multiplying rational expressions.

To Multiply Rational Expressions:

1. Factor all polynomials.
2. Write the product of the numerators over the product of the denominators.
3. Reduce if possible,

or

$$\frac{P}{Q} \cdot \frac{R}{S} = \frac{P \cdot R}{Q \cdot S}$$

where Q and S are not equal to zero.

Example 3 Multiply.

a. $\dfrac{x-3}{5x} \cdot \dfrac{10x^3}{4x^2 - 11x - 3}$ b. $\dfrac{6x^2 + 7x - 5}{x^2 + 2x - 24} \cdot \dfrac{4x^2 + 21x - 18}{12x^2 + 11x - 15}$

c. $\dfrac{4x^2 - 9}{15x - 10x^2} \cdot \dfrac{4 + 4x - 15x^2}{6x^2 + 5x - 6} \cdot \dfrac{x^3}{x+2}$

Solution To multiply rational expressions, first factor all polynomials. Then, express the numerators as a product and the denominators as a product. Finally, simplify.

378 Rational Expressions

3a. $\dfrac{x-3}{5x} \cdot \dfrac{10x^3}{4x^2-11x-3} = \dfrac{\cancel{(x-3)}\cancel{(10x^3)}^{2x^2}}{(5x)\cancel{(x-3)}(4x+1)}$

$= \dfrac{2x^2}{4x+1}$

3b. $\dfrac{6x^2+7x-5}{x^2+2x-24} \cdot \dfrac{4x^2+21x-18}{12x^2+11x-15} = \dfrac{\cancel{(3x+5)}(2x-1)\cancel{(4x-3)}\cancel{(x+6)}}{\cancel{(x+6)}(x-4)\cancel{(3x+5)}\cancel{(4x-3)}}$

$= \dfrac{2x-1}{x-4}$

3c. $\dfrac{4x^2-9}{15x-10x^2} \cdot \dfrac{4+4x-15x^2}{6x^2+5x-6} \cdot \dfrac{x^3}{x+2}$

$= \dfrac{\cancel{(2x+3)}^{-1}\cancel{(2x-3)}^{-1}\cancel{(2-3x)}(2+5x)(\cancel{x^3})^{x^2}}{5 \cdot \cancel{x}(3-2x)\cancel{(2x+3)}\cancel{(3x-2)}(x+2)}$

$= \dfrac{x^2(2+5x)}{5(x+10)}$ or $\dfrac{2x^2+5x^3}{5x+10}$ Multiply remaining factors

Practice Problem 3 *Multiply.*

a. $\dfrac{14x}{y^2-4} \cdot \dfrac{2-y}{77x^3}$ b. $\dfrac{4x^2-4y^2}{12x^3y-6x^2y^2} \cdot \dfrac{2x^2+xy-y^2}{x^2+2xy+y^2}$

IDEA 2

We divide rational expressions in the same manner that we divide rational numbers. In other words, to divide one rational expression by another, we multiply the dividend by the reciprocal of the divisor. The reciprocal is the expression that results from interchanging the numerator and denominator.

$\dfrac{4x}{x^2-4} \div \dfrac{4x^2-12x}{x^2-5x+6} = \dfrac{4x}{x^2-4} \cdot \dfrac{x^2-5x+6}{4x^2-12x}$ Multiply the dividend by the reciprocal of the divisor

$= \dfrac{\cancel{(4x)}\cancel{(x-3)}\cancel{(x-2)}}{\cancel{(x-2)}(x+2)\cancel{(4x)}\cancel{(x-3)}}$ Factor and express the numerators and denominators as a product

$= \dfrac{1}{x+2}$ Divide numerators and denominators by common factors

To find the reciprocal of a rational expression, interchange the numerator and denominator. For example, the reciprocal of

$\dfrac{4x^2}{x-6}$ is $\dfrac{x-6}{4x^2}$.

The rule for dividing one polynomial by another is given below.

To Divide One Rational Expression by Another:

1. Multiply the dividend by the reciprocal of the divisor.
2. Reduce if possible,

or

$$\dfrac{P}{Q} \div \dfrac{R}{S} = \dfrac{P}{Q} \cdot \dfrac{S}{R} = \dfrac{P \cdot S}{Q \cdot R}$$

where Q, R, and S are not equal to zero.

Example 4 Divide.

a. $-\dfrac{14x^2y}{17xy^3} \div \dfrac{77xy}{51x^2y^2}$ b. $\dfrac{24x^2 - 4x}{4x^2 + 16} \div \dfrac{6x^2 + 11x - 2}{x^4 - 16}$

c. $\dfrac{48x^6 - 14x^5 - 12x^4}{64x^5 + 88x^4 + 24x^3} \div (2 - 3x)$

Solution To divide one rational expression by another, multiply the dividend by the reciprocal of the divisor.

4a. $-\dfrac{14x^2y}{17xy^3} \div \boxed{\dfrac{77xy}{51x^2y^2}} = -\dfrac{14x^2y}{17xy^3} \cdot \boxed{\dfrac{51x^2y^2}{77xy}}$

$= -\dfrac{\overset{2}{\cancel{14}} \cdot \overset{3}{\cancel{51}} \cdot \overset{x^2}{\cancel{x^4}} \cdot \overset{1}{\cancel{y^3}}}{\underset{1}{\cancel{17}} \cdot \underset{11}{\cancel{77}} \cdot \underset{1}{\cancel{x^2}} \cdot \underset{y}{\cancel{y^4}}}$ Rearrange factors and multiply powers

$= -\dfrac{6x^2}{11y}$

In step 2, if you did not recognize that the GCF of 14 and 77 is 7 and the GCF of 17 and 51 is 17, then factor.

$-\dfrac{14 \cdot 51 \cdot x^4 \cdot y^3}{17 \cdot 77 \cdot x^2 \cdot y^4} = -\dfrac{2 \cdot \cancel{7} \cdot 3 \cdot \cancel{17} \cdot \cancel{x^2} \cdot x^2 \cdot \cancel{y^3}}{\cancel{17} \cdot \cancel{7} \cdot 11 \cdot \cancel{x^2} \cdot \cancel{y^3} \cdot y} = -\dfrac{6x^2}{11y}$

4b. $\dfrac{24x^2 - 4x}{4x^2 + 16} \div \dfrac{6x^2 + 11x - 2}{x^4 - 16} = \dfrac{24x^2 - 4x}{4x^2 + 16} \cdot \dfrac{x^4 - 16}{6x^2 + 11x - 2}$

$= \dfrac{\cancel{4} \cdot x \cancel{(6x - 1)}\cancel{(x^2 + 4)}(x + 2)(x - 2)}{\cancel{4}\cancel{(x^2 + 4)}\cancel{(6x - 1)}\cancel{(x + 2)}}$

$= x(x - 2)$

4c. $\dfrac{48x^6 - 14x^5 - 12x^4}{64x^5 + 88x^4 + 24x^3} \div (2 - 3x) = \dfrac{48x^6 - 14x^5 - 12x^4}{64x^5 + 88x^4 + 24x^3} \cdot \dfrac{1}{(2 - 3x)}$

$= \dfrac{\overset{x}{\cancel{2}} \cdot \overset{}{\cancel{x^4}}\overset{-1}{\cancel{(3x - 2)}}\cancel{(8x + 3)}}{\underset{4}{\cancel{8}} \cdot \cancel{x^3}\cancel{(8x + 3)}(x + 1)\cancel{(2 - 3x)}}$

$= -\dfrac{x}{4(x + 1)}$

Practice Problem 4 *Divide.*

a. $-\dfrac{28xy^4}{36xy^2} \div \dfrac{35y^2}{24x^2y^3}$ b. $\dfrac{x^3 + 4x^2}{5x^2 + 20x} \div \dfrac{x^2 - 16}{x^2 - x - 12}$

IDEA 3   Let us now consider an example containing the operations of multiplication and division.

Example 5 Simplify.

$$\dfrac{4x^3 + 8x^2}{2x^2 + x - 15} \cdot \dfrac{2x - 5}{x^2 - x - 6} \div \dfrac{32x - 8}{4x^2 + 11x - 3}$$

380 Rational Expressions

Solution $\dfrac{4x^3 + 8x^2}{2x^2 + x - 15} \cdot \dfrac{2x - 5}{x^2 - x - 6} \div \dfrac{32x - 8}{4x^2 + 11x - 3}$

$= \dfrac{4x^3 + 8x^2}{2x^2 + x - 15} \cdot \dfrac{2x - 5}{x^2 - x - 6} \cdot \dfrac{4x^2 + 11x - 3}{32x - 8}$

$= \dfrac{\cancel{4}x^2(x + 2)(2x - 5)(x + 3)(4x - 1)}{(x + 3)(2x - 5)(x + 2)(x - 3)(\cancel{8})(4x - 1)}$

$= \dfrac{x^2}{2(x - 3)}$

11.2 Exercises

Multiply.

1. $\dfrac{6xy}{x + 5} \cdot \dfrac{x^2 - 25}{x^2 - 5x}$

2. $\dfrac{3xy}{2x - 3} \cdot \dfrac{4x^2 - 9}{2x^2 + 3x}$

3. $\dfrac{x^2 - x}{2x^2 + 6} \cdot \dfrac{8x - 4}{x^3 - x^2}$

4. $\dfrac{x^3 - 2x^2}{5x^2 + 20x} \cdot \dfrac{x^2 - 16}{x^2 - x - 12}$

5. $\dfrac{5a^2 + 33a - 14}{10a^2 + 21a - 10} \cdot \dfrac{2a^2 - 3a - 20}{2a^2 + 6a - 56}$

6. $\dfrac{3a^2 + 10a - 48}{3a^2 + 15a - 18} \cdot \dfrac{8a^2 + 2a - 10}{12a^2 - 17a - 40}$

7. $(2x + 3)\dfrac{8x^3 - 12x^4}{6x^2 + 5x - 6}$

8. $(3x + 2)\dfrac{18x - 6x^2}{3x^2 - 7x - 6}$

9. $\dfrac{a^2 + 4a}{a^2 + 4a + 4} \cdot \dfrac{a^2 - 4}{a^2 - 4a + 4}$

10. $\dfrac{x^2 - 4x}{x^2 - 4x - 12} \cdot \dfrac{x^2 + 5x + 6}{x^2 - x - 12}$

11. $\dfrac{x^2 - 4x - 5}{2x - 10} \cdot \dfrac{4x^2}{x^2 - 3x + 2}$

12. $\dfrac{x^2 + 3x - 10}{5x^2} \cdot \dfrac{x^2 + 10x + 25}{x^2 + 5x}$

13. $\dfrac{x^2 + 2x}{x^2 + 3x + 2} \cdot \dfrac{x + 2}{x}$

14. $\dfrac{x^2 - 2x}{x^2 - 3x + 2} \cdot \dfrac{x^2 - 2x}{2x}$

Divide.

15. $-\dfrac{42x^2y}{120xy^4} \div \dfrac{77x^4}{16x^2y^3}$

16. $-\dfrac{86xy^4}{15y^2} \div \dfrac{129x^5y}{10y^4}$

17. $\dfrac{x - 5}{6x} \div \dfrac{3x - 15}{2x^2}$

18. $\dfrac{16x^2}{3x + 2} \div \dfrac{4x}{6x + 4}$

19. $\dfrac{8x}{2x - 12} \div \dfrac{x^3}{x^2 - 36}$

20. $\dfrac{x^2 - 25}{x} \div \dfrac{3x + 15}{9x^2}$

21. $\dfrac{y^2 - 9y}{4y^2 + 28y} \div \dfrac{y^2 - 7y - 18}{y^2 + 9y + 14}$

22. $\dfrac{y^2 - y - 12}{y^2 - 10y + 24} \div \dfrac{5y^2 + 15y}{y^2 - 6y}$

23. $\dfrac{n + 2}{56n} \div \dfrac{n^2 - 2n - 8}{64n^3}$

24. $\dfrac{54x^2}{x + 2} \div \dfrac{72x^3}{x^2 - x - 6}$

25. $\dfrac{6x^2 - x - 2}{2x^2 - 5x + 3} \div \dfrac{3x^2 + x - 2}{2x^2 - 7x + 6}$

26. $\dfrac{6x^2 + 7x - 2}{4x^2 - 3x - 1} \div \dfrac{3x^2 + 7x + 2}{5x^2 - 3x - 2}$

27. $\dfrac{-6x^3 - 8x^2 + 8x}{3x^2 + 7x - 6} \div \dfrac{-2x^4 - 2x^3 + 12x^2}{x^3 + 4x^2 - 9x - 36}$

28. $\dfrac{-4x^4 + 8x^3 - 3x^2}{2x^2 - 11x + 12} \div \dfrac{-2x^3 - 7x^2 + 4x}{x^3 - 16x + 3x^2 - 48}$

Perform the indicated operations.

29. $\dfrac{n^4 - 81}{n^2 - 6n + 9} \div \dfrac{5n^2 + 8n - 21}{6n^2 - 11n - 21}$

30. $\dfrac{10x^3 + 25x}{20x + 10} \div \dfrac{x^5 - x}{2x^2 - x - 1}$

31. $\dfrac{21a^2 + 22a - 8}{5a^2 - 43a - 18} \cdot \dfrac{20a^2 - 7a - 6}{12a^2 + 7a - 12}$

32. $\dfrac{6y^2 - 35y + 25}{4y^2 - 11y - 45} \cdot \dfrac{24y^2 + 74y + 45}{18y^2 + 9y - 20}$

33. $\dfrac{x^2 - 25}{4x^2 - 20xy + 25y^2} \cdot \dfrac{6x^2 - 19xy + 10y^2}{6x^3 - 60x^2 + 150x} \cdot \dfrac{15x^3 - 3x^4}{2x^4 + 10x^3}$

34. $\dfrac{x^2 - 36}{9x^2 - 12xy + 4y^2} \cdot \dfrac{12x^2 - 11xy + 2y^2}{3x^6 - 36x^5 + 108x^4} \cdot \dfrac{12x - 2x^2}{4x^3 + 24x}$

35. $\dfrac{6x^3 + 15x^2 - 9x}{6x^3 + 12x^2 - 18x} \div \left[\dfrac{x^2 + 2x - 35}{x^2 - 6x + 5} \div \dfrac{2x^2 - 18x + 28}{4x^2 - 10x + 4}\right]$

36. $\dfrac{a^2 - 4}{a^2 - a - 6} \div \left[\dfrac{a^2 - a - 2}{2a^2 - 16a + 30} \div \dfrac{3a^2 - 9a - 30}{3a^3 + 9a^2 + 6a}\right]$

Fill in the blanks.

37. To multiply rational expressions, first _____ all polynomials. Then, express the numerators and denominators as a _____. Finally, simplify the resulting expression.

38. To divide one rational expression by another, multiply the dividend times the _____ of the _____.

Answers to Practice Problems 3a. $-\dfrac{2}{11x^2(y + 2)}$ b. $\dfrac{2(x - y)}{3x^2 y}$ 4a. $-\dfrac{8x^2 y^3}{15}$ b. $\dfrac{x(x + 3)}{5(x + 4)}$

11.3 Adding and Subtracting Rational Expressions

IDEA 1 We add and subtract rational expressions in the same manner that we add and subtract fractions. For example,

Fractions

$\dfrac{1}{8} + \dfrac{5}{8} = \dfrac{1 + 5}{8}$
$= \dfrac{6}{8}$
$= \dfrac{3}{4}$

Rational Expressions

$\dfrac{1}{8y} + \dfrac{5}{8y} = \dfrac{1 + 5}{8y}$
$= \dfrac{6}{8y}$
$= \dfrac{3}{4y}$

In other words, to add (or subtract) fractions or rational expressions, perform the indicated operations on the numerators and place this value over the common denominator. If possible, reduce the resulting expression.

The above example suggests the following rule for adding and subtracting rational expressions having a common denominator.

To Add or Subtract Rational Expressions Having a Common Denominator:

1. Perform the indicated operation on the numerators.
2. Place the result over the common denominator.
3. Reduce if possible,

or

$\dfrac{P}{Q} + \dfrac{R}{Q} = \dfrac{P + R}{Q}$ and $\dfrac{P}{Q} - \dfrac{R}{Q} = \dfrac{P - R}{Q}$

where Q is not equal to zero.

382 Rational Expressions

Example 6   Perform the indicated operation.

a. $\dfrac{x^2}{x+5} - \dfrac{25}{x+5}$ b. $\dfrac{11}{117x} + \dfrac{34}{117x}$ c. $\dfrac{x}{x^2-4} + \dfrac{2}{4-x^2}$

Solution   To add or subtract rational expressions having a common denominator, perform the indicated operation on the numerators and place the result over the common denominator. Reduce the resulting fraction.

6a. $\dfrac{x^2}{x+5} - \dfrac{25}{x+5} = \dfrac{x^2-25}{x+5}$
$= \dfrac{\cancel{(x+5)}(x-5)}{\cancel{(x+5)}}$
$= x - 5$

6b. $\dfrac{11}{117x} + \dfrac{34}{117x} = \dfrac{11+34}{117x}$
$= \dfrac{45}{117x}$ Divide numerator and denominator by 9
$= \dfrac{5}{13x}$

6c. $\dfrac{x}{x^2-4} + \dfrac{2}{4-x^2} = \dfrac{x}{x^2-4} + \left[\dfrac{2}{-(x^2-4)}\right]$
$= \dfrac{x}{x^2-4} + \left[-\dfrac{2}{x^2-4}\right]$
$= \dfrac{x-2}{x^2-4}$
$= \dfrac{\cancel{(x-2)}}{(x+2)\cancel{(x-2)}}$
$= \dfrac{1}{x+2}$

Practice Problem 5   **Perform the indicated operation.**

a. $\dfrac{x}{x^2-y^2} - \dfrac{y}{x^2-y^2}$ b. $\dfrac{2x^2}{2x^3-3x^2} + \dfrac{7x-15}{2x^3-3x^2}$

IDEA 2   To add or subtract rational expressions that have different denominators, it is necessary to first find the least common denominator (LCD). The LCD of two or more rational expressions is a polynomial that each denominator divides evenly. The LCD of a set of rational expressions is found in the same manner that we calculated the LCD for fractions. For example, to find the LCD for

$$\dfrac{5}{3x^2} + \dfrac{2}{x^2-7x}$$

we first factor the denominators $3x^2$ and $x^2 - 7x$.

$$3x^2 = 3 \cdot x^2$$
$$x^2 - 7x = x(x-7)$$

Next, we determine the LCD of $3x^2$ and $x^2 - 7x$ by forming the product of our distinct factors. Each factor is used the greatest number of times it occurs in any one factorization. That is,

3 occurs 1 time in any single factorization (3).
x occurs 2 times in any single factorization (x^2).
$x - 7$ occurs 1 time in any single factorization ($x - 7$).

Finally, the LCD is the product of these factors. Thus, the LCD of $3x^2$ and $x^2 - 7x$ is

$$3 \cdot x^2 \cdot (x - 7) = 3x^2(x - 7)$$

This procedure is summarized below.

To Find the LCD of Rational Expressions:

1. Factor each denominator.
2. The LCD is the product of the different factors. Each factor is used the greatest number of times it occurs in any one factorization.

Example 7 Find the LCD.

a. $\dfrac{5}{18x^3y} + \dfrac{1}{10xy^2}$ b. $\dfrac{2x}{x^2 - 4} - \dfrac{1}{x^2 - 3x + 2} + \dfrac{3x^2}{3x^2 - 3x}$

Solution

7a. $18x^3y = \boxed{2 \cdot 3 \cdot 3} \cdot \boxed{x^3} \cdot y$

$10xy^2 = 2 \cdot \boxed{5} \cdot x \cdot \boxed{y^2}$

LCD $= 2 \cdot 3 \cdot 3 \cdot 5 \cdot x^3 \cdot y^2 = 90x^3y^2$

7b. $x^2 - 4 = \boxed{(x + 2)}(x - 2)$

$x^2 - 3x + 2 = \boxed{(x - 2)}(x - 1)$

$3x^2 - 3x = \boxed{3x\,(x - 1)}$

LCD $= 3x(x + 2)(x - 2)(x - 1)$

IDEA 3

Having discussed how to find the LCD of a set of rational expressions, it will be helpful to review how to write equivalent fractions before stating a rule for adding and subtracting rational expressions.

Example 8 Find the missing numerator that makes the expressions equivalent.

a. $\dfrac{5}{18x^3y} = \dfrac{?}{90x^3y^2}$ b. $\dfrac{2x}{x^2 - 4} = \dfrac{?}{(x + 2)(x - 2)(x - 1)}$

Solution To change a rational expression to an equivalent expression having a different denominator, divide the original denominator into the new denominator and then multiply this quotient times the numerator and denominator of the original fraction.

8a. $\dfrac{5}{18x^3y} = \dfrac{5 \cdot \boxed{5y}}{18x^3y \cdot \boxed{5y}} = \dfrac{25y}{90x^3y^2}$ We multiplied the numerator and denominator by $5y$ since $90x^3y^2 \div 18x^3y = 5y$

8b. $\dfrac{2x}{x^2 - 4} = \dfrac{2x}{(x + 2)(x - 2)}$

$= \dfrac{2x \cdot \boxed{(x - 1)}}{(x + 2)(x - 2)\boxed{(x - 1)}}$ We multiplied the numerator and the denominator by $(x - 1)$ since $(x + 2)(x - 2)(x - 1) \div (x + 2)(x - 2) = (x - 1)$

$= \dfrac{2x^2 - 2x}{(x + 2)(x - 2)(x - 1)}$

Alternate Method

$$\frac{2x}{x^2-4} = \frac{2x}{(x+2)(x-2)} = \frac{2x \cdot (x-1)}{(x+2)(x-2)(x-1)} = \frac{2x^2-2x}{(x+2)(x-2)(x-1)}$$

$(x-1)$ is the missing factor

In this case, we multiplied both the numerator and denominator by $(x-1)$ since $(x-1)$ is the only factor of the new denominator $(x+2)(x-2)(x-1)$ that is missing from the original denominator $x^2 - 4 = (x+2)(x-2)$.

Practice Problem 6 **Find the LCD.**

a. $\dfrac{3}{28xy^2} + \dfrac{1}{12x^2}$ b. $\dfrac{4x}{x^2+3x-10} + \dfrac{5}{x^2+x-6}$

Find the missing numerator that makes the expressions equivalent.

c. $\dfrac{3}{28xy^2} = \dfrac{?}{84x^2y^2}$ d. $\dfrac{4x}{x-2} = \dfrac{?}{(x-2)(x+3)}$

IDEA 4

Having reviewed how to find the LCD for rational expressions and how to write equivalent expressions, we can now discuss how to add and subtract rational expressions having different denominators. For example, to find the sum of $\dfrac{2}{x^2-4}$ and $\dfrac{2}{4x+8}$, we first find the LCD for $x^2 - 4$ and $4x + 8$. That is,

$$x^2 - 4 = (x+2)(x-2)$$
$$4x + 8 = 4(x+2)$$
$$\text{LCD} = 4(x+2)(x-2)$$

Next, we rewrite $\dfrac{2}{x^2-4}$ and $\dfrac{2}{4x+8}$ as equivalent fraction having a denominator that equals the LCD.

$$\frac{2}{x^2-4} + \frac{2}{4x+8} = \frac{2}{(x+2)(x-2)} + \frac{2}{4(x+2)}$$
$$= \frac{4 \cdot 2}{4(x+2)(x-2)} + \frac{2 \cdot (x-2)}{4(x+2)(x-2)}$$
$$= \frac{8}{4(x+2)(x-2)} + \frac{2x-4}{4(x+2)(x-2)}$$

Now, we perform the indicated operation and simplify the resulting expression, if possible.

$$\frac{8}{4(x+2)(x-2)} + \frac{2x-4}{4(x+2)(x-2)}$$
$$= \frac{8+2x-4}{4(x+2)(x-2)}$$
$$= \frac{2x+4}{4(x+2)(x-2)}$$
$$= \frac{\overset{1}{\cancel{2}}(x+2)}{\underset{2}{\cancel{4}}\cancel{(x+2)}(x-2)}$$
$$= \frac{1}{2(x-2)}$$

11.3 Adding and Subtracting Rational Expressions   385

Let us now state a rule for adding and subtracting rational expressions having different denominators.

To Add or Subtract Rational Expressions Having Different Denominators:

1. Find the LCD.
2. Change each rational expression to an equivalent expression using the LCD as the new denominator.
3. Perform the indicated operation on the resulting expressions and write the answer in its simplest form.

Example 9 Perform the indicated operation.

a. $\dfrac{1}{39x^2y} + \dfrac{5}{26x^3}$ b. $\dfrac{3x}{x^2 - 9} - \dfrac{x + 2}{x^2 + 3x}$

c. $\dfrac{2a}{a^2 - 4} - \dfrac{1}{a^2 - 3a + 2} + \dfrac{a + 1}{a^2 + a - 2}$

Solution To add or subtract rational expressions, first find the LCD. Next, change each rational expression to an equivalent expression using the LCD as its denominator. Then, perform the indicated operation and simplify.

9a. $\dfrac{1}{39x^2y} + \dfrac{5}{26x^3}$

$= \dfrac{1 \cdot 2x}{39x^2y \cdot 2x} + \dfrac{5 \cdot 3y}{26x^3 \cdot 3y}$

$= \dfrac{2x}{78x^3y} + \dfrac{15y}{78x^3y}$

$= \dfrac{2x + 15y}{78x^3y}$

$39x^2y = 3 \cdot 13 \cdot x^2 \cdot y$
$26x^3 = 2 \cdot 13 \cdot x^3$
LCD $= 2 \cdot 3 \cdot 13 \cdot x^3 \cdot y = 78x^3y$

This can be done mentally

9b. $\dfrac{3x}{x^2 - 9} - \dfrac{x + 2}{x^2 + 3x}$

$= \dfrac{3x}{(x + 3)(x - 3)} - \dfrac{x + 2}{x(x + 3)}$

$= \dfrac{x \cdot 3x}{x(x + 3)(x - 3)} - \dfrac{(x + 2)(x - 3)}{x(x + 3)(x - 3)}$

$= \dfrac{3x^2}{x(x + 3)(x - 3)} - \dfrac{x^2 - x - 6}{x(x + 3)(x - 3)}$

$= \dfrac{3x^2 - (x^2 - x - 6)}{x(x + 3)(x - 3)}$

$= \dfrac{2x^2 + x + 6}{x(x + 3)(x - 3)}$

$x^2 - 9 = (x + 3)(x - 3)$
$x^2 + 3x = x(x + 3)$
LCD $= x(x + 3)(x - 3)$

Use ( ) to ensure that you subtract properly
Think: $3x^2 - x^2 + x + 6$

Since the numerator and denominator have no common factor, the expression cannot be simplified.

To divide a denominator into the LCD, first delete all factors that the denominator and the LCD have in common. Then, the product of the factors remaining in the LCD is the quotient of the LCD and the denominator. For example, in Solution 9b, we have that

$$x^2 - 9 = (x + 3)(x - 3)$$
$$x^2 + 3x = x(x + 3)$$
$$\text{LCD} = x(x + 3)(x - 3)$$

If you delete the factors that $x^2 - 9$ and the LCD have in common, a factor of x remains. This implies that

$$x(x + 3)(x - 3) \div (x^2 - 9) = x$$

9c. $\dfrac{2a}{a^2 - 4} - \dfrac{1}{a^2 - 3a + 2} + \dfrac{a + 1}{a^2 + a - 2}$

$= \dfrac{2a}{(a + 2)(a - 2)} - \dfrac{1}{(a - 2)(a - 1)} + \dfrac{a + 1}{(a - 1)(a + 2)}$ Factor denominators

The LCD $= (a + 2)(a - 2)(a - 1)$, please verify.

$= \dfrac{2a(a - 1)}{(a + 2)(a - 2)(a - 1)} - \dfrac{1(a + 2)}{(a + 2)(a - 2)(a - 1)} + \dfrac{(a + 1)(a - 2)}{(a + 2)(a - 2)(a - 1)}$

$= \dfrac{2a^2 - 2a}{(a + 2)(a - 2)(a - 1)} - \dfrac{a + 2}{(a + 2)(a - 2)(a - 1)} + \dfrac{a^2 - a - 2}{(a + 2)(a - 2)(a - 1)}$

$= \dfrac{2a^2 - 2a - (a + 2) + a^2 - a - 2}{(a + 2)(a - 2)(a - 1)}$

$= \dfrac{2a^2 - 2a - a - 2 + a^2 - a - 2}{(a + 2)(a - 2)(a - 1)}$

$= \dfrac{3a^2 - 4a - 4}{(a + 2)(a - 2)(a - 1)}$

$= \dfrac{(3a + 2)(a - 2)}{(a + 2)(a - 2)(a - 1)}$ Don't forget to simplify

$= \dfrac{3a + 2}{(a + 2)(a - 1)}$

Practice Problem 7   *Perform the indicated operation.*

a. $\dfrac{1}{18x} + \dfrac{2x}{24}$ b. $\dfrac{2x + 11}{x^2 + x - 6} - \dfrac{2}{x + 3} - \dfrac{3}{x - 2}$

11.3 Exercises

Perform the indicated operation.

1. $\dfrac{5}{111y} + \dfrac{28}{111y}$

2. $\dfrac{5}{91x} + \dfrac{9}{91x}$

3. $\dfrac{x}{x^2 - 4} - \dfrac{2}{x^2 - 4}$

4. $\dfrac{x}{x^2 - 4x + 4} - \dfrac{2}{x^2 - 4x + 4}$

5. $\dfrac{5y + 13}{3y^2} + \dfrac{y - 4}{3y^2}$

6. $\dfrac{6x^2 - 5}{10x^3} + \dfrac{2x^2 - 7}{10x^3}$

7. $\dfrac{4x}{x^2 + 5x + 6} - \dfrac{2x - 4}{x^2 + 5x + 6}$

8. $\dfrac{3x}{2x^2 + 5x - 3} - \dfrac{x + 1}{2x^2 + 5x - 3}$

9. $\dfrac{x}{x^2 - 9} + \dfrac{3}{9 - x^2}$

10. $\dfrac{2x}{4x^2 - 9} - \dfrac{3}{9 - 4x^2}$

Given the denominators of several fractions, find the LCD.

11. $21x^2$, $14x$, and 18

12. $42x^3y$, $28x^2$, and $63y^2$

13. $x^2 - 25, x^2 + 5x$
14. $4x^2 + 20x + 25, 2x^2 + 5x$
15. $2x - 3, 6x^2 - 5x - 6$
16. $4x^2 + 9x - 9, 4x + 3$
17. $x^3 + 5x^2 + 6x, x^2 + 4x + 3$
18. $x^3 - x^2 + 6x, x^2 - 6x + 8$
19. $4x^2 + 4x + 2, 2x^2 - 5x - 3$, and $x^4 - 3x^3$
20. $9x^2 - 12x + 4, 6x^2 - x - 2$, and $2x^4 + x^3$

Perform the indicated operation.

21. $\dfrac{x+5}{6} - \dfrac{x-1}{9}$

22. $\dfrac{3a-1}{9} - \dfrac{a+2}{12}$

23. $\dfrac{2x+8}{10} + \dfrac{4x-8}{12}$

24. $\dfrac{2x-6}{12} + \dfrac{4x-8}{16}$

25. $\dfrac{1}{6x} + \dfrac{5}{8x} + \dfrac{7}{18x}$

26. $\dfrac{3}{10y} + \dfrac{7}{15y} + \dfrac{5}{12y}$

27. $\dfrac{5}{36x^3y} - \dfrac{7}{66xy^2}$

28. $\dfrac{1}{18xy^3} - \dfrac{7}{15x^3y^2}$

29. $6x + \dfrac{1}{x^2}$

30. $8x + \dfrac{3}{x^3}$

31. $\dfrac{6x}{x^2 - 8x} + \dfrac{3}{x}$

32. $\dfrac{6}{x^2 - 4x} + \dfrac{4}{x}$

33. $\dfrac{x^2}{x^2 - 5x + 6} - \dfrac{1}{x-3}$

34. $\dfrac{x}{x^2 - x + 12} - \dfrac{2}{x+4}$

35. $\dfrac{1}{x-2} + \dfrac{2}{x+2}$

36. $\dfrac{1}{2x-1} + \dfrac{3}{2x+1}$

37. $\dfrac{4}{x^2 - 2x - 8} - \dfrac{x}{2x^2 + x - 6}$

38. $\dfrac{2x}{2x^2 + 3x - 2} - \dfrac{3}{x^2 - 3x - 10}$

39. $\dfrac{3a}{3a - b} - \dfrac{3a}{3a + b} - \dfrac{2b^2}{9a^2 - b^2}$

40. $\dfrac{x+6}{x^2 - 4} - \dfrac{4}{x+2} - \dfrac{2}{x-2}$

41. $2x - 3 + \dfrac{1}{x+2}$

42. $x + 2 + \dfrac{2}{x-2}$

43. $\dfrac{x}{2-x} + \dfrac{3}{x-2} - \dfrac{3x-2}{x^2 - 4}$

44. $\dfrac{x}{x-y} + \dfrac{y}{x+y} + \dfrac{x^2+y^2}{y^2-x^2}$

45. $\dfrac{n}{n^2 - 1} + \dfrac{2n^2}{n^4 - 1} - \dfrac{1}{n-1}$

46. $\dfrac{a^3}{a^4 - 16} - \dfrac{2}{a+2} + \dfrac{a}{a^2 - 4}$

47. $\dfrac{5}{x^2 + x - 6} - \dfrac{3}{x^2 + 3x} + \dfrac{2}{x^2 - 2x}$

48. $\dfrac{2x}{x^2 - 4} + \dfrac{1}{x^2 + 2x} - \dfrac{1}{x-2}$

49. $\dfrac{2x-7}{x^2 - 5x + 6} - \dfrac{2 - 4x}{x^2 - 6x + 9} + \dfrac{5x+2}{4 - x^2}$

50. $\dfrac{9x+2}{3x^2 - 2x - 8} - \dfrac{7}{4 - x - 3x^2}$

51. $\dfrac{x^2 - 2x - 2}{2x^2 - 13x + 6} - \dfrac{7x - 2}{2x^2 + x - 1} + \dfrac{x - 20}{x^2 - 5x - 6}$

52. $\dfrac{2a-1}{a^2 + a - 6} - \dfrac{3a - 5}{a^2 - 2a - 15} + \dfrac{2a - 3}{a^2 - 7a + 10}$

53. $\dfrac{3}{n^2 + 5n + 6} - \dfrac{2}{n^2 + 4n + 3} + \dfrac{4}{n^2 + n - 2}$

54. $\dfrac{2x + y}{x^2 - 3xy + 2y^2} - \dfrac{x + 4y}{x^2 - 4xy + 3y^2} - \dfrac{x - 7y}{x^2 - 5xy + 6y^2}$

Fill in the blanks.

55. To add or subtract rational expressions having a common denominator, perform the indicated operation on the _____ and place the result over the common denominator.

56. To add or subtract rational expressions having different denominators, first find the _____. Next, change each rational expression to an equivalent expression using the LCD as its _____. Then, perform the indicated operation and simplify.

Answers to Practice Problems **5a.** $\dfrac{1}{x + y}$ **b.** $\dfrac{x + 5}{x^2}$ **6a.** $84x^2y^2$ **b.** $(x - 2)(x + 3)(x + 5)$ **c.** $9x$ **d.** $4x(x + 3)$ **7a.** $\dfrac{2 + 3x}{36x}$ **b.** $-\dfrac{3}{x + 3}$

11.4 Complex Fractions

A **complex fraction** is a rational expression that has a fraction in the numerator, the denominator, or both. For example,

$$\dfrac{\tfrac{3}{5}}{\tfrac{6}{7}}, \quad \dfrac{\tfrac{4}{y}}{\tfrac{3y}{y^2 - 4}}, \quad \text{and} \quad \dfrac{\tfrac{x^2 - 9}{x + 1}}{\tfrac{x^2 + 5x + 6}{x + 3}}$$

are complex fractions.

IDEA 1 There are two procedures used to simplify complex fractions. The first method is to simplify the numerator and denominator of the complex fraction and then divide. For example,

$$\dfrac{\tfrac{5}{6} - \tfrac{1}{4}}{\tfrac{1}{3} + \tfrac{1}{6}} = \dfrac{\tfrac{10}{12} - \tfrac{3}{12}}{\tfrac{2}{6} + \tfrac{1}{6}} \quad \text{Perform the indicated operations in the numerator and denominator}$$

$$= \dfrac{\tfrac{7}{12}}{\tfrac{3}{6}}$$

$$= \dfrac{7}{12} \cdot \dfrac{6}{3}$$

$$= \dfrac{7 \cdot \cancel{6}}{\cancel{12} \cdot 3} \quad \text{Multiply the numerator by the reciprocal of the divisor}$$

$$= \dfrac{7}{6}$$

The first method for simplifying complex fractions is summarized on page 389.

To Simplify a Complex Fraction, Method 1:

1. Simplify the numerator and denominator of the complex fraction.
2. Multiply the numerator by the reciprocal of the denominator.

Example 10   Simplify.

$$\text{a. } \frac{\frac{56x}{15y}}{\frac{42xz}{90y^2}} \qquad \text{b. } \frac{\frac{1}{x} - \frac{1}{y}}{x - y}$$

Solution   To simplify a complex fraction, simplify its numerator and denominator first. Then, apply the division of fractions rule.

10a. $\dfrac{\frac{56x}{15y}}{\frac{42xz}{90y^2}} = \dfrac{56x}{15y} \div \dfrac{42xz}{90y^2}$

$= \dfrac{56x}{15y} \cdot \dfrac{90y^2}{42xz}$

$= \dfrac{56 \cdot 90 \cdot x \cdot y^2}{15 \cdot 42 \cdot x \cdot y \cdot z}$ Divide 15 and 90 by 15
Divide 56 and 42 by 7

$= \dfrac{8y}{z}$

10b. $\dfrac{\frac{1}{x} - \frac{1}{y}}{x - y} = \dfrac{\frac{y}{xy} - \frac{x}{xy}}{x - y}$

$= \dfrac{\frac{y - x}{xy}}{x - y}$

$= \dfrac{y - x}{xy} \cdot \dfrac{1}{x - y}$

$= \dfrac{(y - x)}{xy(x - y)}$

$= -\dfrac{1}{xy}$

Practice Problem 8   **Simplify.**

$$\text{a. } \frac{\frac{14x}{18y^2}}{\frac{42x^2}{15y}} \qquad \text{b. } \frac{\frac{x^2 - y^2}{15xy^2}}{-\frac{1}{6xy} + \frac{1}{6x^2}}$$

IDEA 2

The second method for simplifying complex fractions is an application of the fundamental property of fractions. In this method, we multiply both the numerator and denominator of the complex fraction by the LCD of all the fractions appearing in the

390 *Rational Expressions*

numerator and denominator of the complex fractions. For example, to simplify the complex fraction

$$\frac{\frac{5}{6} - \frac{1}{4}}{\frac{1}{3} + \frac{1}{6}}$$

we multiply the numerator and denominator of the fraction by the LCD of 6, 4, 3, and 6, which is 12.

$$\frac{\frac{5}{6} - \frac{1}{4}}{\frac{1}{3} + \frac{1}{6}} = \frac{12\left(\frac{5}{6} - \frac{1}{4}\right)}{12\left(\frac{1}{3} + \frac{1}{6}\right)} \quad \text{Multiply by LCD} = 12$$

$$= \frac{12\left(\frac{5}{6}\right) - 12\left(\frac{1}{4}\right)}{12\left(\frac{1}{3}\right) + 12\left(\frac{1}{6}\right)} \quad \text{Distributive property}$$

$$= \frac{10 - 3}{4 + 2}$$

$$= \frac{7}{6}$$

The second method for simplifying complex fractions is summarized in the box below.

To Simplify a Complex Fraction, Method 2:

1. Multiply the numerator and denominator of the complex fraction by the LCD of all the fractions appearing in the complex fraction.

2. Simplify the resulting expression.

Example 11 Simplify.

a. $\dfrac{\dfrac{x}{y} - \dfrac{y}{x}}{\dfrac{1}{x} + \dfrac{1}{y}}$ b. $\dfrac{\dfrac{2}{x} + \dfrac{2}{x-1}}{\dfrac{4}{x-1} - \dfrac{2}{x}}$

Solution To simplify a complex fraction, multiply the numerator and denominator of the complex fraction by the LCD of all the fractions appearing in the complex fraction. Then, simplify the results.

11a. $\dfrac{\dfrac{x}{y} - \dfrac{y}{x}}{\dfrac{1}{x} + \dfrac{1}{y}} = \dfrac{xy\left(\dfrac{x}{y} - \dfrac{y}{x}\right)}{xy\left(\dfrac{1}{x} + \dfrac{1}{y}\right)}$ Multiply by LCD = xy

$$= \dfrac{xy\left(\dfrac{x}{y}\right) + xy\left(-\dfrac{y}{x}\right)}{xy\left(\dfrac{1}{x}\right) + xy\left(\dfrac{1}{y}\right)}$$ Distributive property

$$= \frac{x^2 - y^2}{y + x}$$

$$= \frac{(x + y)(x - y)}{(y + x)}$$

$$= x - y$$

11b. The LCD of the fractions is $x(x - 1)$.

$$\frac{\frac{2}{x} + \frac{2}{x - 1}}{\frac{4}{x - 1} - \frac{2}{x}} = \frac{[x(x - 1)]\left[\frac{2}{x} + \frac{2}{x - 1}\right]}{[x(x - 1)]\left[\frac{4}{x - 1} - \frac{2}{x}\right]}$$

$$= \frac{x(x - 1)\left(\frac{2}{x}\right) + x(x - 1)\left(\frac{2}{x - 1}\right)}{x(x - 1)\left(\frac{4}{x - 1}\right) + x(x - 1)\left(-\frac{2}{x}\right)}$$

$$= \frac{(x - 1)(2) + (x)(2)}{(x)(4) + (x - 1)(-2)}$$

$$= \frac{2x - 2 + 2x}{4x - 2x + 2}$$

$$= \frac{4x - 2}{2x + 2}$$

$$= \frac{2(2x - 1)}{2(x + 1)}$$

$$= \frac{2x - 1}{x + 1}$$

Practice Problem 9 **Simplify.**

a. $\dfrac{\dfrac{3}{x} - \dfrac{7}{y^2}}{\dfrac{5}{xy} - \dfrac{2}{y}}$ **b.** $\dfrac{\dfrac{2}{a - b} + \dfrac{3}{a + b}}{\dfrac{5}{a + b} - \dfrac{1}{a^2 - b^2}}$

Example 12 **Simplify.**

$$3 + \frac{2x}{1 - \frac{1}{x}}$$

Solution First, simplify the complex fraction.

$$\frac{2x}{1 - \frac{1}{x}} = \frac{x(2x)}{x\left(1 \cdot \frac{1}{x}\right)} = \frac{2x^2}{x - 1}$$

Now, substitute $\dfrac{2x^2}{x - 1}$ for the complex fraction, and then perform the operation of addition.

$$3 + \frac{2x}{1 - \frac{1}{x}} = 3 + \frac{2x^2}{x - 1}$$

$$= \frac{3(x - 1)}{(x - 1)} + \frac{2x^2}{x - 1}$$

$$= \frac{3x - 3}{x - 1} + \frac{2x^2}{x - 1}$$

$$= \frac{3x - 3 + 2x^2}{x - 1}$$

$$= \frac{2x^2 + 3x - 3}{x - 1}$$

Practice Problem 10 *Simplify.*

$$2 + \frac{1 + \frac{1}{k}}{\frac{2}{k} - \frac{1}{k - 1}}$$

11.4 Exercises

Simplify each complex fraction.

1. $\dfrac{\frac{33}{84x^2y}}{\frac{22}{60xy^2}}$

2. $\dfrac{\frac{35}{16xy}}{\frac{21}{22x}}$

3. $\dfrac{\frac{5}{6} - \frac{3}{8}}{\frac{1}{2} + \frac{3}{4}}$

4. $\dfrac{\frac{3}{8} + \frac{1}{2}}{\frac{5}{28} - \frac{13}{14}}$

5. $\dfrac{\frac{2}{x} + \frac{3}{2x}}{5 + \frac{1}{x}}$

6. $\dfrac{\frac{3}{y} + \frac{1}{2y^2}}{3 - \frac{1}{y}}$

7. $\dfrac{\frac{1}{x} + \frac{1}{y}}{\frac{1}{x}}$

8. $\dfrac{\frac{1}{x}}{\frac{1}{x} - \frac{1}{y}}$

9. $\dfrac{\frac{x + 2y}{y^2}}{\frac{x^2 - 4y^2}{2y}}$

10. $\dfrac{\frac{25p^2 - q^2}{4p}}{\frac{5p + q}{7p}}$

11. $\dfrac{\frac{1}{x} - \frac{4}{y}}{\frac{x^2 - 16y^2}{xy}}$

12. $\dfrac{\frac{2}{a} - \frac{3}{b}}{\frac{4b^2 - 9a^2}{2a}}$

13. $\dfrac{2 + \frac{3}{x + 2}}{4 - \frac{1}{x + 2}}$

14. $\dfrac{5 + \frac{4}{a - 4}}{3 - \frac{2}{a - 4}}$

15. $\dfrac{\frac{x + 3}{x} - \frac{4}{x - 1}}{\frac{x}{x - 1} + \frac{1}{x}}$

16. $\dfrac{\frac{w + 2}{w} + \frac{1}{w + 2}}{\frac{w}{w + 2} + \frac{5}{w}}$

17. $\dfrac{\frac{3}{x} + \frac{2}{x - 1}}{4 - \frac{2}{1 - x}}$

18. $\dfrac{\frac{6}{a - 5} + \frac{5}{5 - a}}{\frac{3}{a} + \frac{2}{a - 5}}$

19. $\dfrac{\frac{x - a}{x + a} - \frac{x + a}{x - a}}{\frac{x^2 + a^2}{x^2 - a^2}}$

20. $\dfrac{\frac{x^2 + y^2}{x^2 - y^2}}{\frac{x + y}{x - y} + \frac{x - y}{x + y}}$

21. $\dfrac{\frac{1}{xy} + \frac{2}{yz} + \frac{3}{xz}}{\frac{2x + 3y + z}{xyz}}$

22. $\dfrac{\frac{x}{yz} - \frac{y}{xz} + \frac{z}{xy}}{\frac{1}{x^2y^2} - \frac{1}{x^2z^2} + \frac{1}{y^2z^2}}$

23. $\dfrac{-\frac{2}{a} - \frac{4}{a + 2}}{\frac{3}{a^2 + 2a} + \frac{3}{a}}$

24. $\dfrac{\frac{3}{c} - \frac{4}{cd - 2c}}{-\frac{1}{d - 2} + \frac{5}{c}}$

25. $\dfrac{-\frac{4}{x - 2y} + \frac{3}{x - y}}{\frac{8}{2y - x} + \frac{1}{x - y}}$

26. $\dfrac{\dfrac{2}{y-5x}+\dfrac{7}{y-x}}{\dfrac{1}{5x-y}-\dfrac{3}{x-y}}$

Simplify.

27. $2-\dfrac{1}{2+\dfrac{1}{y}}$

28. $1+\dfrac{3}{3-\dfrac{1}{2x}}$

29. $3-\dfrac{2}{1-\dfrac{1}{1+a}}$

30. $2+\dfrac{3}{1+\dfrac{2}{1-a}}$

31. $\dfrac{2x^{-1}-y^{-1}}{x^{-1}+5y^{-1}}$

32. $\dfrac{3x^{-1}+4y^{-1}}{6x^{-1}+3y^{-1}}$

Fill in the blanks.

33. A _____ is a rational expression that has a fraction in the numerator, the denominator, or both.

34. To simplify a complex fraction, multiply the numerator and denominator of the complex fraction by the _____ of all fractions appearing in the _____. Then simplify the results.

Answers to Practice Problems 8a. $\dfrac{5}{18xy}$ b. $-\dfrac{2x(x+y)}{5y}$ 9a. $\dfrac{3y^2-7x}{5y-2xy}$ b. $\dfrac{5a-b}{5a-5b-1}$ 10. $\dfrac{K^2+6K-5}{3K-2}$

11.5 Solving Equations Containing Rational Expressions

IDEA 1 In Chapter 6 we solved equations containing fractions whose denominators were constants (numbers). In this section we will solve equations containing fractions whose denominators contain a variable (or variables) by using the same method. In other words, to solve equations containing rational expressions, we multiply both sides of the equation by the LCD of all denominators in the equations. This procedure is used to clear the fractions from the equation and give us an equivalent equation that should be easier to solve. For example, to solve the equation

$$\dfrac{1}{x}+\dfrac{2}{3}=\dfrac{7}{x}$$

we first multiply both sides of the equation by the LCD of x, 3, and x, which is $3x$.

$$\dfrac{1}{x}+\dfrac{2}{3}=\dfrac{7}{x}$$

$3x\left(\dfrac{1}{x}+\dfrac{2}{3}\right)=3x\left(\dfrac{7}{x}\right)$ Multiply both sides by the LCD $= 3x$

$3x\left(\dfrac{1}{x}\right)+3x\left(\dfrac{2}{3}\right)=3x\left(\dfrac{7}{x}\right)$ Distributive property

$3+2x=21$ Multiply

$2x=18$

$x=9$

394 Rational Expressions

$$\text{Check:} \quad \frac{1}{x} + \frac{2}{3} = \frac{7}{x}$$

$$\frac{1}{9} + \frac{2}{3} \stackrel{?}{=} \frac{7}{9}$$

$$\frac{1}{9} + \frac{6}{9} \stackrel{?}{=} \frac{7}{9}$$

$$\frac{7}{9} = \frac{7}{9} \quad \text{True}$$

When solving equations containing rational expressions, always check your solution *(this check is not optional)*. Reject any solution that would make the denominator of an expression in the original equation zero since a denominator of zero implies division by zero, which is not permitted or is said to be undefined.

The procedure for solving equations containing rational expressions is summarized below.

To Solve an Equation Containing Rational Expressions:

1. Multiply both sides of the equation by the LCD of all terms of the equation.
2. Solve the resulting equation.
3. Check the solution obtained in step 2.

In step 1, when we multiply both sides of an equation by the LCD, we will obtain an equation equivalent to the original, provided we do not multiply by zero. However, if we do multiply by zero, we will obtain an equation that is not equivalent to the original and a solution that is not a solution of the original equation. Such solutions are called **extraneous roots,** and when they are substituted into the original equation they produce a denominator of zero, and division by zero is not permitted.

Example 13 Solve.

a. $\dfrac{5}{3x} - \dfrac{1}{9} = \dfrac{1}{x}$ b. $\dfrac{2}{y-5} + 6 = \dfrac{8}{y-5}$

c. $\dfrac{x}{x-2} + \dfrac{2}{3} = \dfrac{2}{x-2}$

Solution To solve an equation containing algebraic fractions, multiply both sides of the equation by the LCD and then solve the resulting equation. Always check your solution to ensure that the solution is not an extraneous root.

13a. The LCD is $9x$.

$$\frac{5}{3x} - \frac{1}{9} = \frac{1}{x}$$

$$9x\left(\frac{5}{3x} - \frac{1}{9}\right) = 9x\left(\frac{1}{x}\right)$$

$$9x\left(\frac{5}{3x}\right) + 9x\left(-\frac{1}{9}\right) = 9x\left(\frac{1}{x}\right)$$

$$15 - x = 9$$

$$-x = -6$$

$$x = 6$$

Check:
$$\frac{5}{3x} - \frac{1}{9} = \frac{1}{x}$$
$$\frac{5}{3 \cdot 6} - \frac{1}{9} \stackrel{?}{=} \frac{1}{6}$$
$$\frac{5}{18} - \frac{1}{9} \stackrel{?}{=} \frac{1}{6}$$
$$\frac{5}{18} - \frac{2}{18} \stackrel{?}{=} \frac{1}{6}$$
$$\frac{3}{18} \stackrel{?}{=} \frac{1}{6}$$
$$\frac{1}{6} = \frac{1}{6}$$

The solution is 6.

13b.
$$\frac{2}{y-5} + 6 = \frac{8}{y-5}$$
$$(y-5)\left(\frac{2}{y-5} + 6\right) = (y-5)\left(\frac{8}{y-5}\right) \quad \text{Multiply by LCD} = y - 5$$
$$(y-5)\left(\frac{2}{y-5}\right) + 6(y-5) = (y-5)\left(\frac{8}{y-5}\right) \quad \text{Distributive property}$$
$$2 + 6y - 30 = 8$$
$$6y - 28 = 8$$
$$6y = 36$$
$$y = 6$$

The solution is 6 (please check it).

13c.
$$\frac{x}{x-2} + \frac{2}{3} = \frac{2}{x-2}$$
$$3(x-2)\left[\frac{x}{x-2} + \frac{2}{3}\right] = 3(x-2)\left[\frac{2}{x-2}\right] \quad \text{Multiply both sides by the LCD} = 3(x-2)$$
$$3(x-2)\left(\frac{x}{x-2}\right) + 3(x-2)\left(\frac{2}{3}\right) = 3(x-2)\left(\frac{2}{x-2}\right)$$
$$3x + 2(x-2) = 6$$
$$3x + 2x - 4 = 6$$
$$5x - 4 = 6$$
$$5x = 10$$
$$x = 2$$

Check:
$$\frac{x}{x-2} + \frac{2}{3} = \frac{2}{x-2}$$
$$\frac{2}{2-2} + \frac{2}{3} \stackrel{?}{=} \frac{2}{2-2}$$
$$\frac{2}{0} + \frac{2}{3} \stackrel{?}{=} \frac{2}{0}$$

When 2 is substituted for x in the original equation, two of the rational expressions have denominators of zero, which implies that these expressions are undefined. Therefore, 2 is an extraneous root and the original equation has no solution.

At this point, you may be wondering how we obtained a value for x (we did not make a mistake) that is not the solution of the equation in Example 13c. The reason is that when we multiplied both sides of the equation by $3(x-2)$ we were actually multiplying both sides of the equation by zero. In other words, when $x = 2$ the expression $3(x-2) = 0$. Therefore, when we multiplied both sides by $3(x-2)$ the resulting equation $3x + 2(x - 2) = 6$ was not equivalent to the original equation, which means that x equal to 2 is not a valid solution.

Example 14 Solve.

$$\frac{2}{x - 1} - \frac{4}{3x} = \frac{1}{x^2 - x}$$

Solution First, factor each denominator to find the LCD. Then multiply both sides by this LCD.

$$\frac{2}{x - 1} - \frac{4}{3x} = \frac{1}{x^2 - x}$$

$$\frac{2}{x - 1} - \frac{4}{3x} = \frac{1}{x(x - 1)} \qquad \text{The LCD} = 3x(x - 1)$$

$$3x(x - 1)\left[\frac{2}{x - 1} - \frac{4}{3x}\right] = 3x(x - 1)\left[\frac{1}{x(x - 1)}\right] \quad \text{Multiply by } 3x(x - 1)$$

$$3x(x - 1)\left[\frac{2}{x - 1}\right] + 3x(x - 1)\left[-\frac{4}{3x}\right] = 3x(x - 1)\left[\frac{1}{x(x - 1)}\right] \quad \text{Distributive property}$$

$$6x + (x - 1)(-4) = 3$$
$$6x - 4x + 4 = 3$$
$$2x + 4 = 3$$
$$2x = -1$$
$$x = -\frac{1}{2}$$

Check:
$$\frac{2}{x - 1} - \frac{4}{3x} = \frac{1}{x^2 - x}$$

$$\frac{2}{-\frac{1}{2} - 1} - \frac{4}{3\left(-\frac{1}{2}\right)} \stackrel{?}{=} \frac{1}{\left(\frac{1}{2}\right)^2 - \frac{1}{2}}$$

$$\frac{4}{-1 - 2} - \frac{4}{-\frac{3}{2}} \stackrel{?}{=} \frac{1}{\frac{1}{4} + \frac{1}{2}}$$

$$\frac{4}{-3} - 4\left(-\frac{2}{3}\right) \stackrel{?}{=} \frac{4}{1 + 2}$$

$$-\frac{4}{3} + \frac{8}{3} \stackrel{?}{=} \frac{4}{3}$$

$$\frac{4}{3} = \frac{4}{3}$$

The solution is $-\frac{1}{2}$.

Practice Problem 11 *Solve.*

a. $\dfrac{1}{2} - \dfrac{6}{y} = \dfrac{3}{2y} - 1$ **b.** $\dfrac{5}{x - 4} - 4 = \dfrac{10}{2x - 8}$

11.5 Exercises

Solve the following equations.

1. $\dfrac{1}{x} + 2 = \dfrac{3}{x}$

2. $\dfrac{1}{x} = \dfrac{3}{x} - 2$

3. $\dfrac{3}{4x} - \dfrac{2}{x} = \dfrac{5}{12}$

4. $\dfrac{5}{2x} - \dfrac{2}{x} = -\dfrac{1}{12}$

5. $\dfrac{1}{y-2} + \dfrac{1}{2} = \dfrac{2}{y-2}$

6. $\dfrac{4}{y-4} - 2 = \dfrac{1}{y-4}$

7. $\dfrac{n}{n-3} - \dfrac{3}{2} = \dfrac{3}{n-3}$

8. $\dfrac{n}{n+5} + 2 = \dfrac{3n}{n+5}$

9. $\dfrac{3}{4x-8} - \dfrac{2}{3x-6} = \dfrac{1}{36}$

10. $\dfrac{4}{x+2} - \dfrac{1}{3x+6} = \dfrac{11}{9}$

11. $\dfrac{5}{6x+14} - \dfrac{2}{3x+7} = \dfrac{1}{56}$

12. $\dfrac{7}{2a+1} + \dfrac{3}{4a+2} = \dfrac{17}{2}$

13. $\dfrac{1}{x+5} + \dfrac{2}{x-5} = \dfrac{11}{x^2-25}$

14. $\dfrac{3}{y+2} - \dfrac{1}{y-2} = \dfrac{2}{y^2-4}$

15. $\dfrac{y+3}{y} - \dfrac{y+4}{y+5} = \dfrac{15}{y^2+5y}$

16. $\dfrac{3x}{x^2+2x-8} = \dfrac{5}{x-2} + \dfrac{2}{x+4}$

17. $\dfrac{5x-22}{x^2-6x+9} - \dfrac{11}{x^2-3x} = \dfrac{5}{x}$

18. $\dfrac{x-12}{x^2-10x+25} - \dfrac{3}{x^2-5x} = \dfrac{1}{x}$

19. $\dfrac{2y}{3y+3} - \dfrac{y+2}{6y+6} - \dfrac{y-6}{8y+8} = \dfrac{5}{12}$

20. $\dfrac{3x}{2x+10} + \dfrac{x}{3x+15} - \dfrac{x+3}{4x+20} = -\dfrac{7}{12}$

21. $\dfrac{2y-1}{y-5} = \dfrac{3y-2}{5-y} - 3$

22. $\dfrac{4-3y}{3y-1} + \dfrac{2y+3}{1-3y} = \dfrac{3}{2}$

23. $\dfrac{1}{x^2-x-2} - \dfrac{3}{x^2-2x-3} = \dfrac{1}{x^2-5x+6}$

24. $\dfrac{5x-22}{x^2-6x+9} - \dfrac{11}{x^2-3x} = \dfrac{5}{x}$

25. $\dfrac{11}{y^2-9} - \dfrac{7}{2y+6} = \dfrac{2}{y+3}$

26. $\dfrac{5}{a^2-2a-3} = \dfrac{4}{a^2-3a-4}$

27. $\dfrac{9}{n^2-3n+2} - \dfrac{2}{n-1} = \dfrac{1}{n-2}$

28. $\dfrac{5}{x^2-3x-10} + \dfrac{3}{x-5} = -\dfrac{1}{4x+8}$

29. $\dfrac{4n}{n-5} - 4 = \dfrac{5}{n}$

30. $\dfrac{5n}{n-3} - 5 = \dfrac{9}{n}$

Fill in the blanks.

31. To solve an equation containing rational expressions, multiply both sides of the equation by the _____, and then solve the resulting equation. Always check your solution to ensure that the solution is not an _____ root.

32. When we substitute a solution that is an extraneous root into the original equation we will obtain rational expressions that are _____.

In electronics, to calculate the resistances in a parallel circuit, we use the formula

$$\dfrac{1}{R_1} + \dfrac{1}{R_2} = \dfrac{1}{R_T}$$

where R_T is the total resistance of the circuit and R_1 and R_2 are the individual resistances. Use this information to solve the following problems.

33. The total resistance of two parallel resistors is 40 ohms. If one resistor has twice the resistance of the other, what is the resistance of each?

34. The total resistance of two parallel resistors is 30 ohms. If one resistor has five times the resistance of the other, what is the resistance of each?

Solve.

35. One number is four times as large as another. If the sum of their reciprocals is $\dfrac{3}{10}$, find the numbers.

36. The ratio of two readings from a gauge is $\frac{4}{5}$. If the reading in the numerator is 10 units above the normal value and the reading in the denominator is five units below the normal value, what is the normal reading of this gauge.

Answers to Practice Problems   **11a.** 5   **b.** no solution

11.6 Applications

IDEA 1   Many word problems involve two or more people or machines working on a task. These types of problems are called work problems. To solve them, we will use the following formula.

Work Formula

> Work done is equal to the rate of work times the time worked,
>
> or
>
> $$W = R \cdot T$$
>
> where W = work done, R = rate of work, and T = time worked.

It is important to note that the rate of work is expressed as work per unit of time. For example, if you can do a job in 5 hours, then

$$\text{rate of work} = \frac{1}{5} \text{ job per hour}$$

Example 15   Determine the rate of work R in each problem.

     a. Jim can paint 1 house in 5 days.

     b. Terry can keypunch 1000 cards in 4 hours.

Solution   Since $W = RT$, the rate of work R is $R = \frac{W}{T}$ or $\frac{\text{Workdone}}{\text{Time work}}$.

15a. $R = \frac{W}{T} = \frac{1 \text{ house}}{5 \text{ day}} = \frac{1}{5}$ house per day

15b. $R = \frac{W}{T} = \frac{1000 \text{ cards}}{4 \text{ hours}} = 250$ cards per hour

We will solve work problems with the same four-step procedure we used to solve word problems in Chapter 7 (see the box on page 399).

Example 16   Solve.

     a. Tommy can paint a barn in 7 hours while Greg can paint the same barn in 5 hours. How long will it take them to paint the barn if they work together?

     b. Working together, Dan and Nora can complete a project in 2 hours and 48 minutes. Working alone Dan can complete this project in 6 hours. How long would it take Nora to complete the project by herself?

To Solve a Work Problem:

1. Use the four-step procedure to solve word problems.

2. In step 2, writing an equation, the equation is usually based on the fact that:

The work A does + work B does = The work A and B do together.
It may be helpful to construct the following table of information:

Person	Rate (R)	Time (T)	Work (W) = $R \cdot T$

Solution

16a. Step 1: x = hours it takes the two people to paint the barn

Step 2: Tommy can paint the barn in 7 hours, which implies that he can paint $\frac{1}{7}$ of the barn in 1 hour. Greg can paint the barn in 5 hours, which suggests that he can paint $\frac{1}{5}$ of the barn in 1 hour. Thus, working together, they can do $\frac{1}{x}$ of the barn in 1 hour.

$$\underbrace{\text{Work done by Tommy in 1 hr.}}_{\frac{1}{7}} + \underbrace{\text{Work done by Greg in 1 hr.}}_{\frac{1}{5}} = \underbrace{\text{Work done together in 1 hr.}}_{\frac{1}{x}}$$

Step 3:
$$\frac{1}{7} + \frac{1}{5} = \frac{1}{x}$$
$$35x\left(\frac{1}{7} + \frac{1}{5}\right) = 35x\left(\frac{1}{x}\right)$$
$$5x + 7x = 35$$
$$12x = 35$$
$$x = \frac{35}{12} \quad \text{or} \quad 2\frac{11}{12}$$

Working together Tommy and Greg can paint the barn in $2\frac{11}{12}$ hours or 2 hours and 55 minutes.

Step 4: The answer checks since:
If Tommy and Greg work together, the entire barn should be painted in 2 hours and 55 minutes $\left(\frac{35}{12} \text{ hr}\right)$.

Work done by Tommy = $\frac{1}{7} \cdot \frac{35}{12} = \frac{5}{12}$ of barn

Work done by Greg = $\frac{1}{5} \cdot \frac{35}{12} = \frac{7}{12}$ of barn

Work done together = $\frac{12}{12}$ of barn (entire barn)

16b. Step 1: x = hours it takes Nora working alone

Step 2: Since Dan can complete the project in 6 hours, he does $\frac{1}{6}$ of the project each hour. Similarly, since Nora can complete the job in x hours, she does $\frac{1}{x}$ of the project each hour. Also, since the *rate of work* is per hour, the *time worked* must be expressed in hours. That is, 2 hours and 48 minutes = $2\frac{4}{5}$ hr.

Person	Rate (R) (in work per hour)	Time (T) (in hours)	Work (W) = R · T
Dan	$\frac{1}{6}$	$2\frac{4}{5} = \frac{14}{5}$	$\frac{1}{6} \cdot \frac{14}{5} = \frac{7}{15}$
Nora	$\frac{1}{x}$	$2\frac{4}{5} = \frac{14}{5}$	$\frac{1}{x} \cdot \frac{14}{5} = \frac{14}{5x}$

Work Dan does + Work Nora does = 1 completed project

$$\frac{7}{15} + \frac{14}{5x} = 1$$

Step 3:
$$\frac{7}{15} + \frac{14}{5x} = 1$$
$$15x\left(\frac{7}{15} + \frac{14}{5x}\right) = 15x(1) \quad \text{Multiply by LCD} = 15x$$
$$7x + 42 = 15x$$
$$42 = 8x$$
$$\frac{42}{8} = x$$
$$\frac{21}{4} = x$$

Nora can complete the project alone in $\frac{21}{4}$ hr or $5\frac{1}{4}$ hours (5 hours and 15 minutes).

Step 4: The answer checks since:

Work done by Dan = $\frac{1}{6} \cdot \frac{14}{5} = \frac{7}{15}$ of the project

Work done by Nora = $\frac{4}{21} \cdot \frac{14}{5} = \frac{8}{15}$ of the project

Work done together = $\frac{15}{15}$ of project (entire project)

Practice Problem 12 *Solve.*

Cazzie can complete a task in 9 hours, but when his brother Guerin helps him, it only takes them 6 hours. How many hours would it take Guerin working alone to complete the task?

IDEA 2

In some uniform motion problems, the equation will contain rational expressions. Recall the distance formula from Chapter 7, $d = r \cdot t$, or distance traveled is equal to the rate of travel times the time traveled.

Example 17   Solve.

Train A travels 630 miles in the same time that Train B travels 420 miles. If Train A travels 20 mph faster than Train B, find the rate of each train.

Solution   **Step 1:** $x = $ rate (mph) of train B
$x + 20 = $ rate (mph) of train A

Step 2: Use the fact that, since $d = r \cdot t$, $t = \dfrac{d}{r}$.

	Rate (in mph)	Time (in hours)	Distance (in miles)
Train A	$x + 20$	$\dfrac{630}{x + 20}$	630
Train B	x	$\dfrac{420}{x}$	420

Time Train A traveled = Time Train B traveled
$$\frac{630}{x + 20} = \frac{420}{x}$$

Step 3:
$$\frac{630}{x + 20} = \frac{420}{x}$$
$$x(x + 20)\left[\frac{630}{x + 20}\right] = x(x + 20)\left[\frac{420}{x}\right]$$
$$630x = 420x + 8400$$
$$210x = 8400$$
$$x = 40$$
$$x + 20 = 60$$

The rate of Train A is 60 mph, and the rate of Train B is 40 mph.

Step 4: The answer checks since:
Time traveled by each train must be the same.
Time Train A travels $= \dfrac{630 \text{ miles}}{60 \text{ mph}} = 10.5$ hr
Time Train B travels $= \dfrac{420 \text{ miles}}{40 \text{ mph}} = 10.5$ hr

Practice Problem 13   *Solve.*

A plane traveled 450 miles in the same time that a train travels 60 miles. If the plane travels 416 mph faster than the train, find the rate of each vehicle.

11.6 Exercises

Determine the rate of work. See Idea 1 and Example 15.

1. Jerry can type 300 words in 5 minutes.
2. Jean can repair 20 scooters in 8 hours.
3. An inlet pipe can fill a pool in 4 hours.
4. A painter can paint a house in 5 days.
5. Val can solve 35 problems in x minutes.
6. Vinnie can catch 8 dogs in x hours.

Solve.

7. Steve can clean a lot in 9 hours, while Tim can clean the same lot in 18 hours. How many hours would it take to clean the same lot if they work together?

9. A window washer can wash the window in a building in 112 hours. With an assistant he can do the job in 63 hours. How many hours would the assistant take to complete the job working alone?

11. Computer A can complete a job in 2 hours and computer B can complete the same job in 3 hours. How long would it take to complete half of the job if both computers worked together?

13. Printing presses A and B working together can print a newspaper in 2 hours. Press A working alone can do the job in 3 hours. How long would it take press B to do the job by itself?

15. Ernie drove 270 miles in the same time that Bert drove 250 miles. If Ernie averaged 4 mph more than Bert, find their rates.

17. An airplane flies 531 miles with the wind. In the same amount of time it can fly 369 miles against the wind. If the speed of the plane is 100 mph in still air, find the speed of the wind?

19. At 2:00 P.M., George started to repair a car, a job that would take him 9 hours to complete. At 3:00 P.M., Chris and Ricky assisted him and the 3 of them completed the job at 5:00 P.M. How many hours would it take Chris and Ricky to repair the car if they worked without George?

21. Ron can complete a project twice as fast as Reggie. When they work together the project can be completed in 15 days. How long would it take each man to complete the project alone?

23. Pipe A takes 30 minutes to fill a tank. After it has been running for 10 minutes, it is shut off and pipe B finishes filling the tank in 15 minutes. How long would it take pipe B to fill the tank alone?

25. The *point of no return* for an airplane flying over water from point A on land to point B on land is the distance for which it takes as much time to fly to point B as it does to return to point A. The distance between two points A and B is 2000 miles. If a plane leaves point A traveling 400 mph (in still air) and there is a tailwind of 50 mph, what is the point of no return.

8. Rochelle can type a term paper in 3 days, whereas Richard would take 5 days to type the same paper. How many days would it take them if they worked together?

10. Plane A can search an area for earthquake victims in 72 hours. If planes A and B can search the same area in 40 hours, how many hours would it take plane B to search the area working alone?

12. Inlet pipe A can fill a vat in 4 hours, while inlet pipe B can do the same job in 6 hours. How many hours would it take to fill three-fourths of the vat if both pipes are open?

14. A landscape contractor and his son can complete a task in 1 hour. The contractor working alone can complete the same task in 3 hours. How long would it take the son to do the job by himself?

16. Eloise covers 30 miles on a moped in the same time that Rita covers 20 miles on her scooter. If Eloise averages 5 mph more than Rita, find their rates.

18. The speed of a stream is 4 mph. If a boat travels 4 miles upstream in the same time it takes to travel 12 miles downstream, find the speed of the boat in still water.

20. Copying machine A can xerox 100 copies of a manuscript in 6 hours. Machine A begins operation at 8:00 A.M. and at 10:00 A.M. machine B is also used to help xerox the material. If both machines working together complete this task at 12:00 P.M., how long would it take machine B to xerox the 100 copies by itself?

22. Tracey can complete a task 3 times as fast as Monica. If they work together, it will take 1 hour to finish this task. How long will it take each of them to complete this job if they work alone?

24. A tank can be filled through pipe A, pipe B, or both pipes together. Pipe B takes three times as long as pipe A to fill the tank. However, if both pipes are used together, the task takes one minute less than if pipe A were used alone. How long does it take pipe A to fill the tank alone?

Answers to Practice Problems   **12.** 18 hours   **13.** plane's rate = 480 mph
                                                                                                               train's rate = 64 mph

Chapter 11 Summary

Important Terms

A **rational expression** or **algebraic fraction** is any expression that can be written in the form $\frac{P}{Q}$, where P and Q are polynomials and $Q \neq 0$. The expressions $\frac{5}{x}$, $\frac{x+2}{x^2+2x}$, and $\frac{3x^2}{6}$ are rational expressions. [Section 11.1/Idea 1]

The **fundamental property of fractions** says that multiplying or dividing the numerator of a fraction by the same non-zero number does not change the value of the fraction. $\frac{a}{b} = \frac{a \cdot k}{b \cdot k}$. [Section 11.1/Idea 1]

A **complex fraction** is a rational expression that has a fraction in the numerator, the denominator, or both. $\frac{\frac{3}{4}}{8}$, $\frac{\frac{x}{x^2+3}}{5}$, and $\frac{\frac{x^2-9}{x}}{\frac{x^2+5x+6}{x+2}}$ are complex fractions. [Section 11.4/Idea 1]

An **extraneous root** is a number that when substituted into the original equation produces a denominator of zero. The number 2 is an extraneous root of the equation $\frac{3}{x-2} = 6$. [Section 11.5/Idea 1]

The **work formula** says that the work done is equal to the rate of work times the time worked. $W = R \cdot T$. [Section 11.6/Idea 1]

Important Skills

Simplifying Rational Expressions

To simplify a rational expression, factor the numerator and denominator. Then, divide both the numerator and denominator by any common factors. [Section 11.1/Idea 1]

Multiplying Rational Expressions
$$\frac{P}{Q} \cdot \frac{R}{S} = \frac{P \cdot R}{Q \cdot S}$$

To multiply rational expressions, factor all polynomials. Then, write the product of the numerators over the product of the denominators and reduce. [Section 11.2/Idea 1]

Dividing Rational Expressions
$$\frac{P}{Q} \div \frac{R}{S} = \frac{P \cdot S}{Q \cdot R}$$

To divide rational expressions, multiply the dividend by the reciprocal of the divisor. [Section 11.2/Idea 2]

Adding or Subtracting Rational Expressions Having a Common Denominator
$$\frac{P}{Q} + \frac{R}{Q} = \frac{P+R}{Q}$$
$$\frac{P}{Q} - \frac{R}{Q} = \frac{P-R}{Q}$$

To add or subtract rational expressions having a common denominator, perform the indicated operation on the numerators and place the result over the common denominator. Reduce if possible. [Section 11.3/Idea 1]

To Find the LCD of Rational Expressions

To find the LCD, factor each denominator. The LCD is the product of the different factors. Each factor is used the greatest number of times it occurs in any one factorization. [Section 11.3/Idea 2]

Adding or Subtracting Rational Expressions Having Different Denominators	To add or subtract rational expressions, find the LCD. Change each rational expression to an equivalent expression using the LCD as denominator. Perform the indicated operation on the resulting expressions, and reduce if possible. [Section 11.3/Idea 4]
Simplifying Complex Fractions	To simplify a complex fraction by Method 1, simplify the numerator and denominator of the complex fraction; then, divide. [Section 11.4/Idea 1]
	To simplify a complex fraction by Method 2, multiply the numerator and denominator of the complex fraction by the LCD of all fractions appearing in the complex fraction; then, simplify. [Section 11.4/Idea 2]
Solving Fractional Equations	To solve an equation containing algebraic fractions, multiply both sides of the equation by the LCD of all terms in the equation. Then, solve the resulting equation and check your answer. [Section 11.5/Idea 1]
Solving Work Problems	To solve work problems, use the four-step procedure used to solve word problems. In step 2, writing an equation, the equation is usually based on the fact that:

The work A does + the work B does = the work A and B do together.

It may be helpful to construct the following table of information:

Person	Rate (R)	Time (T)	Work (W) = $R \cdot T$

[Section 11.6/Idea 1]

Chapter 11 Review Exercises

Simplify.

1. $\dfrac{117x^2y^3}{153x^4y^2}$

2. $\dfrac{x^2 - 3x - 10}{x^2 + x - 2}$

3. $\dfrac{8y^3 - 2y^2 - 3y}{12y^2 - 9y}$

4. $\dfrac{1 - \dfrac{1}{x + 1}}{1 + \dfrac{1}{x - 1}}$

5. $\dfrac{\dfrac{1}{x} + \dfrac{2}{y}}{\dfrac{2}{x} - \dfrac{1}{y}}$

6. $\dfrac{(x - y)^{-2}}{x^{-2} - y^{-2}}$

7. $\dfrac{n^2 + 4n + 4}{n^2 - n - 6} \cdot \dfrac{n^2 - 9}{n^2 + 5n + 6}$

8. $\dfrac{y^2 + y - 12}{y^2 + 8y + 16} \div \dfrac{y^2 - 16}{y + 4}$

9. $\dfrac{3n + 7}{n^2 + n - 12} - \dfrac{2n + 3}{n^2 + n - 12}$

10. $\dfrac{7}{x^2 + x - 12} + \dfrac{2}{x^2 - 8x + 15}$

11. $\dfrac{a + b}{18a^2b^3} \cdot \dfrac{9a^2b + 9ab^2}{a^2 - b^2} \div \dfrac{a^2 + 2ab + b^2}{36b^4}$

12. $\left(\dfrac{4}{5y + 25} + \dfrac{1}{5y - 25}\right) \div \dfrac{y^2 - y - 6}{y + 5}$

13. $\dfrac{y^2 - 5y - 6}{2y - 12} \div \dfrac{y^2 + 2y + 1}{8y^2}$

14. $\dfrac{1}{x} + \dfrac{2}{x + 2} - \dfrac{3}{x + 6}$

15. $\dfrac{5}{x^2 + x - 6} - \dfrac{3}{x^2 + 3x}$

16. $\dfrac{a^2 + ab - 2b^2}{a^2 + 4ab + 4b^2} \div (a^2 - b^2)$

17. $\dfrac{x^2 + 3x - 4}{3x - 15} \cdot \dfrac{x^2 - 25}{6x - 6}$

18. $\left(\dfrac{y}{3} - \dfrac{3}{y}\right) \div \dfrac{2y^2 - 6y}{y^2}$

20. $\dfrac{\dfrac{1}{xy^2} - \dfrac{1}{x^2 y}}{\dfrac{1}{xy}}$

21. $\dfrac{1 - \dfrac{49}{n^2}}{1 + \dfrac{10}{n} + \dfrac{21}{n^2}}$

22. $\dfrac{1 - \dfrac{11}{n} + \dfrac{24}{n^2}}{1 - \dfrac{64}{n^2}}$

23. $2 - \dfrac{y}{y-1}$

24. $\dfrac{x^2 - 4}{x^2 - 1} \cdot \dfrac{1 - x}{2x^2 + 4x}$

25. $\dfrac{8}{3 - 7x} - \dfrac{2}{7x - 3}$

Solve each equation.

26. $\dfrac{5}{6n} - \dfrac{2}{3} = \dfrac{7}{10n}$

27. $\dfrac{2}{x + 5} - \dfrac{16}{x^2 - 25} = \dfrac{1}{x - 5}$

28. $\dfrac{15}{x^2 + 5x} + \dfrac{x + 4}{x + 5} = \dfrac{x + 3}{x}$

29. $\dfrac{2}{n^2 - 4} - \dfrac{3}{n + 2} = \dfrac{4}{n - 2}$

30. $\dfrac{x}{x - 2} + 4 = \dfrac{2}{2 - x}$

31. $\dfrac{6}{x^2 - 9} = \dfrac{3}{x + 3}$

32. $\dfrac{9}{y^2 - 3y + 2} - \dfrac{2}{y - 1} = \dfrac{1}{y - 2}$

33. $\dfrac{6}{x^2 - 3x + 2} - \dfrac{2}{x^2 - 4x + 3} = \dfrac{8}{x^2 - 5x + 6}$

Solve.

34. Mike can complete a report in 14 hours. With the assistance of his wife, the report can be completed in 8 hours. How many hours would it take his wife to complete this job working alone?

35. Train A is traveling 12 mph slower than train B. If train A travels 230 miles in the same time it takes train B to travel 290 miles, find the speed of each train.

36. If an object moves a distance s and has an initial velocity of V_0 and a final velocity of V_f, then its acceleration a is

$$a = \dfrac{1 - \dfrac{V_0^2}{V_f^2}}{\dfrac{2s}{V_f^2}}$$

Find a when $V_0 = 40$, $V_f = 60$, and $s = 100$.

37. Cazzie and Guerin weigh a total of 160 pounds. The ratio of Cazzie's weight to Guerin's weight is 3 to 5. Find the weight of each boy.

Name: _____

Class: _____

Chapter 11 Test

Reduce.

1. $\dfrac{6x^2 - 6x}{12x^3 + 12x}$

2. $\dfrac{x^2 - 16}{x^2 + 8x + 16}$

Simplify.

3. $\dfrac{4}{9x} + \dfrac{7}{6x}$

4. $\dfrac{4a^2 - 12ab}{3x - 6y} \cdot \dfrac{6x - 12y}{24a - 72b}$

5. $\dfrac{5}{4x} - \dfrac{10}{2x^2 + 8x}$

6. $\dfrac{2x - 2y}{6y} \div \dfrac{x^2 - 2xy + y^2}{x^2 - y^2}$

7. $\dfrac{15}{3x - 9} - \dfrac{12x}{9 - 3x}$

8. $\dfrac{a - \dfrac{1}{b}}{b - \dfrac{1}{a}}$

9. $\dfrac{\dfrac{1}{x + 1} - 1}{\dfrac{1}{x + 1} + 1}$

10. $\dfrac{x^2 - 25}{x^2 - x - 12} \cdot \dfrac{3x - 3}{x^2 - x - 20} \div \dfrac{x^2 + 4x - 5}{x^2 - 16}$

Solve.

11. $\dfrac{8}{2y + 5} = 4 + \dfrac{2}{2y + 5}$

12. $\dfrac{2}{x - 2} - \dfrac{4}{x} = \dfrac{1}{2x}$

13. $\dfrac{7}{3x - 1} = \dfrac{2}{x + 2}$

1. _____

2. _____

3. _____

4. _____

5. _____

6. _____

7. _____

8. _____

9. _____

10. _____

11. _____

12. _____

13. _____

14. The denominator of a fraction is 4 less than twice its numerator. If the value of the fraction is three-eights, what are the numerator and denominator of the fraction?

15. Lorie can do a job in 5 hours, and Shirley can do the same job in 4 hours. How long would it take for them to do the job together?

14. numerator _____

    denominator _____

15. _____

12 Square Roots and Quadratic Equations

Objectives The objectives for this chapter are listed below along with sample problems for each objective. By the end of this chapter you should be able to find the solutions to the given problems.

1. Find the square root of a number (*Section 12.1/Idea 1*).
   a. $\sqrt{16}$ b. $-\sqrt{y^8}$

2. Find the decimal approximation of a square root (*Section 12.1/Idea 2*).
   a. $\sqrt{13}$ b. $\sqrt{97}$

3. Multiply square roots (*Section 12.2/Idea 1*).
   a. $\sqrt{2}\,\sqrt{11}$ b. $\sqrt{3}\,\sqrt{27}$

4. Simplify square roots (*Section 12.2/Idea 2*).
   a. $\sqrt{20x^3}$ b. $\sqrt{539}$

5. Divide one square root by another (*Section 12.2/Idea 3*).
   a. $\dfrac{\sqrt{32x}}{\sqrt{2}}$ b. $\dfrac{\sqrt{64x^7}}{\sqrt{x}}$

6. Rationalize the denominator of a fraction (*Section 12.2/Idea 5*).
   a. $\dfrac{1}{\sqrt{3}}$ b. $\dfrac{\sqrt{3}}{\sqrt{20}}$

7. Add and subtract square roots (*Section 12.3/Idea 1*).
   a. $\sqrt{3} - 4\sqrt{3}$ b. $2\sqrt{18} + 3\sqrt{50} - \sqrt{8}$

8. Solve a quadratic equation by using the square root method (*Section 12.4/Idea 1*).
   a. $(x + 3)^2 = 16$ b. $6x^2 - 8 = 1$

9. Solve a quadratic equation by factoring (*Section 12.5/Idea 1*).
   a. $3x^2 = 15x$ b. $6x^2 + 3x = 3$

10. Solve a quadratic equation by using the quadratic formula (*Section 12.6/Idea 1*).
    a. $x^2 + 6x = -9$ b. $2x^2 + 4x = 1$

12.1 Square Roots

IDEA 1

To determine the square root of a given number, our goal is to find a number whose square is the given number. For example, one square root of 25 is 5 since $5^2 = 25$. Another square root of 25 is -5 since $(-5)^2 = 25$. Therefore, 25 has two square roots. In general, every positive number has two square roots (one positive and one negative).

The **positive** or **principal square root** of a number is written with the symbol $\sqrt{\ }$. The negative square root of a number is written $-\sqrt{\ }$. The $\sqrt{\ }$ is called a **radical symbol** and it is also used to represent the square root of zero ($\sqrt{0} = 0$). For example,

$\sqrt{36}$ is read as "the square root of 36"
$-\sqrt{36}$ is read as "the negative square root of 36"

The expression under the radical symbol is called the **radicand**. The entire expression (radical symbol and radicand) is called a **radical**. For example, $\sqrt{36}$ is a radical and 36 is the radicand.

Example 1   Find the square root of each number.

     a. $\sqrt{16}$ **b.** $\sqrt{100}$ **c.** $-\sqrt{36}$ **d.** $-\sqrt{81}$ **e.** $\sqrt{-9}$

Solution   To find the square root of a number x, remember that K is the square root of x if $K^2 = x$.

   1a. $\sqrt{16} = 4$ since $4^2 = 16$ **1b.** $\sqrt{100} = 10$ since $10^2 = 100$

   1c. $-\sqrt{36} = -6$ **1d.** $-\sqrt{81} = -9$

   1e. $\sqrt{-9}$ does not exist. There is no real number that when squared is -9.

Practice Problem 1   **Find the square root of each number.**

     a. $\sqrt{49}$ **b.** $-\sqrt{9}$ **c.** $\sqrt{64}$ **d.** $\sqrt{25}$ **e.** $-\sqrt{4}$

Sometimes the radicand is a number raised to a power or a variable. For example,

$\sqrt{x^2} = x$ because $(x)^2 = x^2$
$\sqrt{x^2} = -x$ because $(-x)^2 = x^2$

Since the variable x could be a positive or a negative number, we cannot determine whether x or $-x$ is the principal square root of x^2. Thus, in general,

$\sqrt{x^2} = |x|$ is the principal square root of x^2

since $|x|$ must be positive, provided $x \neq 0$. However, in this chapter (unless otherwise indicated), we will always assume that all radicands represent positive numbers. Therefore, we will not need to use the absolute value symbol when computing square roots.

Example 2   Find the square root of each number.

     a. $\sqrt{x^6}$ **b.** $-\sqrt{y^8}$ **c.** $\sqrt{y^4}$ **d.** $\sqrt{5^{30}}$ **e.** $-\sqrt{3^{60}}$

Solution   To find the square root of a variable or a number raised to an even power, keep the same base and divide the exponent by 2.

2a. $\sqrt{x^6} = x^{6/2} = x^3$ Since $(x^3)^2 = x^6$

2b. $-\sqrt{y^8} = -y^4$

2c. $\sqrt{y^4} = y^2$

2d. $\sqrt{5^{30}} = 5^{15}$

2e. $-\sqrt{3^{60}} = -3^{30}$

Practice Problem 2   **Find the square root of each number.**

   a. $\sqrt{x^{10}}$ **b.** $-\sqrt{y^6}$ **c.** $\sqrt{7^{20}}$ **d.** $-\sqrt{11^{50}}$

IDEA 2

You can also find the square root of a number by using a table (see Appendix Table A-4). However, if the number is not a perfect square, the square root given in the table is a decimal approximation.

Example 3   Find the decimal approximation of each square root.

   a. $\sqrt{37}$ **b.** $\sqrt{10}$ **c.** $-\sqrt{96}$

Solution   To find the square root of a number by using a table (see Appendix), locate the number in the column labeled n and then find its square root in the column labeled $\sqrt{n}$.

3a. Locate 37 in the column labeled n. Move your finger across this row to the column headed $\sqrt{n}$ and you should see 6.083. Therefore, $\sqrt{37} \doteq 6.083$.

3b. $\sqrt{10} \doteq 3.162$ **3c.** $-\sqrt{96} \doteq -9.798$

Until now, the only real numbers we have considered have been *rational numbers*. **Rational numbers** such as $\sqrt{9}$ and $\sqrt{25}$ are numbers that can be expressed in the form $\frac{a}{b}$, where a and b are integers and $b \neq 0$. Numbers such as $\sqrt{37}$ and $\sqrt{96}$ are also real numbers, but they are called *irrational numbers* (see Figure 12-1).

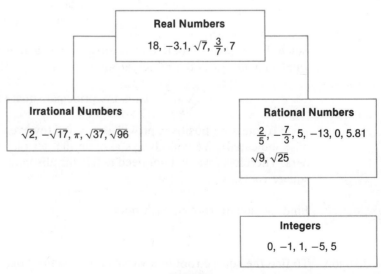

Figure 12-1

Irrational numbers are numbers that cannot be expressed as a fraction $\frac{a}{b}$, where a and b are integers and $b \neq 0$.

Practice Problem 3 **Find the decimal approximation of each square root.**

    a. $\sqrt{79}$ **b.** $-\sqrt{68}$ **c.** $\sqrt{87}$

12.1 Exercises

Find the square root of each number.

1. 4
2. 9
3. 16
4. 25
5. 81
6. 100
7. 1
8. 0
9. 49
10. 36
11. $-\sqrt{25}$
12. $\sqrt{121}$
13. $-\sqrt{1}$
14. $-\sqrt{49}$
15. $\sqrt{x^{16}}$
16. $\sqrt{y^{10}}$
17. $-\sqrt{2^{10}}$
18. $\sqrt{3^6}$
19. $-\sqrt{x^{100}}$
20. $\sqrt{z^8}$
21. $-\sqrt{7^{70}}$
22. $\sqrt{x^{10}}$
23. $-\sqrt{a^{20}}$
24. $-\sqrt{b^{30}}$
25. $\sqrt{5^{70}}$
26. $-\sqrt{c^{66}}$

Find the decimal approximation of each square root.

27. $\sqrt{32}$
28. $\sqrt{92}$
29. $\sqrt{47}$
30. $\sqrt{84}$
31. $\sqrt{14}$
32. $\sqrt{83}$
33. $\sqrt{80}$
34. $-\sqrt{87}$
35. $-\sqrt{13}$
36. $-\sqrt{18}$
37. $-\sqrt{7}$
38. $-\sqrt{15}$

Fill in the blanks.

39. Every positive number has one _____ square root and one _____ square root.
40. The symbol $-\sqrt{}$ is used to indicate the _____ of a number.
41. In the expression $\sqrt{18}$, 18 is called the _____ and $\sqrt{}$ is a _____ symbol.
42. The $\sqrt{-9}$ is not a real number since there is no real number that you can _____ to obtain -9.
43. To determine the square root of a given number, you must find a number whose _____ is that _____.
44. To find the square root of a number raised to an even power, keep the same _____ and divide the _____ by 2.
45. To find the square root of a number by using a table, locate the _____ in a column labeled n and then find its principal square root in the column labeled _____.

Find the square root of each number. Round each answer to the nearest hundredth.

46. $\sqrt{113}$
47. $-\sqrt{175}$
48. $-\sqrt{229}$
49. $\sqrt{0.0684}$
50. $\sqrt{0.0899}$
51. $-\sqrt{7,375,000}$
52. $-\sqrt{2,345,000}$
53. $\sqrt{6.790}$

Answers to Practice Problems   **1a.** 7   **b.** -3 **c.** 8   **d.** 5   **e.** -2 **2a.** x^5 **b.** $-y^3$ **c.** 7^{10} **d.** -11^{25} **3a.** 8.888   **b.** -8.246 **c.** 9.327

12.2 Multiplying, Simplifying, and Dividing Square Roots

IDEA 1   You can multiply square roots by recognizing that the product of two or more square roots is the square root of the product of the radicands. For example,

$$\sqrt{3} \cdot \sqrt{5} = \sqrt{3 \cdot 5} = \sqrt{15}$$

The expression $\sqrt{3} \cdot \sqrt{5}$ is usually written as $\sqrt{3}\sqrt{5}$. In general,

$$\sqrt{a} \cdot \sqrt{b} = \sqrt{a}\sqrt{b}$$

The rule for multiplying square roots is stated below.

To Multiply Square Roots:

1. Multiply the radicands.
2. Take the square root of the product,

or

$$\sqrt{a} \cdot \sqrt{b} = \sqrt{a \cdot b}$$

where a and b are not negative.

Example 4   Multiply.

     a. $\sqrt{2}\,\sqrt{7}$ b. $\sqrt{6}\,\sqrt{x}$ c. $\sqrt{8}\,\sqrt{8}$
     d. $\sqrt{3}\,\sqrt{12}$ e. $\sqrt{3}\,\sqrt{5}\,\sqrt{7}$ f. $\sqrt{x}\,\sqrt{y}$

Solution   To multiply square roots having positive radicands, remember that $\sqrt{a}\sqrt{b} = \sqrt{a \cdot b}$.

     4a. $\sqrt{2}\,\sqrt{7} = \sqrt{2 \cdot 7}$ 4b. $\sqrt{6}\,\sqrt{x} = \sqrt{6 \cdot x}$
             $= \sqrt{14}$ $= \sqrt{6x}$

     4c. $\sqrt{8}\,\sqrt{8} = \sqrt{8 \cdot 8}$ 4d. $\sqrt{3}\,\sqrt{12} = \sqrt{3 \cdot 12}$
             $= \sqrt{64}$ $= \sqrt{36}$
             $= 8$ $= 6$

     4e. $\sqrt{3}\,\sqrt{5}\,\sqrt{7} = \sqrt{3 \cdot 5 \cdot 7}$ 4f. $\sqrt{x}\,\sqrt{y} = \sqrt{x \cdot y}$
                   $= \sqrt{105}$ $= \sqrt{xy}$

Practice Problem 4   Multiply.

     a. $\sqrt{3}\,\sqrt{2}$ b. $\sqrt{2}\,\sqrt{8}$ c. $\sqrt{19}\,\sqrt{x}$

IDEA 2   According to the multiplication rule for square roots, $\sqrt{a}\,\sqrt{b} = \sqrt{a \cdot b}$. However, if you interchange the left and right members of the equation, we have that

$$\sqrt{a \cdot b} = \sqrt{a}\,\sqrt{b}.$$

The above fact can be used to simplify square roots. For example, to simplify $\sqrt{12}$

notice that 12 can be expressed as a product ($12 = 4 \cdot 3$) in which the first factor is a perfect square.

$$\sqrt{12} = \sqrt{4 \cdot 3}$$

Next, we apply the formula $\sqrt{ab} = \sqrt{a}\sqrt{b}$, which states that the square root of a product is the product of the square roots of the factors of the radicand.

$$\sqrt{12} = \sqrt{4 \cdot 3} = \sqrt{4} \cdot \sqrt{3} = 2\sqrt{3}$$

Therefore, $\sqrt{12}$ written in simplified form is $2\sqrt{3}$. In general, a square root is in its simplest form when 1 is the only perfect square factor of the radicand.

The rule for simplifying square roots is stated below.

To Simplify Square Roots:

1. Factor the radicand by factoring out the largest perfect square.
2. Find the square root of each factor and express the result as a product,

or

$$\sqrt{a \cdot b} = \sqrt{a}\sqrt{b}$$

where a is the largest perfect square of the radicand $a \cdot b$.

Example 5 Simplify each square root.

a. $\sqrt{75}$ b. $\sqrt{48x}$ c. $\sqrt{x^7}$ d. $\sqrt{12x^{19}}$ e. $\sqrt{45x^2}$

Solution To simplify a square root, use the formula $\sqrt{a \cdot b} = \sqrt{a}\sqrt{b}$, where a is the largest perfect square factor of the radicand.

5a. $\sqrt{75} = \sqrt{25 \cdot 3}$ 25 is the largest perfect square factor
$= \sqrt{25}\sqrt{3}$
$= 5\sqrt{3}$ $5\sqrt{3} = 5 \cdot \sqrt{3}$

5b. $\sqrt{48x} = \sqrt{16 \cdot 3x}$ **5c.** $\sqrt{x^7} = \sqrt{x^6 \cdot x}$
$= \sqrt{16}\sqrt{3x}$ $= \sqrt{x^6}\sqrt{x}$
$= 4\sqrt{3x}$ $= x^3\sqrt{x}$

5d. $\sqrt{12x^{19}} = \sqrt{4x^{18} \cdot 3x}$ **5e.** $\sqrt{45x^2} = \sqrt{9x^2 \cdot 5}$
$= \sqrt{4x^{18}}\sqrt{3x}$ $= \sqrt{9x^2}\sqrt{5}$
$= 2x^9\sqrt{3x}$ $= 3x\sqrt{5}$

! When simplifying square roots make sure that the first factor in your product is the largest perfect square factor of the radicand.

Practice Problem 5 *Simplify each square root.*

a. $\sqrt{32}$ b. $\sqrt{y^{15}}$ c. $\sqrt{200x^2}$ d. $\sqrt{5x^6}$

When you cannot find the largest perfect square factor of the radicand by inspection (trial-and-error method), it may be helpful to factor by prime factorization.

Example 6 Simplify each square root.

414 Square Roots and Quadratic Equations

a. $\sqrt{567}$ b. $\sqrt{450}$ c. $\sqrt{875x^2}$

Solution To simplify the square root of a large number, write the prime factorization of the radicand and then use the formula $\sqrt{a \cdot b} = \sqrt{a}\sqrt{b}$, where a is the largest perfect square factor of the radicand.

6a. $\sqrt{567} = \sqrt{3^4 \cdot 7}$ $567 = 3 \cdot 3 \cdot 3 \cdot 3 \cdot 7 = 3^4 \cdot 7$
$= \sqrt{3^4}\sqrt{7}$
$= 3^2\sqrt{7}$ Think: $\sqrt{3^4} = 3^{4/2} = 3^2$
$= 9\sqrt{7}$

6b. $\sqrt{450} = \sqrt{3^2 \cdot 5^2 \cdot 2}$ $450 = 2 \cdot 5 \cdot 5 \cdot 3 \cdot 3 = 2 \cdot 5^2 \cdot 3^2$
$= \sqrt{3^2}\sqrt{5^2}\sqrt{2}$
$= 3 \cdot 5 \cdot \sqrt{2}$
$= 15\sqrt{2}$

6c. $\sqrt{875x^2} = \sqrt{5^2 \cdot x^2 \cdot 5 \cdot 7}$ $875x^2 = 5^3 \cdot 7 \cdot x^2$
$= \sqrt{5^2}\sqrt{x^2}\sqrt{5}\sqrt{7}$
$= 5 \cdot x \cdot \sqrt{5} \cdot \sqrt{7}$
$= 5x\sqrt{35}$

Practice Problem 6 *Simplify each square root.*

a. $\sqrt{441}$ b. $\sqrt{360x^6}$ c. $\sqrt{1764}$

Sometimes you can simplify a square root *after* you have multiplied.

Example 7 Multiply and simplify.

a. $\sqrt{2}\sqrt{6}$ b. $\sqrt{8x}\sqrt{10x^5}$

Solution   7a. $\sqrt{2}\sqrt{6} = \sqrt{2 \cdot 6}$ 7b. $\sqrt{8x}\sqrt{10x^5} = \sqrt{8x \cdot 10x^5}$
$= \sqrt{12}$ $= \sqrt{80x^6}$
$= \sqrt{4 \cdot 3}$ $= \sqrt{16x^6 \cdot 5}$
$= \sqrt{4}\sqrt{3}$ $= \sqrt{16x^6}\sqrt{5}$
$= 2\sqrt{3}$ $= 4x^3\sqrt{5}$

Practice Problem 7 *Multiply and simplify.*

a. $\sqrt{2}\sqrt{10}$ b. $\sqrt{3x}\sqrt{15x^3}$ c. $\sqrt{11}\sqrt{33}$

IDEA 3

The *division rule for square roots* is similar to the multiplication rule. In other words, the quotient of two square roots equals the square root of the quotient of the radicands. For example,

$$\frac{\sqrt{6}}{\sqrt{2}} = \sqrt{\frac{6}{2}} = \sqrt{3}$$

The rule for dividing square roots is stated in the box on page 415.

Example 8 Divide and simplify.

a. $\dfrac{\sqrt{30}}{\sqrt{2}}$ b. $\dfrac{\sqrt{48}}{\sqrt{3}}$ c. $\dfrac{\sqrt{49x^9}}{\sqrt{x}}$ d. $\dfrac{\sqrt{160x^6}}{\sqrt{20x}}$

> **To Divide One Square Root by Another:**
>
> 1. Write the radicands as a quotient.
> 2. Take the square root of the quotient,
>
> or
>
> $$\frac{\sqrt{a}}{\sqrt{b}} = \sqrt{\frac{a}{b}}$$
>
> where a and b are not negative and b is not equal to zero.

Solution To find the quotient of two square roots, use the formula $\frac{\sqrt{a}}{\sqrt{b}} = \sqrt{\frac{a}{b}}$

8a. $\frac{\sqrt{30}}{\sqrt{2}} = \sqrt{\frac{30}{2}}$
    $= \sqrt{15}$

8b. $\frac{\sqrt{48}}{\sqrt{3}} = \sqrt{\frac{48}{3}}$
    $= \sqrt{16}$
    $= 4$

8c. $\frac{\sqrt{49x^9}}{\sqrt{x}} = \sqrt{\frac{49x^9}{x}}$
    $= \sqrt{49x^8}$
    $= 7x^4$

8d. $\frac{\sqrt{160x^6}}{\sqrt{20x}} = \sqrt{\frac{160x^6}{20x}}$
    $= \sqrt{8x^5}$
    $= \sqrt{4x^4 \cdot 2x}$
    $= 2x^2\sqrt{2x}$

Practice Problem 8 **Divide and simplify.**

a. $\frac{\sqrt{28}}{\sqrt{4}}$ b. $\frac{\sqrt{80}}{\sqrt{5}}$ c. $\frac{\sqrt{75x^3}}{\sqrt{3x}}$

IDEA 4

The division rule for square roots can also be used to determine the square root of a fraction. That is, if we interchange the left and right members of the equation

$$\frac{\sqrt{a}}{\sqrt{b}} = \sqrt{\frac{a}{b}}$$

we will obtain

$$\sqrt{\frac{a}{b}} = \frac{\sqrt{a}}{\sqrt{b}}$$

This implies that the square root of a quotient (fraction) is the square root of the numerator divided by the square root of the denominator. For example,

$$\sqrt{\frac{16}{25}} = \frac{\sqrt{16}}{\sqrt{25}} = \frac{4}{5}$$

The above statements suggest a rule for finding the square root of a fraction (see the box on page 416).

To Find the Square Root of a Fraction:

1. Write the square root of the numerator over the square root of the denominator.
2. Simplify if possible,

or

$$\sqrt{\frac{a}{b}} = \frac{\sqrt{a}}{\sqrt{b}}$$

where a and b are not negative and b is not equal to zero.

Example 9 Simplify.

a. $\sqrt{\frac{25}{4}}$ b. $\sqrt{\frac{3}{16}}$ c. $\sqrt{\frac{8}{x^4}}$ d. $\sqrt{\frac{x^8}{y^6}}$

Solution To find the square root of a fraction, write the square root of the numerator over the square root of the denominator.

9a. $\sqrt{\frac{25}{4}} = \frac{\sqrt{25}}{\sqrt{4}}$
$= \frac{5}{2}$

9b. $\sqrt{\frac{3}{16}} = \frac{\sqrt{3}}{\sqrt{16}}$
$= \frac{\sqrt{3}}{4}$

9c. $\sqrt{\frac{8}{x^4}} = \frac{\sqrt{8}}{\sqrt{x^4}}$
$= \frac{\sqrt{4}\sqrt{2}}{x^2}$
$= \frac{2\sqrt{2}}{x^2}$

9d. $\sqrt{\frac{x^8}{y^6}} = \frac{\sqrt{x^8}}{\sqrt{y^6}}$
$= \frac{x^4}{y^3}$

Practice Problem 9 *Simplify.*

a. $\sqrt{\frac{16}{9}}$ b. $\sqrt{\frac{15}{81}}$ c. $\sqrt{\frac{32x^5}{2x}}$

IDEA 5

Sometimes when you find the square root of a fraction, there will be a square root left in the denominator. When this occurs, you can eliminate it by using the fundamental property of fractions. For example,

$$\sqrt{\frac{1}{3}} = \frac{\sqrt{1}}{\sqrt{3}}$$
$$= \frac{1}{\sqrt{3}}$$
$$= \frac{1 \cdot \sqrt{3}}{\sqrt{3} \cdot \sqrt{3}}$$ Multiplying by $\sqrt{3}$ will make the radicand in the denominator a perfect square.
$$= \frac{\sqrt{3}}{\sqrt{9}}$$
$$= \frac{\sqrt{3}}{3}$$

12.2 Multiplying, Simplifying, and Dividing Square Roots   417

This process of eliminating a square root from the denominator of a fraction is called *rationalizing the denominator*.

To Rationalize the Denominator of a Fraction:

1. Multiply the numerator and denominator of the fraction by the square root appearing in the denominator, or choose a multiplier that will make the radicand in the denominator a perfect square.
2. Simplify the square roots.

Example 10   Simplify.

a. $\dfrac{3}{\sqrt{5}}$ b. $\dfrac{\sqrt{2}}{\sqrt{7}}$ c. $\sqrt{\dfrac{9}{6}}$ d. $\dfrac{1}{\sqrt{20}}$

Solution   To rationalize the denominator of a fraction, first multiply the numerator and denominator by a square root that will make the radicand of the denominator a perfect square. Then, simplify the square roots.

10a. $\dfrac{3}{\sqrt{5}} = \dfrac{3 \cdot \sqrt{5}}{\sqrt{5} \cdot \sqrt{5}}$

$= \dfrac{3\sqrt{5}}{\sqrt{25}}$

$= \dfrac{3\sqrt{5}}{5}$

10b. $\dfrac{\sqrt{2}}{\sqrt{7}} = \dfrac{\sqrt{2} \cdot \sqrt{7}}{\sqrt{7} \cdot \sqrt{7}}$

$= \dfrac{\sqrt{14}}{\sqrt{49}}$

$= \dfrac{\sqrt{14}}{7}$

Alternate method

10c. $\sqrt{\dfrac{9}{6}} = \dfrac{\sqrt{9}}{\sqrt{6}}$

$= \dfrac{\sqrt{9}\sqrt{6}}{\sqrt{6}\sqrt{6}}$

$= \dfrac{\cancel{3}\sqrt{6}}{\cancel{6}}_{2}$

$= \dfrac{\sqrt{6}}{2}$

Reduce

$\sqrt{\dfrac{9}{6}} = \sqrt{\dfrac{3}{2}}$

$= \dfrac{\sqrt{3}}{\sqrt{2}}$

$= \dfrac{\sqrt{3}\sqrt{2}}{\sqrt{2}\sqrt{2}}$

$= \dfrac{\sqrt{6}}{2}$

10d. $\dfrac{1}{\sqrt{20}} = \dfrac{1 \cdot \sqrt{5}}{\sqrt{20} \cdot \sqrt{5}}$

$= \dfrac{\sqrt{5}}{\sqrt{100}}$

$= \dfrac{\sqrt{5}}{10}$

You could have multiplied by $\sqrt{20}$. However, multiplying by $\sqrt{5}$ will make your computations easier.

Practice Problem 10   *Simplify.*

a. $\dfrac{2}{\sqrt{3}}$ b. $\dfrac{\sqrt{2}}{\sqrt{5}}$ c. $\sqrt{\dfrac{2}{12}}$ d. $\dfrac{\sqrt{8}}{\sqrt{5}}$

IDEA 6   At this point it may be helpful to review what is meant when we say that a square root is in its simplest form. An expression containing radicals is in its simplest form:

418 *Square Roots and Quadratic Equations*

1. if the number 1 is the only perfect square factor of the radicand;
2. if the radicand does not contain a fraction; or
3. if the denominator does not contain a radical.

Example 11   Simplify.

a. $\sqrt{18x^3}$ b. $\sqrt{0.12}$ c. $\dfrac{\sqrt{20}}{\sqrt{3}}$

Solution   11a. $\sqrt{18x^3} = \sqrt{9x^2 \cdot 2x}$
$= \sqrt{9x^2}\sqrt{2x}$
$= 3x\sqrt{2x}$

11b. $\sqrt{0.12} = \sqrt{\dfrac{12}{100}}$ Express 0.12 as a fraction
$= \dfrac{\sqrt{12}}{\sqrt{100}}$ $\sqrt{12} = \sqrt{4}\sqrt{3} = 2\sqrt{3}$
$= \dfrac{2\sqrt{3}}{10}$
$= \dfrac{\sqrt{3}}{5}$

11c. $\dfrac{\sqrt{20}}{\sqrt{3}} = \dfrac{\sqrt{20}}{\sqrt{3}} \dfrac{\sqrt{3}}{\sqrt{3}}$
$= \dfrac{\sqrt{60}}{3}$ $\sqrt{60} = \sqrt{4}\sqrt{15} = 2\sqrt{15}$
$= \dfrac{2\sqrt{15}}{3}$

Practice Problem 11   **Simplify.**

a. $\sqrt{150x^6}$ b. $\sqrt{\dfrac{1}{12}}$ c. $\dfrac{\sqrt{12}}{\sqrt{7}}$

12.2 Exercises

Simplify.

1. $\sqrt{8}$
2. $-\sqrt{32}$
3. $\sqrt{40}$
4. $\sqrt{20}$
5. $-\sqrt{125}$
6. $\sqrt{300}$
7. $\sqrt{28}$
8. $-\sqrt{75}$
9. $\sqrt{x^5}$
10. $-\sqrt{y^7}$
11. $\sqrt{x^{33}}$
12. $-\sqrt{x^{17}}$
13. $\sqrt{54x^4}$
14. $\sqrt{x^{11}}$
15. $\sqrt{48}$
16. $-\sqrt{27x}$
17. $\sqrt{50}$
18. $\sqrt{80}$
19. $\sqrt{72x^4}$
20. $\sqrt{32x^6}$
21. $\sqrt{98}$
22. $\sqrt{60}$
23. $\sqrt{x^{19}}$
24. $\sqrt{x^{21}}$
25. $-\sqrt{x^7}$
26. $-\sqrt{x^{13}}$
27. $-\sqrt{x^{31}}$
28. $-\sqrt{x^{29}}$
29. $\sqrt{25x^5}$
30. $\sqrt{36x^3}$
31. $\sqrt{16x}$
32. $\sqrt{9x}$
33. $\sqrt{c^7}$
34. $\sqrt{x^9}$
35. $-\sqrt{90}$
36. $-\sqrt{20}$
37. $\sqrt{100}$
38. $\sqrt{486}$
39. $-\sqrt{490}$
40. $-\sqrt{425}$
41. $\sqrt{588x}$
42. $\sqrt{720x}$
43. $-\sqrt{2700x^8}$
44. $-\sqrt{7500x^6}$

Multiply and simplify.

45. $\sqrt{3}\,\sqrt{5}$
46. $\sqrt{5}\,\sqrt{6}$
47. $\sqrt{x}\,\sqrt{13}$
48. $\sqrt{3}\,\sqrt{6}$
49. $\sqrt{19}\,\sqrt{19}$
50. $\sqrt{2}\,\sqrt{8}$
51. $\sqrt{7}\,\sqrt{112}$
52. $\sqrt{8x}\,\sqrt{6x}$
53. $\sqrt{10}\,\sqrt{10}$
54. $\sqrt{xy}\,\sqrt{x^3y}$
55. $\sqrt{2}\,\sqrt{27x^3}$
56. $\sqrt{6x}\,\sqrt{6x^9}$
57. $\sqrt{20xy}\,\sqrt{10x^{11}}$
58. $\sqrt{2xy}\,\sqrt{250x^2y^5}$

Divide and simplify.

59. $\dfrac{\sqrt{12}}{\sqrt{3}}$
60. $\dfrac{\sqrt{27}}{\sqrt{3}}$
61. $\dfrac{\sqrt{128}}{\sqrt{2}}$
62. $\dfrac{\sqrt{80x}}{\sqrt{5x}}$
63. $\dfrac{\sqrt{75x^7}}{\sqrt{3x^2}}$
64. $\dfrac{\sqrt{64x^{15}}}{\sqrt{16x^{11}}}$
65. $\dfrac{\sqrt{75x^{13}}}{\sqrt{x^4}}$
66. $\dfrac{\sqrt{x^8y^{11}}}{\sqrt{x^2y}}$
67. $\dfrac{\sqrt{250x^{11}}}{\sqrt{5x}}$
68. $\dfrac{\sqrt{350x^{13}}}{\sqrt{2x}}$

Simplify.

69. $\sqrt{\dfrac{4}{9}}$
70. $\sqrt{\dfrac{16}{81}}$
71. $\sqrt{\dfrac{36}{49}}$
72. $\sqrt{\dfrac{1}{100}}$
73. $\sqrt{\dfrac{5}{9}}$
74. $\sqrt{\dfrac{1}{16}}$
75. $\sqrt{\dfrac{11}{25}}$
76. $\sqrt{\dfrac{3}{49}}$

Rationalize the denominator.

77. $\dfrac{5}{\sqrt{2}}$
78. $\dfrac{5}{\sqrt{3}}$
79. $\dfrac{1}{\sqrt{6}}$
80. $\dfrac{1}{\sqrt{14}}$
81. $\dfrac{\sqrt{2}}{\sqrt{3}}$
82. $\dfrac{\sqrt{5}}{\sqrt{2}}$
83. $\dfrac{\sqrt{3}}{\sqrt{5}}$
84. $\dfrac{\sqrt{5}}{\sqrt{3}}$
85. $\dfrac{1}{\sqrt{2}}$
86. $\dfrac{13}{\sqrt{5}}$
87. $\dfrac{\sqrt{2}}{\sqrt{8}}$
88. $\dfrac{1}{\sqrt{7}}$
89. $\dfrac{1}{\sqrt{x}}$
90. $\dfrac{\sqrt{5}}{\sqrt{18}}$
91. $\dfrac{8x}{\sqrt{6}}$
92. $\dfrac{\sqrt{5}}{\sqrt{20}}$
93. $\dfrac{\sqrt{5}}{\sqrt{8x}}$
94. $\dfrac{\sqrt{3}}{\sqrt{21}}$
95. $\dfrac{\sqrt{8x^5}}{\sqrt{3x}}$
96. $\dfrac{\sqrt{x^5}}{\sqrt{32x}}$

Fill in the blanks.

97. The product of two or more square roots is the _____ of the _____ of the radicands.

98. To simplify a square root, use the formula $\sqrt{a \cdot b} = \sqrt{a}\,\sqrt{b}$, where a is the largest _____ of the radicand.

99. To simplify the square root of a large number, first write the prime factorization of the _____. Then, use the formula $\sqrt{a \cdot b} = \sqrt{a}\,\sqrt{b}$, where a is the largest _____ of the radicand.

100. The quotient of two square roots is the _____ of the quotient of the _____.

101. To find the square root of a fraction, write the square root of the _____ over the square root of the _____.

102. To rationalize the denominator of a fraction, first _____ the numerator and the denominator by a square root that will make the _____ in the denominator a _____. Then, simplify the square roots.

Simplify.

103. $\sqrt{\dfrac{4}{25}}$
104. $\sqrt{\dfrac{9}{16}}$
105. $\sqrt{\dfrac{3}{4}}$
106. $\sqrt{\dfrac{81}{100}}$

107. $\sqrt{\frac{9}{12}}$ 108. $\sqrt{\frac{3}{5}}$ 109. $\sqrt{0.48}$ 110. $\sqrt{90x^8y^6}$

111. $\frac{\sqrt{10}}{\sqrt{80}}$ 112. $\frac{\sqrt{15}}{\sqrt{75}}$ 113. $\frac{\sqrt{8}}{\sqrt{32}}$ 114. $\sqrt{726x^4y^2z^5}$

115. $\sqrt{10}\sqrt{\frac{7}{40}}$ 116. $\sqrt{\frac{9}{50}}\sqrt{\frac{2}{3}}$ 117. $\sqrt{x^{15}y^{20}z^3}$

Answers to Practice Problems 4a. $\sqrt{6}$ b. 4 c. $\sqrt{19x}$ 5a. $4\sqrt{2}$ b. $y^7\sqrt{y}$ c. $10x\sqrt{2}$ d. $x^3\sqrt{5}$
6a. 21 b. $6x^3\sqrt{10}$ c. 42 7a. $2\sqrt{5}$ b. $3x^2\sqrt{5}$ c. $11\sqrt{3}$ 8a. $\sqrt{7}$ b. 4 c. $5x$ 9a. $\frac{4}{3}$ b. $\frac{\sqrt{15}}{9}$
c. $4x^2$ 10a. $\frac{2\sqrt{3}}{3}$ b. $\frac{\sqrt{10}}{5}$ c. $\frac{\sqrt{6}}{6}$ d. $\frac{2\sqrt{10}}{5}$ 11a. $5x^3\sqrt{6}$ b. $\frac{\sqrt{3}}{6}$ c. $\frac{2\sqrt{21}}{7}$

12.3 Adding and Subtracting Square Roots

IDEA 1   Square roots are added and subtracted in the same manner that polynomials are added and subtracted. For example, to combine the like terms of a polynomial we use the distributive property.

$$5x + 4x = (5 + 4)x = 9x$$

Similarly, to combine **like square roots,** square roots having the same radicands, we use the distributive property.

$$5\sqrt{3} + 4\sqrt{3} = (5 + 4)\sqrt{3} = 9\sqrt{3}$$

At this point, we should also note that we cannot combine unlike square roots by using the distributive property. For example, $5\sqrt{3} + 4\sqrt{2}$ cannot be combined or simplified further.

> **To Add or Subtract Square Roots:**
>
> 1. Simplify each square root.
> 2. Use the distributive law to combine like square roots.

Remember, you can only add and subtract like square roots. Like square roots are square roots having the same radicand.

Example 12   Perform the indicated operation.

a. $2\sqrt{3} + 5\sqrt{3}$ b. $7\sqrt{5} - 9\sqrt{5}$
c. $\sqrt{2} + \sqrt{3}$ d. $5\sqrt{7} - 8\sqrt{7} + 14\sqrt{7}$ e. $5\sqrt{x} + 3\sqrt{x}$

Solution   To add or subtract like radicals, use the distributive law $ab + cd = (a + c)b$.

12a. $2\sqrt{3} + 5\sqrt{3} = (2 + 5)\sqrt{3}$
$= 7\sqrt{3}$

12b. $7\sqrt{5} - 9\sqrt{5} = (7 - 9)\sqrt{5}$
$= -2\sqrt{5}$

12c. $\sqrt{2} + \sqrt{3}$ cannot be simplified since $\sqrt{2}$ and $\sqrt{3}$ are not like square roots.

12d. $5\sqrt{7} - 8\sqrt{7} + 14\sqrt{7} = (5 - 8 + 14)\sqrt{7}$
$= 11\sqrt{7}$

12e. $5\sqrt{x} + 3\sqrt{x} = (5 + 3)\sqrt{x}$
$= 8\sqrt{x}$

Practice Problem 12 **Perform the indicated operation.**

    a. $3\sqrt{11} + 5\sqrt{11}$ **b.** $\sqrt{13} - 6\sqrt{13}$ **c.** $\sqrt{2} - 3\sqrt{2} + 2\sqrt{2}$

Example 13   Perform the indicated operation.

    a. $5\sqrt{2} + \sqrt{8}$ **b.** $\sqrt{20} - \sqrt{125}$ **c.** $\sqrt{7} + \sqrt{\frac{1}{7}}$

    d. $\sqrt{12} + 4\sqrt{75}$ **e.** $\sqrt{27} - \sqrt{48} + 2\sqrt{3} + \sqrt{6}$

Solution   To add or subtract unlike square roots, first simplify each square root. Then, use the distributive law to combine like square roots.

13a. $5\sqrt{2} + \sqrt{8} = 5\sqrt{2} + \sqrt{4}\sqrt{2}$ Multiplication rule
$= 5\sqrt{2} + 2\sqrt{2}$
$= (5 + 2)\sqrt{2}$
$= 7\sqrt{2}$

13b. $\sqrt{20} - \sqrt{125} = \sqrt{4}\sqrt{5} - \sqrt{25}\sqrt{5}$
$= 2\sqrt{5} - 5\sqrt{5}$
$= (2 - 5)\sqrt{5}$
$= -3\sqrt{5}$

13c. $\sqrt{7} + \sqrt{\frac{1}{7}} = \sqrt{7} + \frac{1}{\sqrt{7}}$ Division rule
$= \sqrt{7} + \frac{1 \cdot \sqrt{7}}{\sqrt{7} \cdot \sqrt{7}}$ Rationalize denominator
$= \sqrt{7} + \frac{\sqrt{7}}{7}$
$= \sqrt{7} + \frac{1}{7}\sqrt{7}$
$= \left(1 + \frac{1}{7}\right)\sqrt{7}$
$= \frac{8}{7}\sqrt{7}$

13d. $\sqrt{12} + 4\sqrt{75} = \sqrt{4}\sqrt{3} + 4 \cdot \sqrt{25}\sqrt{3}$
$= 2\sqrt{3} + 4 \cdot 5 \cdot \sqrt{3}$
$= 2\sqrt{3} + 20\sqrt{3}$
$= (2 + 20)\sqrt{3}$
$= 22\sqrt{3}$

13e. $\sqrt{27} - \sqrt{48} + 2\sqrt{3} + \sqrt{6} = \sqrt{9}\sqrt{3} - \sqrt{16}\sqrt{3} + 2\sqrt{3} + \sqrt{6}$
$= 3\sqrt{3} - 4\sqrt{3} + 2\sqrt{3} + \sqrt{6}$
$= (3 - 4 + 2)\sqrt{3} + \sqrt{6}$
$= \sqrt{3} + \sqrt{6}$

Practice Problem 13 **Perform the indicated operations.**

a. $5\sqrt{12} - \sqrt{27}$ b. $\sqrt{98} + \sqrt{50}$ c. $\sqrt{\dfrac{5}{3}} + \sqrt{\dfrac{3}{5}}$ d. $\sqrt{18} - \sqrt{27} + 3\sqrt{32}$

12.3 Exercises

Perform the indicated operations.

1. $4\sqrt{2} + 5\sqrt{2}$
2. $3\sqrt{5} + 4\sqrt{5}$
3. $6\sqrt{7} - 8\sqrt{7}$
4. $7\sqrt{11} - \sqrt{11}$
5. $7\sqrt{x} + 5\sqrt{x}$
6. $-\sqrt{3} - 5\sqrt{3}$
7. $8\sqrt{13} + 3\sqrt{13}$
8. $\sqrt{7} + \sqrt{11}$
9. $4\sqrt{2} - 8\sqrt{2}$
10. $\sqrt{18} + \sqrt{32}$
11. $\sqrt{27} - \sqrt{3}$
12. $\sqrt{20} + \sqrt{45}$
13. $\sqrt{72} - \sqrt{98}$
14. $\sqrt{24} - \sqrt{54}$
15. $-\sqrt{3} + \sqrt{12}$
16. $\sqrt{48} + 2\sqrt{27}$
17. $3\sqrt{80} - \sqrt{45}$
18. $4\sqrt{50} - \sqrt{32}$
19. $\sqrt{18} + 3\sqrt{48} - 7\sqrt{28}$
20. $-\sqrt{18} - 3\sqrt{32} - \sqrt{50}$

Fill in the blanks.

21. Like square roots are square roots having the same _____.
22. To add or subtract square roots, first _____ each square root. Then, use the _____ to combine the like square roots.
23. You cannot add $\sqrt{5} + \sqrt{7}$ since $\sqrt{5}$ and $\sqrt{7}$ are not _____.

Perform the indicated operations.

24. $\sqrt{8} + 3\sqrt{18} - 5\sqrt{2}$
25. $\sqrt{2} + \sqrt{3} + \sqrt{5}$
26. $3\sqrt{48} - 2\sqrt{27} + \sqrt{12}$
27. $\sqrt{12} + \sqrt{27} + \sqrt{48}$
28. $\sqrt{\dfrac{1}{3}} + \sqrt{27}$
29. $\sqrt{\dfrac{1}{2}} - \sqrt{2}$
30. $\sqrt{\dfrac{1}{12}} + \sqrt{\dfrac{1}{27}}$
31. $\sqrt{\dfrac{3}{7}} + \sqrt{\dfrac{7}{3}}$
32. $\dfrac{3}{4}\sqrt{6} + \dfrac{1}{4}\sqrt{6}$
33. $3\sqrt{\dfrac{1}{6}} + 5\sqrt{\dfrac{3}{2}}$
34. $\sqrt{48} - \sqrt{\dfrac{1}{3}}$
35. $\sqrt{\dfrac{25}{2}} - \sqrt{\dfrac{9}{2}} + \sqrt{12}$
36. $\sqrt{x^3} - 3\sqrt{4x^3}$
37. $\sqrt{16a} - \sqrt{9a} + \sqrt{36a}$
38. $\sqrt{8x^2} - 4x\sqrt{2}$
39. $\sqrt{40x^4} + x^2\sqrt{90}$
40. $\sqrt{112} + \sqrt{252}$
41. $3\sqrt{275} - 4\sqrt{396} + \sqrt{99}$

Answers to Practice Problems   12a. $8\sqrt{11}$ b. $-5\sqrt{13}$ c. 0   13a. $7\sqrt{3}$ b. $12\sqrt{2}$ c. $\dfrac{8}{15}\sqrt{15}$ d. $15\sqrt{2} - 3\sqrt{3}$

12.4 Quadratic Equations

IDEA 1   A **quadratic** or **second-degree equation** in one variable is an equation that can be written in the form

$$ax^2 + bx + c = 0$$

where a, b, and c are real numbers and $a \neq 0$. A quadratic equation written in this form is said to be in **standard form.** For example, $2x^2 + 7x - 9 = 0$ is a quadratic equation in standard form.

Example 14 Write each equation in standard form.

a. $3x^2 + 5x = 4$ b. $7x = 8 - 2x^2$

Solution To write a quadratic equation in standard form, use the rules for solving linear equations to make the right side of the equation equal to zero.

14a. $3x^2 + 5x = 4$
$3x^2 + 5x - 4 = 0$ Add -4 to both sides
$a = 3, b = 5$, and $c = -4$

14b. $7x = 8 - 2x^2$
$2x^2 + 7x = 8$ Add $2x^2$ to both sides
$2x^2 + 7x - 8 = 0$ Add -8 to both sides
$a = 2, b = 7$, and $c = -8$

Practice Problem 14 *Write each equation in standard form.*

a. $5x^2 = 6x + 4$ b. $-2x = -6x^2 + 2$ c. $6x + 7x^2 = x^2 - 9 + 2x$

IDEA 2

When $b = 0$, the equation $ax^2 + bx + c = 0$ becomes $ax^2 + 0x + c = 0$ or $ax^2 + c = 0$. This type of equation is called a **pure quadratic equation** since the x term is missing. To solve this equation we isolate the x^2 term on one side of the equation and take the square root of both sides of the equation. For example,

$3x^2 - 48 = 0$
$3x^2 = 48$ Add 48 to both sides
$x^2 = 16$ Divide both sides by 3

To solve the equation $x^2 = 16$ we take the square root of both sides of the equation. Remember, the square root of a positive number has two solutions (one positive, one negative), which is indicated by writing $\pm\sqrt{K}$, where K is a positive number.

$x^2 = 16$
$x = \pm\sqrt{16}$ Take square root of both sides
$x = \pm 4$ This means $x = 4$ and $x = -4$

We will check the solutions by using the same procedures used to check linear equations.

Check: For 4   For -4
$3x^2 - 48 = 0$ $3x^2 - 48 = 0$
$3(4)^2 - 48 = 0$ $3(-4)^2 - 48 = 0$
$3(16) - 48 = 0$ $3(16) - 48 = 0$
$0 = 0$ $0 = 0$

This procedure, sometimes called the *square root method*, is summarized in the box on page 424.

To Solve a Pure Quadratic Equation of the Form $ax^2 + c = 0$ (Square Root Method):

1. Rewrite the equation (if necessary) so that x^2 is on one side of the equation and a constant is on the opposite side.
2. Take the square root of both sides. Remember that every positive real number has both a positive and a negative square root.
3. Check both solutions.

Example 15 Solve.

a. $x^2 = 9$ b. $x^2 = 20$ c. $x^2 + 16 = 0$
d. $4x^2 - 9 = 0$ e. $9x^2 - 7 = 25$ f. $2x^2 - 3.5 = -0.5$

Solution To solve a quadratic equation containing no x term, first write the equation in the form $x^2 = K$. Then, take the square root of both sides and simplify.

15a. $x^2 = 9$
$x = \pm\sqrt{9}$ $x = \pm\sqrt{9}$ is shorthand for the statement
$x = \pm 3$ $x = \sqrt{9}$ or $x = -\sqrt{9}$

The solutions are 3 and -3.

Check: For 3   For -3
$x^2 = 9$ $x^2 = 9$
$3^2 = 9$ $(-3)^2 = 9$
$9 = 9$ $9 = 9$

15b. $x^2 = 20$
$x = \pm\sqrt{20}$
$x = \pm 2\sqrt{5}$ Think: $\sqrt{20} = \sqrt{4}\sqrt{5} = 2\sqrt{5}$

The solutions are $2\sqrt{5}$ and $-2\sqrt{5}$.

Check: For $2\sqrt{5}$ For $-2\sqrt{5}$
$x^2 = 20$ $x^2 = 20$
$(2\sqrt{5})^2 = 20$ $(-2\sqrt{5})^2 = 20$
$(2)^2(\sqrt{5})^2 = 20$ $(-2)^2(\sqrt{5})^2 = 20$
$20 = 20$ $20 = 20$

15c. $x^2 + 16 = 0$
$x^2 = -16$ Add -16 to both sides
$x = \pm\sqrt{-16}$

This equation has no real solution since there is no real number that we can square and obtain -16.

15d. $4x^2 - 9 = 0$
$4x^2 = 9$ Add 9 to both sides
$x^2 = \dfrac{9}{4}$ Divide both sides by 4
$x = \pm\sqrt{\dfrac{9}{4}}$
$x = \pm\dfrac{3}{2}$

The solutions are $\frac{3}{2}$ and $-\frac{3}{2}$.

Check: For $\frac{3}{2}$ | For $-\frac{3}{2}$
$$4x^2 - 9 = 0 \qquad\qquad 4x^2 - 9 = 0$$
$$4\left(\frac{3}{2}\right)^2 - 9 = 0 \qquad 4\left(-\frac{3}{2}\right)^2 - 9 = 0$$
$$4\left(\frac{9}{4}\right) - 9 = 0 \qquad 4\left(\frac{9}{4}\right) - 9 = 0$$
$$9 - 9 = 0 \qquad\qquad 9 - 9 = 0$$
$$0 = 0 \qquad\qquad 0 = 0$$

15e. $9x^2 - 7 = 25$
$\qquad 9x^2 = 32$ — Add 7 to both sides
$\qquad x^2 = \frac{32}{9}$ — Divide both sides by 9
$\qquad x = \pm\sqrt{\frac{32}{9}}$
$\qquad x = \pm\frac{4\sqrt{2}}{3}$ — Simplify the square root

The solutions are $\frac{4\sqrt{2}}{3}$ and $-\frac{4\sqrt{2}}{3}$.

Check: For $\frac{4\sqrt{2}}{3}$ | For $-\frac{4\sqrt{2}}{3}$
$$9x^2 - 7 = 25 \qquad\qquad 9x^2 - 7 = 25$$
$$9\left(\frac{4\sqrt{2}}{3}\right)^2 - 7 = 25 \qquad 9\left(-\frac{4\sqrt{2}}{3}\right)^2 - 7 = 25$$
$$9\left(\frac{32}{9}\right) - 7 = 25 \qquad 9\left(\frac{32}{9}\right) - 7 = 25$$
$$32 - 7 = 25 \qquad\qquad 32 - 7 = 25$$
$$25 = 25 \qquad\qquad 25 = 25$$

15f. $2x^2 - 3.5 = -0.5$
$\qquad 2x^2 = 3$ — Add 3.5 to both sides
$\qquad x^2 = \frac{3}{2}$ — Divide both sides by 2.
$\qquad x = \sqrt{\frac{3}{2}}$
$\qquad x = \pm\frac{\sqrt{6}}{2}$ — Simplify the square root

The solutions are $\frac{\sqrt{6}}{2}$ and $-\frac{\sqrt{6}}{3}$.

Check: For $\frac{\sqrt{6}}{2}$ | For $-\frac{\sqrt{6}}{2}$
$$2x^2 - 3.5 = -0.5 \qquad\qquad 2x^2 - 3.5 = -0.5$$
$$2\left(\frac{\sqrt{6}}{2}\right)^2 - 3.5 = -0.5 \qquad 2\left(-\frac{\sqrt{6}}{2}\right)^2 - 3.5 = -0.5$$
$$2\left(\frac{6}{4}\right) - 3.5 = -0.5 \qquad 2\left(\frac{6}{4}\right) - 3.5 = -0.5$$
$$3 - 3.5 = -0.5 \qquad\qquad 3 - 3.5 = -0.5$$
$$-0.5 = -0.5 \qquad\qquad -0.5 = -0.5$$

Square Roots and Quadratic Equations

Practice Problem 15 **Solve and check.**

a. $x^2 = 25$ b. $2x^2 = 16$ c. $5x^2 - 7 = 5$

The square root method for solving pure quadratic equations can also be used to solve equations in which one side of the equation is a binomial squared and the other side is a constant.

Example 16 Solve and check.

a. $(x + 4)^2 = 5$ b. $(3x + 2)^2 = 25$ c. $(2x - 3)^2 - 9 = 23$

Solution

16a. $(x + 4)^2 = 5$
$x + 4 = \pm\sqrt{5}$ Take the square root of both sides
$x + 4 = \sqrt{5}$ or $x + 4 = -\sqrt{5}$ Rewrite as two equations
$x = -4 + \sqrt{5}$ $x = -4 - \sqrt{5}$

The solutions are $-4 + \sqrt{5}$ and $-4 - \sqrt{5}$.

Check: For $-4 + \sqrt{5}$ For $-4 - \sqrt{5}$
$(x + 4)^2 = 5$ $(x + 4)^2 = 5$
$(-4 + \sqrt{5} + 4)^2 = 5$ $(-4 - \sqrt{5} + 4)^2 = 5$
$(\sqrt{5})^2 = 5$ $(-\sqrt{5})^2 = 5$
$5 = 5$ $5 = 5$

16b. $(3x + 2)^2 = 25$
$3x + 2 = \pm 5$ Take the square root of both sides
$3x + 2 = 5$ or $3x + 2 = -5$
$3x = 3$ $3x = -7$
$x = 1$ $x = -\dfrac{7}{3}$

Check: For 1 For $-\dfrac{7}{3}$
$(3x + 2)^2 = 25$ $(3x + 2)^2 = 25$
$(3 \cdot 1 + 2)^2 = 25$ $\left(3\left[-\dfrac{7}{3}\right] + 2\right)^2 = 25$
$(5)^2 = 25$ $(-7 + 2)^2 = 25$
$25 = 25$ $(-5)^2 = 25$
                           $25 = 25$

16c. $(2x - 3)^2 - 9 = 23$
$(2x - 3)^2 = 32$
$2x - 3 = \pm 4\sqrt{2}$ Take the square root of both sides
$2x = 3 \pm 4\sqrt{2}$ Add 3 to both sides
$x = \dfrac{3 \pm 4\sqrt{2}}{2}$ Divide both sides by 2

The solutions are $\dfrac{3 + 4\sqrt{2}}{2}$ and $\dfrac{3 - 4\sqrt{2}}{2}$.

Check: Verify that the solutions are correct.

Practice Problem 16 **Solve and check.**

a. $(x + 3)^2 = 36$ b. $(x + 1)^2 = 6$

12.4 Exercises

Write each equation in standard form.

1. $4x^2 - 7x = 9$
2. $x^2 = 5x - 9$
3. $3x - 1 = 7 - 3x^2$
4. $(2x + 3)(3x - 7) = 7x + 2$
5. $5x^2 - 9x + 2 = 2x^2 + 8x$
6. $2x^2 = 9$

Solve.

7. $x^2 = 9$
8. $x^2 = 16$
9. $x^2 = 36$
10. $x^2 = 81$
11. $x^2 = 12$
12. $x^2 = 18$
13. $x^2 = 7$
14. $x^2 = 10$
15. $x^2 + 8 = 0$
16. $9x^2 = 25$
17. $4x^2 = 1$
18. $x^2 - 80 = 0$
19. $9x^2 - 4 = 0$
20. $4x^2 - 6 = 19$
21. $9x^2 - 4 = 6$
22. $2x^2 - 9 = 40$
23. $8x^2 - 16 = -15$
24. $20x^2 = 3$
25. $16x^2 - 101 = -1$
26. $25x^2 - 7 = 13$
27. $5x^2 - 3 = 5$
28. $81x^2 = 18$
29. $4x^2 - 1 = -7$
30. $32x^2 - 5 = 10$
31. $4x^2 - 15 = 60$
32. $x^2 - 1 = 97$
33. $9x^2 - 4 = 28$
34. $(x + 2)^2 = 7$
35. $(3x + 2)^2 = 36$
36. $(x - 3)^2 = 1$
37. $(x - 3)^2 = 4$
38. $(2x - 1)^2 = 9$
39. $(3x - 7)^2 = 4$
40. $(3x + 5)^2 = 9$
41. $(6x - 2)^2 = 121$
42. $(7x - 10)^2 = 144$
43. $(3x - 2)^2 = 27$
44. $(5x - 3)^2 = 50$
45. $(2x - 5)^2 = 98$
46. $(4x + 1)^2 = 25$

Fill in the blanks.

47. An equation of the form $ax^2 + bx + c = 0$ is called a _____ equation.

48. To solve a pure quadratic equation, first write the equation in the form $x^2 = K$. Then, take the _____ of both sides and _____.

Solve.

49. A ball is dropped from a cliff that is 320 feet above the ground. The distance S (in feet) it falls in t seconds is given by the formula

$$S = 16t^2$$

How many seconds will it take for the ball to hit the ground? (Round the answer to the nearest tenth of a second.)

50. A rock falls from a tower that is 640 feet high. As it is falling, its height h (in feet) is given by the formula

$$h = 640 - 16t^2$$

where the falling time t is measured in seconds. How many seconds will it take for the rock to hit the ground? (Round the answer to the nearest tenth of a second. *Hint:* when the object hits the ground, $h = 0$.)

51. An object is dropped from a tower that is 400 feet above the ground. The distance S (in feet) it falls in t seconds is given by the formula

$$S = 16t^2$$

How many seconds will it take for the object to hit the ground?

52. An object falls from a building 576 feet high. While it is falling, its height h (in feet) is given by the formula

$$h = 576 - 16t^2$$

where the falling time t is measured in seconds. How many seconds will it take for the object to hit the ground?

Answers to Practice Problems **14a.** $5x^2 - 6x - 4 = 0$ **b.** $6x^2 - 2x - 2 = 0$ **c.** $6x^2 + 4x + 9 = 0$
15a. $5, -5$ **b.** $2\sqrt{2}, -2\sqrt{2}$ **c.** $\dfrac{2\sqrt{15}}{5}, -\dfrac{2\sqrt{15}}{5}$ **16a.** $3, -9$ **b.** $-1 + \sqrt{6}, -1 - \sqrt{6}$

12.5 Solving Quadratic Equations by Factoring

IDEA 1 The square root method cannot be used to solve the equation $x^2 - 3x - 10 = 0$ since it is not a pure quadratic equation. To solve this type of equation, you must use the factoring techniques discussed in Chapter 10 and a rule called the **zero-factor property.**

Zero-Factor Property

> If the product of two numbers is zero, then at least one of the numbers must be zero.
>
> If $a \cdot b = 0$, then $a = 0$ or $b = 0$, or both a and b are zero.

For example, to solve the equation $x^2 - 3x - 10 = 0$, we first factor the left side of the equation.

$$x^2 - 3x - 10 = 0$$
$$(x - 5)(x + 2) = 0$$

Next, the zero-factor property tells us that if the product of $(x - 5)$ and $(x + 2)$ is 0, then

$$x - 5 = 0 \quad \text{or} \quad x + 2 = 0$$

We now have two linear equations in one variable, solve them.

$$\begin{aligned} x - 5 &= 0 & \text{or} \quad x + 2 &= 0 \\ x &= 5 & x &= -2 \end{aligned}$$

The solutions are 5 and -2.

Check:
For 5
$x^2 - 3x - 10 = 0$
$5^2 - 3(5) - 10 = 0$
$25 - 15 - 10 = 0$
$0 = 0$

For -2
$x^2 - 3x - 10 = 0$
$(-2)^2 - 3(-2) - 10 = 0$
$4 + 6 - 10 = 0$
$0 = 0$

The above example implies that to solve a quadratic equation by factoring, we can use the procedure shown in the box at the top of page 429.

12.5 Solving Quadratic Equations by Factoring

To Solve a Quadratic Equation by Factoring:

1. Write the equation in standard form, $ax^2 + bx + c = 0$.
2. Factor the polynomial, if possible.
3. Make each factor equal to zero and solve the resulting first-degree equations.
4. The values found in step 3 are the solutions of the equation.
5. Check the solutions in the original equation.

Example 17 Solve by factoring.

a. $x^2 + 15x = -36$ b. $2x^2 - 13x + 20 = 0$ c. $3x^2 = 21x$
d. $4x^2 - 25 = 0$ e. $5(4x - 5) = 4x^2$

Solution To solve a quadratic equation by factoring, first write the equation in standard form. Next, factor the polynomial and then use the zero-factor property. Finally, solve the resulting first-degree equations.

17a.
$$x^2 + 15x = -36$$
$$x^2 + 15x + 36 = 0 \quad \text{Standard form}$$
$$(x + 12)(x + 3) = 0 \quad \text{Factor}$$
$$x + 12 = 0 \qquad x + 3 = 0 \quad \text{Zero-factor property}$$
$$x = -12 \qquad x = -3$$

The solutions are -12 and -3.

Check:

For -12:
$$x^2 + 15x = -36$$
$$(-12)^2 + 15(-12) = -36$$
$$144 + (-180) = -36$$
$$-36 = -36$$

For -3:
$$x^2 + 15x = -36$$
$$(-3)^2 + 15(-3) = -36$$
$$9 + (-45) = -36$$
$$-36 = -36$$

17b.
$$2x^2 - 13x + 20 = 0$$
$$(2x - 5)(x - 4) = 0$$
$$2x - 5 = 0 \qquad x - 4 = 0$$
$$2x = 5 \qquad x = 4$$
$$x = \frac{5}{2}$$

Check:

For $\frac{5}{2}$:
$$2x^2 - 13x + 20 = 0$$
$$2\left(\frac{5}{2}\right)^2 - 13\left(\frac{5}{2}\right) + 20 = 0$$
$$\frac{25}{2} - \frac{65}{2} + \frac{40}{2} = 0$$
$$0 = 0$$

For 4:
$$2x^2 - 13x + 20 = 0$$
$$2(4)^2 - 13(4) + 20 = 0$$
$$32 - 52 + 20 = 0$$
$$0 = 0$$

17c.
$$3x^2 = 21x$$
$$3x^2 - 21x = 0$$
$$3x(x - 7) = 0$$

$$3x = 0 \quad x - 7 = 0$$
$$x = 0 \quad\quad x = 7$$

The solutions are 0 and 7.

Check:
For 0	For 7
$3x^2 = 21x$	$3x^2 = 21x$
$3(0)^2 = 21(0)$	$3(7)^2 = 21(7)$
$0 = 0$	$147 = 147$

17d.
$$4x^2 - 25 = 0$$
$$(2x + 5)(2x - 5) = 0$$
$$2x + 5 = 0 \quad\quad 2x - 5 = 0$$
$$2x = -5 \quad\quad 2x = 5$$
$$x = -\frac{5}{2} \quad\quad x = \frac{5}{2}$$

The solutions are $-\frac{5}{2}$ and $\frac{5}{2}$.

Check:
For $-\frac{5}{2}$	For $\frac{5}{2}$
$4x^2 - 25 = 0$	$4x^2 - 25 = 0$
$4\left(-\frac{5}{2}\right)^2 - 25 = 0$	$4\left(\frac{5}{2}\right)^2 - 25 = 0$
$25 - 25 = 0$	$25 - 25 = 0$
$0 = 0$	$0 = 0$

17e.
$$5(4x - 5) = 4x^2$$
$$20x - 25 = 4x^2$$
$$4x^2 - 20x + 25 = 0 \quad \text{Add } -4x^2. \text{ Then, multiply by } -1.$$
$$(2x - 5)(2x - 5) = 0$$
$$2x - 5 = 0 \quad\quad 2x - 5 = 0$$
$$2x = 5 \quad\quad 2x = 5$$
$$x = \frac{5}{2} \quad\quad x = \frac{5}{2}$$

The solution is $\frac{5}{2}$.

Check:
$$5(4x - 5) = 4x^2$$
$$5\left(4\left[\frac{5}{2}\right] - 5\right) = 4\left(\frac{5}{2}\right)^2$$
$$5(10 - 5) = 4\left(\frac{25}{4}\right)$$
$$25 = 25$$

In the Solution to Example 17e, since the two solutions are equal, $\frac{5}{2}$ is called a **double root**.

Practice Problem 17 *Solve by factoring.*

a. $x^2 + 2x - 8 = 0$ b. $2x^2 = 5 - 3x$

c. $9x^2 - 100 = 0$ d. $5x^2 = 15x$

IDEA 2   Knowing how to solve quadratic equations by factoring will help you in solving some word problems.

Example 18 Solve.

a. The product of two consecutive even integers is 48. What are the integers?

b. The length of a rectangle is five inches more than twice its width. If the area of the rectangle is 33 square inches, find the length and width of the rectangle.

c. Part of a rectangular-shaped field that is six miles by 12 miles is to be used for a recreational area. The City Council decides that the recreational area will cover 40 square miles of the field (see Figure 12-2). The remainder of the field will be used as a picnic area. If the picnic area will have a uniform width, find the width of the picnic area.

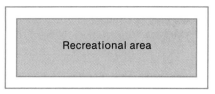

Figure 12-2

Solution To solve a word problem, (1) represent the unknown(s), (2) write an equation, (3) solve the equation written in step 2 and determine the solution(s), and (4) check.

18a. Step 1: x = an even integer
$x + 2$ = the next consecutive even integer

Step 2: Product of two numbers is 48
$$x(x + 2) = 48$$

Step 3:
$$x(x + 2) = 48$$
$$x^2 + 2x = 48$$
$$x^2 + 2x - 48 = 0$$
$$(x + 8)(x - 6) = 0$$
$$x + 8 = 0 \qquad x - 6 = 0$$
$$x = -8 \qquad x = 6$$
$$x + 2 = -6 \qquad x + 2 = 8$$

The solutions are -8 and -6, and 6 and 8.

Step 4: The answer checks since:

For -8 and -6	For 6 and 8
a. -8 and -6 are consecutive even integers	a. 6 and 8 are consecutive even integers
b. $(-8)(-6) = 48$	b. $(6)(8) = 48$

18b. Step 1: x = width of rectangle (in inches)
$2x + 5$ = length of rectangle (in inches)

Step 2: Draw the rectangle (see Figure 12-3)

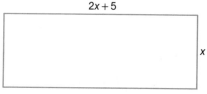

Figure 12-3

Area of rectangle = (length)(width)
33 = (2x + 5)(x)

Step 3:
$$33 = (2x + 5)(x)$$
$$33 = 2x^2 + 5x$$
$$2x^2 + 5x - 33 = 0$$
$$(2x + 11)(x - 3) = 0$$
$$2x + 11 = 0 \qquad x - 3 = 0$$
$$2x = -11 \qquad x = 3 \quad \text{(width)}$$
$$x = -\frac{11}{2} \qquad 2x + 5 = 11 \quad \text{(length)}$$

The solution $-\frac{11}{2}$ is meaningless since a rectangle cannot have a width that is negative. Therefore, the width of the rectangle is 3 inches and the length is 11 inches.

Step 4: The answer checks since:
    a. The length (11 inches) is 5 inches more than twice the width (3 inches).

    b. Area = (11 inches)(3 inches) = 33 square inches.

18c. Step 1: x = width of picnic area

Step 2: Draw the rectangular-shaped field (see Figure 12–4)

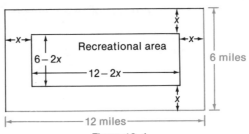

Figure 12–4

From this figure, the recreational area is $6 - 2x$ by $12 - 2x$. Therefore,

Area = (length)(width)
40 = (12 − 2x)(6 − 2x)

Step 3:
$$40 = (12 - 2x)(6 - 2x)$$
$$(12 - 2x)(6 - 2x) = 40$$
$$72 - 36x + 4x^2 = 40$$
$$4x^2 - 36x + 32 = 0$$
$$x^2 - 9x + 8 = 0 \quad \text{Divide both sides by 4}$$
$$(x - 8)(x - 1) = 0$$
$$x - 8 = 0 \qquad x - 1 = 0$$
$$x = 8 \qquad x = 1$$

The solution cannot be 8 miles since one of the dimensions of the field is only 6 miles. Therefore, the width of the picnic area is 1 mile.

Step 4: The answer checks since if the width of the picnic area is 1 mile, then the area of the recreational area is

$$A = (10 \text{ mi})(4 \text{ mi}) = 40 \text{ mi}^2$$

Practice Problem 18 *Solve.*

   a. The product of two consecutive positive integers is 90. What are the integers?

   b. The area of a square is 2.5 times its perimeter. Find the length of each side. (*Hint:* Perimeter = 4s.)

IDEA 3 Sometimes an equation containing rational expressions will lead to a quadratic equation.

Example 19 Solve for x: $x + \dfrac{6x}{x+1} = -\dfrac{6}{x+1}$

Solution First, multiply both sides of the equation by the LCD = $x + 1$ to clear the fractions. Then, solve the resulting equation.

$$x + \dfrac{6x}{x+1} = -\dfrac{6}{x+1}$$

$$(x+1)\left(x + \dfrac{6x}{x+1}\right) = (x+1)\left(-\dfrac{6}{x+1}\right) \quad \text{Multiply by LCD}$$

$$(x+1)(x) + (x+1)\left(\dfrac{6x}{x+1}\right) = (x+1)\left(-\dfrac{6}{x+1}\right) \quad \text{Distributive property}$$

$$x^2 + x + 6x = -6$$
$$x^2 + 7x = -6$$
$$x^2 + 7x + 6 = 0$$
$$(x+6)(x+1) = 0$$

$x + 6 = 0 \quad$ or $\quad x + 1 = 0$
$x = -6 \qquad\qquad x = -1$

Check:

For -6

$$x + \dfrac{6x}{x+1} = -\dfrac{6}{x+1}$$
$$-6 + \dfrac{6(-6)}{-6+1} \stackrel{?}{=} -\dfrac{6}{-6+1}$$
$$-6 + \dfrac{-36}{-5} \stackrel{?}{=} \dfrac{-6}{-5}$$
$$-6 + \dfrac{36}{5} \stackrel{?}{=} \dfrac{6}{5}$$
$$-\dfrac{30}{5} + \dfrac{36}{5} \stackrel{?}{=} \dfrac{6}{5}$$
$$\dfrac{6}{5} = \dfrac{6}{5}$$

For -1

$$x + \dfrac{6x}{x+1} = -\dfrac{6}{x+1}$$
$$-1 + \dfrac{6(-1)}{-1+1} \stackrel{?}{=} -\dfrac{6}{-1+1}$$
$$-1 + \dfrac{-6}{0} \stackrel{?}{=} -\dfrac{6}{0}$$

Clearly -1 is an extraneous root

The solution is -6.

Practice Problem 19 *Solve.*

$$y + \dfrac{4y}{y-1} = \dfrac{4}{y-1}$$

12.5 Exercises

Solve by factoring.

1. $x^2 + 2x - 8 = 0$
2. $2x^2 - 3x = 5$
3. $x^2 + 12x + 35 = 0$
4. $x^2 - 8x - 48 = 0$

5. $x^2 = 5x - 6$
6. $2x^2 + 5x - 3 = 0$
7. $x^2 + 3x - 10 = 0$
8. $3x^2 + 16x = 12$
9. $x^2 - 16 = 0$
10. $9x^2 - 25 = 0$
11. $x^2 - 14x + 49 = 0$
12. $9x^2 = 30x - 25$
13. $10x^2 - 29x + 21 = 0$
14. $6x^2 + 17x - 45 = 0$
15. $x^2 - 6x + 8 = 0$
16. $x^2 + x = 6$
17. $4x^2 + x = 0$
18. $8x^2 - 6x = 0$
19. $x^2 - 9x = 0$
20. $x^2 + 5x = 0$
21. $3x^2 = 2x$
22. $7x^2 = -5x$
23. $8x^2 + 8x = 0$
24. $5x^2 - 6x = 0$
25. $x^2 + 11x = -24$
26. $2x^2 - x - 15 = 0$
27. $(x - 3)(x - 5) = 3$
28. $2(x^2 + 10) = 13x$
29. $x(x - 6) + 9 = 0$
30. $-11x = 3(2x^2 + 1)$
31. $12x^2 - x - 20 = 0$
32. $14x^2 + 29x - 15 = 0$
33. $25x^2 = 36$
34. $6x^2 - 23x = -20$
35. $4x^2 - 5x - 6 = 0$
36. $3x(3x - 4) = -4$

Fill in the blanks.

37. The zero-factor property states that if the _____ of two numbers is zero, then at least one of the numbers must be _____.

38. To solve a quadratic equation by factoring, first write the equation in _____. Next, factor the polynomial and then use the _____. Finally, solve the resulting first degree equations and check your solutions.

39. When the two solutions of a quadratic equation are equal, the solution is called a _____.

Solve.

40. $(3p + 4)(p - 1) = -2$
41. $\dfrac{b^2}{2} + b = -b - 2$
42. $(b - 1)(b + 1) = 12(b - 3)$
43. $\dfrac{a^2}{2} = \dfrac{a}{4} + \dfrac{5}{2}$
44. $6(2x + 1) = x(5x - 1)$
45. $(2y + 9)^2 = 9(2y + 9)$

Solve each word problem.

46. The product of two consecutive odd integers is 63. What are the integers?

47. The product of two consecutive positive integers is 72. What are the integers?

48. The sum of two numbers is 17. Their product is 66. Find the numbers.

49. One number is five more than another. Their product is 24. Find the two numbers.

50. Two times a positive number is 15 less than its square. What is the number?

51. The length of a rectangle is four meters more than its width. If the area of the rectangle is 32 square meters, find the length and the width of the rectangle.

52. A rectangular-shaped lot is 20 yards by 14 yards. Jim decides to plant a garden on this lot with a uniform strip of lawn around it. If the area of the garden is 160 square yards, find the width of the strip of lawn.

53. The height of a triangle is six inches more than the base. If the area of the triangle is 20 square inches, find the base and the height of the triangle. (*Hint:* $A = \dfrac{1}{2}bh$.)

54. A rectangular-shaped floor is 20 feet by 12 feet. Betty decides to place a rug on the floor so that the wooden border surrounding the rug has a uniform width. If the area of the rug is 84 square feet, find the width of the wooden border.

55. The length of a rectangle is twice its width. If the area of the rectangle is numerically 20 more than the perimeter, find the length and the width of the rectangle.

Solve.

56. $y + \dfrac{1}{y} = 2$
57. $y + \dfrac{3}{y} = \dfrac{7}{2}$
58. $-\dfrac{4}{x^2 - 9} = \dfrac{x + 1}{x^2 + 3x}$
59. $\dfrac{6}{x^2 - 16} = \dfrac{5}{x^2 + x - 12}$

60. $\dfrac{2y}{y+2} + 1 = y$

61. $\dfrac{2}{y-1} + y = 4$

62. $x - \dfrac{5x}{x+1} = \dfrac{5}{x+1}$

63. $x + \dfrac{3x}{x-3} = \dfrac{9}{x-3}$

64. $1 + \dfrac{5}{y} = \dfrac{6}{y^2}$

65. $1 - \dfrac{3}{y} = \dfrac{10}{y^2}$

66. $\dfrac{3}{a-2} + \dfrac{7}{a+2} = \dfrac{a+1}{a-2}$

67. $\dfrac{2}{a-1} + \dfrac{3a}{a+2} = \dfrac{10a+18}{a^2+a-2}$

Answers to Practice Problems   17a. $-4, 2$ b. $-\dfrac{5}{2}, 1$ c. $\dfrac{10}{3}, -\dfrac{10}{3}$ d. $0, 3$ 18a. $9, 10$ b. 10 19. -4

12.6 The Quadratic Formula

IDEA 1 The methods that we have discussed so far for solving quadratic equations can be used to solve only some quadratic equations. However, if a quadratic equation is written in standard form, you can always find the solution by using the *quadratic formula*.

Quadratic Formula

> The solutions of the quadratic equation $ax^2 + bx + c = 0$ are given by the formula
>
> $$x = \dfrac{-b \pm \sqrt{b^2 - 4ac}}{2a}$$
>
> where a is the coefficient of the x^2 term; b is the coefficient of the x term; and c is the constant.

In this formula, the radicand $b^2 - 4ac$ is called the **discriminant**. When $b^2 - 4ac = 0$, the equation has a solution that is a double root. When $b^2 - 4ac$ is positive, the equation has two real solutions. When $b^2 - 4ac$ is negative, the equation has no real solutions.

Let us now solve the equation $x^2 - 8x + 7 = 0$ by the quadratic formula. Since the equation is in standard form, $a = 1$, $b = -8$, and $c = 7$. Now, substitute these values into the formula and simplify.

$$\begin{aligned} x &= \dfrac{-b \pm \sqrt{b^2 - 4ac}}{2a} \\ &= \dfrac{-(-8) \pm \sqrt{(-8)^2 - 4(1)(7)}}{2(1)} \\ &= \dfrac{8 \pm \sqrt{64 - 28}}{2} \\ &= \dfrac{8 \pm \sqrt{36}}{2} \\ &= \dfrac{8 \pm 6}{2} \end{aligned}$$

Finally, write the two solutions separately by first using the plus sign and then the minus sign.

$$x = \frac{8+6}{2} \quad \text{or} \quad x = \frac{8-6}{2}$$
$$x = \frac{14}{2} \quad\quad\quad x = \frac{2}{2}$$
$$x = 7 \quad\quad\quad x = 1$$

The solutions are 7 and 1 (the check is left for the student).

NOTE: The above equation could have been solved by factoring. Therefore, when you are asked to solve a quadratic equation, first check to see if the equation can be solved by factoring since this method is usually less time-consuming. If it cannot be solved by factoring, then use the quadratic formula.

The above results suggest that the following procedure can be used to solve quadratic equations by the quadratic formula.

To Solve a Quadratic Equation by Using the Quadratic Formula:

1. Write the equation in standard form, $ax^2 + bx + c = 0$.
2. Substitute the values of a, b, and c into the quadratic formula.
3. Simplify the resulting expression.

Example 20 Solve by using the quadratic formula.

a. $3x^2 - 4x - 2 = 0$ b. $4x^2 - 20x + 25 = 0$

c. $x^2 - 2x + 3 = 0$ d. $\frac{1}{3}x^2 + \frac{5}{6}x = \frac{1}{2}$

e. $4x(x - 3) = 3x^2 - 6x - 1$

Solution To solve a quadratic equation by the quadratic formula, first write the equation in standard form. Then, substitute the values for a, b, and c into the quadratic formula and simplify.

20a. Since $3x^2 - 4x - 2 = 0$ is in standard form, use the values $a = 3$, $b = -4$, and $c = -2$.

$$x = \frac{-b \pm \sqrt{b^2 - 4ac}}{2a}$$
$$x = \frac{-(-4) \pm \sqrt{(-4)^2 - 4(3)(-2)}}{2(3)}$$
$$x = \frac{4 \pm \sqrt{16 + 24}}{6}$$
$$x = \frac{4 \pm \sqrt{40}}{6}$$
$$x = \frac{4 \pm 2\sqrt{10}}{6} \quad\quad \textit{Think:} \;\; \sqrt{40} = \sqrt{4}\sqrt{10} = 2\sqrt{10}$$

$$x = \frac{\cancel{2}(2 \pm \sqrt{10})}{\cancel{6}_3} \qquad \text{Factor}$$

$$x = \frac{2 \pm \sqrt{10}}{3}$$

The solutions are $\frac{2 + \sqrt{10}}{3}$ and $\frac{2 - \sqrt{10}}{3}$.

We can use the square root to find an approximate value for the $\sqrt{10}$. From Appendix Table A-4, $\sqrt{10} \doteq 3.162$. Therefore, the approximate solutions are

$$\frac{2 + \sqrt{10}}{3} \doteq \frac{2 + 3.162}{3} \doteq \frac{5.162}{3} \doteq 1.7$$

$$\frac{2 - \sqrt{10}}{3} \doteq \frac{2 - 3.162}{3} \doteq \frac{-1.162}{3} \doteq -0.4$$

20b. Since $4x^2 - 20x + 25 = 0$, $a = 4$, $b = -20$, and $c = 25$.

$$x = \frac{-b \pm \sqrt{b^2 - 4ac}}{2a}$$

$$x = \frac{-(-20) \pm \sqrt{(-20)^2 - 4(4)(25)}}{2(4)}$$

$$x = \frac{20 \pm \sqrt{400 - 400}}{8}$$

$$x = \frac{20 \pm \sqrt{0}}{8}$$

$$x = \frac{20 \pm 0}{8}$$

$$x = \frac{20 + 0}{8} \qquad x = \frac{20 - 0}{8}$$

$$x = \frac{5}{2} \qquad x = \frac{5}{2}$$

The solution is $\frac{5}{2}$, which is a double root.

20c. Since $x^2 - 2x + 3 = 0$, $a = 1$, $b = -2$, and $c = 3$.

$$x = \frac{-b \pm \sqrt{b^2 - 4ac}}{2a}$$

$$x = \frac{-(-2) \pm \sqrt{(-2)^2 - 4(1)(3)}}{2(1)}$$

$$x = \frac{2 \pm \sqrt{4 - 12}}{2}$$

$$x = \frac{2 \pm \sqrt{-8}}{2}$$

The radical $\sqrt{-8}$ is not a real number. Therefore, the equation has no real solution.

Alternate method

The discriminant of $x^2 - 2x + 3 = 0$ is

$$b^2 - 4ac = (-2)^2 - 4(1)(3)$$
$$= 4 - 12$$
$$= -8$$

Since the discriminant is negative, $x^2 - 2x + 3 = 0$ has no real solution.

20d. To eliminate the fractions, multiply both sides of the equation by 6 (LCD).

$$6\left(\frac{1}{3}x^2 + \frac{5}{6}x\right) = 6\left(\frac{1}{2}\right)$$

$$2x^2 + 5x = 3$$

Now, write the equation in standard form.

$$2x^2 + 5x = 3$$
$$2x^2 + 5x - 3 = 0$$

In standard form, $a = 2$, $b = 5$, and $c = -3$

$$x = \frac{-b \pm \sqrt{b^2 - 4ac}}{2a}$$

$$x = \frac{-5 \pm \sqrt{5^2 - 4(2)(-3)}}{2(2)}$$

$$x = \frac{-5 \pm \sqrt{25 + 24}}{4}$$

$$x = \frac{-5 \pm \sqrt{49}}{4}$$

$$x = \frac{-5 \pm 7}{4}$$

$$x = \frac{-5 + 7}{4} \qquad x = \frac{-5 - 7}{4}$$

$$x = \frac{1}{2} \qquad x = -3$$

The solutions are $\frac{1}{2}$ and -3.

20e. Write the equation in standard form.

$$4x(x - 3) = 3x^2 - 6x - 1$$
$$4x^2 - 12x = 3x^2 - 6x - 1$$
$$x^2 - 6x + 1 = 0 \quad \text{Add } -3x^2, 6x, \text{ and } 1$$

In standard form, $a = 1$, $b = -6$, and $c = 1$.

$$x = \frac{-b \pm \sqrt{b^2 - 4ac}}{2a}$$

$$x = \frac{-(-6) \pm \sqrt{(-6)^2 - 4(1)(1)}}{2(1)}$$

$$x = \frac{6 \pm \sqrt{36 - 4}}{2}$$

$$x = \frac{6 \pm \sqrt{32}}{2}$$

$$x = \frac{6 \pm 4\sqrt{2}}{2}$$

$$x = \frac{\cancel{2}(3 \pm 2\sqrt{2})}{\cancel{2}}$$

$$x = 3 \pm 2\sqrt{2}$$

The solutions are $3 + 2\sqrt{2}$ and $3 - 2\sqrt{2}$.

Practice Problem 20 **Solve using the quadratic formula.**

a. $x^2 - 5x + 4 = 0$ **b.** $5x^2 - 2 = 3x$

c. $x^2 - 10x = -23$ **d.** $4x(x + 3) = 7$

12.6 Exercises

Solve using the quadratic formula.

1. $x^2 - 6x + 8 = 0$
2. $x^2 - 4x + 3 = 0$
3. $5x^2 - 3x + 2 = 0$
4. $9x^2 - 12x + 4 = 0$
5. $2x^2 - 13x = -20$
6. $6x^2 - 7x - 5 = 0$
7. $x^2 - 2x + 1 = 0$
8. $x^2 = 2 - 2x$
9. $2x^2 - 3x + 5 = 0$
10. $x^2 + 2x - 5 = 0$
11. $3x^2 - 2x - 8 = 0$
12. $x^2 = 25$
13. $9x^2 - 16 = 0$
14. $3x^2 + 6x = 0$
15. $9x^2 = 15x$
16. $x^2 = 10x - 22$
17. $3x^2 - 7x = 1$
18. $5x^2 - 8x = 3$
19. $3x^2 + 2x + 7 = 0$
20. $2x^2 = 7x - 3$
21. $x^2 = 6x + 3$
22. $4x^2 - 5x - 2 = 0$
23. $2x^2 = 12x - 2$
24. $2x^2 - 5x = 3$
25. $4 + 4x - x^2 = 0$
26. $x^2 + 2x = 12 - 3x - x$
27. $x^2 = (2x - 2)(x + 2)$
28. $\frac{2}{3}x^2 = \frac{11}{3}x + 2$
29. $\frac{1}{8}x^2 = \frac{1}{2} - \frac{x}{2}$
30. $(2x + 1)^2 - 8x = 0$

Fill in the blanks.

31. In the quadratic formula, a is the _____ of the x^2-term, b is the _____ of the x-term, and c is the _____ term.

32. The radicand $b^2 - 4ac$ is called the _____. When $b^2 - 4ac$ is negative, the quadratic equation has _____ solutions.

33. To solve a quadratic equation by the quadratic formula, first write the equation in _____ form. Then substitute the values for _____, _____, and _____ into the quadratic formula and simplify the resulting expression.

Solve and approximate the solutions to the nearest tenth.

34. $x^2 - 10x + 23 = 0$
35. $3x^2 + 6x - 2 = 0$
36. $4x^2 - 8x - 2 = 3x^2 - 6x$
37. $4x^2 - 2 = 5x$
38. $x^2 = 8$
39. $4x^2 - 3x + 5 = 0$

Answers to Practice Problems   20a. 4, 1   b. $-\frac{2}{5}, 1$ c. $5 + \sqrt{2}, 5 - \sqrt{2}$ d. $\frac{1}{2}, -\frac{7}{2}$

Chapter 12 Summary

Important Terms

The **principal square root** of a number is its positive square root and it is written with the **radical symbol** $\sqrt{}$. The principal square root of 81 is written as $\sqrt{81}$. The principal square root of 100 is written as $\sqrt{100}$. The expression $\sqrt{x}$ is called a **radical** and x is called the **radicand**. The expression $\sqrt{16}$ is a radical and 16 is the radicand. Similarly, $\sqrt{8}$ is called a radical and 8 is the radicand. [Section 12.1/Idea 1]

A **rational number** is a number that can be expressed as a fraction $\frac{a}{b}$, where a and b are integers, and b is not equal to zero. $\frac{17}{12}$ and $\frac{-32}{34}$ are rational numbers. An *irrational number* is a number that cannot be expressed as a fraction $\frac{a}{b}$, where a and b are integers and b is not equal to zero. The numbers $\sqrt{17}$ and $\sqrt{2}$ are irrational numbers. [Section 12.1/Idea 2]

Like square roots are square roots having the same radicand. $5\sqrt{7}$ and $-8\sqrt{7}$ are like square roots. Similarly, $\sqrt{2}$ and $8\sqrt{2}$ are like square roots. [Section 12.3/Idea 1]

A **quadratic** or **second degree equation** in one variable is an equation that can be written in the form $ax^2 + bx + c = 0$ where a, b, and c are real numbers and $a \neq 0$. A quadratic equation written in this form is in **standard form**. The equations $2x^2 + 7x - 8 = 0$ and $x^2 - 7x + 8 = 0$ are quadratic equations that are written in standard form. A **pure quadratic equation** is one in which there is no x term. $3x^2 + 5 = 0$ is a pure quadratic equation. [Section 12.4/Ideas 1-2]

The **zero-factor property** says that if the product of two numbers is zero, then at least one of the numbers is zero. If $a \cdot b = 0$, then $a = 0$, $b = 0$, or both $a = 0$ and $b = 0$. [Section 12.5/Idea 1]

The **quadratic formula** says that the solutions of the quadratic equation $ax^2 + bx + c = 0$ are given by the formula

$$x = \frac{-b \pm \sqrt{b^2 - 4ac}}{2a}.$$

[Section 12.6/Idea 1]

The **discriminant** is the expression $b^2 - 4ac$ in the quadratic formula. The discriminant for the equation $2x^2 - 7x - 4 = 0$ is 81. [Section 12.6/Idea 1]

Important Skills

Multiplying Square Roots
$\sqrt{a}\sqrt{b} = \sqrt{a \cdot b}$

To multiply square roots, multiply the radicands; then take the square root of the product. [Section 12.2/Idea 1]

Simplifying Square Roots
$\sqrt{a \cdot b} = \sqrt{a}\sqrt{b}$

To simplify square roots, factor the radicand by factoring out the largest perfect square. Find the square root of each factor and express the result as a product. [Section 12.2/Idea 2]

Dividing Square Roots
$\frac{\sqrt{a}}{\sqrt{b}} = \sqrt{\frac{a}{b}}$

To divide one square root by another, write the radicands as a quotient; then take the square root of the quotient. [Section 12.2/Idea 3]

Finding Square Roots of Fractions
$\sqrt{\frac{a}{b}} = \frac{\sqrt{a}}{\sqrt{b}}$

To find the square root of a fraction, write the square root of the numerator over the square root of the denominator. Simplify if possible. [Section 12.2/Idea 4]

Rationalizing Denominators

To rationalize the denominator of a fraction, first multiply the numerator and denominator by a square root that will make the radicand in the denominator a perfect square; then, simplify. [Section 12.2/Idea 5]

Adding and Subtracting Square Roots

To add or subtract square roots, first simplify all square roots. Then, add or subtract the like square roots by using the distributive property. [Section 12.3/Idea 1]

Solving Pure Quadratic Equations

To solve a quadratic equation of the form $ax^2 + c = 0$ (square root method), first write the equation in the form $x^2 = K$. Then, take the square root of both sides of the equation. [Section 12.4/Idea 2]

Solving Quadratic Equations

To solve a quadratic equation by factoring, first write the equation in standard form, $ax^2 + bx + c = 0$. Next, factor the polynomial, make each factor each to zero, and solve the resulting first degree equations. [Section 12.5/Idea 1]

To solve a quadratic equation by using the quadratic formula, first write the equation in standard form, $ax^2 + bx + c = 0$. Then, substitute the values for a, b, and c into the quadratic formula $x = \dfrac{-b \pm \sqrt{b^2 - 4ac}}{2a}$ and simplify. [Section 12.5/Idea 1]

Chapter 12 Review Exercises

Perform the indicated operations and simplify.

1. $\sqrt{36}$
2. $-\sqrt{81}$
3. $\sqrt{2}\sqrt{14}$
4. $\sqrt{40x^2}$
5. $\dfrac{\sqrt{80}}{\sqrt{5}}$
6. $\sqrt{\dfrac{12}{49}}$
7. $\sqrt{3} + 4\sqrt{3}$
8. $\sqrt{45} - \sqrt{20}$
9. $\dfrac{1}{\sqrt{8}}$
10. $\sqrt{\dfrac{3}{5}}$
11. $\sqrt{891}$
12. $\sqrt{\dfrac{3}{20}}\sqrt{\dfrac{8}{15}}$
13. $\sqrt{5} + \sqrt{\dfrac{1}{5}}$
14. $\sqrt{5x}\sqrt{15x^5}$
15. $\sqrt{18} - 3\sqrt{32} + 5\sqrt{50}$

Find the decimal approximation of each square root.

16. $\sqrt{43}$
17. $\sqrt{252}$
18. $\sqrt{\dfrac{5}{4}}$

19. Solve by factoring: $6x^2 - 11x = 10$
20. Solve using the quadratic formula: $x^2 = 4x + 1$
21. Solve using the square root method: $4x^2 - 7 = 59$

Solve.

22. $x^2 + 2x - 8 = 0$
23. $x^2 = 49$
24. $8x^2 = 10x$
25. $3x^2 - 4x + 2 = 0$
26. $4x^2 = 7 - 12x$
27. $\dfrac{1}{2}x^2 = \dfrac{x}{6} + \dfrac{1}{3}$
28. $(3x - 8)^2 = 16$
29. $x(x + 16) = 16$

Solve by the quadratic formula.

30. $2x^2 - 7x = 4$
31. $\dfrac{2}{5}x^2 = \dfrac{3}{5}(2x - 1)$
32. $4x^2 + 2 = 5x$

Solve by the square root method.

33. $x^2 + 36 = 0$
34. $(x - 1)^2 = 50$
35. $(x + 8)^2 - 9 = 40$

Solve by factoring.

36. $4x^2 - 9 = 0$
37. $3x^2 = 21x$
38. $12x^2 - 7x - 12 = 0$

Solve.

39. $2x = \dfrac{3}{x} + 4$
40. $x + \dfrac{1}{x} = \dfrac{29}{10}$
41. $\dfrac{x + 6}{x + 3} = \dfrac{x + 4}{x + 2}$

Perform the indicated operation and simplify.

42. $\dfrac{10}{\sqrt{6}}$
43. $\sqrt{54x^8 y^{15} z^{12}}$
44. $3\sqrt{28x^3} + y\sqrt{63x}$

Solve.

45. Find the length and width of a square-shaped room if its area is 196 square yards.

46. The product of two consecutive positive integers is 90. What are the integers?

47. The base of a right triangle is three yards shorter than the height. If the hypotenuse is five yards, find the height and the base.

48. The perimeter of a rectangle is 48 feet. The area of this rectangle is 140 square feet. Find the length and the width of the rectangle. (*Hint:* The length plus the width is 24 feet.)

Chapter 12 Test

Perform the indicated operation.
1. $-\sqrt{81}$
2. $\sqrt{x^{30}}$
3. Find the decimal approximation of $\sqrt{63}$.

Perform the indicated operation and simplify.
4. $\sqrt{2}\ \sqrt{8}$
5. $\sqrt{3}\ \sqrt{x^3}\ \sqrt{24}$
6. $-\sqrt{80}$
7. $\sqrt{3600x^9}$
8. $\dfrac{\sqrt{72}}{\sqrt{2}}$
9. $\sqrt{\dfrac{25}{81}}$
10. $\sqrt{\dfrac{3}{5}}$
11. $8\sqrt{2} + 15\sqrt{2}$
12. $\sqrt{12} - 2\sqrt{75} + \sqrt{12}$

Solve.
13. $x^2 + 2 = 5 - 3x^2$

Solve by factoring.
14. $4x^2 - 15x = 9$
15. $3x^2 = 15x$
16. $x^2 - 6x = 3$
17. $4x^2 + 7x - 2 = 0$

Solve.
18. If the product of two consecutive even numbers is increased by 4, the result is 84. Find the numbers.
19. The length of a rectangle is 4 more than twice its width. If its area is 77 square inches, find its dimensions.
20. The sum of a number and its reciprocal is $\dfrac{17}{4}$. Find the number.

1. _____
2. _____
3. _____
4. _____
5. _____
6. _____
7. _____
8. _____
9. _____
10. _____
11. _____
12. _____
13. _____
14. _____
15. _____
16. _____
17. _____
18. _____
19. _____
20. _____

13 Geometry

Objectives The objectives for this chapter are listed below along with sample problems for each objective. By the end of this chapter you should be able to find the solutions to the given problems.

1. Find the perimeter of a polygon *(Section 13.1/Ideas 1 and 3)*.
   a.
   b.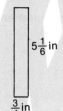

2. Find the circumference of a circle *(Section 13.1/Idea 2)*.
   a.
   b.

3. Find the area of a geometric figure *(Section 13.2/Ideas 1–6)*.
   a.
   b.
   c.
   d.

4. Find the volume of a three dimensional figure *(Section 13.3/Ideas 1–2)*.
   a.
   b.

c.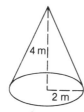

d.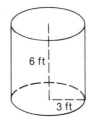

5. Find the length of the unknown side of a right triangle by using the Pythagorean Theorem *(Section 13.4/Ideas 1–2)*.

a.

b.

13.1 Perimeter

IDEA 1 A **polygon** is a closed figure whose sides are straight lines.

Example 1   The figures shown in Figure 13-1 are polygons.

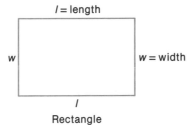
Rectangle

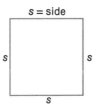
Square

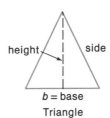
Triangle

Polygon

Figure 13-1

The **perimeter** of a polygon is the total distance around the figure. You can find the perimeter of a polygon by computing the sum of the lengths of its sides.

To Find the Perimeter of a Polygon:

1. Determine the lengths of all sides of the polygon.
2. Compute the sum of the lengths of the sides.

Example 2 Find the perimeter of the polygons shown in Figures 13–2 through 13–5.

a.

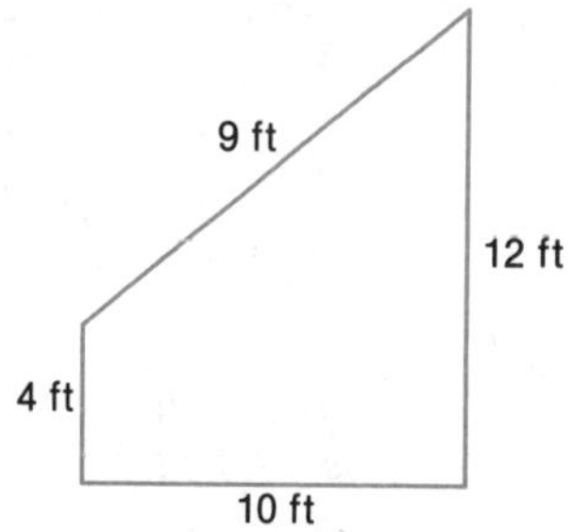

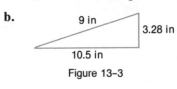

Figure 13–2

b.

Figure 13–3

c.

Figure 13–4

d.

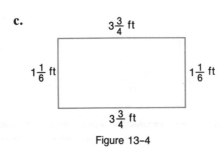

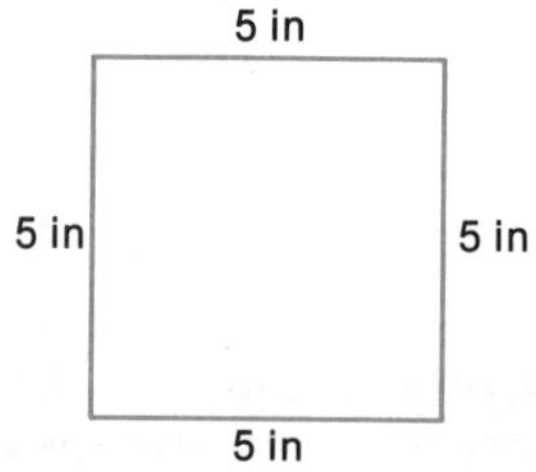

Figure 13–5

Solution To determine the perimeter (P) of a polygon, find the sum of the lengths of the sides.

2a. $P = 4\text{ ft} + 9\text{ ft} + 12\text{ ft} + 10\text{ ft}$
$= (4 + 9 + 12 + 10)\text{ ft}$
$= 35\text{ ft}$ Read as "35 feet"

2b. $P = 9\text{ in} + 3.28\text{ in} + 10.5\text{ in}$
$= 22.78\text{ in}$ Read as "22.78 inches"

2c. $P = 1\frac{1}{6}\text{ in} + 3\frac{3}{4}\text{ in} + 1\frac{1}{6}\text{ in} + 3\frac{3}{4}\text{ in}$
$= 1\frac{2}{12}\text{ in} + 3\frac{9}{12}\text{ in} + 1\frac{2}{12}\text{ in} + 3\frac{9}{12}\text{ in}$
$= 8\frac{22}{12}\text{ in}$
$= 9\frac{5}{6}\text{ in}$

Alternate method
Figure 13–4 is a rectangle. Its perimeter is given by the formula $P = 2l + 2w$, where l is the length of the rectangle and w is the width. Therefore,

$$\begin{aligned} P &= 2l + 2w \\ &= 2\left(3\frac{3}{4}\text{ in}\right) + 2\left(1\frac{1}{6}\text{ in}\right) \\ &= 2\left(\frac{15}{4}\text{ in}\right) + 2\left(\frac{7}{6}\text{ in}\right) \\ &= \frac{15}{2}\text{ in} + \frac{7}{3}\text{ in} \\ &= \frac{45}{6}\text{ in} + \frac{14}{6}\text{ in} \\ &= \frac{59}{6}\text{ in or } 9\frac{5}{6}\text{ in} \end{aligned}$$

2d. $P = 5\text{ in} + 5\text{ in} + 5\text{ in} + 5\text{ in}$
$= 20\text{ in}$

Alternate method

Figure 13-5 is a square. Its perimeter is given by the formula $P = 4s$, where s is the length of a side. Therefore,

$$P = 4s$$
$$= 4(5 \text{ in})$$
$$= 20 \text{ in}$$

Example 3   Find the perimeter of the Figure 13-6.

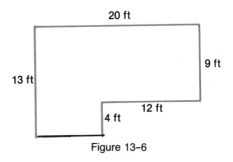

Figure 13-6

Solution   To find the perimeter, first find the length of the missing side. The missing side is $20 \text{ ft} - 12 \text{ ft} = 8 \text{ ft}$ (see Figure 13-7).

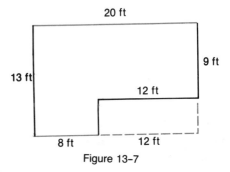

Figure 13-7

$$P = 13 \text{ ft} + 20 \text{ ft} + 9 \text{ ft} + 12 \text{ ft} + 4 \text{ ft} + 8 \text{ ft}$$
$$= 66 \text{ ft}$$

Practice Problem 1   *Find the perimeter of each figure.*

a.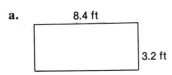

b.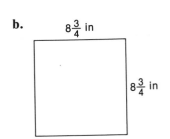

c.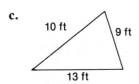

d.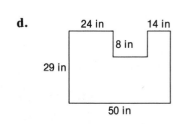

IDEA 2

A **circle** is a geometric figure in which all points are the same distance away from a fixed point called the **center** of the circle (see Figure 13-8). The distance from the center to any point on the circle is called the **radius** (r) of the circle. A line that passes through the center and connects any two points on the circle is called the **diameter** (d) of the circle. It should be noted that the diameter d is twice the radius r. Therefore, we have the relationship

$$d = 2r \quad \text{or} \quad r = \frac{d}{2}$$

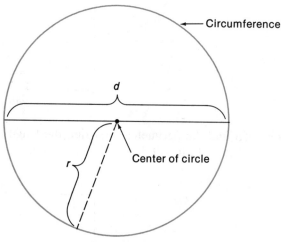

Figure 13-8

The perimeter (or distance around) a circle is called the **circumference**. To determine the circumference of a circle early mathematicians discovered that the ratio of the circumference of any circle to its diameter is the fixed number $3.14159265\ldots$, which is represented by the Greek letter π (pi). Thus,

$$\frac{C}{d} = \pi$$

Now, if we multiply both sides of this equation by d we obtain

$$d \cdot \frac{C}{d} = d \cdot \pi$$
$$C = \pi d$$

However, since $d = 2r$, the formula is sometimes written as

$$C = \pi d$$
$$C = \pi(2r)$$
$$C = 2\pi r$$

Based on the above statements, we have the following result.

13.1 Perimeter 449

Formula for the Circumference of a Circle

> The circumference of a circle is equal to the diameter times the number π,
>
> or
>
>
>
> $C = \pi d \qquad C = 2\pi r$
>
> where C = the circumference, π (pi) $\doteq$ 3.14, r = the radius, and d = the diameter of the circle.

Example 4 Find the circumferences of the circles in Figures 13–9 and 13–10 (units in inches).

a.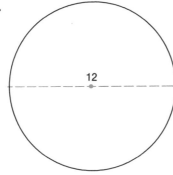

Figure 13–9

b.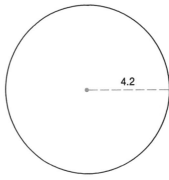

Figure 13–10

Solution To find the circumference of a circle, use the formula $C = \pi d$. Remember that $\pi \doteq 3.14$ and that $d = 2r$.

4a. $C = \pi d$
$\doteq (3.14)(12 \text{ in})$
$\doteq 37.68 \text{ in}$

4b. $C = \pi d$
$\doteq (3.14)(8.4 \text{ in})$
$\doteq 26.376 \text{ in}$

Example 5 Find the lengths of the semicircles in Figures 13–11 and 13–12 (units in feet).

a.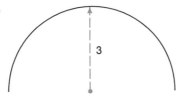

Figure 13–11

b.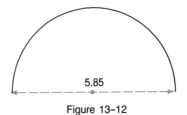

Figure 13–12

Solution Since a semicircle is half a circle, the length l of a semicircle is half the circumference of a circle. Thus,

$$l = \frac{1}{2} \cdot \pi \cdot d = \pi \cdot \frac{d}{2} = \pi r$$

where r is the radius of the circle.

5a. $l = \pi r$
$\doteq (3.14)(3 \text{ ft})$
$\doteq 9.42 \text{ ft}$

5b. $l = \pi r$
$\doteq (3.14)(2.925 \text{ ft})$
$\doteq 9.1845 \text{ ft}$

The radius $r = \frac{d}{2} = \frac{5.85 \text{ ft}}{2} = 2.925 \text{ ft}.$

Practice Problem 2 *Find the circumference of each circle and the length of the semicircle. The units are given in inches.*

a.
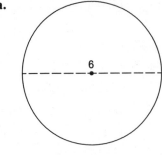

b.
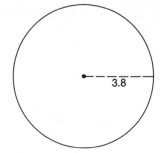

c.
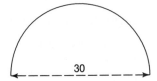

IDEA 3

Sometimes you will need to find the perimeter of some unfamiliar geometric figures.

Example 6 Find the perimeter of Figure 13–13. The units are given in feet.

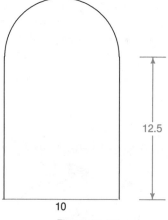

Figure 13–13

Solution To find the perimeter of an unfamiliar figure, first divide the figure into parts that you can recognize such as line segments, rectangles, circles or semicircles. Then find the sum of the lengths of these parts (see Figure 13–14).

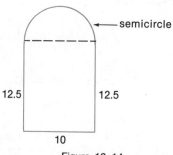

Figure 13–14

P = Length of side 1 + Length of side 2 + Length of side 3
    + Length of semicircle
  = 12.5 ft + 10 ft + 12.5 ft + (3.14)(5 ft)
  = 12.5 ft + 10 ft + 12.5 ft + 15.7 ft
  = 50.7 ft

Practice Problem 3 *Find the perimeter. The units are given in feet. The ends of the figure are semicircles.*

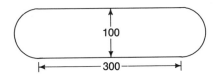

13.1 Exercises

Find the perimeter. The units are given in inches.

1.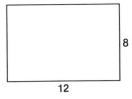

2.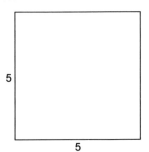

3.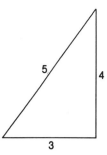

4.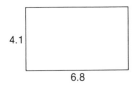

5.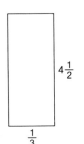

6.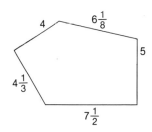

Find the perimeter. The units are given in feet.

7.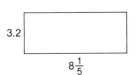

8.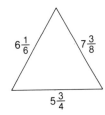

9.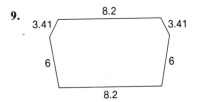

Find the circumferences of each circle and the length of each semicircle. The units are given in feet.

10.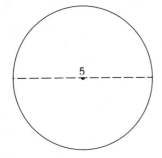

11.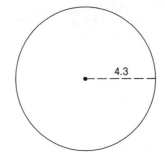

12.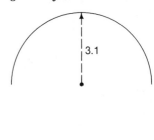

13.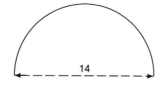

14.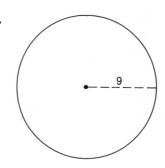

15.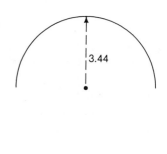

Find the perimeter. The units are given in yards.

16.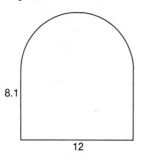

17.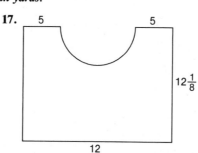

18.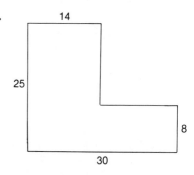

19.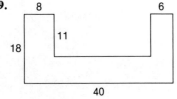

20.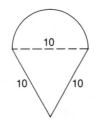

21.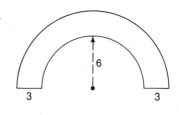

Fill in the blanks.

22. A _____ is a closed figure whose sides are straight lines.

23. The _____ of a polygon is the total distance around the figure.

24. The perimeter of a circle is called the _____.

25. To find the perimeter of a polygon, first determine the lengths of all _____ of the polygon. Next, compute the _____ of the lengths of the _____.

26. You can find the perimeter of a rectangle by using the formula _____.
27. You can find the circumference of a circle by using the formula _____.

Solve.

28. A rectangular-shaped garden is 30 feet by 19 feet. If fencing costs $3.50 a foot and Pete wants to build a fence around this garden, what is the cost of the fencing?

29. If Joe can put one foot of binding on a rug in $1\frac{1}{2}$ minutes, how long will it take him to put binding on a rug that is 16 feet by 12 feet?

30. Betty is planting a circular garden. How many feet of fencing will she need to enclose this garden if the diameter of the garden is 9.2 feet?

31. Cazzie wants to make a magnet by wrapping wire around a metal pole with a three-inch diameter. If he plans to wrap the wire around the pole 11 times, how much wire will he need?

32. The Johnson's den is sketched below. How much floor molding will they need to buy to go around the walls?

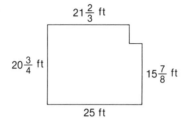

33. Fencing costs $6 per yard. Find the total cost of fencing in a diamond-shaped play area whose sides are six yards each.

Solve.

34. Fencing costs $6.25 per yard. Find the total cost of fencing in a diamond-shaped garden whose sides are 8.35 yards each.

35. Find the perimeter of a rectangle when the length is 26.582 inches and the width is 14.0836 inches?

36. Find the circumference of a circle when the radius is 8.088 centimeters.

Answers to Practice Problems   1a. 23.2 ft   b. 35 in   c. 32 ft   d. 174 in   2a. 18.84 in   b. 23.864 in   c. 47.1 in   3. 914 ft

13.2 Area

IDEA 1   Area is a measure of a surface, and it is expressed in square units. Two such units—a square inch (sq in) and a square centimeter (sq cm)—are shown in Figures 13–15 and 13–16.

Figure 13–15

Figure 13–16

NOTE: 1 inch is equal to approximately 2.54 centimeters.

When you are asked to find the area of a region, you will need to determine the number of square units contained within the region. Consider a rectangle with a length of 3 centimeters and a width of 2 centimeters (see Figure 13-17). Since this rectangle contains six square centimeters, its area is 6 sq cm or 6 cm². It should be noted that the same result can be obtained by multiplying the length (3 cm) times the width (2 cm).

$$A = (3 \text{ cm})(2 \text{ cm}) = 6 \text{ cm}^2$$

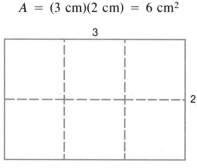

Figure 13-17

The above results suggest the following formula.

Formula for the Area of a Rectangle

> The area of a rectangle is equal to the length times the width,
>
> or 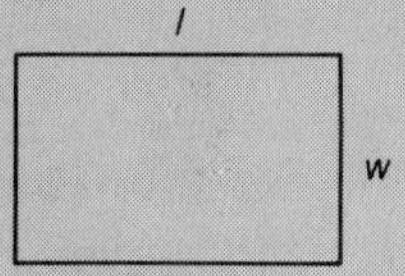
>
> $$A = l \cdot w$$
>
> where A = area, l = length, and w = width of the rectangle.

When using this formula, make sure that you express the length and width in the same units of measurement. Also, since a square is a rectangle in which the length and width are equal, we can also use the formula $A = l \cdot w$ to find the area of a square.

Example 7 Find the area of each rectangle in Figure 13-18 and 13-19. The units are given in feet.

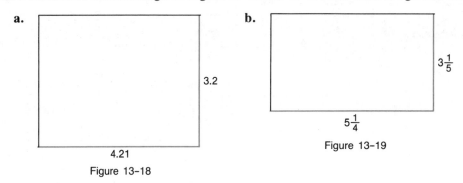

Figure 13-18

Figure 13-19

Solution To find the area of a rectangle, use the formula $A = lw$.

7a. $A = lw$
$ = (4.21 \text{ ft})(3.2 \text{ ft})$
$ = \boxed{13.472 \text{ ft}^2}$ Read as "13.472 square feet"

7b. $A = lw$
$ = \left(3\frac{1}{5} \text{ ft}\right)\left(5\frac{1}{4} \text{ ft}\right)$
$ = \left(\frac{16}{5} \text{ ft}\right)\left(\frac{21}{4} \text{ ft}\right)$
$ = \frac{\overset{4}{\cancel{16}} \cdot 21}{5 \cdot \underset{1}{\cancel{4}}} \text{ ft}^2$
$ = \frac{84}{5} \text{ ft}^2$
$ = 16\frac{4}{5} \text{ ft}^2$

Practice Problem 4 **Find the area of each rectangle. The units are given in inches.**

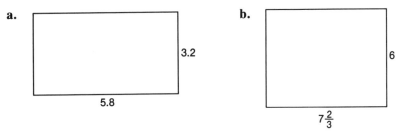

a. (rectangle with sides 5.8 and 3.2)

b. (rectangle with sides $7\frac{2}{3}$ and 6)

c. Find the area of a square whose sides are 8.45 feet.

IDEA 2

Let us now consider how to find the area of some other geometric figures. The first figure is a parallelogram (see Figure 13–20). A **parallelogram** is a four-sided polygon whose opposite sides are equal and parallel. In a parallelogram the base b is the length of one of its sides and the height h is the distance between the base and the side parallel to it.

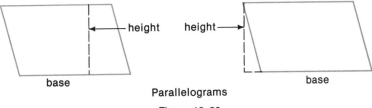

Parallelograms
Figure 13–20

Using the Figure 13–21 below we can show that since a rectangle is a special type of parallelogram, we can find the area of a parallelogram in the same manner that we determined the area of a rectangle.

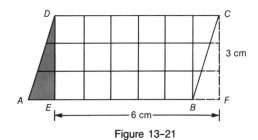

Figure 13–21

In other words, if we cut off the shaded area on the left of the parallelogram *ABCD* and attach it to the right of this parallelogram we will obtain the rectangle *DECF*. Please notice that the area of this resulting rectangle (18 cm²) is the same as the area of the original parallelogram *ABCD*. Also notice that this same result could have been determined by multiplying the length of the base of the parallelogram (6 cm) and its height (3 cm).

$$A = b \cdot h$$
$$= (6 \text{ cm})(3 \text{ cm})$$
$$= 18 \text{ cm}^2$$

The above results are summarized below.

Formula for the Area of a Parallelogram

The area of a parallelogram is equal to the base times the height,

or

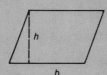

$$A = b \cdot h$$

where A = area, b = the base, and h = the height of the parallelogram.

Example 8 Find the area of each parallelogram in Figures 13–22 and 13–23.

a.
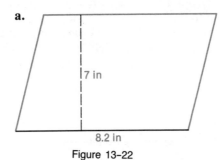

7 in

8.2 in

Figure 13–22

b.
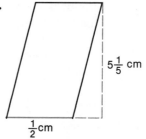

$5\frac{1}{5}$ cm

$\frac{1}{2}$ cm

Figure 13–23

Solution To find the area of a parallelogram, use the formula $A = bh$.

8a. $A = bh$
$= (8.2 \text{ in})(7 \text{ in})$
$= 57.4 \text{ in}^2$

8b. $A = bh$
$= \left(\frac{1}{2} \text{ cm}\right)\left(5\frac{1}{5} \text{ cm}\right)$
$= \left(\frac{1}{2} \text{ cm}\right)\left(\frac{26}{5} \text{ cm}\right)$
$= \frac{1 \cdot \overset{13}{\cancel{26}}}{\underset{1}{\cancel{2}} \cdot 5} \text{ cm}^2$
$= \frac{13}{5} \text{ cm}^2$
$= 2\frac{3}{5} \text{ cm}^2$

Practice Problem 5 *Find the area of each parallelogram. The units are given in centimeters.*

a.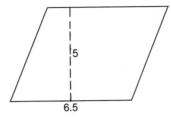

b.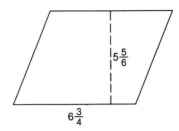

IDEA 3 The second geometric figure we wish to find the area of is a triangle (see Figure 13-24). Recall that a triangle is a three-sided polygon.

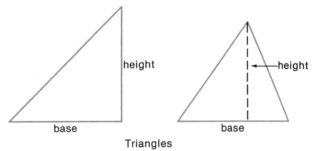

Triangles

Figure 13-24

In a triangle the base b is the length of one of the sides and the height h is the perpendicular distance from the base to the vertex opposite the base. If we carefully examine Figure 13-25 we can see that the area of a triangle ABD is equal to half the area of the parallelogram $ABCD$ that has the same base and height. Since the area of the parallelogram is $b \cdot h$, the area of the triangle is $\frac{1}{2} bh$.

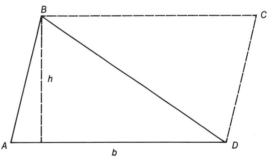

Figure 13-25

The above results are summarized below.

Formula for the Area of a Triangle

The area of a triangle is equal to one-half the product of the base times the height.

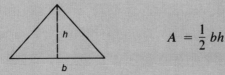

$$A = \frac{1}{2} bh$$

where A = area, b = the base, and h = the height of the triangle.

Example 9 Find the area of each triangle in Figures 13–26 and 13–27. The units are given in feet.

a.
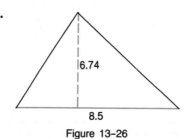
Figure 13–26

b.
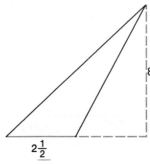
Figure 13–27

Solution To find the area of a triangle, use the formula $A = \frac{1}{2}bh$.

9a. $A = \frac{1}{2}bh$
$= \frac{1}{2}(8.5 \text{ ft})(6.74 \text{ ft})$
$= (0.5)(8.5 \text{ ft})(6.74 \text{ ft})$
$= 28.645 \text{ ft}^2$

9b. $A = \frac{1}{2}bh$
$= \frac{1}{2}\left(2\frac{1}{2} \text{ ft}\right)(8 \text{ ft})$
$= \left(\frac{1}{2}\right)\left(\frac{5}{2} \text{ ft}\right)\left(\frac{8}{1} \text{ ft}\right)$
$= \frac{1 \cdot 5 \cdot \cancel{8}^{\,4\,\,2}}{\cancel{2} \cdot \cancel{2} \cdot 1} \text{ ft}^2$
$= 10 \text{ ft}^2$

Practice Problem 6 *Find the area of each triangle. The units are given in centimeters.*

a.
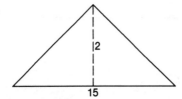

b.
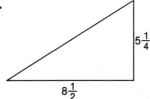

IDEA 4

The third figure you should be able to find the area of is a trapezoid (see Figure 13–28). A **trapezoid** is a four-sided polygon having only two parallel sides.

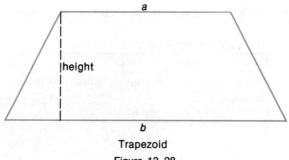
Trapezoid
Figure 13–28

In a trapezoid a and b are the lengths of the parallel sides and the height h is the distance between the parallel sides. The area of a trapezoid can be computed by dividing the trapezoid into two triangles and then finding the sum of these two areas (see Figure 13-29). In other words,

$$\begin{aligned} \text{Area of trapezoid } ABCD &= \text{Area of triangle } ABD + \text{Area of Triangle } BCD \\ &= \qquad A_1 \qquad + \qquad A_2 \\ &= \tfrac{1}{2} bh + \tfrac{1}{2} ah \\ &= \tfrac{1}{2} h(a + b) \qquad \text{Factor out } \tfrac{1}{2}h \end{aligned}$$

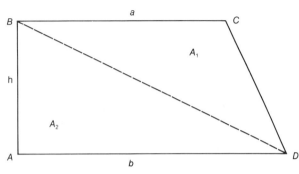

Figure 13-29

The above results are summarized below.

Formula for the Area of a Trapezoid

The area of a trapezoid is equal to one-half the product of the height times the sum of the lengths of the parallel sides,

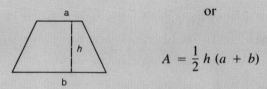

or

$$A = \tfrac{1}{2} h (a + b)$$

where A = area, a and b = lengths of the parallel sides, and h = the height of the trapezoid.

Example 10 Find the area of each trapezoid in Figures 13-30 and 13-31. The units are given in feet.

a.

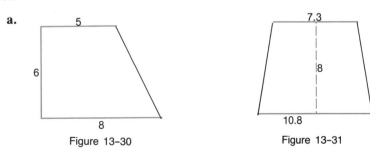

Figure 13-30   Figure 13-31

Solution To find the area of a trapezoid, use the formula $A = \tfrac{1}{2} h (a + b)$.

10a. $A = \frac{1}{2} h (a + b)$

$= \frac{1}{2} (6 \text{ ft})(5 \text{ ft} + 8 \text{ ft})$

$= \dfrac{\overset{3}{\cancel{(6 \text{ ft})}}(13 \text{ ft})}{\underset{1}{\cancel{2}}}$

$= 39 \text{ ft}^2$

10b. $A = \frac{1}{2} h (a + b)$

$= \frac{1}{2} (8 \text{ ft})(7.3 \text{ ft} + 10.8 \text{ ft})$

$= \dfrac{\overset{4}{\cancel{(8 \text{ ft})}}(18 \cdot 1 \text{ ft})}{\underset{1}{\cancel{2}}}$

$= 72.4 \text{ ft}^2$

Practice Problem 7 **Find the area. The units are given in centimeters.**

a.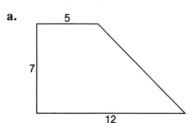

b.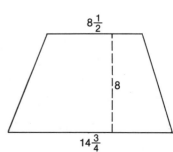

IDEA 5

In Section 13.1 you learned how to find the circumference of a circle. Let us now consider how to find the *area of a circle*.

Formula for the Area of a Circle

The area of a circle is equal to the product of π times the radius squared,

or

$$A = \pi r^2$$

where A = area, r = radius, and $\pi \doteq 3.14$.

Example 11 Find the area of each circle in Figures 13-32 and 13-33. The units are given in meters (m). A meter is approximately 3.3 feet.

a.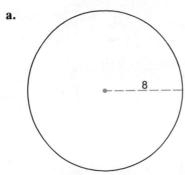

Figure 13-32

b.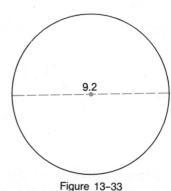

Figure 13-33

Solution To find the area of a circle, use the formula $A = \pi r^2$.

11a. $A = \pi r^2$
$\doteq (3.14)(8 \text{ m})^2$
$\doteq (3.14)(64 \text{ m}^2)$
$\doteq 200.96 \text{ m}^2$

11b. $A = \pi r^2$
$\doteq (3.14)(4.6 \text{ m})^2$
$\doteq (3.14)(21.16 \text{ m}^2)$
$\doteq 66.4424 \text{ m}^2$

$r = \frac{9.2 \text{ m}}{2} = 4.6 \text{ m}$

Example 12 Find the area of each semicircle in Figures 13-34 and 13-35. The units are given in inches.

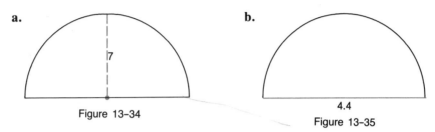

a.

Figure 13-34

b.

4.4

Figure 13-35

Solution Since a semicircle is half a circle, its area is found by using the formula $A = \frac{\pi r^2}{2}$.

12a. $A = \frac{\pi r^2}{2}$
$\doteq \frac{(3.14)(7 \text{ in})^2}{2}$
$\doteq \frac{(3.14)(49 \text{ in}^2)}{2}$
$\doteq \frac{153.86 \text{ in}^2}{2}$
$\doteq 76.93 \text{ in}^2$

12b. $A = \pi r^2$
$\doteq \frac{(3.14)(2.2 \text{ in})^2}{2}$
$\doteq \frac{(3.14)(4.84 \text{ in}^2)}{2}$
$\doteq \frac{15.1976 \text{ in}^2}{2}$
$\doteq 7.5988 \text{ in}^2$

Practice Problem 8 *Find the area of each circle and semicircle. The units are given in centimeters.*

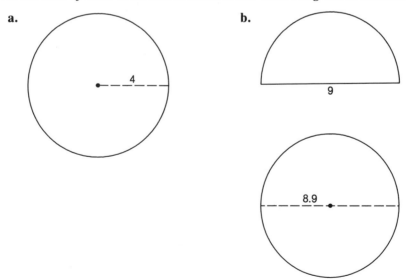

a.

4

b.

9

8.9

IDEA 6

Sometimes you may need to find the area of a geometric figure that is a combination of two or more of the figures we have discussed. When this occurs, use the procedure outlined in the box on page 462.

462 Geometry

> **To Find the Area of a Combined Geometric Figure:**
>
> 1. Divide the figure into rectangles, triangles, circles, line segments, or other familiar figures.
> 2. Find the area of each figure.
> 3. Find the sum or difference of these areas.

Example 13 Find the area of each figure in Figures 13-36 and 13-37. The units are given in feet.

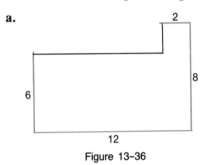

Figure 13-36   Figure 13-37

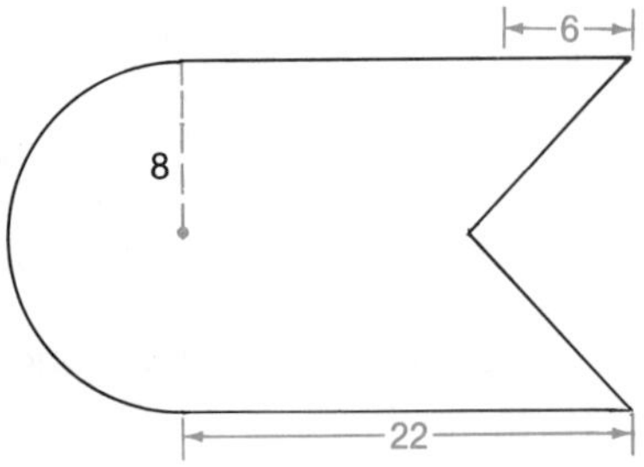

Solution To find the area of an unfamiliar figure, first divide the figure into two or more geometric figures (see Figures 13-38 and 13-39). Next, determine the area of these figures. Finally, compute the sum or differences of the areas.

13a.

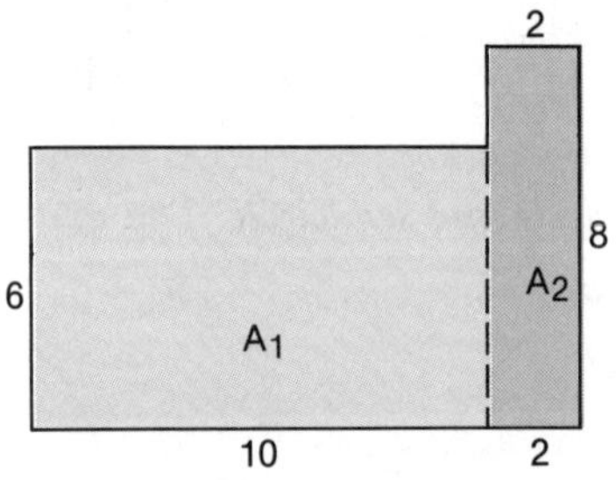

Figure 13-38

Total area is A = Area of rectangle + Area of rectangle

$$A = A_1 + A_2 \qquad A_1 = l \cdot w = (10 \text{ ft})(6 \text{ ft})$$
$$= 60 \text{ ft}^2 + 16 \text{ ft}^2 \qquad \qquad = 60 \text{ ft}^2$$
$$= 76 \text{ ft}^2 \qquad A_2 = l \cdot w = (2 \text{ ft})(8 \text{ ft})$$
$$\qquad \qquad = 16 \text{ ft}^2$$

13b.

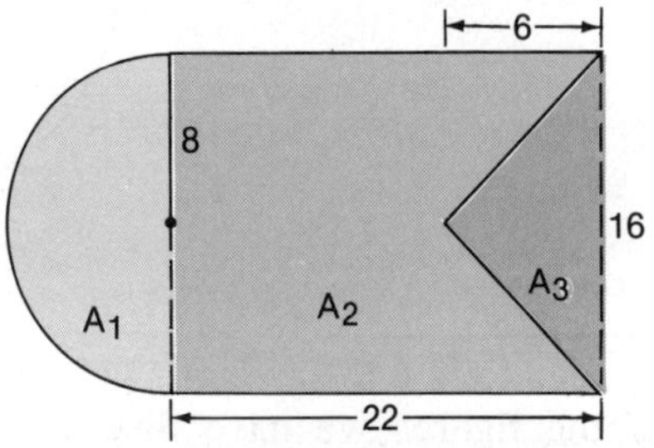

Figure 13-39

The base of the triangle and the width of the rectangle are the same as the diameter of the semicircle.

Total area is $A = $ Area of semicircle $+$ Area of rectangle $-$ Area of triangle

$$A_1 = \frac{\pi r^2}{2} \doteq \frac{(3.14)(8 \text{ ft})^2}{2} \doteq \frac{(3.14)(64 \text{ ft}^2)}{2} \doteq 100.48 \text{ ft}^2$$
$$A_2 = l \cdot w = (22 \text{ ft})(16 \text{ ft}) = 352 \text{ ft}^2$$
$$A_3 = \frac{1}{2} bh = \frac{1}{2}(16 \text{ ft})(6 \text{ ft}) = 48 \text{ ft}^2$$

Thus,

$$\begin{aligned} A &= A_1 + A_2 - A_3 \\ &\doteq 100.48 \text{ ft}^2 + 352 \text{ ft}^2 - 48 \text{ ft}^2 \\ &\doteq 404.48 \text{ ft}^2 \end{aligned}$$

Practice Problem 9 **Find the area of each figure. The units are given in meters.**

a. Find area of the unshaded region.   **b.**

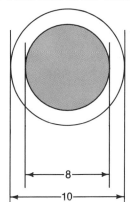

13.2 Exercises

Find the area. The units are given in feet.

1.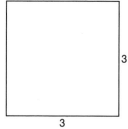

2.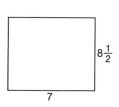

3.

4.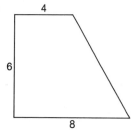

5.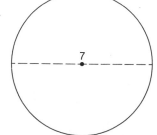

6.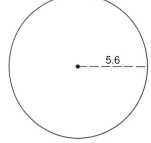

Find the area. The units are given in centimeters.

7.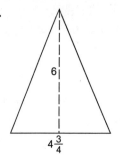

8.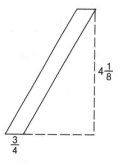

9.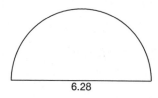

10.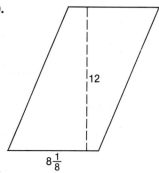

11.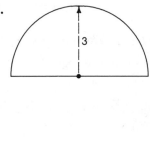

12.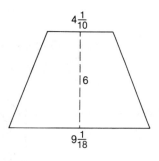

Find the area. The units are given in centimeters.

13.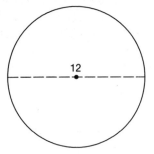

14.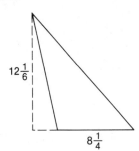

15.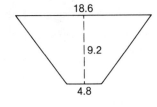

16.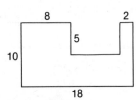

17.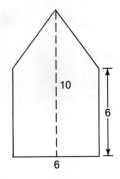

18.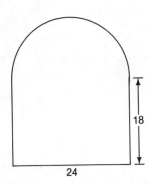

13.2 Exercises   465

19. 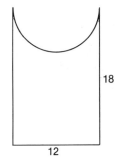   20. 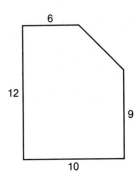   21. Find the area of the shaded region.
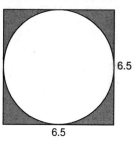

Fill in the blanks.

22. _____ is a measure of a surface, and it is expressed in _____ units.

23. A _____ is a four-sided polygon in which the opposite sides are equal and parallel.
    A _____ is a four-sided polygon having only two parallel sides.

24. The formulas for finding the areas of rectangles, triangles, parallelograms, trapezoids, and circles are as follows:
    a. rectangle: _____
    b. triangle: _____
    c. parallelogram: _____
    d. trapezoid: _____
    e. circle: _____

Solve.

25. Ben has a roll of paper that is 34 inches by 55 inches. How many sheets of paper $8\frac{1}{2}$ inches by 11 inches can he cut from this roll of paper?

26. A lot is 40 meters by 20 meters. If a circular swimming pool with a diameter of 12 meters is constructed on the lot, how much area is left over?

27. Tim wants to carpet a rectangular-shaped room that is 80 feet by 90 feet.
    a. How many square yards of carpeting will he need? (1 yd² = 9 ft²)
    b. If the carpeting costs $8.50 per square yard and Tim must pay 6% sales tax, what is the total cost of the carpeting?

28. A radio station broadcasts over an area with a 40-mile radius. How much area is this?

29. If one ounce of KEM Weed Killer will effectively treat $1\frac{1}{2}$ square yards of a lawn, how many ounces of KEM Weed Killer will Bill need to treat a triangular-shaped lawn with a height of 15 yards and a base of 10 yards?

30. A square tile (one foot on each side) costs 70¢ per tile.
    a. How many square tiles will Vanessa need to cover a rectangular-shaped room that is 19 feet by 23 feet?
    b. If Vanessa must pay $5\frac{3}{4}$% sales tax, what is the total cost of the tile? (Round to the nearest cent.)

Solve (round to nearest hundredth).

31. Find the area of a circle having a radius of 42.35 miles.

32. Find the area of a circle having a diameter of 160.25 meters.

33. Find the area of a semicircle having a diameter of 58.75 feet.

Answers to Practice Problems   4a. 18.56 in²   b. 69 in²   c. 71.4025 ft²   5a. 32.5 cm²   b. $39\frac{3}{8}$ cm²
6a. 15 cm²   b. $22\frac{5}{16}$ cm²   7a. $59\frac{1}{2}$ cm²   b. 93 cm²   8a. 50.24 cm²   b. 31.7925 cm²   c. 62.17985 cm²
9a. 28.26 m²   b. 106.08 m²

13.3 Volume

IDEA 1

Volume is the measure of the capacity within a three-dimensional figure, and it is expressed in cubic units. Two such units—a cubic inch (in³) and a cubic centimeter (cc or cm³)—are shown in Figures 13-40 and 13-41.

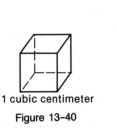
1 cubic centimeter
Figure 13-40

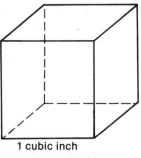
1 cubic inch
Figure 13-41

When you are asked to find the volume of a figure, you will need to determine the number of cubic units contained in that figure. Consider a rectangular box with a length of 4 cm, a width of 3 cm, and a height of 2 cm (see Figure 13-42). You can see that the box contains 12 cubes on the top layer and 12 cubes on the bottom layer. Thus, its volume is 24 cubic centimeters (24 cm³ or 24 cc).

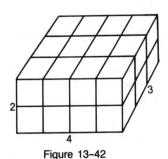
Figure 13-42

IDEA 2

Obviously, it will be inconvenient to mentally divide a geometric figure into unit cubes and then count these cubes. Therefore, to find the volume of a rectangular box, cylinder, cone, or sphere, use the formulas given below.

A **cylinder** is a figure shaped like a can. A cylinder has two circular bases which are joined by perpendicular sides. The radius r of a cylinder is the radius of the circular bases. The height h of a cylinder is the distance between the two bases.

A **sphere** is a figure shaped like a ball. The radius r of a sphere is the distance from its center to any point on the sphere.

Finally, a **cone** is a figure shaped like a funnel. A cone has a circular base and its lateral surface meets at a point directly above or below the center of the base. The radius r of a cone is the radius of the circle that forms the base. The height

13.2 Volume 467

h of a cone is the distance from the center of the base to the point directly above or below the center of the base.

Formulas for Volumes

The volume of a rectangular box is equal to the product of the length times the width times the height,

or

$$V = l \cdot w \cdot h$$

where V = volume, l = length, w = width, and h = height of the box.

The volume of a cylinder is equal to the product of π times the radius squared times the height,

or

$$V = \pi r^2 h$$

where V = volume, $\pi \doteq 3.14$, r = radius, and h = height of the cylinder.

The volume of a cone is equal to one-third the product of π times the radius squared times the height,

or

$$V = \frac{1}{3}\pi r^2 h = \frac{\pi r^2 h}{3}$$

where V = volume, $\pi \doteq 3.14$, r = radius, and h = height of the cone.

The volume of a sphere is equal to four-thirds the product of π times the radius cubed,

or

$$V = \frac{4}{3}\pi r^3 = \frac{4\pi r^3}{3}$$

where V = volume, $\pi \doteq 3.14$, and r = radius of the sphere.

Example 14   Find the volume of each figure in Figures 13–43 through 13–46. The units are given in meters.

a.
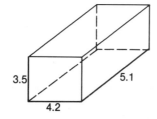

Figure 13–43

b.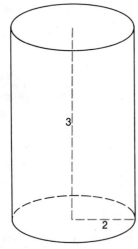

Figure 13-44

c.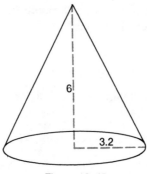

Figure 13-45

d.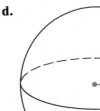

Figure 13-46

Solution **14a.** The volume of the rectangular box (Figure 13–43) is

$$V = lwh$$
$$= (4.2 \text{ m})(5.1 \text{ m})(3.5 \text{ m})$$
$$= 74.97 \text{ m}^3 \text{ (read as 74.97 cubic meters)}.$$

14b. The volume of the cylinder (Figure 13–44) is

$$V = \pi r^2 h$$
$$\doteq (3.14)(2 \text{ m})^2(3 \text{ m})$$
$$\doteq (3.14)(4 \text{ m}^2)(2 \text{ m})$$
$$\doteq 37.68 \text{ m}^3$$

14c. The volume of the cone (Figure 13–45) is

$$V = \frac{\pi r^2 h}{3}$$
$$\doteq \frac{(3.14)(3.2 \text{ m})^2(6 \text{ m})}{3}$$
$$\doteq \frac{(3.14)(10.24 \text{ m}^2)(6 \text{ m})}{3}$$
$$\doteq 64.3072 \text{ m}^3$$

14d. The volume of the sphere (Figure 13–46) is

$$V = \frac{4\pi r^3}{3}$$
$$\doteq \frac{4(3.14)(5 \text{ m})^3}{3}$$
$$\doteq \frac{4(3.14)(125 \text{ m}^3)}{3}$$
$$\doteq 523.33 \text{ m}^3. \quad \text{Round to nearest hundredth}$$

Practice Problem 10 Find the volume of each figure. The units are given in inches.

a.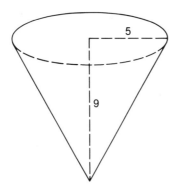

b.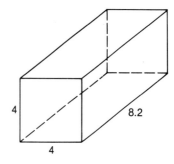

c.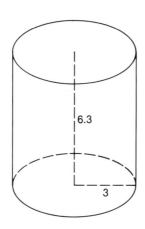

d.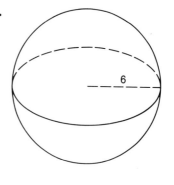

13.3 Exercises

Find the volume of each figure. The units are given in inches.

1.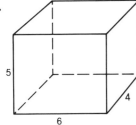

2.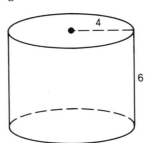

3.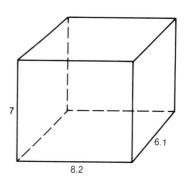

4.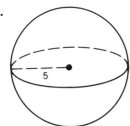

5.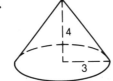

6.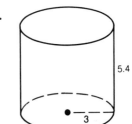

470 Geometry

7.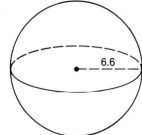

8.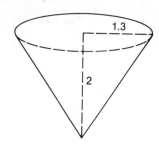

9.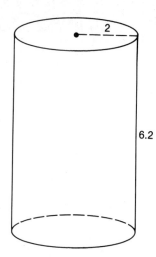

10.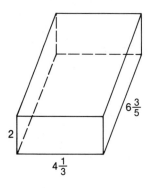

11.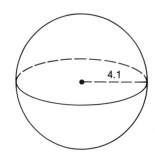

12.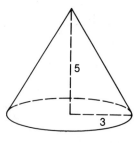

13.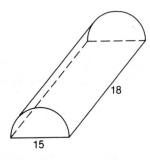

14.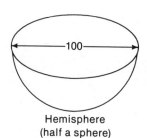

Hemisphere (half a sphere)

15.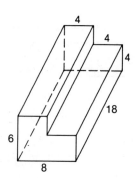

Fill in the blanks.

16. _____ is the measure of the capacity within a three-dimensional figure, and it is measured in _____ units.

17. The formulas for finding the volumes of rectangular boxes, cylinders, cones, and spheres are as follows:
  a. rectangular box: _____   **b.** cylinder: _____
  c. cone: _____   **d.** sphere: _____

Solve.

18. The figure shown at the top of page 471 is a cylinder capped with a hemisphere (half a sphere). Find its volume.

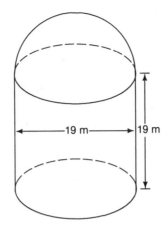

19. Find the volume of air contained in a balloon with a diameter of 8 cm.
20. A hole that is 18 feet wide, 11 feet deep, and six feet long is being made for a storage room. If the bed of a truck holds 11 yd³, how many truckloads of dirt must be hauled away? (one yard = three feet)
21. A cylindrical-shaped water tank is 44 inches in diameter and 21 inches high. If there are 231 in³ in one gallon, how many gallons of water can this tank hold?
22. A truck is eight feet wide, six feet high, and 21 feet long. How much dirt can be loaded into the truck?
23. A swimming pool is 40 feet long, 10 feet wide, and 6 feet deep.
    a. How many cubic feet of water will the pool hold?
    b. If 1 ft³ = 7.5 gallons, how many gallons of water will the pool hold?

Answers to Practice Problems   **10a.** 235.5 in³   **b.** 131.2 in³   **c.** 178.038 in³   **d.** 904.32 in³

13.4 Pythagorean Theorem

IDEA 1   A **right triangle** is a triangle that has a right angle (square corner) (see Figure 13–47). In a right triangle, the side opposite the right angle is called the **hypotenuse,** while the other two sides (*a* and *b*) are called the **legs.** It will be helpful to remember that *the sum of the squares of the legs of a right triangle is equal to the square of the hypotenuse* ($a^2 + b^2 = c^2$). This relationship concerning the sides of a right triangle is called the **Pythagorean theorem.** It was discovered by the famous Greek mathematician—Pythagoras around 500 B.C. This theorem is used to find the length of one side of a right triangle when the lengths of the other two sides are known.

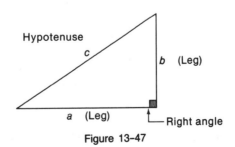

Figure 13–47

Geometry

The Pythagorean theorem is summarized below.

Pythagorean Theorem

For any right triangle, the sum of the squares of the legs is equal to the square of the hypotenuse,

or

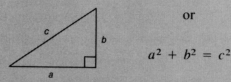

$$a^2 + b^2 = c^2$$

where a and b = the legs and c = the hypotenuse of a right triangle.

Example 15 Solve.

a. Find the length of the hypotenuse of a right triangle whose legs are 3 meters and 4 meters.

b. Find the length of the leg b in the right triangle shown in Figure 13–48.

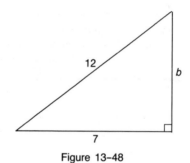

Figure 13–48

c. A ladder that is 15 feet long leans against a house and reaches a point on the house that is 10 feet above the ground. How far is the foot of the ladder from the side of the house?

Solution 15a. Make a sketch (see Figure 13–49) and then use the Pythagorean theorem to solve.

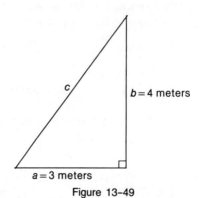

Figure 13–49

$$a^2 + b^2 = c^2$$
$$3^2 + 4^2 = c^2$$
$$9 + 16 = c^2$$
$$25 = c^2 \quad \text{Take the square root of both sides}$$
$$c = 5$$

Only the positive square root is considered since the length of a side of a triangle cannot be negative. Therefore, the hypotenuse is 5 meters.

15b. Sketch the triangle (see Figure 13–50).

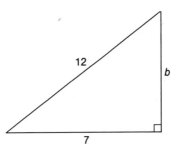

Figure 13-50

$$a^2 + b^2 = c^2$$
$$7^2 + b^2 = 12^2$$
$$49 + b^2 = 144$$
$$b^2 = 95$$
$$b \doteq 9.747 \quad \text{See Appendix Table A–4}$$

15c. Sketch the house and ladder (see Figure 13–51).

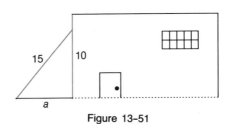

Figure 13-51

Step 1: a = distance (in feet) from foot of the ladder to the house
Step 2: $a^2 + b^2 = c^2$
$a^2 + 10^2 = 15^2$
Step 3: $a^2 + 10^2 = 15^2$
$a^2 + 100 = 225$
$a^2 = 125$
$a = \sqrt{125}$
$a = 5\sqrt{5} \quad \text{Think: } \sqrt{125} = \sqrt{25}\sqrt{5} = 5\sqrt{5}$
$a \doteq 5(2.236) \quad \text{See Appendix Table A–4}$
$a \doteq 11.18$

The foot of the ladder is approximately 11.18 feet from the house.

Practice Problem 11 **Solve.**

a. Find the length of the leg a in the right triangle in the figure shown at the top of page 474.

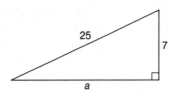

b. A brace is needed to reinforce a rectangular shaped gate. If the gate is 8 feet long and 6 feet high, how long is the brace?

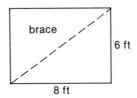

IDEA 2

Sometimes when working a word problem involving a right triangle you will need to use the quadratic formula.

Example 16 One leg of a right triangle is two meters longer than the other leg. If the hypotenuse is six meters long, what are the lengths of the two legs?

Solution   **Step 1:** x = length of one leg (in meters)
$x + 2$ = length of the other leg (in meters)
Step 2: Draw a right triangle (see Figure 13–52).

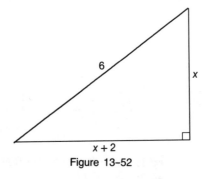

Figure 13–52

$$a^2 + b^2 = c^2$$
$$(x + 2)^2 + x^2 = 6^2$$

Step 3: $(x + 2)^2 + x^2 = 6^2$
$x^2 + 4x + 4 + x^2 = 36$
$2x^2 + 4x - 32 = 0$
$x^2 + 2x - 16 = 0$ Divided by 2

Since you cannot solve this equation by factoring, use the quadratic formula.

$a = 1, b = 2, c = -16$
$$x = \frac{-b \pm \sqrt{b^2 - 4ac}}{2a}$$

$$x = \frac{-2 \pm \sqrt{2^2 - 4(1)(-16)}}{2(1)}$$

$$x = \frac{-2 \pm \sqrt{4 + 64}}{2}$$

$$x = \frac{-2 \pm \sqrt{68}}{2}$$

$$x \doteq \frac{-2 \pm 8.246}{2} \qquad \sqrt{68} \doteq 8.246$$

$$x \doteq \frac{-2 + 8.246}{2} \qquad x \doteq \frac{-2 - 8.246}{2}$$

$$x \doteq \frac{6.246}{2} \qquad x \doteq \frac{-10.246}{2}$$

$$x \doteq 3.123 \qquad x \doteq -5.123$$

$$x + 2 \doteq 5.123$$

A length of -5.123 is meaningless. Therefore, one leg is approximately 3.123 meters and the other is approximately 5.123 meters.

Practice Problem 12 *Solve.*

One leg of a right triangle is three inches longer than the other leg. If the hypotenuse is five inches, find the length of the two legs.

13.4 Exercises

Find the length of the third side of each right triangle.

1.
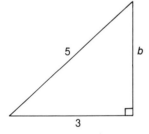

2.
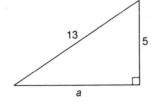

3.
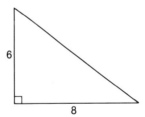

4.
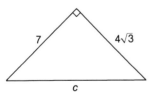

5.
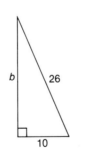

6.
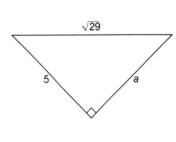

7.
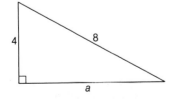

8.
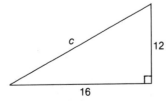

9.
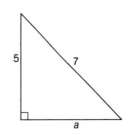

Find the length of the unknown side of each right triangle.

10. $a = 24$, $b = ?$, $c = 25$
11. $a = 12$, $b = ?$, $c = 13$
12. $a = 14$ in, $b = ?$, $c = 18$ in
13. $a = 8$ ft, $b = 6$ ft, $c = ?$
14. $a = 8$ ft, $b = 15$ ft, $c = ?$
15. $a = ?$, $b = 12$ cm, $c = 20$ cm
16. $a = ?$, $b = 12$ ft, $c = 15$ ft
17. $a = 5$ yd, $b = 5$ yd, $c = ?$
18. $a = 7$ m, $b = ?$, $c = 9$ m
19. $a = ?$, $b = 5$ ft, $c = 7$ ft

Fill in the blanks.

20. In a right triangle, the side opposite the right angle is called the _____, and the other two sides are called _____ of the triangle.

21. The Pythagorean theorem states that the sum of the squares of the _____ of a right triangle is equal to the square of the _____.

Solve.

22. A slow-pitch softball diamond (see figure below) is a square whose sides are each 60 feet long. What is the distance from first base to third base?

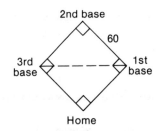

23. How long must a piece of board be to reach from the top of a building 15 feet high to a point on the ground eight feet from the building?

24. A 17-foot ladder is leaning against a wall. If the ladder is 15 feet from the wall, how many feet is it from the top of the ladder to the bottom of the wall?

25. While trying to "square" the corner of a tree house, a carpenter takes the measurements shown in the Figure below. Are the walls of the tree house square?

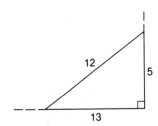

Solve.

26. One leg of a right triangle is 2 cm longer than the other. If the hypotenuse is 3 cm, find the length of the two legs.

27. One leg of a right triangle is four meters shorter than the other. If the hypotenuse is five meters, find the length of the two legs.

28. The length of a rectangle is two feet longer than the width. If the area is 18 ft², find the length and the width of the rectangle.

29. The length of a rectangle is three meters less than the width. If the area is 15 square meters, find the length and the width of the rectangle.

30. The area of a triangle is 14 square inches. If the height of the triangle is three inches more than the base, find the height and the base of the triangle. (*Hint:* $A = \frac{1}{2}bh$.)

31. The length of one leg of a right triangle is one foot longer than the other leg. If the hypotenuse is three feet, find the length of the two legs.

Answers to Practice Problems   **11a.** 24   **b.** 10 ft   **12.** 1.7015 inches, 4.7015 inches

Chapter 13 Summary

Important Terms

A **polygon** is a closed figure whose sides are straight lines.

The **perimeter** of a polygon is the total distance around the figure. Below are two polygons and their perimeters. [Section 13.1/Idea 1]

 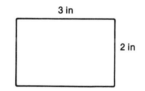

Perimeter = 13 ft   Perimeter = 10 in

A **circle** is a figure in which all points are the same distance away from a fixed point called the **center** of the circle. The distance from any point on the circle to the center is the **radius** of the circle. A line that passes through the center and connects any two points on the circle is the **diameter** of the circle. The perimeter of a circle is called the **circumference**. The **formula for the circumference of a circle** says that the circumference is equal to the product of π times the diameter. $c = \pi d$ or $c = 2\pi r$. [Section 13.1/Idea 2]

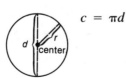

Area is the measure of a surface, and it is expressed in square units. The following figure has an area of one square centimeter (cm²) since it represents the area of a square whose sides are one centimeter in length. [Section 13.2/Idea 1]

1 sq cm

The **formula for the area of a rectangle** says that the area is equal to the product of the length times the width. $A = l \cdot w$. [Section 13.2/Idea 1]

A **parallelogram** is a four-sided polygon whose opposite sides are equal and parallel. The **formula for the area of a parallelogram** says that the area is equal to the product of the base times the height. $A = b \cdot h$. [Section 13.2/Idea 2]

$A = b \cdot h$

The **formula for the area of a triangle** (see top of page 478) says that the area is equal to one-half the product of the base times the height. $A = \frac{1}{2} b \cdot h$. [Section 13.2/Idea 3]

$$A = \tfrac{1}{2} b \cdot h$$

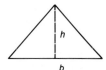

A **trapezoid** is a four-sided polygon having two parallel sides. The **formula for the area of a trapezoid** says that the area is equal to one-half the product of the height times the sum of the lengths of the parallel sides. $A = \tfrac{1}{2} h (a + b)$. [Section 13.2/Idea 4]

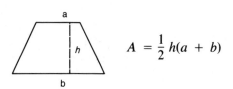

The **formula for the area of a circle** says that the area is equal to the product of π times the radius squared. [Section 13.2/Idea 5]

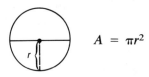

Volume is a measure of the capacity within a three-dimensional figure, and it is expressed in cubic units. The following figure has a volume of one cubic centimeter (cm³) since it represents the volume of a cube whose edges are each 1 centimeter in length. [Section 13.3/Idea 1]

A **cylinder** is a figure shaped like a can (top of next column). The **formula for the volume of a cylinder** says that the volume is equal to the product of π times the radius squared times the height. $V = \pi r^2 h$. [Section 13.3/Idea 2]

$$V = \pi r^2 h$$

A **sphere** is a figure shaped like a ball. The **formula for the volume of a sphere** says that the volume is equal to four-thirds the product of π times the radius cubed. $V = \tfrac{4}{3}\pi r^3$. [Section 13.3/Idea 2]

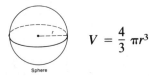

The **formula for the volume of a rectangular box** says that the volume is equal to the product of the length times the width times the height. $V = l \cdot w \cdot h$. [Section 13.3/Idea 2]

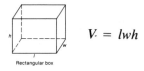

A **cone** is a figure shaped like a funnel. The **formula for the volume of a cone** says that the volume is equal to one-third the product of π times the radius squared times the height. $V = \tfrac{1}{3}\pi r^2 h$. [Section 13.3/Idea 2]

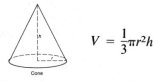

A **right triangle** is a triangle that has a right angle (square corner). The side opposite the right angle is called the **hypotenuse**, while the other two sides are called **legs**. The **Pythagorean theorem** says that for any right triangle the sum of the squares of the legs is equal to the square of the hypotenuse. $a^2 + b^2 = c^2$. [Section 13.4/Idea 1]

$$a^2 + b^2 = c^2$$

Important Skills

Finding Perimeters of Polygons To find the perimeter of a polygon, compute the sum of the length of the sides. [Section 13.1/Idea 1]

Finding Areas of Combined Geometric Figures To find the area of an unfamiliar figure, first divide the figure into two or more familiar geometric figures. Next, determine the area of these figures. Finally, compute the sum or difference of the areas. [Section 13.2/Idea 6]

Chapter 13 Review Exercises

Find the perimeter. The units are given in inches.

1.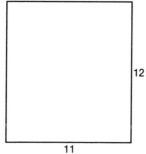

2.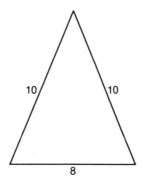

3.

Find the area. The units are given in feet.

4.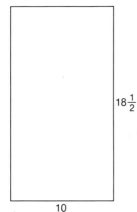

5.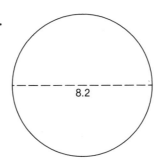

6.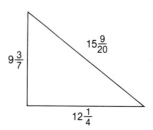

7. Find the circumference of the circle in Problem 5.

8. Find the area of the figure in Problem 3.

Find the volume. The units are given in inches.

9.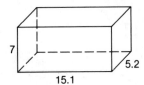
10.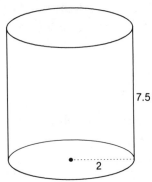
11.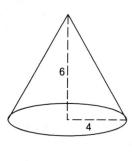

Find the perimeter. The units are given in feet.

12.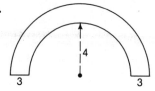
13.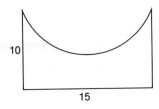
14.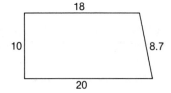

15. Find the area of the figure in Problem 12.

16. Find the area of the figure in Problem 13.

17. Find the area of the figure in Problem 14.

18. A storage tank 8.3 feet long and 4.1 feet wide is filled with water to a depth of 6.9 feet. How many cubic feet of water are in the tank?

19. How many square yards of carpeting will Gilly need to carpet a rectangular-shaped floor that measures 33 feet by 18 feet? (one yard = three feet)

20. A rectangular-shaped playground is $20\frac{1}{4}$ yards by $30\frac{5}{6}$ yards. How many yards of fencing will be needed to build a fence around the playground?

Find the length of the unknown side for each right triangle.

21. $a = 5$ in, $b = 4$ in, $c = ?$
22. $a = 15$ m, $b = 9$ m, $c = ?$
23. $a = ?$, $b = 5$ yd, $c = 13$ yd
24. $a = 12$ m, $b = ?$, $c = 15$ m
25. $a = 900$ in, $b = 1200$ in, $c = ?$
26. $a = ?$, $b = 800$ mi, $c = 1000$ mi

Solve.

27. The length of one leg of a right triangle is one yard longer than the other leg. If the hypotenuse is 3 yards, find the length of the two legs.

28. A ladder is leaning against the top of a 15 foot wall. If the bottom of the ladder is 20 feet from the wall, how long is the ladder?

29. Given a triangle with legs of 6 ft and 8 ft and a hypotenuse of 12 ft, is the triangle a right triangle? (*Hint:* use the Pythagorean theorem.)

30. Find the value of a sphere having a radius of 6 meters.

31. Find the area of a circle whose diameter is 4 meters.

32. Find the volume of a cylinder with a diameter of 12 meters and a height of 30 meters.

33. Find the area of a rectangle having a length of 82.5 feet and a width of 25 feet.

34. Find the area of a triangle having a base of 24 ft and a height of 12 ft.

35. Find the perimeter of a square in which the sides are 68.375 meters.

Chapter 13 Test

Name: _____

Class: _____

Find the perimeter. The units are in feet.

1.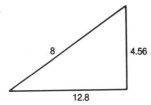

2.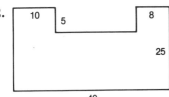

3.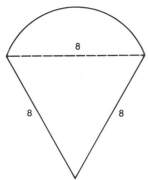

Find the circumference.

4.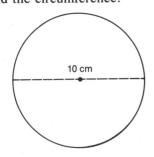

1. _____

2. _____

3. _____

4. _____

482   Geometry

Find the area. The units are in meters.

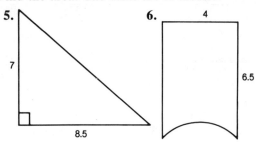

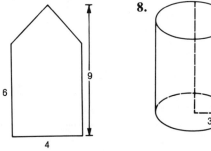

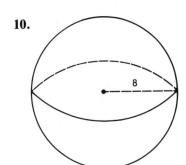

5. _____
6. _____
7. _____
8. _____
9. _____
10. _____

Solve.

11. How many feet of caulking are needed to insulate a rectangular window 6 feet long and 4.5 feet wide?

12. How many square feet of tile is needed to cover a floor that is 16.8 feet long and 10.5 feet wide?

13. A dome on a building is shaped like a hemisphere. If the diameter of the dome is 30 feet. What is its volume?

14. A 25 foot ladder is placed against a building at a point 24 feet from the ground. Find the distance from the base of the building to the base of the ladder.

15. The hypotenuse of a right triangle is 2 inches longer than its shortest leg. The other leg is 1 inch longer than the shortest leg. Find the lengths of the legs and the hypotenuse of the triangle.

11. _____
12. _____
13. _____
14. _____
15. _____

A Measurement

IDEA 1 The two major systems of measurement used today are the **English system** and the **metric system**. The relationships *within* the two systems are given in Tables A–1 and A–2. Table A–3 summarizes the basic relationships *between* the two systems.

Table A–1   Common English Units of Measure

Volume
1 gallon (gal) = 4 quarts (qt)
1 quart (qt) = 2 pints (pt)

Length
1 foot (ft) = 12 inches (in)
1 yard (yd) = 3 feet (ft)
1 mile (mi) = 5280 feet (ft)

Weight
1 pound (lb) = 16 ounces (oz)
1 ton = 2000 pounds (lb)

Time
1 minute (min) = 60 seconds (sec)
1 hour (hr) = 60 minutes (min)
1 day (da) = 24 hours (hr)

Table A–2   Common Metric Units of Measure

Prefixes
kilo (1000 times)
milli $\left(\frac{1}{1000} \text{ of}\right)$
centi $\left(\frac{1}{100} \text{ of}\right)$

Volume
1 liter (ℓ) = 1000 milliliters (ml)
1 milliliter (ml) = 1 cubic centimeter (cc)

Weight
1 kilogram (kg) = 1000 grams (g)
1 gram (g) = 1000 milligrams (mg)

Length
1 kilometer (km) = 1000 meters (m)
1 meter (m) = 100 centimeters (cm)
1 meter (m) = 1000 millimeters (mm)
1 centimeter (cm) = 10 millimeters (mm)

Table A-3 Approximate English-Metric Conversions

Volume
1.06 quarts (qt) $\doteq$ 1 liter (ℓ)
1 gallon (gal) $\doteq$ 3.8 liters (ℓ)

Weight
2.2 pounds (lb) $\doteq$ 1 kilogram (kg)
1 ounce (oz) $\doteq$ 28.4 grams (g)

Length
1 inch (in) $\doteq$ 2.54 centimeters (cm)
1 foot (ft) $\doteq$ 0.3 meters (m)
1.1 yards (yd) $\doteq$ 1 meter (m)
1 mile (mi) $\doteq$ 1.6 kilometers (km)

IDEA 2

To make conversions between the English and metric systems (or within the systems), we will use unit fractions. A **unit fraction** is a fraction that is equal to 1. To form unit fractions, use the tables. For example, Table A-1 states that 1 ft = 12 in. Therefore, the ratio of these units $\frac{1 \text{ ft}}{12 \text{ in}}$ or $\frac{12 \text{ in}}{1 \text{ ft}}$ is equal to 1, and $\frac{1 \text{ ft}}{12 \text{ in}}$ and $\frac{12 \text{ in}}{1 \text{ ft}}$ are unit fractions.

To Convert From One Unit of Measure to Another:

1. Express the given unit of measure as a fraction with a denominator of 1.
2. Write a unit fraction that has the same unit of measure in the denominator that appears in the numerator of the preceding fraction.
3. Repeat step 2 until the desired units appear in the numerator of a unit fraction.
4. Multiply and divide out common units of measure.

Example 1 Fill in the missing numbers.

    a. 6 gal = ? ℓ **b.** 3.5 m = ? cm   **c.** 3 tons = ? oz

Solution

Table A-3
↓

1a. $6 \text{ gal} = \frac{6 \text{ gal}}{1} \cdot \frac{3.8 \ell}{1 \text{ gal}}$
$= \frac{6(3.8 \ell)}{1}$
$= 22.8 \ell$

1b. $3.5 \text{ m} = \frac{3.5 \text{ m}}{1} \cdot \frac{100 \text{ cm}}{1 \text{ m}}$
$= \frac{3.5 (100 \text{ cm})}{1}$
$= 350 \text{ cm}$

1c. $3 \text{ tons} = \frac{3 \text{ tons}}{1} \cdot \frac{2000 \text{ lb}}{1 \text{ ton}} \cdot \frac{16 \text{ oz}}{1 \text{ lb}}$
$= \frac{3(2000)(16 \text{ oz})}{1}$
$= 96{,}000 \text{ oz}$

B Table of Squares, Square Roots, and Primes

n	n²	√n	Prime Factorization	n	n²	√n	Prime Factorization	Prime Numbers Through 200
1	1	1.000	—	51	2,601	7.141	3 · 17	2
2	4	1.414	Prime	52	2,704	7.211	2 · 2 · 13	3
3	9	1.732	Prime	53	2,809	7.280	Prime	5
4	16	2.000	2 · 2	54	2,916	7.348	2 · 3 · 3 · 3	7
5	25	2.236	Prime	55	3,025	7.416	5 · 11	11
6	36	2.449	2 · 3	56	3,136	7.483	2 · 2 · 2 · 7	13
7	49	2.646	Prime	57	3,249	7.550	3 · 19	17
8	64	2.828	2 · 2 · 2	58	3,364	7.616	2 · 29	19
9	81	3.000	3 · 3	59	3,481	7.681	Prime	23
10	100	3.162	2 · 5	60	3,600	7.746	2 · 2 · 3 · 5	29
11	121	3.317	Prime	61	3,721	7.810	Prime	31
12	144	3.464	2 · 2 · 3	62	3,844	7.874	2 · 31	37
13	169	3.606	Prime	63	3,969	7.937	3 · 3 · 7	41
14	196	3.742	2 · 7	64	4,096	8.000	2 · 2 · 2 · 2 · 2 · 2	43
15	225	3.873	3 · 5	65	4,225	8.062	5 · 13	47
16	256	4.000	2 · 2 · 2 · 2	66	4,356	8.124	2 · 3 · 11	53
17	289	4.123	Prime	67	4,489	8.185	Prime	59
18	324	4.243	2 · 3 · 3	68	4,624	8.246	2 · 2 · 17	61
19	361	4.359	Prime	69	4,761	8.307	3 · 23	67
20	400	4.472	2 · 2 · 5	70	4,900	8.367	2 · 5 · 7	71
21	441	4.583	3 · 7	71	5,041	8.426	Prime	73
22	484	4.690	2 · 11	72	5,184	8.485	2 · 2 · 2 · 3 · 3	79
23	529	4.796	Prime	73	5,329	8.544	Prime	83
24	576	4.899	2 · 2 · 2 · 3	74	5,476	8.602	2 · 37	89
25	625	5.000	5 · 5	75	5,625	8.660	3 · 5 · 5	97
26	676	5.099	2 · 13	76	5,776	8.718	2 · 2 · 19	101
27	729	5.196	3 · 3 · 3	77	5,929	8.775	7 · 11	103
28	784	5.292	2 · 2 · 7	78	6,084	8.832	2 · 3 · 13	107
29	841	5.385	Prime	79	6,241	8.888	Prime	109
30	900	5.477	2 · 3 · 5	80	6,400	8.944	2 · 2 · 2 · 2 · 5	113
31	961	5.568	Prime	81	6,561	9.000	3 · 3 · 3 · 3	127
32	1,024	5.657	2 · 2 · 2 · 2 · 2	82	6,724	9.055	2 · 41	131
33	1,089	5.745	3 · 11	83	6,889	9.110	Prime	137
34	1,156	5.831	2 · 17	84	7,056	9.165	2 · 2 · 3 · 7	139
35	1,225	5.916	5 · 7	85	7,225	9.220	5 · 17	149
36	1,296	6.000	2 · 2 · 3 · 3	86	7,396	9.274	2 · 43	151
37	1,369	6.083	Prime	87	7,569	9.327	3 · 29	157

n	n^2	$\sqrt{n}$	Prime Factorization	n	n^2	$\sqrt{n}$	Prime Factorization	Prime Numbers Through 200
38	1,444	6.164	2 · 19	88	7,744	9.381	2 · 2 · 2 · 11	163
39	1,521	6.245	3 · 13	89	7,921	9.434	Prime	167
40	1,600	6.325	2 · 2 · 2 · 5	90	8,100	9.487	2 · 3 · 3 · 5	173
41	1,681	6.403	Prime	91	8,281	9.539	7 · 13	179
42	1,764	6.481	2 · 3 · 7	92	8,464	9.592	2 · 2 · 23	181
43	1,849	6.557	Prime	93	8,649	9.644	3 · 31	191
44	1,936	6.633	2 · 2 · 11	94	8,836	9.695	2 · 47	193
45	2,025	6.708	3 · 3 · 5	95	9,025	9.747	5 · 19	197
46	2,116	6.782	2 · 23	96	9,216	9.798	2 · 2 · 2 · 2 · 2 · 3	199
47	2,209	6.856	Prime	97	9,409	9.849	Prime	
48	2,304	6.928	2 · 2 · 2 · 2 · 3	98	9,604	9.899	2 · 7 · 7	
49	2,401	7.000	7 · 7	99	9,801	9.950	3 · 3 · 11	
50	2,500	7.071	2 · 5 · 5	100	10,000	10.000	2 · 2 · 5 · 5	

perfect squares are in color

Answers to Selected Exercises

CHAPTER 1

CHAPTER 1 Objectives

1a. 10 **b.** 30 **c.** 15 **2a.** −45 **b.** −30 **c.** 9 **3a.** 10
b. −30 **c.** −15 **4a.** −45 **b.** 35 **5a.** −29 **b.** −10
6a. −50 **b.** 30 **7a.** −6 **b.** 6 **8a.** −2 **b.** −90
9. 22

Section 1.1

1. 300 **3.** 100 **5.** −39 **7.** −300 **9.** 35 **11.** 9
13. 95 **15.** 35 **17.** 104 **19.** 724 **21.** 0 **23.** 5
25. 324 **27.** negative

Section 1.2

1. −26 **3.** 6 **5.** −14 **7.** −129 **9.** 70 **11.** −91
13. −105 **15.** 95 **17.** −134 **19.** −23 **21.** 384
23. −500 **25.** 17 **27.** 30 **29.** 670 **31.** 170
33. −421 **35.** 56 **37.** −50 **39.** 0 **41.** −32
43. −10 **45.** −3 **47.** −15 **49.** −300 **51.** 50
53. −6 **55.** −19 **57.** −15 **59.** 103 **61.** −100
63. 35 **65.** negative **67.** commutative **69.** −60
71. 185 **73.** −34,135 **75.** 9,538

Section 1.3

1. 10 **3.** 4 **5.** −8 **7.** 0 **9.** −100 **11.** −21
13. −3 **15.** −25 **17.** 3 **19.** −205 **21.** −48
23. −18 **25.** −68 **27.** 20 **29.** 185 **31.** −15
33. 189 **35.** 1900 **37.** −284 **39.** −31 **41.** 5
43. −8 **45.** 32 **47.** 0 **49.** −500 **51.** 45
53. −45 **55.** zero **57.** 14,540 feet **59.** −34
61. −5 **63.** 5 **65.** −109,140 **67.** −27,582

Section 1.4

1. −27 **3.** −26 **5.** 33 **7.** −40 **9.** −34
11. −160 **13.** −30 **15.** −49 **17.** −33 **19.** −15
21. −18 **23.** −20 **25.** −25 **27.** 22 **29.** −23
31. −41 **33.** −17 **35.** 32 **37.** 1 **39.** −25
41. −110 **43.** subtraction, addition, subtraction, addition
45. subtraction **47.** account is overdrawn $4
49. $8,820 **51.** −12,108 **53.** −15,992

Section 1.5

1. 10 **3.** −15 **5.** −32 **7.** −24 **9.** 20 **11.** −24
13. 0 **15.** 30 **17.** −55 **19.** −28 **21.** 120 **23.** 24
25. −8 **27.** −105 **29.** 28 **31.** −8 **33.** 16
35. −60 **37.** −12 **39.** 120 **41.** addition
43. negative **45.** negative **47.** absolute values, sign
49. −30 **51.** 2700 **53.** 14,070 **55.** −46,800

Section 1.6

1. 8 **3.** −3 **5.** −5 **7.** −50 **9.** 15 **11.** 11 **13.** 1
15. −1 **17.** −13 **19.** −10 **21.** 22 **23.** −4
25. 8 **27.** −8 **29.** −100 **31.** 2 **33.** −13
35. 311 **37.** 0 **39.** undefined **41.** positive
43. absolute values, sign **45.** undefined **47.** −1,000
49. −45 **51.** 104 **53.** −100

Section 1.7

1. 4 **3.** −29 **5.** −65 **7.** 0 **9.** 6 **11.** 46
13. −23 **15.** 4 **17.** −56 **19.** 10 **21.** 7 **23.** −12
25. −19 **27.** 24 **29.** −11 **31.** 16 **33.** −30
35. 213 **37.** −64 **39.** −40 **41.** multiply, divide, add, subtract **43.** parentheses, multiply, divide, add, subtract
45. innermost grouping symbol **47.** 280 **49.** −17,315
51. −1,745

Section 1.8

1. 6 **3.** −4 **5.** 36 **7.** formula **9.** letters **11.** −23
13. −34 **15.** 48 **17.** 1 **19.** 31 **21.** −2 **23.** −36
25. 47 **27.** −28 **29.** −123 **31.** 64 feet **33.** 20
35. 130 **37.** 30 **39.** 7 **41.** 3 **43.** −10 **45.** 8
47. −4° **49.** 59° **51.** −22° **53.** 50° **55.** −3
57. 85

CHAPTER 1 Review Exercises

1. 20 *(Section 1.1)* **2.** 60 *(1.1)* **3.** 95 *(1.1)* **4.** 0 *(1.1)*
5. 20 *(1.3)* **6.** −60 *(1.3)* **7.** −95 *(1.3)* **8.** 0 *(1.3)*
9. −50 *(1.2)* **10.** 39 *(1.3)* **11.** 68 *(1.7)* **12.** −40 *(1.5)*
13. −5 *(1.6)* **14.** 6 *(1.6)* **15.** 53 *(1.2)* **16.** −120 *(1.5)*
17. 67 *(1.4)* **18.** −50 *(1.3)* **19.** 1 *(1.4)*
20. undefined *(1.6)* **21.** 220 *(1.7)* **22.** $131 *(1.2)*
23. 45 *(1.3)* **24.** 196 *(1.7)* **25.** −14 *(1.6)*
26. $100 *(1.4)* **27.** 8 *(1.2)* **28.** −10 *(1.6)*
29. −120 *(1.5)* **30.** 30 *(1.2)* **31.** −10 *(1.3)*
32. 10 *(1.3)* **33.** 48 *(1.7)* **34.** −4 *(1.7)*
35. False, 161 = 6 *(1.1)* **36.** True *(1.3)* **37.** False, the absolute value of every nonzero integer is a positive integer. *(1.1)* **38.** True *(1.3)* **39.** −7 *(1.6)* **40.** 13 *(1.6)*
41. 143 *(1.6)* **42.** −11 *(1.6)* **43.** −15 *(1.3)*
44. 16 *(1.7)* **45.** −556 *(1.7)* **46.** −7857 *(1.7)*
47. −704 *(1.4)* **48.** 20 *(1.7)* **49.** 50 *(1.7)*
50. 174 *(1.8)* **51.** −44 *(1.8)* **52.** −7 *(1.8)*

CHAPTER 1 Test

1. 85 **2.** 362 **3.** 6 **4.** −12 **5.** −56 **6.** −20
7. −74 **8.** 49 **9.** −45 **10.** 8 **11.** 35 **12.** −35
13. −120 **14.** 80 **15.** −154 **16.** −401 **17.** 3

18. undefined **19.** −13 **20.** −39 **21.** 37
22. 18 feet **23.** 100 **24.** 20° **25.** $1388

CHAPTER 2

CHAPTER 2 Objectives

1a. $2 \cdot 2 \cdot 2 \cdot 3$ **b.** $7 \cdot 11$ **c.** $3 \cdot 3 \cdot 3 \cdot 11$ **2a.** $\frac{7}{8}$ **b.** $\frac{3}{2}$
3a. $\frac{10}{3}$ **b.** $\frac{77}{8}$ **4a.** $7\frac{1}{2}$ **b.** $12\frac{1}{8}$ **5a.** $\frac{21}{10}$ **b.** $91\frac{1}{2}$
6a. $\frac{9}{8}$ **b.** $3\frac{279}{427}$ **7a.** $\frac{113}{198}$ **b.** $\frac{125}{132}$ **8a.** $12\frac{2}{7}$ **b.** $28\frac{23}{24}$
9a. $\frac{23}{40}$ **b.** $3\frac{5}{42}$ **10a.** $\frac{8}{15}$ **b.** $-\frac{5}{44}$ **10c.** $-\frac{2}{5}$

Section 2.1

1. numerator = 3; denominator = 14 **3.** numerator = 10; denominator = 9 **5.** $\frac{2}{5}$ **7.** $\frac{3}{4}$ **9.** 3 **11.** 3
13. 3 and 5 **15.** 2, 3, and 5 **17.** $2 \cdot 7$ **19.** $2 \cdot 2 \cdot 3 \cdot 5$
21. $3 \cdot 13$ **23.** $2 \cdot 3 \cdot 7$ **25.** $7 \cdot 11$ **27.** $2 \cdot 2 \cdot 2 \cdot 11$
29. $3 \cdot 3 \cdot 3 \cdot 3$ **31.** $5 \cdot 5$ **33.** $3 \cdot 5 \cdot 11$
35. $3 \cdot 3 \cdot 3 \cdot 5$ **37.** $5 \cdot 23$ **39.** $2 \cdot 3 \cdot 7 \cdot 11$
41. $7 \cdot 11 \cdot 13$ **43.** $2 \cdot 17 \cdot 47$ **45.** $2 \cdot 2 \cdot 2 \cdot 2 \cdot 3 \cdot 5 \cdot 5$
47. $2 \cdot 2 \cdot 2 \cdot 2 \cdot 53$ **49.** $3 \cdot 3 \cdot 3 \cdot 11$ **51.** $11 \cdot 13 \cdot 17$
53. prime **55.** prime, quotient, product, prime, quotient
57. $\frac{17}{30}$

Section 2.2

1. not equal **3.** equal **5.** $\frac{2}{3}$ **7.** $\frac{3}{5}$ **9.** $\frac{3}{4}$ **11.** $\frac{2}{3}$
13. $\frac{1}{9}$ **15.** $\frac{5}{7}$ **17.** $\frac{2}{13}$ **19.** $\frac{7}{11}$ **21.** $\frac{5}{6}$ **23.** $\frac{2}{5}$ **25.** $\frac{32}{55}$
27. $\frac{11}{35}$ **29.** $\frac{5}{6}$ **31.** $\frac{121}{169}$ **33.** $\frac{19}{23}$ **35.** $\frac{26}{37}$ **37.** $\frac{23}{27}$
39. $\frac{2}{3}$ **41.** $\frac{19}{33}$ **43.** $\frac{21}{22}$ **45.** $\frac{6}{35}$ **47.** $\frac{1}{3}$ **49.** $\frac{25}{66}$
51. $\frac{31}{43}$ **53.** cross-products **55.** prime, numerator, denominator, common, numerator, denominator **57.** $\frac{1}{8}$
59. $\frac{1}{2}$ **61.** $\frac{3}{7}$ **63.** $\frac{1}{3}, \frac{2}{3}$

Section 2.3

1. $\frac{5}{2}$ **3.** $\frac{47}{8}$ **5.** $\frac{92}{9}$ **7.** $\frac{47}{5}$ **9.** $\frac{135}{11}$ **11.** $\frac{227}{16}$ **13.** $\frac{316}{9}$
15. $\frac{662}{9}$ **17.** $\frac{86}{17}$ **19.** $\frac{172}{13}$ **21.** $7\frac{1}{2}$ **23.** $8\frac{4}{7}$ **25.** $11\frac{2}{3}$
27. $6\frac{8}{9}$ **29.** $5\frac{1}{4}$ **31.** $2\frac{1}{8}$ **33.** $3\frac{3}{5}$ **35.** $5\frac{2}{3}$ **37.** $13\frac{9}{11}$
39. $2\frac{3}{5}$ **41.** mixed number **43.** denominator, numerator, quotient, remainder **45.** $\frac{34}{5}$

Section 2.4

1. $\frac{3}{10}$ **3.** $\frac{3}{10}$ **5.** $\frac{5}{42}$ **7.** $\frac{15}{32}$ **9.** $\frac{15}{44}$ **11.** $\frac{270}{7}$ **13.** $\frac{25}{21}$

15. 21 **17.** $\frac{6}{65}$ **19.** $\frac{9}{16}$ **21.** 1 **23.** $\frac{4}{9}$ **25.** $\frac{12}{7}$ **27.** 1
29. 4 **31.** $\frac{891}{200}$ **33.** $\frac{5}{14}$ **35.** $\frac{1}{55}$ **37.** 600 **39.** $\frac{44}{28175}$
41. 2 **43.** 54 **45.** $35\frac{5}{8}$ **47.** $12\frac{1}{2}$ **49.** $1\frac{8}{21}$ **51.** $6\frac{1}{2}$
53. $6\frac{2}{3}$ **55.** $9\frac{1}{6}$ **57.** 1 **59.** $1\frac{13}{20}$ **61.** $5\frac{5}{6}$ **63.** $1\frac{5}{39}$
65. $1\frac{5}{28}$ **67.** $91\frac{1}{2}$ **69.** $6\frac{2}{3}$ **71.** 8 **73.** $3\frac{3}{4}$ **75.** $\frac{N}{1}$
77. $562\frac{1}{2}$ **79.** $\frac{1}{6}$ ton **81.** $195\frac{1}{2}$ miles **83.** 300
85. $\frac{10086}{3773}$ **87.** $\frac{1}{2}$ **89.** 576 **91.** 154 miles **93.** $25,125
95. $\frac{1}{15}$ ton

Section 2.5

1. 3 **3.** $\frac{5}{4}$ **5.** $\frac{225}{196}$ **7.** $\frac{9}{4}$ **9.** $\frac{150}{7}$ **11.** $\frac{45}{8}$ **13.** $\frac{5}{3}$
15. $\frac{83}{108}$ **17.** $\frac{14}{3}$ **19.** $\frac{1}{81}$ **21.** $\frac{55}{24}$ **23.** 28 **25.** $\frac{15}{14}$
27. 1 **29.** 0 **31.** undefined **33.** 15 **35.** $\frac{21}{4}$
37. 900 **39.** $\frac{205}{224}$ **41.** $\frac{38}{39}$ **43.** $1\frac{8}{125}$ **45.** $1\frac{3}{7}$ **47.** 1
49. $\frac{1}{2}$ **51.** 100 **53.** $\frac{3}{8}$ **55.** 66 **57.** 4
59. undefined **61.** $2\frac{4}{5}$ **63.** $\frac{7}{9}$ **65.** 1 **67.** $7\frac{5}{7}$
69. $14\frac{17}{20}$ **71.** $\frac{27}{35}$ **73.** $2\frac{59}{273}$ **75.** $3\frac{45}{56}$ **77.** $\frac{8}{15}$
79. reciprocal **81.** mixed numbers, improper fractions
83. $\frac{55}{91}$ **85.** $\frac{2}{3}$ **87.** $\frac{1}{12}$ pound **89.** 6 **91.** $6\frac{4}{5}$ **93.** 5
95. $19\frac{5}{13}$

Section 2.6

1. $\frac{5}{7}$ **3.** $\frac{16}{7}$ **5.** $\frac{2}{3}$ **7.** $\frac{13}{20}$ **9.** $\frac{1}{2}$ **11.** 9 **13.** 3
15. 15 **17.** 7 **19.** 77 **21.** 3 **23.** 15 **25.** 75
27. 35 **29.** 126 **31.** 24 **33.** 105 **35.** 18 **37.** 40
39. 90 **41.** 180 **43.** 600 **45.** 72 **47.** $\frac{17}{18}$ **49.** $\frac{13}{12}$
51. $\frac{47}{40}$ **53.** $\frac{19}{48}$ **55.** $\frac{41}{60}$ **57.** $\frac{109}{60}$ **59.** $\frac{61}{144}$ **61.** $\frac{1802}{7007}$
63. $\frac{29}{150}$ **65.** $\frac{19}{55}$ **67.** $\frac{43}{84}$ **69.** $\frac{3}{16}$ **71.** $\frac{59}{105}$ **73.** $\frac{16}{21}$
75. $\frac{49}{80}$ **77.** $\frac{79}{65}$ **79.** $\frac{29}{84}$ **81.** $\frac{71}{72}$ **83.** $\frac{59}{40}$ **85.** $\frac{274}{645}$
87. $\frac{589}{588}$ **89.** numerators, common denominator **91.** $\frac{241}{120}$
93. $\frac{97}{168}$ **95.** $\frac{3}{4}$ pound **97.** $\frac{31}{60}$ **99.** $\frac{17}{24}$ teaspoon
101. $\frac{1}{77}$ **103.** $\frac{7}{24}$ **105.** $\frac{19}{40}$ inch

Section 2.7

1. $5\frac{2}{3}$ **3.** $10\frac{3}{8}$ **5.** 9 **7.** $15\frac{3}{5}$ **9.** $5\frac{31}{36}$ **11.** $18\frac{7}{12}$
13. $14\frac{9}{20}$ **15.** $15\frac{41}{60}$ **17.** $8\frac{4}{7}$ **19.** $6\frac{2}{15}$ **21.** $7\frac{23}{42}$

23. $24\frac{25}{72}$ 25. $13\frac{31}{48}$ 27. $24\frac{29}{72}$ 29. $189\frac{23}{60}$ 31. $72\frac{1}{20}$
33. $11\frac{2}{45}$ 35. $10\frac{19}{20}$ 37. $40\frac{31}{72}$ 39. $135\frac{191}{588}$ 41. $263\frac{19}{35}$
43. $8\frac{3}{4}$ pounds 45. $346\frac{1}{10}$ miles 47. $29\frac{1}{24}$ inches
49. $571\frac{1}{24}$ pounds

Section 2.8

1. $\frac{1}{2}$ 3. $\frac{3}{5}$ 5. $\frac{1}{15}$ 7. $\frac{3}{10}$ 9. $\frac{1}{12}$ 11. $\frac{19}{30}$ 13. $\frac{11}{20}$
15. $\frac{13}{24}$ 17. $\frac{17}{36}$ 19. $\frac{1}{15}$ 21. $\frac{1}{4}$ 23. $\frac{1}{48}$ 25. $\frac{23}{48}$
27. $\frac{13}{15}$ 29. $\frac{1}{70}$ 31. $\frac{9}{20}$ 33. $\frac{1}{84}$ 35. $\frac{23}{90}$ 37. $\frac{271}{504}$
39. $\frac{77}{200}$ 41. $\frac{28}{99}$ 43. $\frac{8}{5}, 6\frac{8}{5}$ 45. $\frac{16}{9}, 4\frac{16}{9}$ 47. 13
49. 39 51. 9 53. 110 55. $3\frac{2}{7}$ 57. $3\frac{2}{9}$ 59. $2\frac{2}{5}$
61. $5\frac{1}{3}$ 63. $3\frac{2}{3}$ 65. $2\frac{2}{3}$ 67. $2\frac{3}{4}$ 69. $3\frac{1}{8}$ 71. $1\frac{1}{24}$
73. $2\frac{23}{36}$ 75. $2\frac{11}{14}$ 77. $27\frac{13}{48}$ 79. $1\frac{53}{90}$ 81. $58\frac{65}{96}$
83. $2\frac{1}{4}$ 85. $4\frac{1}{3}$ 87. $8\frac{5}{8}$ 89. $5\frac{31}{63}$ 91. $12\frac{7}{90}$
93. numerators, denominator 95. whole, fractional, whole, minuend, fractional 97. $3\frac{7}{12}$ 99. $\frac{17}{24}$ 101. $1\frac{1}{2}$
103. $7\frac{19}{24}$ 105. $35\frac{3}{4}$ 107. $\frac{67}{396}$ 109. $49\frac{1}{2}$ yards

Section 2.9

1. $\frac{3}{5}$ 3. $\frac{13}{8}$ 5. $\frac{5}{8}$ 7. $-\frac{5}{9}$ 9. $-\frac{3}{40}$ 11. $-\frac{5}{24}$ 13. $\frac{2}{7}$
15. $\frac{13}{48}$ 17. $-\frac{9}{70}$ 19. $\frac{1}{2}$ 21. $-\frac{32}{5}$ 23. $-\frac{2}{25}$
25. $-\frac{5}{3}$ 27. -15 29. $\frac{6}{5}$ 31. $-\frac{1}{98}$ 33. $-\frac{24}{5}$
35. $\frac{3}{40}$ 37. $-\frac{7}{8}$ 39. -2 41. $-\frac{13}{42}$ 43. $\frac{2}{9}$ 45. $-\frac{71}{60}$
47. $-\frac{17}{36}$ 49. $-\frac{13}{12}$ 51. $-\frac{25}{3}$ 53. $-\frac{10}{21}$ 55. $\frac{13}{72}$
57. -72 59. $-\frac{3}{10}$ 61. $\frac{17}{48}$ 63. $-\frac{2}{5}$ 65. $\frac{1}{36}$ 67. 0
69. $2\frac{31}{40}$ 71. $-\frac{15}{4}$ 73. -24 75. $-3\frac{19}{40}$ 77. $80\frac{23}{72}$

CHAPTER 2 Review Exercises

1. $2 \cdot 2 \cdot 3 \cdot 3$ (Section 2.1) 2. $11 \cdot 11$ (2.1)
3. $2 \cdot 3 \cdot 3 \cdot 29$ (2.1) 4. $7 \cdot 11 \cdot 13$ (2.1) 5. $\frac{3}{5}$ (2.2)
6. $\frac{5}{13}$ (2.2) 7. $\frac{3}{7}$ (2.2) 8. $\frac{7}{9}$ (2.2) 9. $3\frac{2}{5}$ (2.3)
10. $12\frac{5}{6}$ (2.3) 11. $13\frac{4}{5}$ (2.3) 12. $\frac{15}{4}$ (2.3) 13. $\frac{61}{8}$ (2.3)
14. $\frac{331}{4}$ (2.3) 15. $\frac{14}{15}$ (2.4) 16. $-\frac{41}{60}$ (2.9) 17. $\frac{29}{18}$ (2.6)
18. $-\frac{15}{14}$ (2.9) 19. $5\frac{1}{4}$ (2.4) 20. $\frac{23}{45}$ (2.8) 21. $3\frac{11}{36}$ (2.8)
22. $\frac{4}{27}$ (2.5) 23. $12\frac{11}{15}$ (2.7) 24. $5\frac{5}{6}$ (2.5) 25. $8\frac{1}{3}$ (2.8)
26. -210 (2.9) 27. $\frac{1}{4}$ (2.8) 28. $\frac{2}{9}$ (2.5) 29. $\$750$ (2.4)

30. $\$3\frac{5}{8}$ (2.8) 31. $9\frac{5}{24}$ cups (2.7) 32. 8 (2.5)
33. $5\frac{7}{8}$ inches (2.8) 34. $\frac{61}{96}$ (2.8) 35. $9\frac{1}{4}$ pies (2.3)
36. 110 miles (2.4) 37. $176\frac{11}{20}$ miles (2.7) 38. $\frac{8}{187}$ (2.4)
39. $138\frac{2}{3}$ (2.4) 40. $-7\frac{1}{2}$ (2.4) 41. $-1\frac{5}{27}$ (2.5)
42. $5\frac{20}{39}$ (2.5) 43. $-\frac{5}{13}$ (2.9) 44. $\frac{23}{90}$ (2.6) 45. $16\frac{1}{4}$ (2.7)
46. $3\frac{11}{126}$ (2.8) 47. $\frac{19}{78}$ (2.8) 48. $\frac{5}{22}$ (2.4) 49. 75 (2.9)
50. $-\frac{1}{60}$ (2.9) 51. $\frac{63}{80}$ (2.9) 52. 9 (2.9) 53. $\frac{33}{56}$ (2.9)
54. $\frac{2}{5}$ (2.9) 55. $\frac{101}{504}$ (2.9) 56. $3\frac{19}{33}$ (2.9)

CHAPTER 2 Test

1. $11 \cdot 11$ 2. $\frac{3}{5}$ 3. $\frac{26}{37}$ 4. $\frac{80}{11}$ 5. $\frac{316}{9}$ 6. $7\frac{1}{2}$ 7. $3\frac{3}{5}$
8. $\frac{1}{35}$ 9. 75 10. $\frac{9}{4}$ 11. $\frac{21}{26}$ 12. $\frac{109}{60}$ 13. $72\frac{1}{20}$
14. $\frac{1}{2}$ 15. $\frac{2}{35}$ 16. $223\frac{1}{4}$ 17. $-\frac{53}{70}$ 18. -72
19. $27\frac{13}{18}$ inches 20. $3\frac{7}{12}$ miles 21. $-\frac{1}{20}$ 22. $\frac{5}{6}$ 23. $\frac{11}{15}$
24. $-\frac{7}{18}$ 25. -29

CHAPTER 3

CHAPTER 3 Objectives

1a. three hundredths **b.** two and four hundred fifteen thousandths **2a.** 3.41 **b.** 52.668 **3a.** 1.771 **b.** 10.672
4a. 0.16 **b.** 28.397 **5a.** 0.32 **b.** 8.0
6a. 33.1 **b.** 2.03 **7a.** 0.6 **b.** 2.67 **8a.** $\frac{7}{20}$ **b.** $\frac{25}{8}$
9a. 0.45 **b.** 10.56 **c.** -3.68 **10a.** $\frac{13}{20}$ **b.** $\frac{1}{1250}$ **c.** $\frac{3}{800}$
11a. 0.65 **b.** 0.0008 **c.** 0.00375
12a. 45% **b.** 0.95% **c.** 320%
13a. 75% **b.** 12.5% **c.** $66\frac{2}{3}\%$

Section 3.1

1. ones 3. tenths 5. thousandths 7. hundred-thousandths 9. thousands 11. eight hundred thirteen thousandths 13. one thousand three hundred fifty seven ten-thousandths 15. thirty-four millionths 17. six and thirty-four thousandths 19. thirty-one thousandths 21. six and seven hundred three thousandths 23. seven and seven tenths 25. 8.17 27. 0.0046 29. 10.000005
31. 16.006 33. 3000.003 35. 20009.00056
37. decimal 39. whole, and, whole, place value

Section 3.2

1. 0.95 3. 1.3 5. 17.72 7. 35.34 9. 4.288
11. 7.689 13. 25.234 15. 110.6628 17. 14.9691
19. 128.171 21. 7.2 23. 2.12 25. 0.1 27. 4.077
29. 0.1 31. 62.36 33. 73.92 35. 16.88 37. 80.315
39. 1.31 41. 20.851 43. 0.19 45. 4.36 47. 8.605

49. 24.66 **51.** 8988 **53.** 1984 **55.** 339.49
57. 1195.05 **59.** 0.0022 **61.** 22.79 **63.** 13.398
65. 5.1264 **67.** 20 **69.** 151.05 **71.** 2099.88
73. 23.81 **75.** 216.7 **77.** 4206.09 **79.** 14.7461
81. 35.82 **83.** decimal points, whole, decimal points
85. place value **87.** 114.516 **89.** 28.725 inches
91. $18.91 **93.** 142.6 **95.** $25.92 **97.** 27.08
99. $318.43 **101.** $111.58 **103.** $702.52 **105.** $133.50

Section 3.3

1. 0.5742 **3.** 5.742 **5.** .05742 **7.** 0.96 **9.** 1.088
11. 0.0016 **13.** 79.56 **15.** 0.10104 **17.** 23.63
19. 0.00016399 **21.** 1676.47 **23.** 133.375 **25.** 670.5
27. 0.6272 **29.** 1.0798137 **31.** 5012.596 **33.** 13.86
35. 3.75 **37.** 21.3444 **39.** 0.2606225 **41.** digits, right
43. zeros **45.** $42.75 **47.** $28.12 **49.** 250.6
51. 0.621 **53.** 0.28608 **55.** $1671 **57.** 334,800
59. $178.50 **61.** 107.1

Section 3.4

1. 0.2 **3.** 0.3 **5.** 18.1 **7.** 0.9 **9.** 15.8 **11.** 0.85
13. 0.17 **15.** 1.38 **17.** 2.75 **19.** 0.146 **21.** 0.679
23. 5.678 **25.** 95.333 **27.** 10.000 **29.** 1335.375
31. 1300 **33.** 1335.374590 **35.** 1335.3746
37. approximately equal to **39.** hundredth **41.** 75.78
43. 0.739 **45.** 5.0

Section 3.5

1. 33.2 **3.** 6.81 **5.** 3.5 **7.** 5.3 **9.** 3.29 **11.** 0.0042
13. 0.0005 **15.** 0.2 **17.** 5.46 **19.** 1.8 **21.** 0.04
23. 65.8 **25.** 5780 **27.** 125,000 **29.** 0.178
31. 35.58 **33.** 0.04 **35.** 0.49 **37.** 0.27 **39.** 45.1
41. dividend, quotient **43.** $3.42 **45.** 21.2 **47.** 8.5
49. 5.1 **51.** 8.5 **53.** 3.68 **55.** $2.15 **57.** 20.4
59. 24 months

Section 3.6

1. 0.6 **3.** 0.25 **5.** 1.125 **7.** 0.28 **9.** 0.33 **11.** 0.71
13. 0.425 **15.** 0.62 **17.** 0.9 **19.** 0.256 **21.** 0.75
23. 3.5 **25.** 0.125 **27.** 0.06 **29.** 0.01 **31.** 0.02
33. 6.2 **35.** 0.19 **37.** 1.17 **39.** 0.014 **41.** $\frac{3}{10}$
43. $\frac{31}{1000}$ **45.** $\frac{1}{20}$ **47.** $\frac{5}{2}$ **49.** $\frac{8}{1}$ **51.** $\frac{128}{25}$ **53.** $\frac{61103}{10,000}$
55. $\frac{165}{1}$ **57.** $\frac{1}{8}$ **59.** $\frac{5}{16}$ **61.** $\frac{9}{20}$ **63.** $\frac{111}{1000}$ **65.** $\frac{13}{1}$
67. $\frac{3}{10}$ **69.** $\frac{7}{2}$ **71.** $\frac{9}{40}$ **73.** $\frac{701}{100}$ **74.** $\frac{9}{25}$ **77.** $\frac{3}{20}$
79. $\frac{1234}{5}$ **81.** denominator, numerator, round
83. multiply, place value **85.** $\frac{33}{8}$, 4.125
87. $\frac{25}{2}$, 12.5 **89.** $\frac{1}{1}$, 1 **91.** $0.43 **93.** $\frac{9}{16}$ **95.** 3.25
97. 7.5 **99.** 0.0017 **101.** $\frac{1}{8}$

Section 3.7

1. 3.21 **3.** 8.231 **5.** 0.85 **7.** 0.96 **9.** 2.06
11. 8.58 **13.** 0.521 **15.** 5.7 **17.** −0.91 **19.** 7.09
21. 0.069 **23.** −0.93 **25.** 13.54 **27.** −0.219
29. −0.15 **31.** −0.024 **33.** −1.83 **35.** −22.2

37. −100 **39.** −4.34 **41.** 0.271 **43.** 11.31
45. −3.03 **47.** −1.82 **49.** 20.2 **51.** −0.09
53. −9.44 **55.** 5.55 **57.** 0.009 **59.** −1.08
61. 0.05 **63.** 4 **65.** −8.85 **67.** 0.21 **69.** −14.4
71. 0.012 **73.** −0.024 **75.** −6 **77.** 44.4 **79.** −20
81. −4.7 **83.** −3.9752 **85.** −22.11 **87.** 0.0012
89. −52.1 **91.** 22.11 **93.** −0.034 **95.** −20.3
97. 2.9 **99.** −21.48 **101.** −33 **103.** 62.05
105. 32.402 **107.** 621.2

Section 3.8

1. $\frac{13}{100}$ **3.** $\frac{31}{100}$ **5.** $\frac{11}{20}$ **7.** $\frac{1}{50}$ **9.** $\frac{19}{300}$ **11.** 1 **13.** $\frac{3}{8}$
15. $\frac{3}{200}$ **17.** $\frac{3}{2}$ **19.** $\frac{1}{25,000}$ **21.** $\frac{17}{100}$ **23.** $\frac{9}{50}$ **25.** $\frac{23}{20}$
27. $\frac{13}{1250}$ **29.** $\frac{1}{12}$ **31.** $\frac{3}{400}$ **33.** 5 **35.** $\frac{441}{800}$
37. $\frac{3}{50,000}$ **39.** $\frac{13}{20}$ **41.** 0.13 **43.** 0.31 **45.** 0.55
47. 0.02 **49.** 0.0625 **51.** 1 **53.** 0.00125 **55.** 2.5
57. 0.1075 **59.** 0.034 **61.** 0.17 **63.** 0.16 **65.** 1.15
67. 0.0104 **69.** 0.0833 **71.** 0.0075 **73.** 5
75. 0.55125 **77.** 0.00006 **79.** 0.65 **81.** percent
83. 0.01, percent symbol **85.** 23, 100 **87.** $\frac{9}{40}$
89. 0.125 **91.** $\frac{131}{100}$ **93.** 0.00375 **95.** $\frac{297}{800}$

Section 3.9

1. 65% **3.** 230% **5.** 0.6% **7.** 81% **9.** 75%
11. 60% **13.** 40% **15.** 5120% **17.** 0.7% **19.** $33\frac{1}{3}$%
21. 13% **23.** 800% **25.** 0.8% **27.** 50%
29. 11.1% **31.** 810% **33.** 0.45% **35.** 60%
37. $55\frac{1}{2}$% **39.** $80\frac{1}{2}$% **41.** 60% **43.** 25%
45. 112.5% **47.** 28% **49.** $33\frac{1}{3}$% **51.** $71\frac{3}{7}$%
53. 42.5% **55.** 62% **57.** 90% **59.** 75% **61.** 10%
63. $66\frac{2}{3}$% **65.** 27.5% **67.** 70% **69.** 150% **71.** 0.7%
73. 500% **75.** 2.5% **77.** $31\frac{1}{4}$% **79.** 110%
81. decimal point, right **83.** fraction, decimal, decimal, percent **85.** 60% **87.** 14.55% **89.** 15.5% **91.** $114\frac{2}{7}$%
93. $52\frac{1}{8}$% **95.** 18.7%

CHAPTER 3 Review Exercises

1. five and nine thousandths *(Section 3.1)* **2.** two thousand three hundred eighteen ten-thousandths *(3.1)* **3.** thirty-five hundredths *(3.1)* **4.** 3.05 *(3.6)* **5.** $\frac{31}{100}$ *(3.6)*
6. 0.33 *(3.6)* **7.** 0.58 *(3.6)* **8.** $\frac{13}{20}$ *(3.6)* **9.** $\frac{1}{8}$ *(3.6)*
10. 1751.859 *(3.4)* **11.** 1800 *(3.4)* **12.** 14.84 *(3.2)*
13. 0.915 *(3.2)* **14.** 0.003945 *(3.3)* **15.** 453.47 *(3.5)*
16. −11.915 *(3.7)* **17.** −110 *(3.7)* **18.** −8.466 *(3.7)*
19. 8.925 *(3.3)* **20.** 0.6625 *(3.2)* **21.** 0.568 *(3.2)*
22. −0.05 *(3.7)* **23.** 17.881 *(3.2)* **24.** $15.08 *(3.3)*
25. $30.75 *(3.5)* **26.** $6.52 *(3.2)* **27.** $314.60 *(3.6)*
28. −18.85 *(3.7)* **29.** −27.2 *(3.7)* **30.** 2.5 *(3.6)*

31. 115.4 *(3.7)* **32.** 1004.9665 *(3.7)* **33.** −0.4 *(3.7)*
34. $\frac{1}{2}$, 50% *(3.9)* **35.** 0.2, 20% *(3.9)* **36.** 0.25, $\frac{1}{4}$ *(3.8)*
37. $\frac{61}{100}$, 61% *(3.9)* **38.** 0.25, 25% *(3.9)* **39.** 0.8, $\frac{4}{5}$ *(3.8)*
40. $\frac{1}{8}$, $12\frac{1}{2}$% *(3.9)* **41.** $0.33\frac{1}{3}$, $33\frac{1}{3}$% *(3.9)*
42. $0.66\frac{2}{3}$, $\frac{2}{3}$ *(3.8)* **43.** $\frac{3}{5}$, 60% *(3.9)*
44. 0.75, 75% *(3.9)* **45.** 0.375, $\frac{3}{8}$ *(3.8)*
46. $\frac{5}{1}$, 500% *(3.9)* **47.** 0.006, $\frac{3}{500}$ *(3.8)*
48. 0.014, 1.4% *(3.9)*

CHAPTER 3 Test

1. three and five thousandths **2.** 33.5165 **3.** 91.9994
4. 544.407 **5.** 4.44 **6.** 8.76 **7.** 150 **8.** 18.36
9. −3.92 **10.** 0.25 **11.** 0.67 or $0.\overline{6}$ **12.** $\frac{9}{250}$
13. 0.15 **14.** −1.92 **15.** $393.98 **16.** 36 months
17. $18.81 **18.** $\frac{7}{20}$ **19.** $\frac{21}{2000}$ **20.** 0.35 **21.** 0.00375
22. 70% **23.** 91.8% **24.** 75% **25.** $11\frac{1}{9}$%

CHAPTER 4

CHAPTER 4 Objectives

1a. 8 **b.** $\frac{1}{8}$ **c.** −8 **2a.** 7^{50} **b.** x^{12} **3a.** 7^{13} **b.** $\frac{1}{x^{18}}$
4a. x^8 **b.** $\frac{1}{7^{15}}$ **5a.** $3^{11} \cdot 6^{11}$ **b.** $81x^4$ **6a.** $\frac{8}{27}$ **b.** $-\frac{x^3}{125}$
7a. 6.8×10^4 **b.** 6.8×10^{-5}

Section 4.1

1. $3 \cdot 3 \cdot 3 \cdot 3 \cdot 3$ **3.** $5 \cdot 5 \cdot 5 \cdot 5 \cdot 5 \cdot 5$
5. $7 \cdot 7 \cdot 7 \cdot 7 \cdot 7 \cdot 7$ **7.** 9^2 **9.** 6^5 **11.** $(.5)^2$ **13.** 4^6
15. 9 **17.** 36 **19.** $\frac{1}{49}$ **21.** $\frac{1}{1000}$ **23.** $\frac{1}{16}$ **25.** 1
27. −1 **29.** 0.09 **31.** 1.331 **33.** 81 **35.** −81
37. $\frac{1}{36}$ **39.** $-\frac{1}{343}$ **41.** $-\frac{1}{32}$ **43.** 0.0000000001
45. base **47.** fraction, one, positive

Section 4.2

1. 3^{11} **3.** 8^{15} **5.** $\frac{1}{4^7}$ **7.** $\frac{1}{2^{11}}$ **9.** 1 **11.** m^7 **13.** $\frac{1}{6^7}$
15. x^{19} **17.** $\frac{1}{5^9}$ **19.** $\frac{1}{3^6}$ **21.** $\frac{1}{3^8}$ **23.** a^{21} **25.** $\frac{1}{m^{41}}$
27. $\frac{1}{x^{16}}$ **29.** $\frac{1}{7^6}$ **31.** n^2 **33.** 1 **35.** $\frac{1}{8}$ **37.** 32
39. $\frac{1}{x^7}$ **41.** $\frac{1}{n^{20}}$ **43.** 7^{15} **45.** x^4 **47.** x^4 **49.** $\frac{1}{x^{55}}$
51. base, add **53.** exponential expression, operation
55. 49 **57.** $\frac{1}{4}$ **59.** 63 **61.** 4

Section 4.3

1. 2^{27} **3.** y^8 **5.** $\frac{1}{3^{18}}$ **7.** $\frac{1}{5^{16}}$ **9.** $\frac{1}{x^{27}}$ **11.** x^{ab}

13. $\frac{1}{x^{12}}$ **15.** x^6 **17.** 2^{80} **19.** 16 **21.** $3^7 \cdot 5^7$
23. $5^{10}n^{10}$ **25.** $-27p^3$ **27.** $16x^2y^2$ **29.** $a^8b^8c^8d^8$
31. $49x^2$ **33.** $\frac{16}{81}$ **35.** $\frac{81}{n^4}$ **37.** $-\frac{x^3}{8}$ **39.** $\frac{b^8}{5^8}$ **41.** $\frac{x^n}{y^n}$
43. $-\frac{n^3}{125}$ **45.** $32x^5$ **47.** $-8x^3$ **49.** $x^9y^9z^9$
51. $49x^2$ **53.** $\frac{1}{36}$ **55.** $\frac{n^4}{81}$ **57.** $-\frac{x^3}{343}$ **59.** 25
61. factor, power **63.** numerator, denominator, power
65. $16x^8y^{12}$ **67.** $\frac{1}{3^{ab}}$ **69.** $\frac{1}{m^7}$ **71.** $\frac{a^{2n}}{b^{4n}}$ **73.** $\frac{1}{x^{22}}$
75. $\frac{y^6}{x^3z^{12}}$ **77.** $\frac{x^{14}}{64}$ **79.** $\frac{64b^3c^6}{a^6}$ **81.** $\frac{x^6}{64y^9z^9}$

Section 4.4

1. 372 **3.** 0.0000000857 **5.** 0.875 **7.** 300,000
9. 0.00000007 **11.** −0.000677 **13.** 81,000
15. 0.000535 **17.** −3,120,000 **19.** 810,000,000,000
21. 2.75×10^2 **23.** 2.75×10^{-4} **25.** 3.18×10^1
27. 1.1×10^4 **29.** 7.5×10^8 **31.** 5×10^{-4}
33. 6.19×10^{-6} **35.** 3.78×10^{17} **37.** 6.25×10^2
39. 8.75×10^{-2} **41.** 3.5×10^{10} **43.** 5.1×10^{-8}
45. n, right **47.** one, ten, decimal point, one, ten
49. 30.8 **51.** 32,000,000 **53.** 0.0000000004
55. 1,920,000,000,000 **57.** 30,000,000,000 **59.** 0.000007
61. −0.000000000155 **63.** 9.3×10^7 **65.** 5.19×10^{-5}
67. 2.25×10^8

CHAPTER 4 Review Exercises

1. 81 *(Section 4.1)* **2.** $\frac{1}{81}$ *(4.1)* **3.** 81 *(4.1)*
4. −81 *(4.1)* **5.** $\frac{1}{x^{20}}$ *(4.2)* **6.** $\frac{121}{16}$ *(4.3)* **7.** x^3 *(4.2)*
8. x^{30} *(4.3)* **9.** $-8p^3$ *(4.3)* **10.** 5^{53} *(4.2)* **11.** $16\frac{1}{8}$ *(4.1)*
12. $\frac{1}{9^{24}}$ *(4.2)* **13.** $\frac{9}{16}$ *(4.3)* **14.** 2^{15} *(4.3)*
15. $2^{15}x^{15}$ *(4.3)* **16.** $\frac{8}{9}$ *(4.1)* **17.** $81x^8$ *(4.3)* **18.** x^{20} *(4.2)*
19. $\frac{1}{5^{29}}$ *(4.3)* **20.** 347,000,000 *(4.4)*
21. 0.000000406 *(4.4)* **22.** −2,470 *(4.4)*
23. −0.00862 *(4.4)* **24.** 1.5×10^7 *(4.4)*
25. 3.12×10^{-7} *(4.4)* **26.** 3.25×10^{-4} *(4.4)*
27. 8.7×10^{11} *(4.4)* **28.** 1.1552×10^{12} *(4.4)*
29. 18,600 *(4.4)* **30.** 1,230,000 *(4.4)*
31. 10^{19} *(4.4)* **32.** $\frac{1}{x^{10}}$ *(4.3)* **33.** $\frac{x^6}{64}$ *(4.3)* **34.** $\frac{1}{x^{26}}$ *(4.3)*
35. $\frac{81}{4}$ *(4.3)* **36.** $\frac{x^4}{9y^2z^8}$ *(4.3)* **37.** $-8x^{12}y^6$ *(4.3)*
38. $\frac{x^{22}}{y^8}$ *(4.3)* **39.** $\frac{49}{4}$ *(4.3)* **40.** $\frac{x^{20}}{y^{10}}$ *(4.3)*

CHAPTER 4 Test

1. 81 **2.** $\frac{1}{125}$ **3.** −0.027 **4.** −1 **5.** $\frac{3}{8}$ **6.** $\frac{1}{36}$
7. $\frac{1}{x^{22}}$ **8.** n^{42} **9.** $\frac{1}{5^{31}}$ **10.** m^{56} **11.** $\frac{1}{x^{15}}$ **12.** x^{56}
13. $0.0016x^4$ **14.** $a^8b^8c^8$ **15.** $\frac{x^3}{64}$ **16.** $-\frac{n^3}{343}$
17. $-125x^6y^9$ **18.** 234,000 **19.** 2.46×10^{-4} **20.** 4

CHAPTER 5

CHAPTER 5 Objectives
1a. $-7x$ and $-8x$ **b.** none **2a.** $-17x$ **b.** $-7x + 7$
3a. 69 **b.** 2.088 **4a.** $6x - 13$ **b.** x
5a. $x + 1$ **b.** $-3.6x^2 - 2.2x - 1.3$ **6a.** $21x^9y$ **b.** $-\frac{9}{56}x^6$
7a. $15x^3y^3 + 9x^2y^3 - 6xy^2$ **b.** $-0.06x^5 + 0.63x^4 - 3x^3$
8a. $2x^3 + 5x^2 - 27x + 5$ **b.** $x^3 - y^3$
9a. $6x^2 - 11x - 10$ **b.** $4x^2 - 9$ **c.** $4x^2 - 4x + 1$
10a. $-\frac{7x^3y}{z}$ **b.** $-\frac{50}{x^3}$ **11a.** $5x^3 - 6x + \frac{4}{x^2}$ **b.** $\frac{3}{5}x - \frac{6}{5y}$
12a. $3x^4 - 2$ **b.** $n^2 + 3n + 9$

Section 5.1
1. polynomial **3.** not a polynomial **5.** $7x^3, -7x^2, 8$; coefficients: 7, -7, 8 **7.** $-3x^2, -7x, 5$; coefficients: $-3, -7, 5$ **9.** trinomial **11.** monomial **13.** monomial
15. $7x^3$ and $3x^3$ **17.** $3x^3$ and $5x^3$; $-7x^2$ and $-x^2$
19. $3xy^3, -6xy^3$, and xy^3 **21.** $7x^2$ and $-6x^2$, 9 and -2
23. $5x^3$ and $-x^3$, $3x^2$ and x^2, 5 and -9 **25.** $3y^2$
27. $-2x^5$ **29.** $-6x$ **31.** $6x^3y + 6xy^3$ **33.** $9x^3 + 7x$
35. $2xy + 8y$ **37.** $18x^2$ **39.** $-2xy$ **41.** $0.81x + x^2$
43. $-0.5x^3 + 3x^2$ **45.** $\frac{11}{72}x^2 + \frac{1}{3}x$ **47.** $-3xy$
49. $-0.21x$ **51.** $\frac{1}{24}x^2y$ **53.** $-9y$ **55.** $\frac{19}{48}x^2 - \frac{1}{8}y^2$
57. $-0.08a$ **59.** $4.2b$ **61.** $\frac{16}{45}x^3$ **63.** $0.5x$ **65.** $\frac{1}{4}x$
67. $-6.1y$ **69.** degree of terms: 3, 2, 1, 0; degree of polynomial $= 3$ **71.** monomial **73.** terms **75.** like, similar

Section 5.2
1. 2 **3.** -15 **5.** 76 **7.** -71 **9.** -27 **11.** 1.59
13. -3.73 **15.** 3.93 **17.** $-\frac{4}{27}$ **19.** $\frac{49}{8}$ **21.** 23
23. 8.14 **25.** $\frac{2}{7}$ **27.** evaluating **29.** \$9500
31. \$10,750 **33.** 10 **35.** \$104,950 **37.** 4.35 **39.** $-\frac{133}{90}$

Section 5.3
1. $5x^2 + 6x - 7$ **3.** $6n^3 - 7n^2 + n - 5$
5. $-8x^6 - 9x^5 + 7x^4 + 7x^3 + 6x + 5$ **7.** $5n - 2$
9. $-3x - 5$ **11.** $-x - 4y$ **13.** $-3n^3 - 10n^2 - 16$
15. $-2x - 15y + 10z$ **17.** $4s^3 + 2s^2 + 7s$
19. $7a^2b + 2ab - b^2$ **21.** $8x^4 - 3x^3 - 11x^2 - 16x - 3$
23. $-3x^3 + 4x^2 + 3x$ **25.** $-3a^2 - 12ac + 9c^2$
27. $5.3x - 1.5$ **29.** $3.3x^2 + 4.3xy - 3.81y^2$
31. $-0.81x^2 + 5x - 2.9$ **33.** $\frac{9}{7}x - \frac{69}{8}$
35. $\frac{97}{396}x^2 + \frac{19}{90}x + \frac{28}{11}$ **37.** $\frac{2}{5}x^2 - \frac{7}{30}x + \frac{1}{6}$
39. alphabetical, first **41.** $5x^2 - 3x + 1$
43. $\frac{19}{12}x^2 - \frac{17}{40}$ **45.** $1.45x + 3.18y$ **47.** $-2.1x + 5z$

Section 5.4
1. $-2x + 6$ **3.** $4x + 5$ **5.** $-x + 3y - 4z$
7. $6x^2 - 2$ **9.** $4a + 12$ **11.** $-3x + 2$
13. $-2x^2 + x - 4$ **15.** $2x^2 + 2y^2$

17. $10n^2 - 2n - 3$ **19.** $-2y^2 - 2y + 2$
21. $3x^3 - 12x^2 + 10$ **23.** $2xy^2 - 2xy + 10$
25. $-2a - 10b - c$ **27.** $0.9x - 2.3$
29. $-3.6x^2 - 0.8x + 0.85$ **31.** $\frac{2}{5}x - \frac{2}{3}$
33. $\frac{5}{24}a^2 + \frac{2}{5}a - \frac{19}{10}$ **35.** $\frac{9}{35}x^2 + \frac{13}{60}x + \frac{1}{18}$
37. additive inverse, subtrahend, minuend **39.** $2x + 1$
41. $-x - 6$ **43.** $-3.8x^2 - 7y + 4.79$
45. $7x^3 - 10x^2 - 2x + 11$ **47.** $-x^2 - \frac{11}{40}x + \frac{19}{5}$

Section 5.5
1. $-15x^9$ **3.** $-15n^9$ **5.** $-45x^4y^2$ **7.** $56x^5$ **9.** $9xy$
11. $30x^6$ **13.** $0.93x^7$ **15.** $-12n^{12}$ **17.** $-0.407n^7$
19. $-\frac{3}{20}x^7$ **21.** $-21x^7$ **23.** $2x^2 + 10x$
25. $7x^5 - 49x^4 + 14x^2$ **27.** $-15x^4 + 12x^3 - 6x^2$
29. $12n^7 - 28n^6 - 32n^5$ **31.** $0.06x^6 - 0.62x^5$
33. $-1.55x^3 + 0.1x^2 - 1.05x$ **35.** $\frac{20}{81}x^3 - \frac{10}{21}x^2$
37. $x^2 + 8x + 15$ **39.** $21x^2 - 41xy + 10y^2$
41. $x^3 + 3x^2 - 8x - 4$ **43.** $24n^3 - 46n^2 + 37n - 12$
45. $7n^4 + 23n^3 - 68n^2 + 39n - 7$
47. $x^4 - 2x^3 - 12x^2 + x + 2$ **49.** numerical coefficients, exponential expressions, numerical coefficient
51. multiplied, sum, like **53.** $-32x^4y^2$ **55.** $72xy$
57. $-0.06x^8$ **59.** $-\frac{12}{55}x^6y^2$ **61.** $24y^5 - 32y^4 + 20y^3$
63. $-16x^3 + 48x^2y - 8xy^2$
65. $30x^3y^2 + 24x^2y^2 - 36xy$
67. $10x^2y^3 + 12x^2y^4 + 6xy^4$ **69.** $0.06x^5 - 0.62x^3$
71. $24.6c^4 - 1.23c^3 + 0.246c^2$ **73.** $\frac{3}{10}x^6 - 8x^4$
75. $\frac{3}{28}a^3b^3 - \frac{3}{8}a^3b^4$ **77.** $x^2 - 16$
79. $x^2 + 8x + 16$ **81.** $x^2 + 10x + 25$
83. $12x^3 - 22x^2 - 3x + 3$ **85.** $8x^2 - 1.4x - 0.15$
87. $0.6x^3 - 2.34x^2 - 8.02x + 15.5$
89. $x^8 + 2x^4 - x^2 + 2$
91. $2x^2 - 3xy + 9xz - 9y^2 + 18yz - 5z^2$
93. $x^3 + y^3$ **95.** $3x^3 - 4x^2 - 11x + 14$
97. $-12x^3 + 15x^2 - 3x$ **99.** $\frac{9}{25}x^4$ or $0.36x^4$

Section 5.6
1. $6x^2 + 13x + 6$ **3.** $4x^2 + 19x - 5$
5. $5x^2 - 44x - 9$ **7.** $x^2 - 0.1x - 0.06$
9. $2x^2z^2 - 8xz - 10$ **11.** $x^2 + xy - 6y^2$
13. $6x^2 - 11x + 3$ **15.** $x^2 + 8x + 16$
17. $6x^2 - 25x + 25$ **19.** $x^2 - 25$
21. $x^2 + 2xy - 3y^2$ **23.** $9x^2 - 16$
25. $6x^2 - 17xy + 5y^2$ **27.** $5x^2 + 14xy - 3y^2$
29. $18x^2 - 9x - 2$ **31.** $x^2 - 0.4x + 0.04$
33. $x^2 - 0.25y^2$ **35.** $x^2 - \frac{11}{6}x - \frac{5}{3}$
37. $x^2 + 4x + 4$ **39.** $x^2 - 14x + 49$
41. $4x^2 + 4xy + y^2$ **43.** $4x^2 + x + \frac{1}{16}$
45. $25x^2 + 20xy + 4y^2$ **47.** $x^2 + 18x + 81$
49. $x^2 - 6x + 9$ **51.** $x^2 + 14x + 49$
53. $x^2 - 12x + 36$ **55.** $x^2 + 6xy + 9y^2$
57. $x^2 - 10xy + 25y^2$ **59.** $4x^2 - 12xy + 9y^2$
61. $x^2 + 0.4x + 0.04$ **63.** $4x^2 + 2x + \frac{1}{4}$

65. $9x^2 + 0.6x + 0.01$ **67.** $4x^2 + 28xy + 49y^2$
69. $x^4 + 26x^2 + 169$ **71.** $x^2 - 9$ **73.** $25x^2 - 1$
75. $9x^4 - 100$ **77.** $a^2b^2 - 1$ **79.** $x^2 - \frac{9}{16}$
81. $\frac{x^2}{4} - 25$ **83.** square, twice, square
85. $20x^2 - 19x + 3$ **87.** $x^2 + 18x + 81$
89. $x^2 + 6x + 8$ **91.** $9x^2 - 12x + 4$
93. $x^2 - 4$ **95.** $x^2 - \frac{3}{2}x + \frac{9}{16}$ **97.** $x^2 - \frac{1}{4}$
99. $9x^2 - 25y^2$ **101.** $2x^2 + 3x - 35$
103. $x^2 - 16x + 64$ **105.** $6x^6 + x^3y^2 - y^4$

Section 5.7

1. $4x^3$ **3.** $-\frac{2}{3}x^3y^3$ **5.** $\frac{7x}{y}$ **7.** $\frac{mn^4}{5}$ **9.** $-\frac{3y^2}{x^2}$ **11.** $70x^3$
13. $\frac{xy^6}{5}$ **15.** $\frac{11}{13}x^2y$ **17.** $\frac{3}{4}x^2$ **19.** $3x^3 + 4x$
21. $4x^5 + 3x^3 - 2x$ **23.** $-5x^5 + 3x^4$ **25.** $\frac{3a}{b^2} - 2b^2$
27. $4x - 3 + \frac{5}{3x}$ **29.** $\frac{3}{5}n - \frac{6}{5m}$
31. $-\frac{15a^3}{b} - 5a^2b + \frac{32b^3}{a}$
33. $\frac{3}{2}x^4y^2 - 2x^2 + \frac{1}{y^2} + \frac{3}{x^2y^3} + \frac{9}{2x^3y^4}$
35. polynomial, monomial, sum   **37.** $5x^6$ **39.** $-\frac{111}{x^4}$
41. $-3x$ **43.** $\frac{2}{3}y^2$ **45.** $-130x^6$ **47.** $\frac{3}{x^4}$ **49.** $\frac{5a}{7b}$
51. $-\frac{11}{13}a^2b^2$ **53.** $\frac{5b}{4a}$ **55.** $0.13x^7$ **57.** $3x^4 - 5x^2 + 7x$
59. $5x^4 - 4x^2$ **61.** $-5x^5 + 6x^3 - 7$ **63.** $5a - \frac{2}{a} + \frac{1}{a^2}$
65. $\frac{2x}{y^3} - 3y^2$ **67.** $5x - 4 + \frac{3}{2x^2}$ **69.** $\frac{6}{5}x - \frac{8}{5y}$
71. $-\frac{10x^4}{y} - 4x^3y^2 + \frac{11x^3}{y}$ **73.** $-x^{12} - x^{10} - x^8 + x^6$
75. $2x + 3y - 5z$ **77.** $-\frac{6}{x} + \frac{8}{x^3} - \frac{4}{x^4}$ **79.** $-\frac{4}{5x^7}$
81. $3x^2 - 4yz - 5xy$

Section 5.8

1. $x + 3$ **3.** $3x + 2 - \frac{1}{3x - 5}$ **5.** $3x + 1$
7. $5x - 3y$ **9.** $5x^2 - 2xy - 2y^2$ **11.** $4x^2 + 3x + 6$
13. $2x - 3y$ **15.** $x + 6 + \frac{13}{x - 3}$
17. $3x - 2 - \frac{12}{2x - 3}$ **19.** $3a^2 + 6a - 5$
21. $x + y$ **23.** $x^2 + xy + y^2$
25. $a^3 - a^2 + 2 + \frac{-3a - 3}{2a^2 - a - 1}$ **27.** $3x^2 - 1$
29. $x + 2$ **31.** $a^2 + 2a + 4$
33. $3x^2 - 8x + 7 - \frac{12}{2x + 2}$ **35.** $5x - 7 + \frac{14}{2x + 1}$
37. $3x^4 - 2$ **39.** $3t + 2 + \frac{1}{3t + 1}$ **41.** descending, long division, whole   **43.** $x - 2$ **45.** $a - b$

CHAPTER 5   Review Exercises

1. $6x^2$ and $-16x^2$, $-7x$ and $3x$ *(Section 5.1)*   **2.** $-5x^2y$ and x^2y *(5.1)*   **3.** $-3x^2 + 3x$ *(5.1)*   **4.** $5x^2 + 0.91x$ *(5.1)*
5. $\frac{29}{24}x^3y$ *(5.1)*   **6.** $-\frac{61}{396}x^2 + \frac{1}{36}y^2$ *(5.1)*   **7.** 35 *(5.2)*
8. $\frac{68}{9}$ *(5.2)*   **9.** $\frac{17}{30}x - \frac{49}{60}$ *(5.3)*   **10.** $-\frac{5x^2}{y^2}$ *(5.6)*
11. $-36.4x^6y^5$ *(5.5)*   **12.** $-2x^3 + 7x^2 + y^2$ *(5.4)*
13. $2x^3 - 2x^2 - 4x - 2$ *(5.3)*   **14.** $x^2 + 2x + 4$ *(5.8)*
15. $-1.5x^4 + 0.05x^3 - 2x^2$ *(5.5)*
16. $-3a^2 + 4a - \frac{5}{2a}$ *(5.7)*
17. $3x^3 - 26x^2 + 25x - 6$ *(5.5)*   **18.** $-\frac{3x^5}{4z}$ *(5.7)*
19. $8x^2 - 10xy - 11y^2$ *(5.4)*   **20.** 39 *(5.2)*
21. $\$5150$ *(5.2)*   **22.** -0.0125 *(5.2)*
23. $9x^2 - 21x + 10$ *(5.6)*   **24.** $4x^2 + 20x + 25$ *(5.6)*
25. $x^2 - 81$ *(5.6)*   **26.** $x^2 - 0.4x + 0.04$ *(5.6)*
27. $x^2 - x - 72$ *(5.6)*   **28.** $\frac{1}{16}x^2 - 4$ *(5.6)*
29. $x^2 - 0.2x - 0.15$ *(5.6)*   **30.** $x^2 - \frac{5}{12}x - \frac{1}{4}$ *(5.6)*
31. $4x^2 - 2x + \frac{1}{4}$ *(5.6)*   **32.** $a^2 + 7a + 27 + \frac{100}{a - 5}$ *(5.8)*
33. $\frac{5x}{6y^2} + \frac{1}{3y^3} + \frac{1}{2xy}$ *(5.7)*   **34.** $-10n - 13$ *(5.4)*
35. $30x^3 - 155x^2 + 200x$ *(5.6)*   **36.** $-40x^3 - 14x^2 - 10x$ *(5.6)*
37. $2x^2 + 50$ *(5.6)*   **38.** $4.02x^3 - x^2 + 0.92$ *(5.3)*
39. $-\frac{1}{14}x^3 + \frac{23}{28}x^2 - \frac{13}{3}$ *(5.4)*
40. $-0.6x^2y^7 + 1.8x^3y^8 - 15xy^6$ *(5.5)*
41. $\frac{2138}{189}x^5$ *(5.1)*   **42.** $0.125x^3$ *(5.1)*

CHAPTER 5   Test

1. $3x^3$ and $-6x^3$ **2.** $-0.21x^3 - 5x$ **3.** $-\frac{8}{45}x^2 + \frac{2}{3}x$
4. -29 **5.** $\frac{14}{9}$ **6.** $7x^2 + 2x - 5$
7. $0.4x^2 - 8.44x + 1.08$ **8.** $-5x^2 + \frac{5}{24}x - \frac{17}{10}$
9. $4x^4y^3$ **10.** $\frac{10}{33}x^{13}y^2$ **11.** $2x^4 - 1.2x^3 + 8x^2$
12. $-35x^5 + 10x^3y - 15x^2y^2$ **13.** $6x^2 - 5x - 6$
14. $15x^3 - x^2 - 27x + 14$ **15.** $-9x^6$ **16.** $-\frac{130}{x^3}$
17. $7x^2 - 6x + 4$ **18.** $-5xy + 4y^2 - \frac{3}{2xy^3}$
19. $3x^2 - 2x + 10 - \frac{35}{x + 3}$ **20.** $x^2 + x + 1 + \frac{2}{x - 1}$
21. $4x^2 - 20x + 25$ **22.** $9x^2 - 25y^2$
23. $\frac{1}{12}x^2 - \frac{1}{12}xy - y^2$ **24.** $x^2 + xy + 0.25y^2$
25. $2x^2 - 1.4x + 0.2$

CHAPTER 6

CHAPTER 6   Objectives

1a. yes   **b.** no   **2a.** 4 **b.** -2.25 **3a.** -18 **b.** $\frac{5}{3}$
4a. 18 **b.** $\frac{1}{2}$ **5a.** 15.1 **b.** 8 **6a.** $\frac{77}{100}$ **b.** 3
7a. $\frac{c}{a + b}$ **b.** $\frac{P - ad}{a}$ **8a.** $x < -19$ **b.** $x \leq -9$
9a. $x < -30$ **b.** $x \geq 15$ **10a.** $x \geq -\frac{5}{3}$ **b.** $x \leq \frac{9}{2}$

c. $x \geq \frac{4}{23}$

Section 6.1

1. not a linear equation in one variable   3. not a linear equation in one variable   5. not a linear equation in one variable   7. 4 is not the solution   9. -2 is not the solution   11. 1.9 is the solution   13. $-\frac{3}{8}$ is not the solution   15. 4 is not the solution   17. 11.9 is not the solution   19. 3 is not the solution   21. -3 is not the solution   23. -2 is not the solution   25. 5 is not the solution   27. 5 is the solution   29. solution   31. -2.8 is the solution   33. 4 is not the solution   35. -1.64 is the solution   37. $-4\frac{1}{3}$ is the solution   39. -10 is not the solution   41. $-\frac{5}{2}$ is the solution   43. -5 is not the solution   45. -3 is not the solution   47. identity   49. inconsistent   51. inconsistent   53. identity

Section 6.2

1. 5   3. -6 5. -5 7. 5   9. -18 11. -1.1
13. -2.71 15. -1 17. $-\frac{37}{40}$ 19. -44 21. $\frac{1}{42}$
23. 2   25. -7 27. $-\frac{3}{2}$ 29. 28   31. -0.09
33. -24 35. $-\frac{2}{5}$ 37. -3 39. 2.5   41. -40
43. additive inverse   45. 5.5   47. -24 49. 6
51. -2 53. -15 55. 15   57. 21   59. 1   61. $\frac{5}{6}$
63. $-\frac{3}{2}$ 65. 35   67. 6   69. $-\frac{21}{2}$ 71. 21
73. -0.015 75. -41 77. -110 79. -3.7
81. -0.7 or $-\frac{7}{10}$ 83. $-\frac{2}{5}$ or -0.4 85. $\frac{7}{6}$ 87. $\frac{15}{2}$
89. -332.99 91. -24.14

Section 6.3

1. 90   3. 5.85   5. -6 7. $\frac{35}{3}$ 9. 4   11. 2   13. -6
15. 16.5   17. $\frac{1}{2}$ 19. $\frac{15}{2}$ 21. 6   23. $\frac{2}{7}$ 25. $-\frac{5}{3}$
27. -3.8 29. $-\frac{7}{4}$ 31. 0.4   33. -1.2 35. 6
37. -1 39. 3.2   41. -9 43. 4   45. -20
47. -4 49. 2   51. 3   53. $\frac{5}{3}$ 55. 2   57. $\frac{11}{7}$ 59. 5
61. $-\frac{9}{2}$ 63. 1   65. -50 67. 0.09   69. -3.2
71. -0.7 73. -4 75. $-\frac{4}{3}$
77. 20   79. -4.2 81. 0.1   83. combine, addition
85. -5.5 87. 2.1

Section 6.4

1. 1   3. 0   5. -5 7. 1   9. 3   11. $\frac{7}{3}$ 13. 7   15. $\frac{32}{3}$
17. 4   19. 1   21. 17   23. $-\frac{5}{3}$ 25. $\frac{1}{2}$ 27. -3
29. $\frac{1}{4}$ 31. $\frac{2}{3}$ 33. -18 35. 12   37. 20   39. 36
41. 4   43. 0   45. 30   47. -3 49. -17 51. -4

53. 14   55. -4 57. $-\frac{60}{11}$ 59. -12 61. $-\frac{8}{5}$
63. perform the indicated operation   65. -6 67. 6
69. $-\frac{3}{2}$ 71. $-\frac{21}{16}$ 73. $-\frac{1}{2}$ 75. -8 77. $\frac{9}{8}$ 79. 43
81. $\frac{17}{8}$ 83. 3   85. -72 87. 2   89. 1.39

Section 6.5

1. $\frac{d}{r}$ 3. $\frac{2A}{h}$ 5. $\frac{2h - 440}{11}$ 7. $\frac{2S - an}{n}$ 9. $\frac{2}{a - b}$
11. $\frac{E - IR}{I}$ 13. $\frac{A - P}{pt}$ 15. $\frac{d + 4a}{4}$
17. $\frac{V}{wh}$ 19. $(a - c)x - cd$ 21. $\frac{1 + a}{ac - a}$ 23. $5 - y$
25. $\frac{12 - 3y}{4}$ 27. $\frac{c + 5a}{3}$ 29. $\frac{5a + 10}{2}$ 31. $\frac{ac}{a - c}$
33. $\frac{I}{Pt}$ 35. $\frac{6}{a + b}$ 37. $\frac{3 - ab}{a}$ 39. $\frac{2A - bh}{h}$
41. literal, constant

Section 6.6

1. $x < -9$

3. $x \leq -6$

5. $x \leq -0.59$

7. $x < 4.7$

9. $x \geq -10$

11. $x \geq \frac{14}{15}$

13. $x < 2$

15. $x \leq 2$

17. $x \geq -\frac{11}{2}$

19. $x \leq 10$

21. $x < 1.2$

23. $x \geq -30$

25. $x < -15$

27. $x < -\frac{8}{3}$

29. $x < -3$

31. $x > 0.02$

33. $x \geq -1.7$

35. $x \geq -20$

37. $x > 12$

39. $x > -10$

41. $x \geq 15$

43. $x < \frac{13}{24}$

45. $x \leq 1.7$

47. $x < -3$

25. $x \leq -4$

27. $x > -2$

49. $x > -1$

51. $x \leq 0.2$

29. $x \leq -3$

31. $x < -\frac{1}{4}$

53. $x \leq -5$

55. $x < \frac{10}{7}$

33. $x \leq \frac{42}{5}$

35. no solution

57. $x \geq -1.2$

59. $x > -10.8$

37. $x \geq \frac{9}{8}$

39. $x < \frac{14}{17}$

61. $x \geq -10$

63. $x > 0.636$

41. $x < -72$

43. $x \leq -\frac{39}{10}$

65. $x > 12$

67. $x \leq -8.6$

45. $x < -\frac{12}{5}$

47. $x > -\frac{3}{5}$

69. $x < -0.12$

71. $x < 14$

49. $x \leq -2$

51. $x \leq 100$

73. $x > 8$

75. $x \leq -8$

53. $x \leq 4$

55. $x > 2$

77. $x \leq \frac{15}{14}$

79. $x \geq -\frac{2}{7}$

57. $x < 3$

59. $x > -2$

81. $x \leq 1$

83. $x \geq -1$

61. $x \geq 10$

63. $x < -6$

85. inequality, greater

87. not, one, solutions

65. infinite number of solutions

Section 6.7

1. $x \leq 2$

3. $x > 2$

5. $x < 6$

7. $x \leq 8$

67. $y \leq 1.33$

9. $x < 10$

11. $x \leq -5$

69. $x \leq -1.5$

13. $x \leq -2$

15. $x < 2$

CHAPTER 6 Review Exercises

1. yes *(Section 6.1)* **2.** no *(6.1)* **3.** yes *(6.1)*
4. -7 *(6.2)*
5. $x > -2$ *(6.6)*

17. $x \geq 4$

19. $x > -2.5$

21. $x \geq 2$

23. $x < 2$

6. -27 *(6.2)* **7.** -3 *(6.2)* **8.** 33 *(6.3)* **9.** 15 *(6.4)*
10. 0.39 *(6.2)*

11. $x > 9$ (6.7)

12. $\dfrac{c - b}{a}$ (6.5) **13.** -20 (6.3) **14.** $-\dfrac{5}{2}$ (6.4)

15. $\dfrac{9}{128}$ (6.4) **16.** $\dfrac{yz}{y + z}$ (6.5)

17. $x < -6$ (6.6)

18. $x \geq -25$ (6.6)

19. $x \geq 3.62$ (6.6)

20. $\dfrac{7}{3}$ (6.4)

21. $x \leq \dfrac{11}{2}$ (6.7)

22. $x < 24$ (6.7)

23. $\dfrac{b}{a - c}$ (6.5) **24.** $-\dfrac{5}{29}$ (6.4)

25. no solution (6.1)
26. infinite number of solutions (6.1)
27. identity (6.7) **28.** no solution (6.7)
29. $x < -21$ (6.7)

30. $x \geq -\dfrac{15}{14}$ or $-1\dfrac{1}{14}$ (6.6)

31. $-\dfrac{8}{5}$ (6.4) **32.** -3.8 (6.3)
33. $x \geq 45$ (6.7)

34. $\dfrac{36}{5}$ (6.4) **35.** no solution (6.4)

CHAPTER 6 Test

1. 8 is not the solution **2.** $-\dfrac{9}{2}$ is the solution **3.** -4

4. $-\dfrac{8}{45}$ **5.** 10 **6.** -6 **7.** -15 **8.** $-\dfrac{4}{9}$ **9.** 5.85

10. $\dfrac{13}{7}$ **11.** $-\dfrac{5}{3}$ **12.** $-\dfrac{60}{11}$ **13.** 20 **14.** $\dfrac{b + ad}{a - c}$

15. $\dfrac{2h - 440}{11}$

16. $x \leq 1.25$

17. $x > -\dfrac{8}{3}$ or $-2\dfrac{2}{3}$

18. $y \geq \dfrac{7}{6}$ or $1\dfrac{1}{6}$

19. $x < 10$

20. $x > -\dfrac{20}{9}$ or $-2\dfrac{2}{9}$

CHAPTER 7

CHAPTER 7   Objectives

1. x, $12 - x$ **2.** 30 **3.** $\dfrac{1}{12}$ **4.** \$25 **5.** 5 **6.** \$3500

7. 25 liters **8.** $3\dfrac{1}{2}$

Section 7.1

1. $x + 10$ **3.** $\dfrac{2}{3}x$ **5.** $2x$ **7.** $x - 6$ **9.** $6x$ **11.** $x - 8$
13. $x - 50$ **15.** $x - 7$ **17.** $x - 80$ **19.** $\dfrac{x}{12}$ **21.** $\dfrac{10}{x}$
23. $x - 2$ **25.** $x + 90$ **27.** $9x$ **29.** $3 + \dfrac{x}{2}$ **31.** $4x + 7$
33. $\dfrac{2}{4} - 2x$ **35.** $2x + 8$ **37.** $\dfrac{x}{2} - 10$ **39.** $9(x + 7)$
41. $\dfrac{1}{4}x + 5$ **43.** $x + \dfrac{2}{6}$ **45.** $x + 130$ **47.** $l - 13$
49. $n + 2$ **51.** $d - 35$ **53.** $s + 75$ **55.** $10 - x$
57. $40 - x$ **59.** $\dfrac{x}{2}$ **61.** x, $2x$
63. $x =$ John's height, $x + 8 =$ Joe's height **65.** $x = $ \$ in Dan's account, $x - 1025 = $ \$ in Tom's account
67. $x =$ length of rectangle, $x + 3 =$ width of rectangle
69. x, $60 - x$ **71.** x, $1000 - x$ **73.** $2b - 5$ **75.** x, $4x$
77. x, $50 - x$ **79.** $300 + x$ **81.** x, $5000 - x$
83. $10 - x$ **85.** x, $x + 1$, $x + 2$ **87.** 25, 35
89. 17, 18 **91.** -5 **93.** 14 **95.** 10 **97.** \$425
99. 39, 40 **101.** 93, 95 **103.** 10, 12, 14 **105.** 23, 33
107. Gene received 122 votes; Howard received 262 votes
109. 30 **11.** 20°, 100°, 60° **113.** 171 **115.** 625
117. 37, 39, 41 **119.** 10 teachers, 11 tutors

Section 7.2

1. $\dfrac{2}{3}$ **3.** $\dfrac{73}{110}$ **5.** $\dfrac{15}{7}$ **7.** $\dfrac{1}{15}$ **9.** $7\dfrac{\text{feet}}{\text{minute}}$
11. $15\dfrac{\text{dollars}}{\text{hour}}$ **13.** 5:9 **15.** 3:7 **17.** 2.5¢ **19.** 80¢
21. extremes: 5, 30; means: 15, 10
23. extremes: 12, 65; means: 60, 13 **25.** equal
27. not equal **29.** 9 **31.** 27 **33.** 60 **35.** 4 **37.** $\dfrac{27}{5}$
39. 0.5 **41.** 3.5 **43.** 5 **45.** $-\dfrac{10}{3}$ **47.** 4 **49.** 78
51. same, units **53.** proportions, product, product

55. 20, 25 **57.** $63.75 **59.** 105 men, 60 women
61. $3750 in stocks, $5000 in bonds, $6250 in real estate
63. Phyllis received $8000, Pam received $14,000
65. $24,000; $40,000; $56,000 **67.** 20 **69.** 595
71. 230.4 **73.** 15 **75.** 42 **77.** 52 **79.** 966 **81.** 96
83. 112.24274 **85.** 6146

Section 7.3

1. 60 **3.** 25% **5.** 25 **7.** 226 **9.** 70% **11.** 1300
13. $66\frac{2}{3}\%$ **15.** 1400 **17.** 1.44 **19.** 73.77% **21.** 20
23. $30 **25.** 25 **27.** $2,800 **29.** 8% **31.** 20
33. markup is $90, selling price is $240 **35.** $16,000
37. 10% **39.** original cost is $50,000; new cost is $70,000
41. $11,200 **43.** $10 **45.** $80,000 **47.** $\frac{4}{3}$ **49.** $3\frac{1}{3}$
51. $731.45 **53.** $5126.78 **55.** $1231562.50

Section 7.4

1. $80 **3.** $240 **5.** $30 **7.** 6.25% **9.** 15%
11. $15,000 **13.** 12 **15.** 4 **17.** $16.20 **19.** $20,000
21. 9 **23.** $2,000 at 15%; $3,000 at 16% **25.** $850
27. $3,000 **29.** $2,000 at 10%; $4,000 at 12%

Section 7.5

1. 9 gallons of 15% solution; 3 gallons of 75% solution
3. 3.6 ounces of 12% solution; 2.4 ounces of 7% solution
5. 17 **7.** $8\frac{1}{3}$ ounces **9.** 400 gallons **11.** $\frac{2}{3}$ quarts
13. 5 **15.** 17 pints **17.** 12 ounces **19.** 20 barrels at $80 per barrel; 30 barrels at $100

Section 7.6

1. 110 **3.** 50 mph **5.** 3 **7.** 110 mph **9.** 12.54 mph
11. $\frac{1}{2}$ hour **13.** 10:24 A.M. **15.** 3
17. Tim's rate is 34 mph; Sam's rate is 42 mph
19. 10:40 A.M. **21.** faster boat travels 30 miles; slower boat travels 20 miles

CHAPTER 7 Review Exercises

1. $x - 5$ (Section 7.1) **2.** $x, 50 - x$ (7.1)
3. 131,132 (7.1) **4.** 60% (7.3) **5.** $\frac{2}{9}$ (7.2) **6.** $600 (7.4)
7. 20 (7.5) **8.** $80,000 (7.3) **9.** $12,480 (7.2)
10. 34 mph, 42 mph (7.6)
11. $5000 at 11%, $3000 at 10% (7.4)
12. 40 (7.2) **13.** 80 mph (7.6)
14. 50 ounces (7.5) **15.** 80.5 (7.1)
16. $5600; $7840; $8960 (7.2) **17.** broccoli (7.2)
18. 32% (7.3) **19.** 7.55 liters (7.2) **20.** 23 student tickets and 37 senior citizen tickets (7.1) **21.** 30 liters (7.5)
22. $4200 (7.4) **23.** 6:20 P.M. (7.6) **24.** 90% (7.3)
25. 3.06 gallons (7.2)

CHAPTER 7 Test

1. $x, 15 - x$ **2.** 18, 20, and 22 **3.** $0.04 per oz.
4. $5625 = Cazzie's share $3375 = Guerin's share
5. $7.82 **6.** 80% **7.** $5300
8. women: $1800; men: $4200 **9.** 25 gallons **10.** Jill: 20 feet per second; Joe: 15 feet per second

CHAPTER 8

CHAPTER 8 Objectives

1.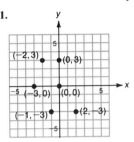

2. $A = (2, 2)$, $B = (5, -3)$, $C = (-5, 0)$, $D = (-5, -4)$, $E = (0, -5)$

3a. 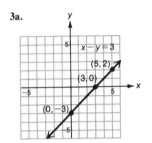 **3b.**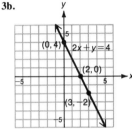

3c. 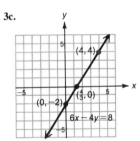 **3d.**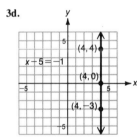

4a. 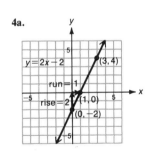 **4b.**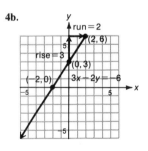

5a. 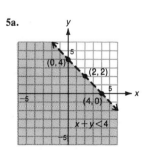 **5b.**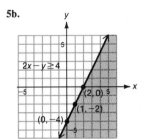

498 Answers to Selected Exercises

5c. 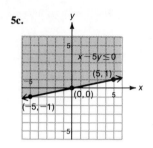 **5d.**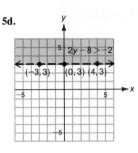

5. $(0, -2), (1, -2), (-3, -2)$

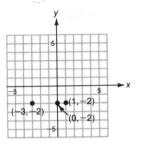

Section 8.1

1. 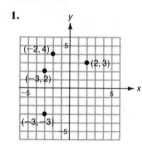 **3.**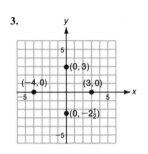

5.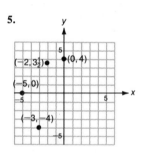

7. $A = (2, 3), B = (-5, -2), C = (0, -5), D = (-6, 5),$
$E = (5, -4), F = (5, 6)$
9. $A = (-6, -2), B = (-4, -6), C = (0, 0), D = (5, 0),$
$E = (-4, 3), F = (3, 6)$ **11.** quadrants
13. ordered pairs **15.** x-coordinate, abscissa, y-coordinate, ordinate **17.** vertical, x, coordinate, horizontal, y, second, coordinates, ordered pair

7. x-intercept $= (2, 0)$, y-intercept $= (0, -6)$
9. x-intercept $= (-2, 0)$, y-intercept $= (0, 6)$
11. x-intercept $= (3, 0)$, y-intercept $= (0, 2)$
13. x-intercept $= (12, 0)$, y-intercept $= (0, -6)$

15. 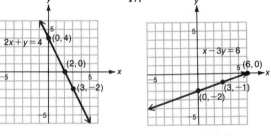 **17.**

19. 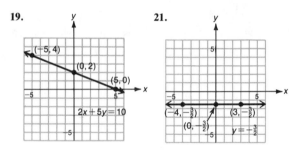 **21.**

23. 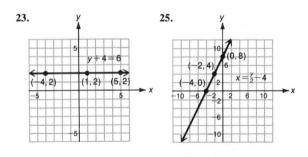 **25.**

Section 8.2

1. $(0, 1), (-2, -5), (1, 4)$ **3.** $(0, -2), (7, 1), \left(5, \dfrac{1}{7}\right)$

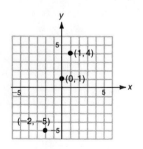 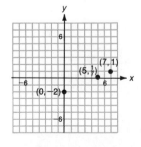

27. 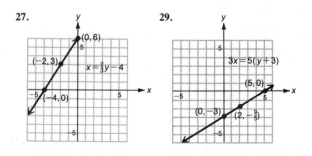 **29.**

31. 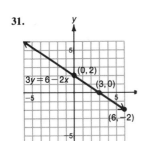 **33.**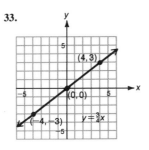

17. 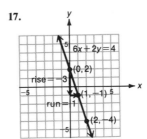 **19.**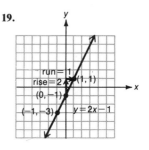

35. 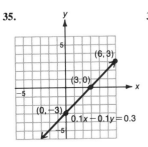 **37.**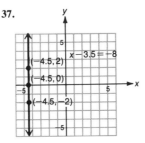

21. 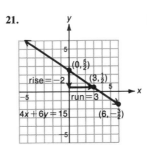 **23.**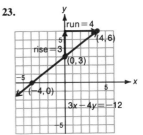

39.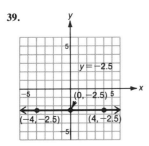

25. 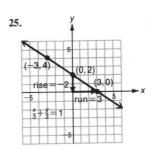 **27.**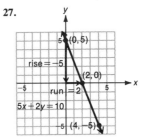

41. straight line **43.** ordered pairs, satisfy, ordered pairs, straight line **45a.** $y = 10x + 10$
45b. (2, 30), (4, 50), (6, 70)
c. 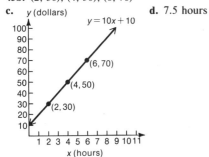   **d.** 7.5 hours

29. 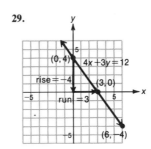 **31.**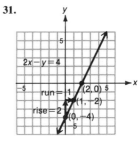

33. 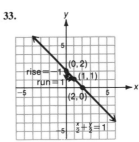 **35.**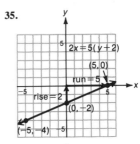

Section 8.3

1. $y = -2x + 6$, slope $= -2$, y-intercept $= 6$
3. $y = \frac{2}{5}x - 6$, slope $= \frac{2}{5}$, y-intercept $= -6$
5. $y = \frac{1}{4}x - 2$, slope $= \frac{1}{4}$, y-intercept $= -2$
7. $y = -\frac{3}{2}x + 3$, slope $= -\frac{3}{2}$, y-intercept $= 3$
9. $y = -3x + 6$, slope $= -3$, y-intercept $= 6$
11. $y = 5x - 8$, slope $= 5$, y-intercept $= -8$
13. $y = -\frac{3}{2}x + \frac{5}{2}$, slope $= -\frac{3}{2}$, y-intercept $= \frac{5}{2}$
15. $y = \frac{3}{5}x - \frac{11}{5}$, slope $= \frac{3}{5}$, y-intercept $= -\frac{11}{5}$

37. 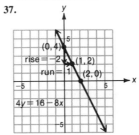 **39.**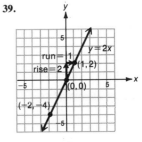

41.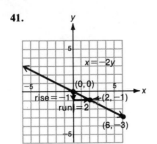

43. $y = mx + b$ **45.** right   **47.** rise, run

Section 8.4

1. $(2, 0)$ is a solution   **3.** $\left(-2, \dfrac{3}{4}\right)$ is not a solution

5. 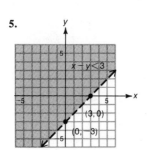   **7.**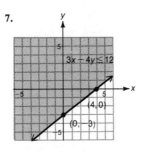

9. 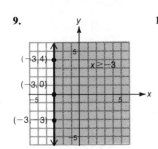   **11.**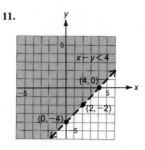

13. 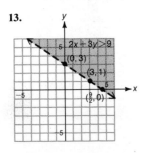   **15.**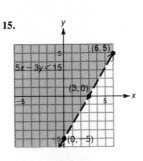

17. 19.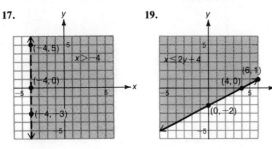

21.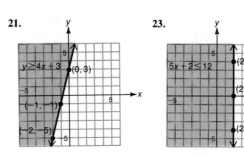

23.

25. 27.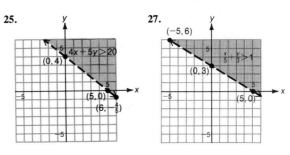

29.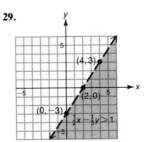

31. equation, equal   **33.** dotted, not part   **35.** boundary line, line, solution, shade

CHAPTER 8   Review Exercises

1. *(Section 8.1)*   **2.** *(8.1)*
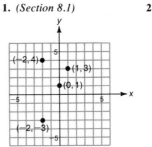   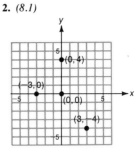

3. $A = (0, 4)$, $B = (-5, 0)$, $C = (3, -2)$, $D = (0, -5)$, $E = (-4, -5)$ *(9.1)*   **4.** $A = (4, 1)$, $B = (-3, 1)$, $C = (0, -2)$, $D = (-6, -2)$, $E = (3, -5)$ *(9.1)*

5. *(8.2)*   **6.** *(8.2)*
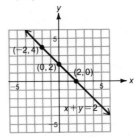   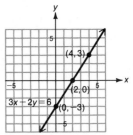

Answers to Selected Exercises   501

7. (8.4)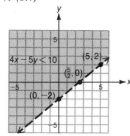

8. (8.4)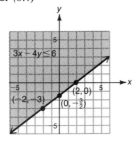

9. (8.2)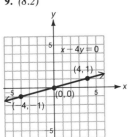

10. (8.2)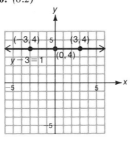

11. (8.4)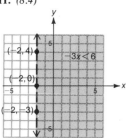

12. (8.2)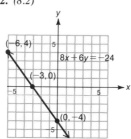

13. (8.4)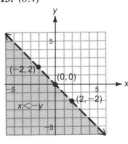

14. (8.2)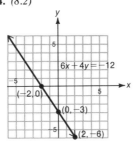

15. (8.3)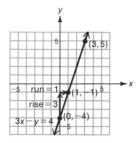

16. (8.3)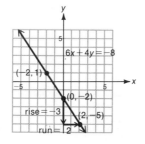

17. (8.3)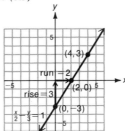

18. (8.3)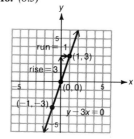

19a. when $n = 2$, $b = 3200$; when $n = 3$, $b = 2800$; when $n = 5$, $b = 2000$ (8.2)

b. (8.2)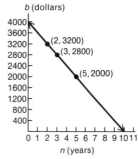

c. After 1 year the book value is $3600 (8.1)   **d.** when $n = 8$ or in 8 years (8.1)

20. (8.2)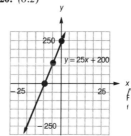

21. (8.2)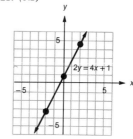

22. (8.2)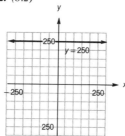

23. (8.4)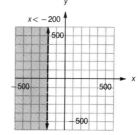

24. (8.2)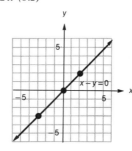

25. (8.2)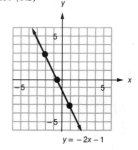

26. (8.2)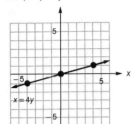

27. (8.4)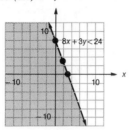

6.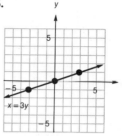

7.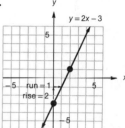

28. (8.4)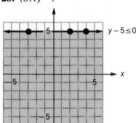

29. (8.4)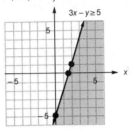

8.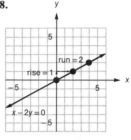

9.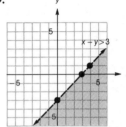

30. (8.4)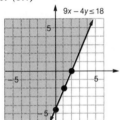

31. (8.4)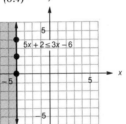

10.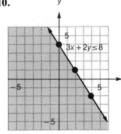

11.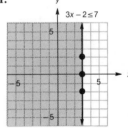

CHAPTER 8 Test

1.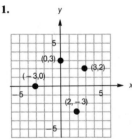

12.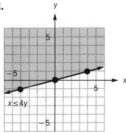

CHAPTER 9

CHAPTER 9 Objectives

2. $A = (4, 0)$, $B = (0, -4)$, $C = (7, -2)$, $D = (-2, 4)$, $E = (-7, -6)$, $F = (-5, 0)$
3. x-intercept $= 2$; y-intercept $= -4$
4.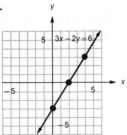

5.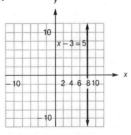

1a.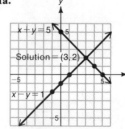

1b.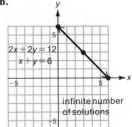

1c. 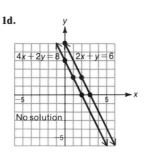 **1d.**

2a. no solution **b.** (1, 3) **c.** $\left(\frac{3}{2}, -2\right)$ **d.** $\left(\frac{40}{7}, \frac{2}{7}\right)$
3a. no solution **b.** infinite number of solutions
c. $\left(\frac{15}{7}, -\frac{3}{7}\right)$ **d.** $(-4, 0)$ **e.** Guerin earns $205, Cazzie earns $410

Section 9.1

1. solution **3.** not the solution **5.** not the solution
7. not the solution
9. **11.**
13. **15.**
17. **19.**
21. 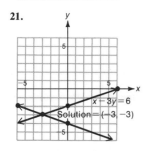 **23.**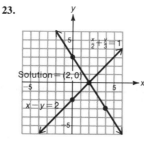

25. consistent, one solution **27.** inconsistent, no solution
29. dependent, infinite number of solutions
31. consistent, one solution **33.** dependent, infinite number of solutions **35.** consistent, one solution
37. consistent, one solution **39.** dependent, an infinite number of solutions **41.** equation, interpret
43.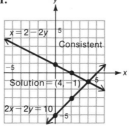

45. inconsistent **47.** dependent **49.** inconsistent
51. **53.** dependent

55. 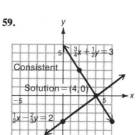 **57.** dependent

59.

Section 9.2

1. (1, 6) **3.** no solution **5.** $\left(\frac{1}{2}, 1\right)$ **7.** (3, 0)
9. $\left(-\frac{45}{34}, -\frac{38}{17}\right)$ **11.** (4, 2) **13.** infinite number of solutions **15.** (2, −1) **17.** $\left(-\frac{2}{5}, \frac{12}{5}\right)$ **19.** $\left(\frac{122}{39}, -\frac{110}{39}\right)$
21. (6, 2) **23.** $\left(\frac{69}{34}, \frac{31}{34}\right)$ **25.** (3, 1) **27.** $\left(\frac{2}{3}, 4\right)$
29. infinite number of solutions **31.** $ax + by = c$, no solution, infinite number of solutions **33.** inconsistent, no
35. 2.4 gallons of the 7% solution, 3.6 gallons of the 12% solution **37.** $1000 for the Ford, $3000 for the BMW

39. $-\frac{82}{7}$ is the first number, $\frac{146}{7}$ is the second number
41. $10 for typesetting, 50¢ per brochure **43.** 6.4 gallons of the 15% solution, 9.6 gallons of the 40% solution
45. width is 38.7 feet, length is 32.3 feet **47.** $\left(\frac{285}{13}, -\frac{304}{13}\right)$
49. $(1.5, -2.5)$

Section 9.3

1. $(6, 0)$ **3.** $(0, 6)$ **5.** $(3, 5)$ **7.** $(3, 4)$ **9.** $\left(\frac{17}{6}, \frac{25}{6}\right)$
11. infinite number of solutions **13.** $\left(-\frac{44}{7}, \frac{20}{7}\right)$ **15.** no solution **17.** $\left(\frac{8}{15}, -\frac{18}{5}\right)$ **19.** $\left(\frac{1}{5}, -\frac{4}{3}\right)$ **21.** $\left(-\frac{11}{3}, -\frac{4}{3}\right)$
23. $(2, -5)$ **25.** $(-4, 0)$ **27.** $(-2, 0)$ **29.** $(3, -10)$
31. $(1.8, -2.2)$ **33.** eliminate, eliminate, substitution, addition **35a.** one, one **b.** substitute, other equation **c.** one, one **d.** substituting, equation **e.** ordered pair **f.** check **37.** Gene received 122 votes, Howard received 262 votes **39.** the lot costs $20,000; the house costs $35,000 **41.** $400 at 6.25%, $1100 at 13%
43. 9 gallons of the 15% solution; 3 gallons of the 75% solution **45.** 50 mph on the highway, 30 mph on the unpaved road **47.** the faster boat traveled 30 miles, the slower boat traveled 20 miles **49.** $(-8, -4)$ **51.** $(2.5, 4.5)$

CHAPTER 9   Review Exercises

1. *(Section 9.1)* **2.** *(9.1)*

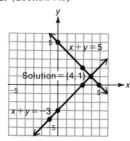 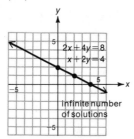

3. *(9.1)*

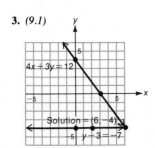

4. $\left(\frac{1}{2}, 4\right)$ *(9.2)* **5.** no solution *(9.2)*
6. $(0, 2)$ *(9.2)* **7.** $(1, 4)$ *(9.3)* **8.** $(1, 1)$ *(9.3)*
9. $\left(\frac{1}{2}, -\frac{11}{6}\right)$ *(9.3)* **10.** no solution *(9.1)*
11. $\left(1, \frac{5}{2}\right)$ *(9.2)* **12.** $(-2, 1)$ *(9.2)* **13.** $(5, -1)$ *(9.3)*
14. $(1.5, -2)$ *(9.2)* **15.** $\left(\frac{51}{4}, -\frac{13}{2}\right)$ *(9.2)* **16.** 38, 32 *(9.2)*
17. 10 pounds of pumpkin seeds, 20 pounds of sunflower seeds *(9.2)* **18.** $15,000 at 15%; $10,000 at 25% *(9.2)*
19. Ed's rate was 27 feet per second; Mike's rate was 24 feet per second *(9.3)* **20.** interest rate for the real estate was 12%, interest rate for stocks was 15% *(9.2)*
21. $(10, 20)$ *(9.2)* **22.** $(5, -4)$ *(9.2)* **23.** no solution *(9.3)*
24. $\left(\frac{35}{8}, \frac{11}{8}\right)$ *(9.2)*

CHAPTER 9   Test

1. no solution **2.** infinite number of solutions

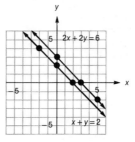 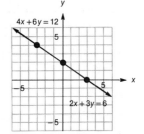

3. $(12, -6)$ is the solution

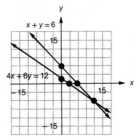

4. $(3, -5)$ **5.** no solution **6.** $\left(2, \frac{1}{3}\right)$ **7.** $\left(\frac{3}{2}, -1\right)$
8. no solution **9.** $(6, 5)$ **10.** $\left(\frac{1}{2}, \frac{2}{3}\right)$ **11.** $(4, 9)$
12. $(3, -1)$ **13.** $1800 at 10%, $1200 at 14%

CHAPTER 10

CHAPTER 10   Objectives

1a. $2x^5(8x^8 - 9)$ **b.** $x^2(15x^4 - 18x^3 + 10)$
2a. $(x - 4)(x - 2)$ **b.** $(x + 8)(x - 3)$
3a. $(6x - 7)(x + 1)$ **b.** $2x(3x - 2)(x + 5)$
4a. $(x - 4)(x + 4)$ **b.** $(4x - 5)(4x + 5)$ **5a.** $(x + 3)^2$
b. $(3x - 5)^2$ **6a.** $(x - y)(a + b)$
b. $(x - 9y - 1)(x + 9y - 1)$

Section 10.1

1. $2(x + 4)$ **3.** $3xy(3x + 5)$ **5.** $5y(5x - 7y)$
7. $5(2x^3 - 5x^2 + 4)$ **9.** $50x^2(x^3 + 2x - 3)$
11. $8xy(2x^2 + x + 3)$ **13.** $19x^2(x + 2y^2)$
15. $6a^2d^2(2a - 3d)$ **17.** $-9b(3b^3 + 2b^2 - 4)$
19. $14xy^6(x^4 - 3x^2y + 2y^2)$ **21.** $6a^2b^2(2a - 3b)$
23. $11a^{10}b^5(4a^4b^2 - 3)$ **25.** $2xy^3(6x^7y^6 + 9x^4y - 10)$
27. $ab(15a^{11} - 8b^{11} + 9)$
29. $-15a^2b^2(2ab^2 + 3a^6b^5 + 1)$
31. $6(3x^3 - 2y^2 - 8x^4)$ **33.** $8xy^2(2x^2 + 3y - 5x^3)$

35. $-5x^2(4x^3 + 2x - 1)$ 37. $11x^2y^2z(4xy^2 - 5)$
39. $40a^2b^2c^3(bc - 2a)$ 41. 12 43. 13 45. 21 47. 3
49. 15 51. 26 53. 77 55. $13x^2(5x - 7)$
57. $35xy(2x^2 - 3xy + 4y^2)$ 59. $11(11x^4 - 13x^3 + 17)$
61. $-12x(4x^2y - 5x + 7y)$
63. greatest common factor 65. greatest common monomial factor, term, greatest common monomial factor

Section 10.2

1. $(x + 7)(x + 5)$ 3. $(x - 5)(x - 1)$
5. $(x - 6)(x + 5)$ 7. $(x - 8)(x - 3)$
9. $(x - 12)(x + 3)$ 11. $(x + 7)(x - 5)$
13. $(x - 7)(x - 4)$ 15. not factorable
17. $(5x + 2)(x - 3)$ 19. $(6x + 5)(x - 1)$
21. $(2x - 5)(x - 4)$ 23. $(3x + 4)(x - 2)$
25. $(5x - 1)(4x - 3)$ 27. $(6x + 7)(2x - 5)$
29. $(9x - 1)(4x - 1)$ 31. $(y + 5)(y - 3)$
33. $(x + 3)(x - 2)$ 35. $(a - 6)(a + 3)$
37. $(7x - 5)(3x - 2)$ 39. $3(x - 18)(x + 2)$
41. $3(3x + 4)(3x + 2)$ 43. $(7x^2 - 1)(x^2 + 10)$
45. $(x - 9y)(x + 2y)$ 47. not factorable
49. $2(3x + 2)(2x - 5)$ 51. $x^5(x - 7)(x + 2)$
53. $(x + 7)(x - 6)$ 55. $(4x^2 + 3)(2x^2 - 3)$
57. $(8x + 5)(x + 1)$ 59. $(3x + 4)(2x - 5)$
61. $(5x + 3)(3x - 2)$ 63. $(5x - 8)(2x + 3)$
65. $(5x^2 - 4)(3x^2 - 2)$ 67a. same, first term
b. product, sum, coefficient c. last terms
d. product, same 69. common monomial
71. $(5x - 2y)(5x + y)$ 73. $(4s + 3)(s - 2)$
75. $(8x - 3)(2x - 3)$ 77. $2(a + 5b)(a - 3b)$
79. $(4a^2 - 5b^2)(3a^2 + 2b^2)$ 81. $(9c + 2)(7c - 5)$

Section 10.3

1. $(x - 3)(x + 3)$ 3. $(3x - 2)(3x + 2)$
5. $(x - 8)(x + 8)$ 7. $(3s + 1)(3s - 1)$
9. $(x^2 + 4)(x - 2)(x + 2)$ 11. not factorable
13. $(9x + 1)(9x - 1)$ 15. $(5y + 2)(5y - 2)$
17. $(x + 2)^2$ 19. $(a - 5)^2$ 21. $(x - 4)^2$
23. not factorable 25. not factorable
27. not factorable 29. $(x + 8)^2$ 31. not factorable
33. $(x + y)(c + d)$ 35. $(x - 1)(b + a)$
37. $(y + 3)(y^2 + 4)$ 39. $(y - 3)(y - 3)(y + 3)$
41. $(-x + 2y + z)(x - 2y + z)$ 43. $(2b - c)(3b - 7)$
45. $(x - 5y + 4)(x + 5y + 4)$ 47. $(x + y)(b + c)$
49. $(c - d)(w - z)$ 51. $(b + 1)(y - 1)$
53. $(x - y)(x + y + 5)$ 55. $(s - 3t)(4s + 7)$
57. $(x - 6y - 3)(x + 6y - 3)$
59. $(-x + y - 6)(x + y + 6)$
61. $(x + y)(x - y + 1)$ 63. perfect squares, twice, perfect squares, additive inverse 65. common, factor
67. $(3x + 8)^2$ 69. $(x - a)(y + p)$
71. $2(2x - 3)^2$ 73. $2(4x + 1)^2$ 75. $3(x - 5)(x + 5)$

CHAPTER 10 Review Exercises

1. $(2x + 3)^2$ *(Section 10.3)* 2. $(x - 5)(x + 2)$ *(10.2)*
3. $(x - 6)(x - 5)$ *(10.2)* 4. $(x^2 + 9)(x - 3)(x + 3)$ *(10.3)*
5. $2(x + 7)(x - 1)$ *(10.2)* 6. $(5x - 1)^2$ *(10.3)* 7. not factorable *(10.3)* 8. $(b - 2)(b + 2)(b - 2)$ *(10.3)*
9. not factorable *(10.3)* 10. $2(4x - 1)(4x + 1)$ *(10.3)*
11. $(2x - 1)(2x + 3)$ *(10.2)* 12. $6x^3(3x^2 - 4x + 6)$ *(10.1)*
13. $12x^4(5x^5 + 7x^4 - 8x^3 + 2)$ *(10.1)*
14. $(2x - 5y - 3)(2x + 5y - 3)$ *(10.3)*
15. $(2x - 1)(x + 3)$ *(10.2)* 16. $(4x + 3)(3x - 4)$ *(10.2)*
17. $(6x + 5)(x - 2)$ *(10.2)* 18. $2(x^2 + 11x + 14)$ *(10.1)*
19. $(8x^2 + 1)(8x^2 - 1)$ *(10.3)*
20. $(6x - 7y)(4x + 3y)$ *(10.2)*
21. $3(3x^2 - 14x + 7)$ *(10.1)* 22. $3(5x^4 - 9x^3 - 3)$ *(10.1)*
23. not factorable *(10.2)* 24. $(x^3 + 2)(x^3 - 2)$ *(10.3)*
25. $(3x + 2)(6x - 5)$ *(10.2)* 26. $(2a + 5)(a + 6)$ *(10.2)*
27. $(2x + 1)(5x + 2)$ *(10.2)* 28. $(3x + 5y)(x - 4y)$ *(10.2)*
29. $(a + b)(3x + y)$ *(10.3)* 30. $(x + y)(9 - 4a)$ *(10.3)*
31. $(x + 9)(x + y)$ *(10.3)* 32. $(x - y)(x - 4)$ *(10.3)*
33. $(2x - y)(4x - 3)$ *(10.3)* 34. not factorable *(10.1)*
35. $2x(4x + 3)(4x - 5)$ *(10.2)*
36. $2(3x + 4)(6x + 5)$ *(10.2)* 37. $4x(x + 4)(x + 4)$ *(10.2)*
38. $ab(a + 5b)(a - 5b)$ *(10.3)* 39. $2z(80z + 1)$ *(10.1)*
40. $(e - f)(5x + 3y - 7)$ *(10.3)*

CHAPTER 10 Test

1. $5x^2y^3(3xy - 2)$ 2. $3xy^2z^2(2x^2z^2 - 5xy + 4z)$
3. $26x(3x^2 - 5x + 7)$ 4. not factorable
5. $(x - 4)(x + 3)$ 6. $(x - 10)(x + 2)$
7. $(2x - 1)(x - 3)$ 8. $(2x + 3)(5x - 4)$
9. $2x(2x + 5)(3x - 4)$ 10. $(3x - 8y)(4x - 5y)$
11. not factorable 12. $(5y + 2)(5y - 2)$
13. $2x(x - y)(x + y)(x^2 + y^2)$ 14. $(2x - 3)^2$
15. $(x + 3y)^2$ 16. $(x - 2y)(3x + 2)$
17. $(y + 1)(x - 1)$ 18. $(x + 3y - 5)(x - 3y - 5)$
19. $2(2x + y)(3x + 1)$ 20. $(3x^2 + 2)(2x + 1)(2x - 1)$

CHAPTER 11

CHAPTER 11 Objectives

1a. $\frac{1}{5}$ b. $\frac{x-4}{x-3}$ 2a. $\frac{2}{x}$ b. $\frac{x^2 - 2xy + y^2}{x^2 + 2xy + y^2}$
3a. 4 b. $\frac{2x-3}{x-3}$ 4a. $\frac{x(x-1)}{x+1}$ b. $\frac{x-7}{(x+2)(x+2)(x-1)}$
5a. -2 b. $\frac{4}{5}$ 6a. $\frac{12}{13}$ hr (55 minutes)
b. Car A travels 120 mph, Car B travels 100 mph

Section 11.1

1. $\frac{1}{2(2x+1)}$ 3. $x + 5$ 5. $\frac{x+2}{x(3x+1)}$ 7. $\frac{1}{2x-3}$
9. $\frac{x-6}{x+6}$ 11. $\frac{5b(a-1)}{a+4}$ 13. $\frac{3x+1}{x+4}$ 15. $\frac{x-2}{x+2}$
17. $\frac{y(y-4)}{3(y+1)}$ 19. $\frac{1}{x+8}$ 21. $\frac{x-8}{x-6}$ 23. $\frac{x-3}{x+8}$
25. $\frac{x+6}{x-7}$ 27. $\frac{n-5}{2}$ 29. $\frac{9}{(a-3)(a^2+9)}$
31. -1 33. $-\frac{1}{5}$ 35. $-\frac{n}{2n+3}$ 37. $-\frac{x+4}{x}$
39. $\frac{3m+4n}{3m-4n}$ 41. $\frac{2a(a+2b)}{3(a+b)}$ 43. $\frac{y+3}{y+4}$
45. $\frac{x-2y+5}{x-2y}$ 47. rational, algebraic fractions

Section 11.2

1. $6y$ 3. $\frac{2(2x-1)}{x(x^2+3)}$ 5. $\frac{1}{2}$ 7. $-4x^3$
9. $\frac{a(a+4)}{(a+2)(a-2)}$ 11. $\frac{2x^2(x+1)}{(x-2)(x-1)}$ 13. $\frac{x+2}{x+1}$
15. $-\frac{4}{55x}$ 17. $\frac{x}{9}$ 19. $\frac{4(x+6)}{x^2}$ 21. $\frac{1}{4}$

23. $\dfrac{8n^2}{7(n-4)}$ 25. $\dfrac{(2x+1)(x-2)}{(x-1)(x+1)}$
27. $\dfrac{(x+2)(x+4)(x-3)}{x(x+3)(x-2)}$ 29. $\dfrac{(n^2+9)(6n+7)}{5n-7}$
31. $\dfrac{7a-2}{a-9}$ 33. $-\dfrac{(3x-2y)}{4x(2x-5y)}$ 35. $\dfrac{x-7}{2(x+7)}$
37. factor, product

Section 11.3

1. $\dfrac{11}{37y}$ 3. $\dfrac{1}{x+2}$ 5. $\dfrac{2y-3}{y^2}$ 7. $\dfrac{2}{x+3}$
9. $\dfrac{1}{x+3}$ 11. $126x^2$ 13. $x(x+5)(x-5)$
15. $(2x-3)(3x+2)$ 17. $x(x+1)(x+2)(x+3)$
19. $2x^3(2x+1)(x-3)(2x^2+2x+1)$ 21. $\dfrac{x+7}{18}$
23. $\dfrac{2(4x+1)}{15}$ 25. $\dfrac{85}{72x}$ 27. $\dfrac{55y-42x^2}{396x^3y^2}$ 29. $\dfrac{6x^3+1}{x^2}$
31. $\dfrac{9x-24}{x(x-8)}$ 33. $\dfrac{x^2-x+2}{(x-3)(x-2)}$
35. $\dfrac{3x-2}{(x-2)(x+2)}$ 37. $\dfrac{-x^2+12x-12}{(x+2)(x-4)(2x-3)}$
39. $\dfrac{2b}{3a+b}$ 41. $\dfrac{2x^2+x-5}{x+2}$ 43. $-\dfrac{x+4}{x+2}$
45. $\dfrac{1}{n^2+1}$ 47. $\dfrac{4}{x(x-2)}$ 49. $\dfrac{x^3+17x^2-54x+32}{(x-3)(x-3)(x-2)(x+2)}$
51. $\dfrac{x^3-6x^2-x+6}{(2x-1)(x-6)(x+1)}$
53. $\dfrac{5n^2+14n+13}{(n+3)(n+2)(n+1)(n-1)}$ 55. numerators

Section 11.4

1. $\dfrac{15y}{14x}$ 3. $\dfrac{11}{30}$ 5. $\dfrac{7}{10x+2}$ 7. $\dfrac{y+x}{y}$
9. $\dfrac{2}{y(x-2y)}$ 11. $\dfrac{y-4x}{x^2-16y^2}$ 13. $\dfrac{2x+7}{4x+7}$
15. $\dfrac{x^2-2x-3}{x^2+x-1}$ 17. $\dfrac{5x-3}{4x^2-2x}$ 19. $-\dfrac{4ax}{x^2+a^2}$
21. 1   23. $\dfrac{-6a-4}{3a+9}$ 25. $\dfrac{-x-2y}{-7x+6y}$ 27. $\dfrac{3y+2}{2y+1}$
29. $\dfrac{a-2}{a}$ 31. $\dfrac{2y-x}{y+5x}$ 33. complex fraction

Section 11.5

1. 1   3. -3 5. 4   7. no solution   9. 5   11. 7
13. 2   15. no solution   17. -4 19. 0   21. $\dfrac{9}{4}$
23. $\dfrac{2}{3}$ 25. 5   27. $\dfrac{14}{3}$ 29. $-\dfrac{5}{3}$ 31. LCD, extraneous
33. 60 ohms and 120 ohms   35. $\dfrac{25}{6}$ and $\dfrac{50}{3}$

Section 11.6

1. 60 words per minute   3. $\dfrac{1}{4}$ pool per hour
5. $\dfrac{35}{x}$ problems per minute   7. 6 hours   9. 144 hours
11. $\dfrac{3}{5}$ hr   13. 6 hours   15. Bert traveled 50 mph, Ernie traveled 54 mph   17. 18 mph   19. 3 hours   21. Ron: 22.5 hours, Reggie: 45 hours   23. 22.5   25. 1125 miles

CHAPTER 11   Review Exercises

1. $\dfrac{13y}{17x^2}$ (Section 11.1)   2. $\dfrac{x-5}{x-1}$ (11.1)   3. $\dfrac{2y+1}{3}$ (11.1)
4. $\dfrac{x-1}{x+1}$ (11.4)   5. $\dfrac{y+2x}{2y-x}$ (11.4)
6. $\dfrac{x^2y^2}{(x-y)^2(y^2-x^2)}$ (11.4)   7. 1 (11.2)
8. $\dfrac{y-3}{(y+4)(y-4)}$ (11.2)   9. $\dfrac{1}{n-3}$ (11.3)
10. $\dfrac{9}{(x+4)(x-5)}$ (11.3)   11. $\dfrac{18b^2}{a(a-b)(a+b)}$ (11.2)
12. $\dfrac{1}{(y-5)(y+2)}$ (11.3)   13. $\dfrac{4y^2}{y+1}$ (11.2)
14. $\dfrac{14x+12}{x(x+2)(x+6)}$ (11.3)   15. $\dfrac{2}{x(x-2)}$ (11.3)
16. $\dfrac{1}{(a+2b)(a+b)}$ (11.2)
17. $\dfrac{(x+4)(x+5)}{18}$ (11.2)   18. $\dfrac{y+3}{6}$ (11.2)
19. $\dfrac{x-y}{xy}$ (11.4)   20. $\dfrac{n-7}{n+3}$ (11.4)   21. $\dfrac{n-3}{n+8}$ (11.4)
22. $\dfrac{y-2}{y-1}$ (11.3)   23. $-\dfrac{x-2}{2x(x+1)}$ (11.2)
24. $-\dfrac{10}{7x-3}$ (11.3)   25. $\dfrac{1}{5}$ (11.5)   26. 31 (11.5)
27. no solution (11.5)   28. 0 (11.5)   29. $\dfrac{6}{5}$ (11.5)
30. 5 (11.5)   31. $\dfrac{14}{3}$ (11.5)   32. $-\dfrac{3}{2}$ (11.5)
33. $18\dfrac{2}{3}$ hr (11.6)   34. Train A: 46 mph, Train B: 58 mph
35. 10 (11.4)   36. Cazzie: 60 pounds, Guerin: 100 pounds (11.6)

CHAPTER 11   Test

1. $\dfrac{x-1}{2(x^2+1)}$ 2. $\dfrac{x-4}{x+4}$ 3. $\dfrac{29}{18x}$ 4. $\dfrac{a}{3}$
5. $\dfrac{5}{4(x+4)}$ 6. $\dfrac{x+y}{3y}$ 7. $\dfrac{4x+5}{x-3}$ 8. $\dfrac{a}{b}$
9. $-\dfrac{x}{x+2}$ 10. $\dfrac{3}{x+3}$ 11. $-\dfrac{7}{4}$ 12. $\dfrac{18}{5}$ 13. -16
14. numerator: -6, denominator: -16 15. $2\dfrac{2}{9}$ hr

CHAPTER 12

CHAPTER 12   Objectives

1a. 4   b. $-y^4$ 2a. 3.606   b. 9.849   3a. $\sqrt{22}$ b. 9
4a. $2x\sqrt{5x}$ b. $7\sqrt{11}$ 5a. $4\sqrt{x}$ b. $8x^3$
6a. $\dfrac{\sqrt{3}}{3}$ b. $\dfrac{\sqrt{15}}{10}$ 7a. $-3\sqrt{3}$ b. $19\sqrt{2}$
8a. 1, -7 b. $\dfrac{\sqrt{6}}{2}, -\dfrac{\sqrt{6}}{2}$ 9a. 0, 5   b. $-1, \dfrac{1}{2}$
10a. -3 b. $\dfrac{-2+\sqrt{6}}{2}, \dfrac{-2-\sqrt{6}}{2}$

Section 12.1

1. 2, -2 3. 4, -4 5. 9, -9 7. 1, -1 9. 7, -7
11. -5 13. -1 15. x^8 17. -2^5 19. $-x^{50}$
21. -7^{35} 23. $-a^{10}$ 25. 5^{35} 27. 5.657   29. 6.856

31. 3.742   **33.** 8.944   **35.** −3.606   **37.** −2.646
39. positive, negative   **41.** radicand, radical   **43.** square, given number   **45.** number, $\sqrt{n}$ **47.** −13.23   **49.** 0.26
51. −2715.70   **53.** 2.61

Section 12.2
1. $2\sqrt{2}$ **3.** $2\sqrt{10}$ **5.** $-5\sqrt{5}$ **7.** $2\sqrt{7}$ **9.** $x^2\sqrt{x}$
11. $x^{16}\sqrt{x}$ **13.** $3x^2\sqrt{6}$ **15.** $4\sqrt{3}$ **17.** $5\sqrt{2}$
19. $6x^2\sqrt{2}$ **21.** $7\sqrt{2}$ **23.** $x^9\sqrt{x}$ **25.** $-x^3\sqrt{x}$
27. $-x^{15}\sqrt{x}$ **29.** $5x^2\sqrt{x}$ **31.** $4\sqrt{x}$ **33.** $c^3\sqrt{c}$
35. $-3\sqrt{10}$ **37.** 10   **39.** $-7\sqrt{10}$ **41.** $14\sqrt{3x}$
43. $-30x^4\sqrt{3}$ **45.** $\sqrt{15}$ **47.** $\sqrt{13x}$ **49.** 19   **51.** 28
53. 10   **55.** $3x\sqrt{6x}$ **57.** $10x^6\sqrt{2y}$ **59.** 2   **61.** 8
63. $5x^2\sqrt{x}$ **65.** $5x^4\sqrt{3x}$ **67.** $5x^5\sqrt{2}$ **69.** $\frac{2}{3}$ **71.** $\frac{6}{7}$
73. $\frac{\sqrt{5}}{3}$ **75.** $\frac{\sqrt{11}}{5}$ **77.** $\frac{5\sqrt{2}}{2}$ **79.** $\frac{\sqrt{6}}{6}$ **81.** $\frac{\sqrt{6}}{3}$
83. $\frac{\sqrt{15}}{5}$ **85.** $\frac{\sqrt{2}}{2}$ **87.** $\frac{1}{2}$ **89.** $\frac{\sqrt{x}}{x}$ **91.** $\frac{4x\sqrt{6}}{3}$
93. $\frac{\sqrt{10x}}{4x}$ **95.** $\frac{2x^2\sqrt{6}}{3}$ **97.** square root, product
99. radicand, perfect square factor   **101.** numerator, denominator   **103.** $\frac{2}{5}$ **105.** $\frac{\sqrt{3}}{2}$ **107.** $\frac{\sqrt{3}}{2}$ **109.** $0.4\sqrt{3}$
111. $\frac{\sqrt{2}}{4}$ **113.** $\frac{1}{2}$ **115.** $\frac{\sqrt{7}}{2}$ **117.** $x^7y^{10}z\sqrt{xz}$

Section 12.3
1. $9\sqrt{2}$ **3.** $-2\sqrt{7}$ **5.** $12\sqrt{x}$ **7.** $11\sqrt{13}$ **9.** $-4\sqrt{2}$
11. $2\sqrt{3}$ **13.** $-\sqrt{2}$ **15.** $\sqrt{3}$ **17.** $9\sqrt{5}$
19. $3\sqrt{2} + 12\sqrt{3} - 14\sqrt{7}$ **21.** radicand   **23.** like square roots   **25.** $\sqrt{2} + \sqrt{3} + \sqrt{5}$ **27.** $9\sqrt{3}$ **29.** $-\frac{1}{2}\sqrt{2}$
31. $\frac{10\sqrt{21}}{21}$ **33.** $3\sqrt{6}$ **35.** $\sqrt{2} + 2\sqrt{3}$ **37.** $7\sqrt{a}$
39. $5x^2\sqrt{10}$ **41.** $-6\sqrt{11}$

Section 12.4
1. $4x^2 - 7x - 9 = 0$ **3.** $3x^2 + 3x - 8 = 0$
5. $3x^2 - 17x + 2 = 0$ **7.** 3, −3   **9.** 6, −6
11. $2\sqrt{3}, -2\sqrt{3}$ **13.** $\sqrt{7}, -\sqrt{7}$ **15.** no real solution
17. $\frac{1}{2}, -\frac{1}{2}$ **19.** $\frac{2}{3}, -\frac{2}{3}$ **21.** $\frac{\sqrt{10}}{3}, -\frac{\sqrt{10}}{3}$
23. $\frac{\sqrt{2}}{4}, -\frac{\sqrt{2}}{4}$ **25.** $\frac{5}{2}, -\frac{5}{2}$ **27.** $\frac{2\sqrt{10}}{5}, -\frac{2\sqrt{10}}{5}$ **29.** no real solutions   **31.** $\frac{5\sqrt{3}}{2}, -\frac{5\sqrt{3}}{2}$ **33.** $\frac{4\sqrt{2}}{3}, -\frac{4\sqrt{2}}{3}$
35. $\frac{4}{3}, -\frac{8}{3}$ **37.** 1, 5   **39.** $\frac{5}{3}, 3$ **41.** $-\frac{3}{2}, \frac{13}{6}$
43. $\frac{2 - 3\sqrt{3}}{3}, \frac{2 + 3\sqrt{3}}{3}$ **45.** $\frac{5 + 7\sqrt{2}}{2}, \frac{5 - 7\sqrt{2}}{2}$
47. quadratic   **49.** 4.5   **51.** 5

Section 12.5
1. −4, 2   **3.** −7, −5   **5.** 2, 3   **7.** −5, 2   **9.** 4, −4
11. 7   **13.** $\frac{3}{2}, \frac{7}{5}$ **15.** 2, 4   **17.** $-\frac{1}{4}, 0$ **19.** 0, 9
21. $0, \frac{2}{3}$ **23.** −1, 0   **25.** −8, −3   **27.** 2, 6   **29.** 3
31. $-\frac{5}{4}, \frac{4}{3}$ **33.** $\frac{6}{5}, -\frac{6}{5}$ **35.** $-\frac{3}{4}, 2$ **37.** product, zero
39. double root   **41.** −2   **43.** $-2, \frac{5}{2}$ **45.** $-\frac{9}{2}, 0$

47. 8, 9   **49.** −8, −3 or 3, 8   **51.** length is 8 meters, width is 4 meters   **53.** base is 4 inches, height is 10 inches
55. width is 5, length is 10   **57.** $\frac{3}{2}, 2$ **59.** −2   **61.** 3, 2
63. 3, −3   **65.** 5, −2   **67.** $\frac{14}{3}, -1$

Section 12.6
1. 2, 4   **3.** no real solutions   **5.** $\frac{5}{2}, 4$ **7.** 1   **9.** no real solutions   **11.** $-\frac{4}{3}, 2$ **13.** $-\frac{4}{3}, \frac{4}{3}$ **15.** $\frac{5}{3}, 0$
17. $\frac{7 - \sqrt{61}}{6}, \frac{7 + \sqrt{61}}{6}$ **19.** no real solutions
21. $3 + 2\sqrt{3}, 3 - 2\sqrt{3}$ **23.** $3 + 2\sqrt{2}, 3 - 2\sqrt{2}$
25. $2 + 2\sqrt{2}, 2 - 2\sqrt{2}$ **27.** $-1 + \sqrt{5}, -1 - \sqrt{5}$
29. $-2 - 2\sqrt{2}, -2 + 2\sqrt{2}$ **31.** coefficient, coefficient, constant   **33.** standard, a, b, c,   **35.** −2.3, 0.3
37. −0.3, 1.6   **39.** no real solutions

CHAPTER 12   Review Exercises
1. 6 (Section 12.1)   **2.** −9 (12.1)   **3.** $2\sqrt{7}$ (12.2)
4. $2x\sqrt{10}$ (12.2)   **5.** 4 (12.2)   **6.** $\frac{2\sqrt{3}}{7}$ (12.2)
7. $5\sqrt{3}$ (12.3)   **8.** $\sqrt{5}$ (12.3)   **9.** $\frac{\sqrt{2}}{4}$ (12.2)
10. $\frac{\sqrt{15}}{5}$ (12.2)   **11.** $9\sqrt{11}$ (12.2)   **12.** $\frac{\sqrt{2}}{5}$ (12.2)
13. $\frac{6}{5}\sqrt{5}$ (12.3)   **14.** $5x^3\sqrt{3}$ (12.2)   **15.** $16\sqrt{2}$ (12.3)
16. 6.557 (12.1)   **17.** 15.876 (12.2)   **18.** 1.118 (12.2)
19. $-\frac{2}{3}, \frac{5}{2}$ (12.5)   **20.** $2 - \sqrt{5}, 2 + \sqrt{5}$ (12.6)
21. $\frac{\sqrt{66}}{2}, -\frac{\sqrt{66}}{2}$ (12.4)   **22.** −4, 2 (12.5)   **23.** 7, −7 (12.4)
24. $0, \frac{5}{4}$ (12.5)   **25.** no real solutions (12.6)
26. $-\frac{7}{2}, \frac{1}{2}$ (12.5)   **27.** $-\frac{2}{3}, 1$ (12.5)   **28.** $\frac{4}{3}, 4$ (12.4)
29. $-8 + \sqrt{70}, -8 - \sqrt{70}$ (12.6)   **30.** $4, -\frac{1}{2}$ (12.6)
31. $\frac{3 + \sqrt{3}}{2}, \frac{3 - \sqrt{3}}{2}$ (12.6)   **32.** no real solutions (12.6)
33. no real solutions (12.4)   **34.** $1 + 5\sqrt{2}, 1 - 5\sqrt{2}$ (12.4)
35. −1, −15 (12.4)   **36.** $\frac{3}{2}, -\frac{3}{2}$ (12.5)   **37.** 0, 7 (12.5)
38. $\frac{4}{3}, -\frac{3}{4}$ (12.5)   **39.** no real solutions (12.6)
40. $\frac{5}{2}, \frac{2}{5}$ (12.5)   **41.** 0 (12.4)   **42.** $\frac{5\sqrt{6}}{3}$ (12.2)
43. $3x^4y^7z^6\sqrt{6y}$ (12.2)   **44.** $(6x + 3y)\sqrt{7x}$ (12.3)
45. 14 yd (12.4)   **46.** 9, 10 (12.5)   **47.** height is 4.7015 yd; base is 1.7015 (12.6)   **48.** 10 feet, 14 feet (12.5)

CHAPTER 12   Test
1. −9   **2.** x^{15} **3.** 7.937   **4.** 4   **5.** $6x\sqrt{2x}$ **6.** $-4\sqrt{5}$
7. $60x^4\sqrt{x}$ **8.** 6   **9.** $\frac{5}{9}$ **10.** $\frac{\sqrt{15}}{5}$
11. $23\sqrt{2}$ **12.** $-6\sqrt{3}$ **13.** $\frac{\sqrt{3}}{2}, -\frac{\sqrt{3}}{2}$ **14.** $\frac{3}{4}, 3$

15. 0, 5   **16.** $3 + 2\sqrt{3}, 3 - 2\sqrt{3}$ **17.** $-\frac{1}{4}, -2$
18. -8 and -10, 8 and 10   **19.** width: 5.2849 inches, length: 14.5698 inches   **20.** $4, \frac{1}{4}$

CHAPTER 13

CHAPTER 13   Objectives

1a. 18.65 in   **b.** $11\frac{5}{6}$ in   **2a.** 25.12 ft   **b.** 14.444 ft
3a. 22.5 m²   **b.** $13\frac{17}{96}$ m²   **c.** $4\frac{1}{6}$ m²   **d.** 17.5053 m²
4a. 267.95 m³   **b.** 64 yd³   **c.** 16.75 m³   **d.** 169.56 ft³
5a. 7.9373 m   **b.** 6 and 8

Section 13.1

1. 40 inches   **3.** 12 inches   **5.** $9\frac{2}{3}$ inches   **7.** 22.8 feet
9. 35.22 feet   **11.** 27.004 feet   **13.** 21.98 feet
15. 10.8016 feet   **17.** 49.39 yards   **19.** 138 yards
21. 53.1 yards   **23.** perimeter   **25.** sides, sum, sides
27. $C = \pi d$ **29.** 84 minutes   **31.** 103.62 inches
33. $144   **35.** 81.3312 inches

Section 13.2

1. 9 ft²   **3.** 24.6 ft²   **5.** 38.465 ft²   **7.** $14\frac{1}{4}$ cm²
9. 15.479572 cm²   **11.** 14.13 cm²   **13.** 113.04 cm²
15. 107.64 cm²   **17.** 48 cm²   **19.** 159.48 cm²
21. 9.08375 cm²   **23.** parallelogram, triangle, trapezoid
25. 20   **27a.** 800 yd²   **b.** $7,208   **29.** 50
31. 5631.6606 mi²   **33.** 1354.7382 ft²

Section 13.3

1. 120 in³   **3.** 350.14 in³   **5.** 523.33 in³
7. 1203.6499 in³   **9.** 77.872 in³   **11.** 288.55 in³
13. 1589.625 in³   **15.** 720 in³   **17a.** $V = lwh$
b. $V = \pi r^2 h$ **c.** $V = \frac{\pi r^2 h}{3}$ **d.** $V = \frac{4\pi r^3}{3}$

19. 267.95 cm³   **21.** 138.16   **23a.** 2400   **b.** 18,000

Section 13.4

1. 4   **3.** 10   **5.** 24   **7.** $4\sqrt{3}$ **9.** $2\sqrt{6}$ **11.** 5
13. 10 ft   **15.** 16 cm   **17.** 7.0711 yd   **19.** 4.9 ft
21. sides, hypotenuse   **23.** 17 feet   **25.** no   **27.** 4.9155 meters and 0.9155 meters   **29.** width: 5.6535 meters, length: 2.6535 meters   **31.** 1.5615 feet, 2.5615 feet

CHAPTER 13   Review Exercises

1. 46 in *(Section 13.1)*   **2.** 28 in *(13.1)*   **3.** 120 in *(13.1)*
4. 185 ft² *(13.2)*   **5.** 52.7834 ft² *(13.2)*   **6.** $57\frac{3}{4}$ ft² *(13.2)*
7. 25.748 ft *(13.1)*   **8.** 502 in² *(13.2)*   **9.** 549.64 in³ *(13.3)*
10. 94.2 in³ *(13.3)*   **11.** 100.48 in³ *(13.3)*
12. 40.54 ft *(13.1)*   **13.** 58.55 ft *(13.1)*   **14.** 56.7 ft *(13.1)*
15. 51.81 ft² *(13.2)*   **16.** 61.6875 ft² *(13.2)*
17. 190 ft² *(13.2)*   **18.** 234.807 *(13.3)*   **19.** 66 *(13.2)*
20. $102\frac{1}{6}$ *(13.1)*   **21.** 6.403 in *(13.4)*   **22.** 17.493 m *(13.4)*
23. 12 yd *(13.4)*   **24.** 9 m *(13.4)*   **25.** 1500 in *(13.4)*
26. 600 mi *(13.4)*   **27.** 1.5615 yd, 2.5615 yd *(13.4)*
28. 25 ft *(13.4)*   **29.** no *(13.4)*   **30.** 904.32 m³ *(13.3)*
31. 50.24 m² *(13.2)*   **32.** 13564.8 m³ *(13.3)*   **33.** 2062.5 ft² *(13.2)*
34. 144 ft² *(13.2)*   **35.** 273.5 m *(13.1)*

CHAPTER 13   Test

1. 25.36 feet   **2.** 140 feet   **3.** 28.56 feet   **4.** 31.4 cm
5. 29.75 m²   **6.** 19.72 m²   **7.** 30 m²   **8.** 163.908 ft³
9. 18.84 ft³   **10.** 2143.5733 ft³   **11.** 21 feet
12. 176.4 ft²   **13.** 7065 ft³   **14.** 7 ft
15. hypotenuse: 5 inches, legs: 3 inches and 4 inches

APPENDIX

1. 10   **3.** 2.5   **5.** 2.5   **7.** 88000   **9.** 10   **11.** 5
13. 15,000,000   **15.** 10,000   **17.** 120   **19.** 2.25
21. 0.75   **23.** 35   **25.** 3600

Index

A

Abscissa, 287
Absolute value, 3
Addition
  of decimals, 103-4
  of fractions, 69-74
  of integers, 5-8
  of mixed numbers, 78-9
  of polynomials, 169-70
  of rational expressions, 381-6
  of signed decimals, 122
  of signed fractions, 88
  of squared roots, 420-1
Addition principle
  of equality, 201, 326
  of inequality, 221
Additive inverse, 11
Algebraic fraction, 373
Amount, 256
Area
  meaning of, 453
  of parallelograms, 456
  of rectangles, 454-5
  of semicircles, 461
  of trapezoids, 459
  of triangles, 457-8
  of unfamiliar figures, 462-3
Associative law
  for addition, 8
  for multiplication, 22

B

Base, 141, 256
Binomials, 161
Borrowing, 83
Boundary line, 303
Braces, 27
Brackets, 27

C

Circle, 448
Circumference, 448
Commutative law
  for addition, 7
  for multiplication, 19
Complex fraction, 388-93
Composite number, 43
Cones, 466
Consistent equations, 321-4
Coordinates, 287-90
Cross-products, 48, 179

Cubic centimeter, 466
Cubic inch, 466
Cylinders, 466

D

Decimals
  addition of, 103-5
  division of, 114-6
  expressing as fractions, 119
  expressing as a percent, 131
  meaning of, 101
  multiplication of, 108-9
  problems containing fractions and decimals, 119-20
  repeating, 118
  rounding, 10-12
  signed decimals, 122-4
  subtraction of, 104-5
  writing word name, 102
Denominator, 42
Dependent equations, 322-4
Descending order, 164
Diameter, 448
Difference, 12
Difference of two squares, 182
Discriminant, 435
Distributive law, 162
Dividend, 23
Divisibility tests, 44
Division
  of decimals, 114-6
  of exponential expressions, 145
  of fractions, 64-5
  of integers, 24
  of mixed numbers, 65-6
  of polynomials, 189
  of rational expressions, 378-9
  of signed decimals, 124
  of fractions, 91-92
  of square roots, 414-5
  using scientific notation, 153
Division principle
  of equality, 202
  of inequality, 223
Divisor, 23, 42
Double root, 430

E

Equations
  conditional, 198
  consistent, 321-4
  dependent, 322-4

Equations, *cont'd.*
  equivalent, 200
  identity, 198
  inconsistent, 199
  linear, 197, 292
  literal, 217-8
  meaning of, 197
  no solution, 199
  pure quadratic, 423
  quadratic, 422
  system of, 319
  writing equations, 240-1
Exponential expressions
  division of, 145
  finding value of, 142
  multiplication of, 144
  raising a fraction to a power, 149-50
  raising a power to a power, 147
  raising a product to a power, 148
  sums and differences, 145
Exponents, 141
Extraneous roots, 394
Extremes of a proportion, 249

F

Factor, 42
Factoring
  by grouping, 366-7
  completely, 364
  difference of two squares, 364
  meaning of, 351
  perfect square trinomial, 365
  polynomials, 352-3
  trinomials, 355-61
FOIL method, 179
Formulas
  area of circles, 460
  area of parallelograms, 456
  area of rectangles, 454
  area of semicircles, 461
  area of trapezoids, 459
  area of triangles, 457
  circumference of a circle, 449
  distance, 274
  evaluation of, 32-3
  length of semicircle, 449
  meaning of, 31
  mixture, 270
  percent, 256
  percent proportion, 260
  simple interest, 265
  volume of cones, 467

Formulas, *cont'd.*
  volume of cylinders, 467
  volume of rectangular box, 467
  volume of spheres, 467
  work, 398
Fractional part, 42
Fractions
  addition of, 69–74
  algebraic, 373
  complex, 388–91
  division of, 64–5
  equivalent, 48–9
  expressing as decimals, 117–8
  expressing as a percent, 131–2
  fundamental property of, 373
  meaning of, 42
  multiplication of, 56–60
  problems containing fractions and decimals, 119–20
  proper, 42
  reducing, 50
  signed, 87–8
  square root of, 416
  subtraction of, 81–4
Fundamental property of proportion, 343

G

Graphing
  inequalities, 303–7
  intercept method, 294–6
  points, 288
  slope-intercept method, 301–2
Greater than, 220
Greatest common factor (GCF), 351–3
Grouping symbols, 27–28

H

Hypotenuse, 471

I

Identity, 198
Improper fractions
  expressed as mixed numbers, 54
  meaning of, 52–3
Inconsistent equations, 322–4
Inequalities
  graphing, 221
  solving, 220–29
  symbols, 220
Integers
  addition of, 5–8
  addition and subtraction of, 16–7
  division of, 24
  meaning of, 2
  multiplication of, 20–1
  order of operations, 26–8
  subtraction of, 12–3
Intercept method, 294
Interest, 265
Irrational numbers, 410–1

L

Least common multiple (LCM), 71
Legs of a triangle, 471
Less than, 220
Like square roots, 420
Like terms, 162–3
Linear equations, 292
Lowest common denominator (LCD), 71, 383–4

M

Means of a proportion, 249
Meter, 460
Minuend, 12
Mixed numbers
  addition of, 78–9
  division of, 65–6
  expressed as improper fractions, 53
  meaning of, 52
  multiplication of, 58–60
  subtraction of, 83–4
Monomials, 161
Multiplication
  binomials by binomials, 180
  of decimals, 108–9
  of exponential expressions, 144
  of fractions, 56–60
  of integers, 20–2
  of mixed numbers, 58–60
  of polynomials, 176–7
  by power of ten, 152
  of rational expressions, 377–8
  of signed decimals, 123
  of signed fractions, 89–90
  of square roots, 412
  squaring binomials, 181
  sum and difference, 182
  using scientific notation, 153–4
Multiplication principle
  of equality, 203
  of inequality, 224

N

Natural numbers, 2
Negative square root, 409
Number line, 2
Numerator, 42
Numerical coefficients, 161

O

Order of operations, 26–8
Order relations, 220
Ordered pairs, 287
Ordinate, 287
Origin, 287

P

Parallel lines, 319–22
Parallelograms, 455
Parentheses, 5
Percent, 256
Percents
  expressed as decimals, 128
  expressed as fractions, 127
  meaning of, 126
  perfect square, 363
Perfect square trinomial, 181, 365
Perimeter
  meaning of, 445
  of polygons, 445–51
Pi(π), 449
Place values, 101
Polygon, 445
Polynomials
  addition of, 169–70
  degree of, 164
  division of, 189
  evaluating, 166–7
  meaning of, 161
  multiplication of, 176–7
  subtraction of, 172–3
  terms of, 161
Prime
  with respect to the integers, 357
Prime factorization
  meaning of, 43
  procedure for, 43–5
Prime number, 43
Principal, 265
Principal square root, 409
Principles of equality
  addition, 201
  division, 202
  multiplication, 203
Principles of inequality
  addition, 221
  division, 223
  multiplication, 224
Proper fraction, 42
Proportion, 249
Pure quadratics, 423
Pythagorean Theorem, 471–5

Q

Quadrants, 287
Quadratic equations
  meaning of, 422
  pure, 423
  solution by factoring, 429–30
  solution by formula, 436–8
  solution by square root method, 424–6
  word problems, 431–2
Quotient, 23

R

Radical, 409
Radical symbol, 409
Radicand, 409
Radius, 448
Rate, 247, 265
Ratio, 246

Rational expressions
  addition of, 381-6
  division of, 378-9
  equations involving, 394-6
  lowest common denominator of, 383-4
  multiplication of, 377-9
  simplifying, 374-5
  subtraction of, 381-6
Rational numbers, 373
Rationalizing the denominator, 416-7
Real numbers, 410
Reciprocal, 62-3
Rectangles, 446
Rectangular coordinate system, 287
Repeating decimals, 118
Right triangle, 471
Rise, 299
Root, 197
Rounding decimals, 110-11
Run, 299

S

Satisfies the equation, 319
Scientific notation, 151-2
Signed decimals, 122-4
Signed fractions, 87-92
Slope, 299-300
Slope-intercept form, 299-300
Slope-intercept method, 301-2
Solutions
  of linear equations, 197-8, 292
  of linear inequalities, 303
  of system of linear equations, 319-20
Solving
  equations containing rational expressions, 394-6
  inequalities, 220-9
  linear equations, 201-15
  literal equations, 217-8
  proportions, 251
  system of equations, 321-40
  word problems, 241-79, 333-4, 341-3, 431-2

Special products
  squaring binomials, 181
  sum times difference, 182
Spheres, 466
Square centimeter, 453
Square inch, 453
Square roots
  addition of, 420-1
  division of, 414-5
  of fractions, 416
  like, 420
  meaning of, 363-4, 409
  multiplication of, 412
  rationalizing the denominator, 416-7
  simplest form, 413-4
  simplification of, 417-8
  subtraction of, 420-1
Subtraction
  of decimals, 104-5
  difference between a and b, 13
  of fractions, 81-2
  of integers, 12-3
  of mixed numbers, 83-4
  of rational expressions, 381-6
  of signed decimals, 123
  of signed fractions, 89
  of square roots, 420-1
Subtrahend, 12
System of equations
  solving of, addition method, 328-32
  solving of, graphing, 321-4
  solving of, substitution, 337-40
  using, 333-4, 341-3

T

Terms
  degree of, 163
  like, 162-3
  of a polynomial, 161
Tests of divisibility, 44
Time, 265
Trapezoids, 458
Triangles, 457
Trinomials, 161

U

Undefined, 23
Unit price, 248

V

Variable, 30
Volume
  of cones, 466
  of cylinders, 466
  meaning of, 466
  of rectangular box, 468
  of spheres, 466

W

Whole numbers, 2
Word problems
  general, 241-2
  mixture, 270-3
  percent, 257-63
  proportion, 252-3
  ratio, 248-9
  simple interest, 265-8
  strategy for solving, 9, 241
  uniform motion, 275-9
  using quadratics, 430-32
  using systems of equations, 333-4, 341-3
  work problems, 399-401

X

x-axis, 287
x-intercept, 294

Y

y-axis, 287
y-intercept, 293

Z

Zero-factor property, 428